T0353440

Applied Categorical and Count Data Analysis

Developed from the authors' graduate-level biostatistics course, **Applied Categorical and Count Data Analysis, Second Edition** explains how to perform the statistical analysis of discrete data, including categorical and count outcomes. The authors have been teaching categorical data analysis courses at the University of Rochester and Tulane University for more than a decade. This book embodies their decade-long experience and insight in teaching and applying statistical models for categorical and count data. The authors describe the basic ideas underlying each concept, model, and approach to give readers a good grasp of the fundamentals of the methodology without relying on rigorous mathematical arguments.

The second edition covers classic concepts and popular topics, such as contingency tables, logistic regression models, and Poisson regression models, along with modern areas that include models for zero-modified count outcomes, parametric and semiparametric longitudinal data analysis, reliability analysis, and methods for dealing with missing values. As in the first edition, R, SAS, SPSS, and Stata programming codes are provided for all the examples, enabling readers to immediately experiment with the data in the examples and even adapt or extend the codes to fit data from their own studies.

Designed for a one-semester course for graduate and senior undergraduate students in biostatistics, this self-contained text is also suitable as a self-learning guide for biomedical and psychosocial researchers. It will help readers analyze data with discrete variables in a wide range of biomedical and psychosocial research fields.

Features

- Describes the basic ideas underlying each concept and model.
- Includes R, SAS, SPSS and Stata programming codes for all the examples.
- Features significantly expanded Chapters 4, 5, and 8 (Chapters 4-6, and 9 in the second edition.
- Expands discussion for subtle issues in longitudinal and clustered data analysis such as time varying covariates and comparison of generalized linear mixed-effect models with GEE.

CHAPMAN & HALL/CRC
Texts in Statistical Science Series

Joseph K. Blitzstein, *Harvard University, USA*
Julian J. Faraway, *University of Bath, UK*
Martin Tanner, *Northwestern University, USA*
Jim Zidek, *University of British Columbia, Canada*

Recently Published Titles

Applied Categorical and Count Data Analysis

Second Edition

Wan Tang, Hua He, Xin M. Tu

CRC Press
Taylor & Francis Group
Boca Raton London New York

CRC Press is an imprint of the
Taylor & Francis Group, an **informa** business

A CHAPMAN & HALL BOOK

Designed cover image: © Wan Tang, Hua He, Xin M. Tu

Second edition published 2023
by CRC Press
6000 Broken Sound Parkway NW, Suite 300, Boca Raton, FL 33487-2742

and by CRC Press
4 Park Square, Milton Park, Abingdon, Oxon, OX14 4RN

CRC Press is an imprint of Taylor & Francis Group, LLC

Library of Congress Cataloging-in-Publication Data

Names: Tang, Wan, author. | He, Hua, author. | Tu, Xin M., author.
Title: Applied categorical and count data analysis / Wan Tang, Hua He, Xin M. Tu.
Description: Second edition. | Boca Raton : Taylor and Francis, 2023. | Series: Chapman & Hall/CRC texts in statistical science | Includes bibliographical references and index. |Identifiers: LCCN 2022050188 (print) | LCCN 2022050189 (ebook) | ISBN 9780367568276 (hardback) | ISBN 9780367625832 (paperback) | ISBN 9781003109815 (ebook)
Subjects: LCSH: Regression analysis. | Categories (Mathematics)
Classification: LCC QA278.2 .T365 2023 (print) | LCC QA278.2 (ebook) | DDC 519.5/3--dc23/eng/20221020
LC record available at https://lccn.loc.gov/2022050188
LC ebook record available at https://lccn.loc.gov/2022050189

ISBN: 978-0-367-56827-6 (hbk)
ISBN: 978-0-367-62583-2 (pbk)
ISBN: 978-1-003-10981-5 (ebk)

DOI: 10.1201/9781003109815

Typeset in CMR10 font
by KnowledgeWorks Global Ltd.

Publisher's note: This book has been prepared from camera-ready copy provided by the authors.

Contents

Preface to the Second Edition

It has been ten years since this book was first published in 2012. Although the material originally covered in this book is comprehensive, there have been some new developments since then. Additionally, we have received a large amount of feedback from professors and students who used this book. Thus, we feel that a second edition would be necessary and helpful to the audience.

In the second edition we have thoroughly revised the book by adding sizeable new material as well as incorporating the feedback we received. However, the main purpose of the book remains the same: to provide an applied textbook focusing on applications and interpretations of statistical models for a broad audience. For this purpose, the theoretical interpretation is often carried out in the scalar situation so that students unfamiliar with matrix algebra may still capture the underlying intuition while ignoring the technical complicities. For students uninterested in the theory at all, they may totally skip the theoretical explanation without loss of continuity on reading the application part. On the other hand, students interested in the theoretical background are encouraged to dive deeper into theoretical arguments and more advanced theoretical exercises in the second edition. For students who are interested in the theory, but lack the necessary math background, they can download from the book website "A Primer on Linear Algebra" developed by Dr. Marlene Edger at the University of Utah to help to bridge the gap.

While all the identified typos have been corrected, we highlight the major revisions below so that the audience of the first edition may be aware of what to expect in the second edition. The main structure of the book remains the same. For space consideration, we separated the chapter on regression models for categorical outcomes into two chapters, one for binary and the other for polytomous responses. For binary responses, the linear probability model is added and more model diagnoses and assessment tools are introduced. A more systematic treatment for regression analyses for polytomous responses is provided in the second edition. In addition to cumulative models, adjacent models and continuous ratio models are also introduced, together with model evaluation tools. Please note that for balance among the chapters, we postponed the discussion of model selection to the chapter on log-linear models for contingency table as we did in the first edition. Instructors may cover this part earlier when talking about logistic regression models. For count responses studied in Chapter 6, we added the general negative binomial, generalized Poisson, and Conway–Maxwell–Poisson models for count data with overdispersion. We also discussed some recent developments on testing for zero inflation. The chapter for longitudinal data analysis has been thoroughly rewritten. A more systematic treatment of the mixed-effects models is provided. For generalized estimating equation (GEE), we added alternating logistic approaches for binary responses. More model diagnoses and assessment tools are also introduced. Also added are discussions on subtle issues such as the validity of GEE models when there are time-varying covariates and special features for clustered data analysis. For the chapter on the Evaluation of Instruments, we added a section to discuss the added predictability in terms of difference in area under the curves (AUCs), net reclassification improvement, and integrated discrimination improvement, which have become popular in recent years as indices for assessing the added value when a new risk factor is added to a prediction model.

We would like to express our sincerest appreciation to Dr. Marlene Egger for her detailed errata of the first edition, constructive comments, and careful review of the entire second edition manuscript. We would also like to thank Dr. Peng Ye for proofreading part of the book and the students and teaching assistants at Tulane University who took the course and provided helpful comments and feedbacks. We are also thankful to editor David Grubbs for his patience and continuing support.

Preface to the First Edition

This book focuses on statistical analysis of discrete data, including categorical and count outcomes. Discrete variables are abundant in practice, and knowledge about and ability to analyze such data is important for professionals and practitioners in a wide range of biomedical and psychosocial research areas. Although there are some excellent books on this general subject such as those by Agresti (2002, 2007), Long (1997), Long and Freese (2006), and Stokes et al. (2009), a book that includes models for longitudinal data, real data examples with detailed programming codes, as well as intuitive explanations of the models and their interpretations and differences thereupon will complement the repertoire of existing texts. Motivated by the lack of such a text, we decided to write this book five years ago when preparing a graduate-level biostatistics course on this topic for students within a medical school setting at the University of Rochester New York. The lecture notes from which this book has evolved have been used for the course over the past five years.

In addition to the classic concepts such as contingency tables and popular topics such as logistic and Poisson regression models, as covered by most available textbooks on categorical data analysis, this book also includes many modern topics. These include models for zero modified count outcomes, longitudinal data analysis (both parametric and semiparametric), reliability analysis, and popular methods for dealing with missing values. More importantly, programming codes are provided for all the examples in the book for the four major software packages, R, SAS, SPSS, and Stata, so that when reading the examples readers can immediately put their knowledge into practice by trying out the codes with the data in the examples using the statistical packages of their choice and/or adapt and even extend them to fit settings arising from their own studies.

We view effective learning as a process of "reverse engineering" in the sense that one develops an in-depth appreciation of a concept, model, or approach by tracing its bumble beginnings that motivate its development in the first place. With this philosophy in mind, we try to describe the basic ideas underlying each concept, model, and approach introduced in this book so that even without rigorous mathematical arguments, readers can have a good grasp of the fundamentals of the concept and methodology. For the rather technical-savvy audience, we have also included a section in Chapter 1 to review some key results on statistical inference to help facilitate the discussion and understanding of the theoretical aspects of the models and inference methods introduced in the subsequent chapters, complemented by theory-oriented exercises at the end of each chapter. Readers should not be discouraged by such theoretical materials and exercises, since skipping such theoretical justifications will not hamper understanding of the concepts and models and principles of applying them in practice. The book is pretty much self-contained, with no prerequisite for using this book, although knowledge on statistics in general is helpful. Fundamental concepts such as confidence intervals, hypothesis tests, and p-values are briefly introduced as they first appear in the text so that people without former exposure to statistics may still benefit from the book.

The outline of the book is as follows. In addition to the review section mentioned above, Chapter 1 also presents various types of discrete random variables, together with an introduction of the study data that will be used throughout the book.

In Chapter 2, we first study individual random variables and introduce the popular discrete distributions including the binomial, multinomial, and Poisson models. Next we concentrate on the study of relationship between two categorical variables, i.e., the study of two-way contingency tables. This is followed in Chapter 3 by stratified two-way tables, controlling for potential categorical confounding variables.

When there are more than two categorical variables or there are continuous variables present, regression analysis becomes necessary to study the relationship between such variables. In Chapter 4, we introduce regression models for categorical responses. We first discuss logistic regression for binary responses in detail, including methods to reduce bias for relatively small samples such as exact logistic models. Less popular models for binary responses such as the Probit and complementary log-log models are then discussed, followed by the models for general polytomous categorical outcomes to conclude this chapter.

Chapter 5 focuses on regression analysis of count responses. As the most commonly used models in this setting, the Poisson log-linear regression is first studied in detail, followed by a discussion on overdispersion, a common violation of the Poisson model, along with its detection and correction within the confines of this model using robust inference methods, such as the sandwich variance estimate. Alternative models that explicitly account for the sources of overdispersion and structural zero, another common violation of the Poisson, such as the negative binomial, hurdle, and zero-modified models, are then introduced to formally address such deviations from the Poisson. This chapter concludes with a systematic guide to modeling count responses using the different models introduced. Chapter 6 illustrates a major application of the Poisson log-linear regression, as it applies to general contingency tables to facilitate inference about the relationship between multiple variables, which is algebraically too complex using the classic methods discussed in Chapters 2 and 3. Also included in Chapter 6 is a section on model selection that introduces popular criteria for deriving optimal models within a given context.

Chapter 7 discusses analyses for discrete survival times. Survival analysis is widely used in statistical applications involving time to occurrence of some event of interest such as heart attacks and suicide attempts. We discuss nonparametric life table methods as well as regression approaches.

The statistical methods covered in Chapters 2–7 are mainly for cross-sectional studies, where the data only include a single assessment point for every subject. This is not the case for longitudinal studies, where the same set of outcomes such as disease status is repeatedly measured from the same subject over time. Methods for longitudinal data must address the within-subject correlations in repeatedly measured outcomes over time. In Chapter 8, we introduce longitudinal data and models for such data and focus on the popular parametric mixed-effects models and semiparametric generalized estimating equations.

Chapter 9 discusses validity and reliability analysis for diagnostic tests and measuring instruments. We discuss how to assess the accuracy of an ordinal test when the true status is known, using the theory of receiver operating characteristic (ROC) curves. We introduce measurement error models for assessing latent constructs and discuss popular indices for addressing inter-rater agreement and instrument validity and reliability such as Cronbach's alpha coefficient and Kappa.

In Chapter 10, we discuss how to deal with missing values. Common approaches such as multiple imputation and inverse probability weighting methods are introduced. Since applications of the missing value concept really go beyond addressing the problem of missing values in study outcomes, we also illustrate how to apply the principles of such methods to a range of seemingly unrelated issues such as causal inference and survey sampling.

This book can serve as a primary text for a course on categorical and count data analysis for senior undergraduate, beginning as well as senior graduate students in biostatistics. It also serves well as a self-learning text for biomedical and psychosocial researchers interested in this general subject. Based on our own experiences, Chapters 1 through 7 can be covered in a one-semester course.

We would like to express our appreciation to all who have contributed to this book. We would like to thank the students at the University of Rochester who took the course in the past five years, many of whom have provided countless helpful comments and feedbacks. We would also like to thank Dr. Yinglin Xia and Dr. Guoxin Zuo, who proofed many parts of the book and offered numerous valuable comments and suggestions; Dr. Naiji Lu, who helped with some of the examples in Chapter 9 whose analyses are not supported by standard software packages; and Dr. Jun Hu, who proofread the entire book multiple times to help eradicate errors and typos. We are grateful to Drs. Linda Chaudron, Steve Lamberti, Jeffrey Lyness, Mary Caserta, and Paul Duberstein from the University of Rochester and Dr. Dianne Morrison-Beedy from the University of South Florida for graciously sharing their study data for use in the book as real data examples. We are also thankful to editor David Grubbs for his patience and continuing support this project despite multiple delays on the project on our part, to one anonymous reviewer for his/her critical comments and constructive suggestions that have led to an improved presentation, and to staff at CRC who carefully proofread the manuscript and helped with some technical issues and numerous corrections. And, last but not least, we thank all the faculty and staff in the department for their support.

Author biographies

Wan Tang (Ph.D.) is a Clinical Professor in the Department of Biostatistics and Data Science, Tulane University School of Public Health and Tropical Medicine. Dr. Tang's research interests include longitudinal data analysis, missing data modeling, structural equation models, causal inference, and nonparametric smoothing methods. He has co-edited a book on modern clinical trials.

Hua He (Ph.D.) is an Associate Professor in Biostatistics in the Department of Epidemiology at Tulane University School of Public Health and Tropical Medicine. Dr. He is a highly experienced biostatistician with expertise in longitudinal data analysis, structural equation models, potential outcome based causal inference, semiparametric models, ROC analysis and their applications to observational studies, and randomized controlled trials across a range of disciplines, especially in the behavioral and social sciences. She has co-authored a series of publications in peer-reviewed journals, one textbook on categorical data analysis and co-edited a book on statistical causal inference and their applications in public health research.

Xin Tu (Ph.D.) is a Professor in the Division of Biostatistics and Bioinformatics, Department of Family Medicine and Public Health, UCSD. Dr. Tu is well versed in statistical methods and their applications to a range of disciplines, particularly within the fields of biomedical, behavioral and social sciences. He has co-authored over 300 peer-reviewed publications, two textbooks on categorical data and applied U-statistics, and co-edited books on modern clinical trials and social network data analysis. He has done important work in the areas of longitudinal data analysis, causal inference, U-statistics, survival analysis with interval censoring and truncation, pooled testing, semiparametric efficiency, and has successfully applied his novel development to addressing important methodological problems in biomedical and psychosocial research.

Chapter 1

Introduction

This book focuses on the analysis of data containing discrete outcomes. *Discrete variables* are abundant in practice, and many familiar outcomes fall into this category. For example, gender and race are discrete outcomes and are present in many studies. Count variables recording the frequency of some events of interest such as strokes and heavy drinking days are also common discrete outcomes in clinical studies. Statistical models for continuous outcomes such as the popular linear regression are not applicable to discrete variables.

In Section 1.1, we describe discrete variables and their different subtypes. In Section 1.2, we provide a brief description of some clinical studies that will be used as real data examples throughout the book. Questions typically asked for such variables, as well as an outline of the book, are given in Section 1.3. This chapter finishes with a review of some important technical results that underlie the foundation of inference for the statistical models discussed in the book in Section 1.4 and a note on statistical software packages to implement such models in Section 1.5.

1.1 Discrete Outcomes

Discrete variables are those outcomes that are only allowed to acquire finitely or countably many values. This is in contrast to continuous outcomes, which may take on any real number in an interval of an either finite or infinite range. Because of the fundamental differences between continuous and discrete outcomes, many methods developed for continuous variables such as the popular linear regression do not apply to discrete outcomes.

There are several subtypes within discrete outcomes. Different types of discrete data may require different methods. A discrete outcome with only finitely many possible values is called a *categorical* variable. In particular, if there are only two possible values, the variable is called *binary*. Binary outcomes are quite popular in clinical studies. For example, gender is often a variable of interest for most studies, with the categories of "male" or "female." In many questionnaires, "yes" and "no" are often the only possible answers to an item. Even when there is no binary outcome planned at the design stage, they may occur in data analysis. For example, if an outcome of interest is subject to missing values, then a binary variable may be created to indicate the missingness of the outcome. In such a setting, it may be important to model the binary missing data indicator for valid inferences about the outcome of interest, even though the binary variable itself is not of primary interest.

Categorical outcomes with more than two levels are also called *polytomous* variables. There are two common types of polytomous variables. If the levels of a polytomous variable are ordered, then it is also called *ordinal*. Many polytomous variables are ordinal in nature; for example, the five-point Likert scale—strongly disagree, disagree, neutral, agree, and strongly agree—is frequently used in survey questionnaires. In practice, natural numbers are often used to denote the ordinal levels. For example, numbers such as 1–5 are often used to denote the five-point Likert scale. Although the numbers may be chosen arbitrarily,

DOI: 10.1201/9781003109815-1

it is important to select the ones that convey different degrees of discrepancy among the categories if applicable. For example, in many situations involving the Likert scale, 1–5 are used since the difference (or disagreement) between strongly disagree and disagree is viewed to be similar to that between disagree and neutral, etc.

Such equidistant ordinal levels also arise often from discretizing (latent) continuous outcomes, either because of failure to observe the original scale directly or for the purpose of modeling and interpretation. For example, time to death is a continuous variable, but is often recorded (grouped) in units of month, quarter, or year in many large epidemiological and survey studies. This kind of ordinal variable whereby each category represents an interval of an underlying continuous variable is also called the *interval scale*.

If the levels of an ordinal variable only represent the ordered structure, it is not appropriate to consider or compare between-level differences. For example, disease diagnoses are often classified into ordinal levels such as severe, moderate, and none. For many diseases, it is difficult to compare the differences from transitions from one level to the next such as from severe to moderate and from moderate to none. Although numbers may still be used to represent the different levels, they only convey the order rather than the degree of differences across the different levels.

If there is no ordering in the levels, the variable is called *nominal*; for example, gender, ethnicity, and living situation are all nominal polytomous variables. The order of the levels of a variable may be important in selecting appropriate statistical models. Further, the treatment of a polytomous outcome as being an ordinal or nominal variable may also depend on the study; for example, race is usually considered a nominal variable. But for studies in which darkness of skin tone is important, race may become ordinal.

Discrete variables may have an infinite range. For practical importance and modeling convenience, we will mostly consider count variables. A *count variable* records the number of occurrences of an event of interest such as heart attacks, suicide attempts, and abortions and thus has a theoretical range that includes all natural numbers.

Note that many authors use the terms *categorical* and *discrete* interchangeably. In this book, we use categorical only for those variables with a finite range to distinguish them from count variables.

1.2   Data Source

In this section, we give a brief overview of the various study data and the associated variables that will be used throughout the book. More details about the data sources can be found in the references for each of the studies.

The Metabolic Syndrome Study. Metabolic syndrome (MS) is a collection of risk factors associated with increased morbidity and mortality due to cardiovascular diseases. Ninety-three outpatients at the University of Rochester Medical Center Department of Psychiatry received clozapine for at least six months. One of the interests is the incidence of MS and whether it is higher than comparable people who did not take clozapine (Lamberti et al., 2006).

The Postpartum Depression Study (PPD). Postpartum depression affects an average of one out of every seven new mothers in the United States with rates as high as one out of four among poor and minority women. To increase the potential for early intervention, primary care providers, including pediatric practitioners, are encouraged to screen and refer mothers for care. One of the research interests is to study the accuracies of different screening tools. In the study, 198 women were screened with the *Edinburgh Postnatal Depression Scale* (EPDS), *Beck Depression Inventory - II* (BDI-II), and *Postpartum Depression Screening Scale* (PDSS) and underwent a comprehensive clinician-assisted diagnosis based on the *Structured Clinical Interview for DSM-IV-TR* (SCID). See Chaudron et al. (2010) for details.

The Sexual Health Study. Adolescence is the only age category where the number of females infected with human immunodeficiency virus (HIV) outnumbers the number of males. A large controlled randomized study was conducted to evaluate the short term and longer term efficacy of an HIV prevention intervention for adolescent girls residing in a high-risk urban environment. Adolescent girls accessing urban reproductive and general healthcare clinics and youth development programs in western New York, as well as those who heard about our program through word of mouth, were recruited for a sexual risk reduction randomized clinical trial. A total of 640 girls were randomized into either the intervention or a control condition containing only nutritional materials.

One of the primary outcomes of this longitudinal study is the number of sexual experiences reported over the three-month period assessed at 3, 6, and 12 months following the intervention. Other intermediate or mediating variables of the study include HIV knowledge, motivation, and behavioral skills assessed at the three follow-up visits. See Morrison-Beedy et al. (2011) for details.

This R01 study was preceded by a pilot study which examined the accuracy of reporting of sexual behaviors using methods with different modes (contemporaneous daily diary vs. retrospective recall) and frequency (every month vs. 3 months) of reporting (see Morrison-Beedy et al. (2008) for details). We also use data from this Sexual Health pilot study on several occasions.

The Depression of Seniors Study (DOS). The data are from a study that examined the 3–4-year course of depression in over 700 older primary care patients. Depression is a common and disabling mental disorder in older persons. This study collected psychiatric, medical, functional, and psychosocial variables. The focus is on major, minor, and subsyndromal depressions in these patients to test theoretical models (e.g., the cerebrovascular model) by a risk factor approach as well as to identify those most at risk for chronicity. Findings based on the study are published in several articles. Interested readers may check Lyness et al. (2007), Cui et al. (2008), and Lyness et al. (2009) for details.

The Detection of Depression in a Primary Care Setting Study (DDPC). This study recruited friends and relatives of older adults enrolled in the DOS. Of the friends and relatives (informants) of the 589 patients (probands) enrolled in the parent DOS who were approached and asked to participate in this study, the informants of 212 probands consented and provided information on depression for the probands. Depression diagnoses for the probands based on their own assessments and from the informants are compared. See Duberstein et al. (2011) for details.

The Stress and Illness Association in Child Study (SIAC). The data are from a longitudinal study which tried to identify potential links between family stress and health in children. One hundred and sixty-nine children between 5 and 10 years of age and one of their primary caregivers were recruited from an ambulatory population already participating in a study of pediatric viral infections at the University of Rochester School of Medicine. The children were initially identified by visits to the emergency department or other pediatric services. Children with chronic diseases affecting the immune system (e.g., receiving chronic corticosteroid therapy) were excluded. One child per family was enrolled, and all children were well at the first visit. See Caserta et al. (2008) for details.

Self Management in Urinary Catheter Users (Catheter Study). This is a randomized clinical trial on indwelling urinary catheter users, teaching awareness and self-monitoring skills to indwelling urinary catheter users in New York state. As one of the largest clinical trials on urinary catheter usage, a total of 202 adult long-term urinary catheter users were randomized to the intervention, which involved learning catheter-related self-monitoring and self-management skills during home visits by a study nurse, and the control group, which provided usual care. The primary outcomes of interest are whether the subjects experienced urinary tract infections (UTIs), catheter blockages, and catheter displacements during the last two months, as well as the corresponding counts of these experiences. Thus, all the primary outcomes are either categorical or count variables. The subjects were measured every 2 months, up to a year. See Wilde et al. (2015) for details.

1.3   Outline of the Book

In this section, we describe questions that are often asked about discrete variables and outline the chapters that discuss the statistical models to address them.

1.3.1   Distribution of random variables

For a single random variable X, its distribution is all we need to know to understand and describe the outcome. For a discrete variable, it ranges over either a finite number of levels or a set of natural numbers (for count responses). In either case, let v_j $(j = 1, 2, \ldots)$ denote the distinct values comprising the range of the random variable. Then, the distribution of X is described by the probability distribution function (PDF):

$$p_j = \Pr(X = v_j), \quad j = 1, 2, \ldots, \quad \sum p_j = 1.$$

Categorical variables have only a finite number of levels, say J, and their distributions are determined by finitely many p_j's $(1 \leq j \leq J)$. These distributions are called multinomial distributions. Because of the constraint imposed on p_j, the multinomial distribution is determined by a subset of $J - 1$ p_j's. An important special case is when there are only two possible levels, i.e., binary responses, and the resulting distribution is known as the *Bernoulli* distribution, *Bernoulli* (p). Inference about multinomial distribution involves estimating and testing hypothesis about the parameter vector $\mathbf{p} = (p_1, \ldots, p_{J-1})^\top$.

Count variables have infinitely many levels, making it difficult to interpret the associated p_j's. A common approach is to impose some constraints among the p_j's using parametric models such as the Poisson distribution. We describe how to estimate and make inferences about such models in Section 2.1.3.

1.3.2   Association between two random variables

Given two or more random variables, one would be interested in their relationships. For continuous outcomes, the two most popular types of relationships of interest are correlation between the two variables and regression analysis with one designated as a response (dependent variable) and the rest as a set of explanatory variables, or independent variables, predictors, and covariates. For correlation analysis, no analytic relationship is assumed, while for regression analysis, an analytic model such as a linear relationship is posited to relate the response to the explanatory variables. In that sense, correlation analysis is less structured or "nonparametric" and regression analysis is more structured or "parametric."

Similar approaches are employed for modeling relationships among variables involving discrete outcomes. We use nonparametric methods for assessing association between two discrete outcomes or between a discrete and a continuous outcome and parametric regression models for relationships for a discrete response with a set of explanatory variables (mixed continuous and discrete variables).

Note that the terms *nonparametric* and *parametric* are also widely used to indicate whether a certain analytic form is postulated for the data distribution of the response. For example, semiparametric regression models for continuous responses in the literature often refer to those models that posit an analytic form for relating a continuous response to explanatory variables, but assume no analytic model for the data distribution. We will refer to these as *distribution-free* models to distinguish them from the nonparametric models that do not impose structural relationships such as linear as in linear regression among a set of outcomes.

Chapter 2 will focus on methods for assessing the association between two categorical outcomes. As such outcomes are usually displayed in *contingency tables*, these are also called contingency table analysis. Associations between two variables can be very different in nature. The most common question is whether they are independent.

Example 1.1
For the Metabolic Syndrome study, we want to see whether the MS rates differ between men and women among patients taking clozapine.

TABLE 1.1:   Gender by MS for the Metabolic Syndrome study

	MS		
Gender	Present	Absent	Total
Male	31	31	62
Female	17	14	31
Total	48	45	93

Shown in the 2 × 2 contingency table above (Table 1.1) are the numbers of MS cases broken down by gender. The percentage of MS is different between males and females in the study sample. However, to arrive at a conclusion about such a difference at the population level, we need to account for sampling variability. This can be accomplished by appropriate statistical tests, which will be the focus of Chapter 2. In the above example, if one or both outcomes have more than two categories, we will have an $s \times r$ contingency table, where s is the number of levels of the row variable and r is the number of levels of the column variable. For example, MS is defined as having three or more of the following: 1. Waist circumference > 102 cm in men and > 88 cm in women 2. Fasting blood triglycerides > 150 mg/dL 3. HDL cholesterol level < 40 mg/dL in men and < 50 mg/dL in women 4. Blood pressure > 130 mm Hg systolic or > 85 mm Hg diastolic 5. Fasting blood glucose

> 100 mg/dL. We may define subsyndromal MS if a person has one or two of the above, and the newly defined MS variable has three levels, MS, subsyndromal MS, and none. In this case, we would have a 2×3 table. We will also discuss methods for such general $s \times r$ contingency table analysis in Chapter 2.   □

In many situations, we know that two variables are related, and thus our goal is to assess how strong their association is and the nature of their association. For example, in pre-post treatment studies, subjects are assessed before and after an intervention, and thus the paired outcomes are correlated. In reliability research, studies typically involve ratings from multiple observers, or judges, on the same subjects, also creating correlated outcomes. In pre-post treatment studies, we are interested in whether the intervention is effective, i.e., if the distribution of the posttreatment outcome will be shifted to indicate changes in favor of the treatment. Thus, the objective of the study is to test whether the distributions are the same between such correlated outcomes. McNemar's test, introduced in Chapter 2, can test such a hypothesis for binary outcomes. This approach is generalized in Chapter 3 to stratified tables to account for heterogeneity in the study subjects.

In reliability studies, the focus is different, the goal being to assess the extent to which multiple judges' ratings agree with one another. Kappa coefficients are commonly used indices of agreement for categorical outcomes between two raters. Kappa coefficients are introduced in Chapter 2, but a systematic treatment of reliability studies is given in Chapter 10.

1.3.3   Regression analysis

Chapter 4 focuses on models for regression analysis when the response is a binary variable taking only two values. For example, in the Metabolic Syndrome study in Example 1.1, we also have other risk factors for MS, such as a family history of diabetes. We can test the association between each risk factor and MS, but this approach does not do justice to the data since we have information for both gender and family history of diabetes. A regression can address this weakness by including multiple risk factors in predicting MS.

The logistic regression is the most popular model for such a relationship involving a binary response and multiple explanatory variables, so we first discuss this model in detail in Chapter 4. Under a logistic model, the probability of the binary outcome, p_i, is assumed to be related to a linear predictor, $\mathbf{x}_i^\top \boldsymbol{\beta}$, through a logistic function, where x_i is the covariate vector. More precisely, we assume $\log \frac{p_i}{1-p_i} = \mathbf{x}_i^\top \boldsymbol{\beta}$. The logistic model is an example of generalized linear models. Under the framework of generalized linear models, one may change the logit link function to other link functions to obtain different regression models. We discuss the commonly used link functions including probit, complementary log-log, and identity functions. Model assessment and diagnosis are also discussed in the chapter.

The regression analysis under the framework of generalized linear models can be generalized to polytomous outcomes where the response variables have more than two, but still limited number of, levels. Depending on the nature of the responses (nominal, ordinal, interval, or aggregated binary), different models are commonly used in practice. However, most of the modeling approaches are based on some transformation of the question to binary and then apply models for binary responses discussed in Chapter 4. We will discuss regression models for polytomous outcomes in Chapter 5.

Regression analysis can also be applied to count responses. Count outcomes such as number of abortions, birth defects, heart attacks, sexual activities, and suicide attempts all have an unbounded range, though in practice they are observed within a finite range. For example, in the Sexual Health study, one of the primary outcomes is the count of protected sex behaviors in a three-month period. At baseline, the observed frequencies ranged from 0 to 65. One way to analyze the data is to group frequencies larger than a threshold into one category as in Table 2.2 and then apply the methods discussed in Chapters 2 to 5 to test the association. Such an approach not only yields results that depend on the choice of cut-point, but also incurs loss of information due to grouping the data. Furthermore, if we use a cutoff larger than 6 to minimize loss of information, it would be difficult to interpret results from the resulting multinomial model with many possible levels. Methods have been developed for modeling count response. In Chapter 6, we first discuss the most popular approach for count outcomes, the Poisson regression. The Poisson-based log-linear models allow us to model the relationship using the natural range of the outcome without imposing any cut-point. However, they often suffer from the overdispersion issue—the observed variation is higher than would be expected under the models. In this chapter, we also discuss how to detect and address the overdispersion issue. In this chapter, we also discuss generalizations of Poisson regression such as negative binomial and zero-modified Poisson and negative binomial regression models to address limitations of the Poisson distribution.

1.3.4 Log-linear methods for contingency tables

Chapter 7 discusses log-linear models for contingency tables. In the study of association between two random discrete variables (Chapter 2), different statistics are developed for different questions. As more variables get involved, the questions and the statistics to address them will become more complex. More importantly, we need to derive the asymptotic distribution for each of the statistics individually. For example, suppose we are interested in the relationship among the variables gender, family history of diabetes, and MS status in the Metabolic Syndrome study discussed in Section 1.2. We may use methods for stratified tables to study the independence between two of them conditional on the third. To study this conditional independence, we need to develop an appropriate statistic following the discussion in Chapter 2. We may further study pairwise independence among the three variables, in which case we need a different statistic. In the latter case, we test pairwise independence, and as such we need to put the three related pairs together in one statistic. It is not easy to construct such a statistic.

The log-linear methods for contingency tables approach such complex associations among variables through formal models. In such models, the frequency of subjects in each cell (a combination of all the variables involved) is used as the response or dependent variable of the log-linear model. The association of the variables is reflected in the model.

Suppose we are interested in whether the variables gender (x), family history of diabetes (y), and MS status (z) are jointly independent in the Metabolic Syndrome study. By definition, joint independence of the three variables corresponds to

$$\Pr\left(X = i, Y = j, Z = k\right) = \Pr\left(X = i\right)\Pr\left(Y = j\right)\Pr\left(Z = k\right),$$

for all levels i, j, and k of the three categorical variables. Thus,

$$N^2 m\left(X, Y, Z\right) = m\left(X\right) m\left(Y\right) m\left(Z\right),$$

where N is the sample size and m stands for expected frequencies. Taking logarithms on both sides, we obtain

$$\log m\left(X, Y, Z\right) = c + \log m\left(X\right) + \log m\left(Y\right) + \log m\left(Y\right),$$

where $c = -2\log N$. Hence, the logarithm of the expected frequency of each cell follows an additive model. In other words, joint independence corresponds to an additive model with no interaction.

Similarly, independence between two variables conditional on the third corresponds to a model with neither three-way interaction nor the interaction between the two variables. Under the log-linear model approach, we only need to find an appropriate model for the question, and the calculations of the statistics for hypothesis testing can be carried out in a systematic fashion following methods for regression analysis without being bogged down with messy algebra.

1.3.5   Discrete survival data analysis

Chapter 8 discusses analyses for discrete survival times. Survival analysis is widely used in statistical applications involving time to occurrence of some event of interest. For example, in studies involving seriously ill patients such as those with cancer and cardiovascular and infectious diseases, death is often a primary outcome, and we are interested in comparing different treatments by testing differences in the patients' survival times (time to death). Time of death can be accurately observed in such studies, and methods for such continuous survival time outcomes can be applied. In many other studies, the occurrence of the event may not be observed "instantaneously" or the event itself is not an "instantaneous" occurrence. For example, depression is not an instantaneous event. Thus, occurrence of depression is usually measured by a coarse scale such as week or month, yielding discrete outcomes. Discrete survival times also arise quite often in large survey studies and surveillance systems. The sample size is typically huge in such databases, and it may not be computationally practical to apply methods for continuous survival time to such large samples. Thus, it is necessary to group the time of occurrence of events into discrete time intervals such as weeks and months.

Methods in Chapters 1 to 7 do not apply to survival time outcomes because of *censoring*; if the event of interest is not observed, then the survival time is only partially observed.

Example 1.2

In the DOS, a total of 370 adults with no depression at baseline were followed up for up to four years. Shown in Table 1.2 are the numbers of adults with depression as well as dropout (no depression) due to death or loss of follow-up of participants over each year, broken down by gender.

TABLE 1.2:   First major depression diagnosis (dropout) post-baseline

	Yr 1	Yr 2	Yr 3	Yr 4	Total
Men	12 (17)	4 (38)	6 (71)	0 (12)	160
Women	29 (28)	12 (40)	10 (67)	2 (22)	210

Since depression was not observed for all the dropouts over time, one may be tempted to group them together. Then, the above becomes a 2×5 contingency table, and the methods in Chapter 2 may be applied to compare the depression diagnosis cases between the male and female. However, this approach ignores the effect of ordered sequence of dropout and gives rise to biased estimates and invalid inference. For example, a subject who dropped out in year 2 could hardly be treated the same as someone who was censored at the end of study in terms of the subject's susceptibility to depression over the study period.

TABLE 1.3: First major depression diagnosis for men (women)

	Yr 1	Yr 2	Yr 3	Yr 4
At risk	160 (210)	131 (153)	89 (101)	12 (24)
Depression	12 (29)	4 (12)	6 (10)	0 (2)
Censored	17 (28)	38 (40)	71 (67)	12 (22)

By conditioning on subjects at risk in each year as in Table 1.3, we can calculate rates of depression and compare the two groups by using such conditional (hazards) rates of depression.

Note that like censoring, a related but fundamentally distinct concept often encountered in survival analysis is *truncation*. For example, in the early years of the AIDS epidemic, interest was centered on estimating the latency distribution between HIV infection and AIDS onset. Data from CDC and other local (state health departments) surveillance systems were used for this purpose. Since the time of HIV infection is usually unknown and observation is limited by chronological time, only those who became infected and came down with AIDS within the time interval 0–T can be observed (see Figure 1.1). Thus, AIDS cases are underreported because of truncation.

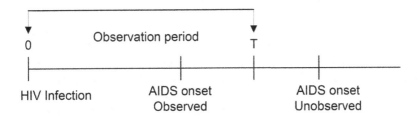

FIGURE 1.1: HIV infection to AIDS onset.

1.3.6 Longitudinal data analysis

The statistical methods covered in Chapters 2 to 8 are mainly for cross-sectional studies. For example, with the exception of McNemar's test and Kappa coefficients, multiple variables of interest such as age, gender, MS, and depression represent different characteristics of an individual and a snapshot of this subject's disease status at a single assessment point. This is not the case for longitudinal studies, where the same set of outcomes such as disease status is repeatedly measured from the same subject over time. In that regard, the paired outcomes from a pre-post treatment study are an example of data from a longitudinal study design. Another example is the Sexual Health study where each sexual behavior of interest such as unprotected vaginal sex and oral sex was assessed repeatedly at 3, 6, and 12 months following the intervention.

Longitudinal data are important in many areas of research including epidemiology, psychology, sociology, economics, and public health. Data from longitudinal studies in clinical trials and cohort studies with long-term follow-ups are a primary example of such data. By taking advantage of multiple assessments over time, data from longitudinal studies capture both between-individual differences and within-individual dynamics, offering the opportunity to study more complicated biological, psychological, and behavioral hypotheses than those that can be addressed using cross-sectional or time series data. For example, if we want to test whether exposure to some chemical agent can cause some disease of interest such as cancer, the between-subject difference observed in cross-sectional data can only provide evidence for an association or correlation between the exposure and disease. The supplementary within-individual dynamics in longitudinal data allow for the inference of a causal nature for such a relationship. Although providing much richer information about the relationship, especially of a causal nature among different outcomes, longitudinal data also present many methodological challenges in study designs and data analyses, the most prominent being correlated responses. As a result, classic models for cross-sectional data analysis such as multiple linear and logistic regressions do not apply to such data.

For example, if we measure an individual's blood pressure twice, the two readings are correlated since they reflect the health condition of this particular individual; if he or she has high blood pressure, both readings tend to be higher than the normal range (positively correlated) despite the variations over repeated assessments. The existence of such within-subject correlations invalidates the independent sampling assumption required for most classic models, precluding applications of statistical methods developed for cross-sectional data based on such an independence assumption. In the blood pressure example, if we ignored the correlations between the two readings and modeled the mean blood pressure using linear regression, then for a sample of n subjects, we would claim to have $2n$ independent observations. However, if the two readings were collected within a very short time span, say 5 seconds apart, they would be almost identical and would certainly not represent independent data comparable to blood pressure readings taken from two different people. In other words, the variation between two within-subject readings would be much smaller than any two between-subject observations, invalidating the model assumption of independent observations and yielding underestimated error variance in this case. Although assessments in most real studies are not spaced as closely as in this extreme example, the within-subject correlation still exists, and ignoring such correlations may yield incorrect inferences.

Thus, methods for longitudinal data must address the within-subject correlations in repeatedly measured outcomes over time. We discuss how to address this issue and popular models for longitudinal data in Chapter 9.

1.3.7   Validity and reliability data analysis

Diagnostic and screening tools are commonly used in clinical and research studies. When a new diagnostic device or instrument (or questionnaire) is introduced for detecting certain medical conditions or latent constructs such as personality, it is important to assess their validity and reliability. For disease diagnosis, even when the true condition can be ascertained by some gold-standard methods such as surgery, it may still be important to use such tests to obtain a timely diagnosis without invasive and costly procedures. For example, SCID is generally considered as the gold standard for diagnosis of depression and other related mental disorders. However, as it involves quite a lengthy (typically several hours) interview of the patient by a clinician, SCID is not normally administered in primary care facilities and large health surveys. Less time-consuming and inexpensive screening tools are often used in large-scale screening studies such as health surveys.

For example, examined in the PPD was the accuracy of several popular screening tools for depression among postpartum women. The outcome of each screening test is the total score obtained by summing binary or Likert-scale responses to individual items in the instrument. Such scores are dimensional, with higher or lower scores indicating higher likelihood of depression. For clinical purposes, the score is dichotomized (or categorized) to indicate diagnosis of depression (or levels of depression severity). To provide valid diagnoses, the cut-points must be selected carefully so that diagnoses based on such instruments correspond to the true disease status. For example, if higher scores indicate higher likelihood of depression, a higher cut-point would give rise to more false negatives, whereas a lower value would lead to more false positives. Thus, the choice of the cut-point needs to balance the false negative and positive rates. Receiver operating characteristic (ROC) curves (Green and Swets, 1966) are commonly used to investigate such a relationship and help determine the choice of optimal cut-point.

For diagnostic instruments that measure latent constructs in mental health and psychosocial research such as depression, eating disorders, personality traits, social support, and quality of life, it is important to assess their internal consistency or validity. Outcomes from such an instrument are typically derived from totaling scores over the items of either the entire instrument or subsets of items, called the *domains*, of the *instrument*. For the total score of an instrument (domain) to be a meaningful dimensional measure, the item scores within the instrument (domain) must be positively correlated, or internally consistent.

Another important consideration for diagnostic instruments is the *test-retest reliability*. As measurement errors are random, scores will vary from repeated administrations of the instrument to the same individual. Like validity and internal consistency, large variations will create problems for replicating research results using the instrument, giving rise to spurious findings. Thus, it is important to assess the test-retest reliability before using an instrument for diagnostic purposes.

In Chapter 10, we discuss measures for validity, internal consistency, test-retest reliability, and other related concepts.

1.3.8   Incomplete data analysis

A common feature in most modern clinical trials but an issue we have avoided in the discussion thus far is the missing value. For example, patients may refuse to answer sensitive private questions such as income and the number of sex partners. The missing value issue is especially common in longitudinal studies where subjects are followed over a period of time. It is common that subjects may drop out of the study for various reasons such as problems with transportation and relocation which may not be related with health outcomes. In many analyses, missing values are simply ignored. For example, in regression analysis, if an observation has missing values in the response and/or in any of the explanatory variables, this observation is excluded from the analysis. This common approach may not be valid if the missing values occur in some systematic way.

For example, in clinical trials, the patient's health condition may be associated with missing visits, with more critical patients more likely to drop out. In treatment control studies, patients may drop out of a study if they do not feel any improvement. It is also possible that they feel better and see no need to continue to receive treatment. Ignoring such treatment-related missing data may give rise to severely biased study outcomes.

Methods for missing values may be necessary even when there are no missing values in the sense that we collected all the data we planned to collect. For example, it is well known that statistical association is different from causal relation. Are the observed differences between two treatment groups the causal effect of the treatment or due to some unobserved confounders? The most popular framework of causal inference is based on the concept of counterfactual outcomes. We assume that there is a potential outcome for each of the subjects for each of the treatments. For the same person, since everything is the same except for the treatment, the difference in the potential outcome represents the causal effect. Of course, we will only be able to observe the one that is realized, i.e., the one corresponding to the treatment the subject actually receives. Thus, the potential outcomes corresponding to the other treatment cannot be observed. We may treat it as a missing value and apply methods for missing values for the causal inference.

Common approaches for dealing with missing values include imputation and weighting. The single imputation approach attempts to fill each missing outcome with a plausible value based on statistical models or other reasonable assumptions. Subsequent analysis then treats the data as if it were really observed. Such an approach usually underestimates the variance of model estimate, because it ignores the sampling variability of the imputed outcome. One way to overcome this problem is to adjust the underestimated variance using approaches such as the mean score method (Reilly and Pepe, 1995). Alternatively, one may simulate the sampling variability in the missing outcome by imputing several plausible values, the so-called multiple imputation procedure (Rubin, 1987). These methods try to complete the missing outcomes so that complete-data methods can be applied.

Another approach for dealing with missing values is to assign weights to observed subjects. This approach has its root in sample survey studies. For cost and efficiency considerations, some underrepresented populations are often oversampled in sample survey studies so reliable estimates can be obtained with a reasonable large sample size. To obtain valid inference for the population as a whole, the sampled subjects are each assigned a weight that is the inverse of the sampling probability, the so-called inverse probability weighting (IPW) method. By treating observed subjects as those sampled, this IPW method can be applied to a wide context such as urn randomization, adaptive sampling, and missing values arising in longitudinal clinical trials.

We discuss these different approaches to missing values as well as their applications to complex sampling designs such as survey studies in Chapter 11.

1.4   Review of Key Statistical Results

Statistical models can be broadly classified into three major categories: parametric, semiparametric, and nonparametric. Under the parametric approach, the distribution of the outcome is assumed to follow some mathematical distributional model defined by a set of parameters such as the normal distribution for continuous outcomes. The method of maximum likelihood is the most popular to provide inference about model parameters. On the other hand, if no such modeling assumption is imposed on the data distribution, then the approach is nonparametric. The parametric approach is generally more efficient, with more power to detect a difference in hypothesis testing. A major weakness of this approach is its dependence on the assumed model; biased estimates may arise if the model assumptions are violated. Inference for nonparametric models is generally more complex, as it involves a parameter vector of an infinite dimension.

The semiparametric approach lies between the two; it assumes a parametric model for some, but not all, of the relationships present in the data. A popular class of semiparametric models for regression analysis, especially for longitudinal data, involves positing a mathematical model for the relationship between the response and the set of explanatory variables, but leaving the data distribution completely unspecified. As no likelihood function can be constructed due to the lack of parametric assumptions for the data distribution, inference for such models usually relies on a set of estimating equations.

Regardless of the approach taken, we must estimate the model parameters and determine the sampling distribution of the estimates, in order to make inferences about the parameters of interest. Except for a few special cases, sampling distributions are generally quite complex and difficult to characterize. The most effective approach approximating the sampling distribution of an estimate is statistical asymptotics or large sample theory. In essence, the theory asserts that the sampling distribution of the estimates of most statistical models can be well approximated by some theoretical distributions such as the normal distribution, with the approximation becoming more accurate as the sample size increases.

In this section, we review the fundamental concepts and techniques in statistical asymptotics that are used to derive inference procedures for the models considered in this book. These concepts and techniques are covered in standard courses on statistical inference. However, to benefit those who may not yet have a formal course on this topic, we will also describe the roles played by these concepts and techniques in the investigation of statistical properties of model estimates. Readers who are familiar with the material or not interested in the theoretical background can safely skip this part.

1.4.1 Central limit theorem and law of large numbers

The basis for statistical asymptotics is the central limit theorem (CLT). For an i.i.d. random sample X_i, the CLT asserts that the sampling distribution of the sample mean, $\overline{X}_n = \frac{1}{n} \sum_{i=1}^n X_i$, becomes more similar to the normal distribution as the sample size gets larger, regardless of the distribution of X_i. Thus, for large n, statistical inference can be based on the approximating normal distribution, and the larger the n, the better the approximation.

For example, if X_i is an i.i.d. sample and $X_i \sim N\left(\mu, \sigma^2\right)$, i.e., X_i follows a normal with mean $\mu = E\left(X_i\right)$ and variance $\sigma^2 = Var\left(X_i\right)$, then it is well known that the sample mean $\overline{X}_n \sim N\left(\mu, \frac{\sigma^2}{n}\right)$. In this special case, the distribution of $\overline{X}_n$ is a normal with mean μ and variance $\frac{\sigma^2}{n}$. However, if X_i is not a normal variate, the sampling distribution of $\overline{X}_n$ is generally non normal. For example, if $X_i \sim Bernoulli\left(p\right)$, a Bernoulli variable with the probability of success p, then $\overline{X}_n$ no longer follows a normal distribution. In fact, the distribution of $\overline{X}_n$ is not even a continuous function. For large n, however, we can approximate the distribution of $\overline{X}_n$ of any random variate X_i using a normal distribution according to the CLT.

To formally characterize such a tendency in statistical asymptotics, let $\mu = E\left(X_i\right)$ and $\sigma^2 = Var\left(X_i\right)$ denote the mean and variance of the i.i.d. sample X_i $(i = 1, 2, \ldots)$. Let $F_n\left(x\right)$ be the cumulative distribution function of the centered sample mean $\sqrt{n}\left(\overline{X}_n - \mu\right)$ and let $\Phi\left(x\right)$ be the cumulative distribution function of a standard normal. Then, for each x, as n approaches infinity $F_n\left(x\right)$ converges to $\Phi\left(x\right)$:

$$F_n\left(x\right) = \Pr\left[\sqrt{n}\left(\overline{X}_n - \mu\right) \leq x\right] \to \Phi\left(\frac{x}{\sigma}\right), \quad n \to \infty, \tag{1.1}$$

where $\Pr\left(\cdot\right)$ denotes probability and "$\to$" denotes convergence. In other words, $\sqrt{n}\left(\overline{X}_n - \mu\right)$ *converges in distribution* to the normal distribution $N\left(0, \sigma^2\right)$. In general, for a sequence of random variables, $X_1, X_2, \ldots$, let F_n be the cumulative distribution function (CDF) of the random variable X_n. We call that the sequence converges to a random variable X in distribution if for each continuous point, t, of the CDF of X, $F(X)$, we have that $F_n(t) \to F(t)$ as $n \to \infty$. Problem 1.3 illustrates that for the meaningful definition of convergence in distribution, we cannot require convergence of the CDFs at discontinuous points of F. However, we will only face the cases where the limiting distributions are normal or chi-square distributions, both of which have continuous CDFs; thus, this does not concern us.

For convenience, the asymptotic distribution is often expressed in several different forms. In this book, we will sometimes denote the asymptotic normal distribution in (1.1) by

$$\sqrt{n}\left(\overline{X}_n - \mu\right) \to_d N\left(0, \sigma^2\right), \quad n \to \infty, \tag{1.2}$$

or by

$$\overline{X}_n - \mu \sim_a N\left(0, \frac{\sigma^2}{n}\right), \quad \text{or} \quad \overline{X}_n \sim_a N\left(\mu, \frac{\sigma^2}{n}\right), \tag{1.3}$$

where "$\to_d$" denotes convergence in distribution and "$\sim_a$" indicates an approximate rather than exact relationship.

Example 1.3
Let $X_i \sim Bernoulli\,(p)$, then $\Pr\,(X_i = 1) = p$, and $\Pr\,(X_i = 0) = 1 - p = q$. Let $\widehat{p}_n = \frac{1}{n}\sum_{i=1}^n X_i$. Then, by the CLT,

$$\sqrt{n}\,(\widehat{p}_n - p) \to_d N\,(0, pq)\,, \quad n \to \infty.$$

So, for large n, $\widehat{p}_n - p \sim_a N\left(0, \frac{pq}{n}\right)$ or $\widehat{p}_n \sim_a N\left(p, \frac{pq}{n}\right)$. ☐

The above example shows that the sample proportion $\widehat{p}_n$, centered around the true proportion p, follows approximately a normal distribution with variance $\frac{pq}{n}$ for large n. More generally, based on the CLT, we can obtain an approximate distribution of the sample mean of an i.i.d. sample for large n no matter how complex its distribution, and we can use the approximate normal distribution to make inference about p (point estimate, confidence intervals, and p-values, etc.). To use such a distribution for inference, however, we must estimate the parameters of the distribution, namely the mean and variance of the random variable. For example, in the Bernoulli example above, the asymptotic distribution $N\left(p, \frac{pq}{n}\right)$ depends on p, which is unknown and must be estimated from the data. In general, there may be many different estimates for the parameters, and optimal estimates are generally difficult to characterize without specific model assumptions. However, we can readily distinguish and thus rule out bad estimates from a pool of good ones by the "consistency" criterion.

Conceptually, an estimate $\widehat{\theta}_n$ of a parameter θ is *consistent* if its error diminishes to 0 as the sample size n increases to infinite. However, we cannot just look at the difference $\left|\widehat{\theta}_n - \theta\right|$ of a realized sample directly and see whether it gets smaller as n increases because $\widehat{\theta}_n$ is random and $\left|\widehat{\theta}_n - \theta\right|$ generally does not exhibit monotonic behavior. For example, let X_i denote a binary outcome with 1 if a head turns up and 0 otherwise on the ith toss of a fair coin. Then, $X_i \sim Bernoulli\left(\frac{1}{2}\right)$. It is quite possible that $\widehat{p}_2 = \frac{1}{2}$ and $\widehat{p}_{100} = \frac{45}{100}$, and thus $\left|\widehat{p}_2 - \frac{1}{2}\right| < \left|\widehat{p}_{100} - \frac{1}{2}\right|$. However, as n gets larger, there will be more stability in $\widehat{p}_n$, and $\left|\widehat{\theta}_n - \theta\right|$ will be more likely to stay small.

Thus, the operational criterion for a consistent estimate is that the probability for the error $\left|\widehat{\theta}_n - \theta\right|$ to exceed any threshold value $\delta > 0$ decreases to 0 when n grows unbounded, i.e.,

$$d_{n,\delta} = \Pr\left(\left|\widehat{\theta}_n - \theta\right| > \delta\right) \to 0, \quad n \to \infty. \tag{1.4}$$

Thus, by using probability, we turn the random sequence $\left|\widehat{\theta}_n - \theta\right|$ into a deterministic one $d_{n,\delta}$, allowing us to investigate the behavior of $\left|\widehat{\theta}_n - \theta\right|$ using convergence criteria for deterministic sequences. The criterion defined in (1.4) is known as *convergence in probability*, or $\widehat{\theta}_n \to_p \theta$. Thus, an estimate $\widehat{\theta}_n$ is *weak consistent* if $\widehat{\theta}_n \to_p \theta$. There is a stronger version of consistency based on *almost sure convergence*. For a sequence of random variables, $X_1, X_2, ...$, defined on the same sample space Ω, we call that it converges to random variable

X almost surely if

$$\Pr\left(w \in \Omega : X_n\left(w\right) \to X\left(w\right) \text{ as } n \to \infty\right) = 1.$$

An estimate $\widehat{\theta}_n$ is strong consistent if $\widehat{\theta}_n \to \theta$ almost surely. For our book, weak consistency is sufficient and we use consistent and weak consistent interchangeably.

Let X_i be an i.i.d. sample with mean μ and variance σ^2 and $\overline{X}_n = \frac{1}{n}\sum_{i=1}^n X_i$. Then, by CLT, $\overline{X}_n - \mu \sim_a N\left(0, \frac{\sigma^2}{n}\right)$ for large n. For any $\delta > 0$, as $n \to \infty$, we have

$$d_{n,\delta} = \Pr\left(\left|\overline{X}_n - \mu\right| > \delta\right) = \Pr\left(\left|\frac{\overline{X}_n - \mu}{\sqrt{\frac{\sigma^2}{n}}}\right| > \frac{\delta}{\sqrt{\frac{\sigma^2}{n}}}\right)$$

$$\approx 2\Phi\left(-\frac{\sqrt{n}\delta}{\sigma}\right) \to 2\Phi\left(-\infty\right) = 0. \tag{1.5}$$

Thus, the sample mean $\overline{X}_n$ is a consistent estimate of the population mean μ. This result is known as the *weak law of large numbers* (LLN). There is also a stronger version of the LLN, based on almost sure convergence.

Note that although we utilized the CLT in the above derivation, the LLN can be proved without using the CLT (see Problem 1.6 for more details). Note also that the CLT and LLN can be generalized to random vectors. Let $\mathbf{X}_i = \left(X_{i1}, X_{i2} \ldots, X_{ik}\right)^\top$ denote an i.i.d. sample of $k \times 1$ random vectors from the k-dimensional Euclidean space $\mathbf{R}^k$ with mean $\boldsymbol{\mu}$ and covariance Σ. Then, we have

$$\sqrt{n}\left(\overline{\mathbf{X}}_n - \boldsymbol{\mu}\right) \to_d N\left(\mathbf{0}, \Sigma\right), \quad n \to \infty, \tag{1.6}$$

or

$$\overline{\mathbf{X}}_n - \boldsymbol{\mu} \sim_a N\left(\mathbf{0}, \frac{1}{n}\Sigma\right), \quad \overline{\mathbf{X}}_n \sim_a N\left(\boldsymbol{\mu}, \frac{1}{n}\Sigma\right), \tag{1.7}$$

where $N\left(\boldsymbol{\mu}, \Sigma\right)$ denotes a $k \times 1$ multivariate normal distribution with mean $\boldsymbol{\mu}$ (a $k \times 1$ vector) and covariance Σ (a $k \times k$ matrix). It follows that $\overline{\mathbf{X}}_n \to_p \boldsymbol{\mu}$, i.e., the random variable $\left\|\mathbf{X}_n - \boldsymbol{\mu}\right\| \to_p 0$, where $\|\cdot\|$ denotes the Euclidean distance in $\mathbf{R}^k$.

1.4.2 Delta method and Slutsky's theorem

In many inference problems in applications, we encounter much more complex statistics than sample means. For example, the sample variance of an i.i.d. sample X_i, $s_n^2 = \frac{1}{n-1}\sum_{i=1}^n \left(X_i - \overline{X}_n\right)^2$, can be used to estimate the populate variance. Since the terms $X_i - \overline{X}_n$ are not independent, we cannot apply LLN or CLT directly to determine the consistency or asymptotic distribution of the sample variance. As will be seen in the subsequent chapters of the book, we often need to determine the consistency and asymptotic distribution of a function of some statistic $g\left(\widehat{\theta}_n\right)$, where $\widehat{\theta}_n$ is some statistic such as the sample mean and $g\left(\cdot\right)$ is some smooth function such as log. The delta method and Slutsky's theorem are the two most popular techniques to facilitate such tasks.

Let $\widehat{\boldsymbol{\theta}}_n$ be a vector-valued statistic following an asymptotic normal distribution, i.e., $\widehat{\boldsymbol{\theta}}_n \sim_a N\left(\boldsymbol{\theta}, \frac{1}{n}\Sigma\right)$. Let $\mathbf{g}\left(\boldsymbol{\theta}\right) = \left(g_1\left(\boldsymbol{\theta}\right), \ldots, g_m\left(\boldsymbol{\theta}\right)\right)^\top$ be a continuous vector-valued function from $\mathbf{R}^k$ to $\mathbf{R}^m$. If $\mathbf{g}\left(\boldsymbol{\theta}\right)$ is differentiable at $\boldsymbol{\theta}$, then the function of the statistic $\mathbf{g}\left(\widehat{\boldsymbol{\theta}}_n\right)$ is also asymptotically normal: $\mathbf{g}\left(\widehat{\boldsymbol{\theta}}_n\right) \sim_a N\left(\mathbf{g}\left(\boldsymbol{\theta}\right), \frac{1}{n}D^\top \Sigma D\right)$, where $D = \frac{\partial}{\partial \boldsymbol{\theta}}\mathbf{g}$ is an $k \times m$

derivative matrix defined as

$$
\frac{\partial}{\partial \boldsymbol{\theta}} \mathbf{g} = \begin{pmatrix} \frac{\partial g_1}{\partial \theta_1} & \cdots & \frac{\partial g_m}{\partial \theta_1} \\ \vdots & \ddots & \vdots \\ \frac{\partial g_1}{\partial \theta_k} & \cdots & \frac{\partial g_m}{\partial \theta_k} \end{pmatrix}_{k \times m} . \tag{1.8}
$$

This *delta method* is quite useful for finding asymptotic distributions of functions of statistics.

Similar to the relationship between CLT and LLN, we also have a version of the delta method for functions of consistent estimates. Let $\widehat{\boldsymbol{\theta}}_n$ be a vector-valued consistent estimate of some vector-valued parameter $\boldsymbol{\theta}$. Let $\mathbf{g}(\boldsymbol{\theta}) = (g_1(\boldsymbol{\theta}), \ldots, g_m(\boldsymbol{\theta}))^\top$ be a continuous vector-valued function from $\mathbf{R}^k$ to $\mathbf{R}^m$. Then, the function $\mathbf{g}\left(\widehat{\boldsymbol{\theta}}_n\right)$ is a consistent estimate of $\mathbf{g}(\boldsymbol{\theta})$. This result helps to find a consistent estimate of the variance of the asymptotic distribution of $\mathbf{g}\left(\widehat{\boldsymbol{\theta}}_n\right)$ above.

Example 1.4

Let $\widehat{\theta}_n$ be a consistent and asymptotically normal estimate of a parameter θ, i.e., $\widehat{\theta}_n \to_p \theta$ and $\widehat{\theta}_n \sim_a N\left(\theta, \frac{1}{n}\sigma^2\right)$. Let $g(\theta) = \exp(\theta)$. Then, $\frac{d}{d\theta}g(\theta) = \exp(\theta)$. By the delta method, the estimate $\exp\left(\widehat{\theta}_n\right)$ for $\exp(\theta_n)$ is also consistent, with the asymptotic distribution $N\left(\exp(\theta), \frac{1}{n}\sigma^2 \exp(2\theta)\right)$. $\Box$

Sometimes, some functions of statistics of interest may involve different modes of convergence, and the following *Slutsky's theorem* is quite useful for finding the asymptotic distributions of such functions of statistics. Let $X_n \to_d X$ and $Y_n \to_p c$, where c is a constant. Then,

1. $X_n + Y_n \sim_a X + c$

2. $X_n Y_n \sim_a cX$

3. If $c \neq 0$, $X_n/Y_n \sim_a X/c$.

Example 1.5

Consider the sample variance $s_n^2 = \frac{1}{n-1}\sum_{i=1}^n \left(X_i - \overline{X}_n\right)^2$ of an i.i.d. sample X_i with mean $\mu = E(X_i)$ and variance $\sigma^2 = Var(X_i)$. We show that s_n^2 has an asymptotic normal distribution if the fourth centered moment of X_i exists. First reexpress s_n^2 as

$$
\frac{n-1}{n}s_n^2 - \sigma^2 = \frac{1}{n}\sum_{i=1}^n \left[(X_i - \mu)^2 - \sigma^2\right] - \left(\overline{X}_n - \mu\right)^2 .
$$

By CLT,

$$
\frac{\sqrt{n}}{n}\sum_{i=1}^n \left[(X_i - \mu)^2 - \sigma^2\right] \sim_a N\left(0, Var\left((X_i - \mu)^2\right)\right)
$$

and $\sqrt{n}\left(\overline{X}_n - \mu\right) \sim_a N\left(0, \sigma^2\right)$. By LLN, $\overline{X}_n - \mu = \frac{1}{n}\sum_{i=1}^n (X_i - \mu) \to_p 0$. Thus, by Slutsky's theorem,

$$
\sqrt{n}\left(\overline{X}_n - \mu\right)^2 \sim_a 0, \text{ and } s_n^2 - \sigma^2 \sim_a N\left(0, \frac{1}{n}Var\left((X_i - \mu)^2\right)\right).
$$

Since $Var\left((X_i - \mu)^2\right) = \mu_4 - \sigma^4$, where $\mu_4 = E\left(X_1 - \mu\right)^4$ is the fourth centered moment of X_i, we can estimate the asymptotic variance of s_n^2 by $\frac{1}{n}\left(\widehat{\mu}_4 - \left(s_n^2\right)^2\right)$, with $\widehat{\mu} = \frac{1}{n}\sum_{i=1}^n X_i$ and $\widehat{\mu}_4 = \frac{1}{n}\sum_{i=1}^n \left(X_i - \widehat{\mu}\right)^4$. $\qquad\qquad\qquad\qquad\qquad\qquad\qquad\qquad\quad$ ☐

1.4.3 Maximum likelihood estimate

One of the most popular inference approaches for parametric models is the maximum likelihood. Let $f(x, \theta)$ denote either the probability density function for a continuous X_i or the probability distribution function for a discrete outcome; i.e., $f(x, \theta)$ is the likelihood that $X_i = x$, where $\theta \in \mathbf{R}^m$ is the parameter vector. Given an i.i.d. sample of X_i $(1 \leq i \leq n)$, the likelihood function is $L(\theta) = \prod_{i=1}^n f(X_i, \theta)$. In most applications, there is some constraint on θ so it can only vary within a subset of $\mathbf{R}^m$. For example, for $X_i \sim N(\mu, \sigma^2)$, we assume that $\sigma^2 > 0$. Thus, the domain of the parameter space of $\theta = \left(\mu, \sigma^2\right)^\top$ is a subset of $\mathbf{R}^2$.

For inference, when the X_i are i.i.d., we always use the logarithm of $L(\theta)$, or the log-likelihood function:

$$l(\theta) = \sum_{i=1}^n l_i(\theta, X_i) = \sum_{i=1}^n \log f(X_i, \theta), \quad \theta \in D, \qquad (1.9)$$

where $l_i(\theta, X_i) = \log f(X_i, \theta)$ is the likelihood for the ith observation and D denotes the domain of θ. Given an i.i.d. sample X_i $(1 \leq i \leq n)$, the log-likelihood in (1.9) is a function of θ. If the maximum of the (log-)likelihood is achieved at a point $\widehat{\theta}_n$ in D, then $\widehat{\theta}_n$ is the *maximum likelihood estimate* (MLE) of θ. If the MLE is an interior point of D, then the derivative of $l(\theta)$ at $\widehat{\theta}_n$ must be 0, and $\widehat{\theta}_n$ is obtained by solving the following *score equations*:

$$\frac{\partial}{\partial \theta} l(\theta) = \sum_{i=1}^n \frac{\partial}{\partial \theta} l_i(\theta, X_i) = \sum_{i=1}^n \frac{1}{f(X_i, \theta)} \frac{\partial}{\partial \theta} f(X_i, \theta) = \mathbf{0}. \qquad (1.10)$$

Note that $\mathbf{U}_i(\theta, X_i) = \frac{\partial}{\partial \theta} l_i(\theta, X_i)$ is called the *score function* for the ith observation. In general, we have $E\left(\mathbf{U}_i(\theta, X_i)\right) = \mathbf{0}$, and this provides another argument of solving the score equations to find an estimate. The variance of the score function, $Var\left(\mathbf{U}_i(\theta, X_i)\right) = -E\left[\frac{\partial^2}{\partial\theta\partial\theta^\top}\log f(X_i, \theta)\right]$, is known as the *Fisher information* matrix (see Problem 1.8).

In most applications, the score equations (1.10) cannot be solved in closed form, but the MLE $\widehat{\theta}_n$ can be obtained numerically using the Newton-Raphson method (Kowalski and Tu, 2008). In general, the MLE is consistent and asymptotically normal, $\widehat{\theta}_n \sim_a N\left(\theta, \frac{1}{n}\Sigma\right)$, where $\Sigma = \mathbf{I}^{-1}(\theta)$ and $\mathbf{I}(\theta) = -E\left[\frac{\partial^2}{\partial\theta\partial\theta^\top}\log f(X_i, \theta)\right]$ is the Fisher information matrix. If $\mathbf{I}(\theta)$ cannot be evaluated in closed form, it can be estimated by $\frac{1}{n}\mathbf{I}_n(\theta)$, where $\mathbf{I}_n(\theta) = -\sum_{i=1}^n \frac{\partial^2}{\partial\theta\partial\theta^\top}\log f(X_i, \theta)$ is the observed Fisher information. Thus, $\widehat{\theta}_n \sim_a N\left(\theta, \mathbf{I}_n^{-1}(\theta)\right)$. When $\mathbf{I}(\theta)$ and $\mathbf{I}_n(\theta)$ involve θ, we may simply substitute $\widehat{\theta}_n$ for θ in $\mathbf{I}(\theta)$ and $\mathbf{I}_n(\theta)$ to estimate the asymptotic variance of MLE; however, the one based on the observed information is generally preferred (Efron and Hinkley, 1978).

Further, the MLE is generally asymptotically *efficient*. If $\widehat{\theta}_n$ is a consistent estimate of a parameter θ for a parametric model defined by $f(X_i, \theta)$ with an asymptotic normal distribution $N\left(\theta, \frac{1}{n}\Sigma\right)$, then $\Sigma - \mathbf{I}^{-1}(\theta)$ is nonnegative definite, or simply $\Sigma \geq \mathbf{I}^{-1}(\theta)$.

Example 1.6

If $X_i \sim N(\mu, \sigma^2)$, let us compute the MLE of μ and its asymptotic variance.

The likelihood is

$$L\left(\mathbf{\theta}\right) = \prod_{i=1}^{n} \left[\frac{1}{\sqrt{2\pi}\sigma} \exp\left(-\frac{(X_i - \mu)^2}{2\sigma^2} \right) \right] = \frac{1}{(\sqrt{2\pi}\sigma)^n} \exp\left(-\frac{\sum_{i=1}^{n}(X_i - \mu)^2}{2\sigma^2} \right).$$

Thus, the log-likelihood is $l\left(\mathbf{\theta}\right) = -\log\left[(2\pi)^{n/2}\sigma^n\right] - \frac{1}{2\sigma^2}\sum_{i=1}^{n}(X_i - \mu)^2$. By taking the derivative with respect to μ, we obtain the score equation $\frac{1}{\sigma^2}\sum_{i=1}^{n}(X_i - \mu) = 0$. The solution is $\widehat{\mu} = \overline{X}_n$. The Fisher information is $I\left(\mathbf{\theta}\right) = -E\left[\frac{\partial^2}{\partial\theta\partial\theta^\top}\log f\left(X_i, \mathbf{\theta}\right)\right] = \frac{1}{\sigma^2}$. Thus, the asymptotic variance of $\widehat{\mu}$ is $\frac{\sigma^2}{n}$. In the special case, $\widehat{\mu}$ has an exact rather than asymptotic normal distribution. ▯

In regression analysis, we have one outcome designated as the response or dependent variable Y and a set of other variables specified as explanatory variables or independent variables predictors, and covariates, $\mathbf{X}$. We are interested in the change of the response as a function of the explanatory variables. To account for the random variability in the response for the given values of the explanatory variables, we assume a distribution of the response conditioning on the explanatory variables, $f\left(Y \mid \mathbf{X}, \mathbf{\theta}\right)$ ($\mathbf{\theta} \in D$). Thus, given an independent sample of Y_i and $\mathbf{X}_i$ ($1 \leq i \leq n$), the likelihood or log-likelihood function is constructed based on the conditional probability or distribution function:

$$L\left(\mathbf{\theta}\right) = \prod_{i=1}^{n} f\left(Y_i \mid \mathbf{X}_i, \mathbf{\theta}\right), \quad l\left(\mathbf{\theta}\right) = \sum_{i=1}^{n} \log f\left(Y_i \mid \mathbf{X}_i, \mathbf{\theta}\right). \tag{1.11}$$

For example, in linear regression, we assume the response Y_i conditional on the explanatory variables $\mathbf{X}_i$ follows a normal distribution $N\left(\mathbf{X}_i^\top\mathbf{\beta}, \sigma^2\right)$ with mean $\mathbf{X}_i^\top\mathbf{\beta}$ and variance σ^2, where $\mathbf{X}_i^\top$ denotes the transpose of the vector $\mathbf{X}_i$ and $\mathbf{\beta}$ is the vector of parameters relating $\mathbf{X}_i$ to the mean of Y_i. In this case, $f\left(Y_i \mid \mathbf{X}_i, \mathbf{\theta}\right)$ is the probability density function of $N\left(\mathbf{X}_i^\top\mathbf{\beta}, \sigma^2\right)$. We can write the model as

$$Y_i = \mathbf{X}_i^\top\mathbf{\beta} + \epsilon_i, \quad \epsilon_i \sim N\left(0, \sigma^2\right), \quad 1 \leq i \leq n. \tag{1.12}$$

Thus, in regression models, interest lies in the relationship between Y_i and $\mathbf{X}_i$, while accounting for random variation of Y_i given the values of $\mathbf{X}_i$ and the distribution of $\mathbf{X}_i$ is of no interest.

Example 1.7

For the linear regression in (1.12), the log-likelihood function is given by

$$l_n\left(\mathbf{\theta}\right) = -\frac{n}{2}\log\left(2\pi\sigma^2\right) - \frac{1}{2\sigma^2}\sum_{i=1}^{n}\left(Y_i - \mathbf{X}_i^\top\mathbf{\beta}\right)^2.$$

Using the properties of differentiation of vector-valued function (see Problem 1.11), the vector score equation is readily calculated (see Problem 1.12):

$$\frac{\partial}{\partial\mathbf{\beta}}l_n = \frac{1}{\sigma^2}\sum_{i=1}^{n}\left(\mathbf{X}_iY_i - \mathbf{X}_i\mathbf{X}_i^\top\mathbf{\beta}\right) = \mathbf{0}, \tag{1.13}$$

where $\frac{\partial}{\partial\mathbf{\beta}}l_n = \left(\frac{\partial}{\partial\beta_1}l_n, \ldots, \frac{\partial}{\partial\beta_m}l_n\right)^\top$ is a column vector. By solving for $\mathbf{\beta}$, we obtain the MLE $\widehat{\mathbf{\beta}} = \left(\sum_{i=1}^{n}\mathbf{X}_i\mathbf{X}_i^\top\right)^{-1}\left(\sum_{i=1}^{n}\mathbf{X}_iY_i\right)$. The second derivative of $\log f\left(Y_i \mid \mathbf{X}_i, \mathbf{\theta}\right)$ is

$$\frac{\partial^2}{\partial\mathbf{\beta}\partial\mathbf{\beta}^\top}\log f\left(y_i \mid \mathbf{\theta}\right) = -\frac{1}{\sigma^2}\mathbf{X}_i\mathbf{X}_i^\top. \tag{1.14}$$

Thus, the observed Fisher information matrix is $I_n\left(\boldsymbol{\theta}\right) = \sum_{i=1}^{n} \frac{1}{\sigma^2} \mathbf{X}_i \mathbf{X}_i^\top$ and the variance of $\widehat{\boldsymbol{\beta}}$ is $V = \sigma^2 \left(\sum_{i=1}^{n} \mathbf{X}_i \mathbf{X}_i^\top\right)^{-1}$. A consistent estimate of V is given by $\widehat{V} = \widehat{\sigma}^2 \left(\sum_{i=1}^{n} \mathbf{X}_i \mathbf{X}_i^\top\right)^{-1}$ with $\widehat{\sigma}^2 = \frac{1}{n} \sum_{i=1}^{n} \left(Y_i - \mathbf{X}_i^\top \widehat{\boldsymbol{\beta}}\right)^2$. $\qquad\square$

1.4.4 Estimating equations

Under the parametric approach, a parametric distribution is assumed for the data. For example, in the linear regression model discussed in Section 1.4.3, the response Y conditional on $\mathbf{X}$ is assumed to follow a normal distribution $N\left(\mathbf{X}^\top \boldsymbol{\beta}, \sigma^2\right)$. Such a normality assumption may be violated by the data at hand, so that inference about the parameters of interest may be wrong. For example, if the response Y is positive, inference based on the normal assumption is likely to be incorrect. One approach is to apply some transformation such as the logarithmic function to help reduce the skewness so that the normal distribution can approximate the data distribution. Another approach is to remove the distributional assumption completely and base inference on a different paradigm. The method of estimating equations based on the method-of-moments estimate is one of the most popular procedures for inference for the latter distribution-free models.

A set of estimating equations for a vector of parameters of interest $\boldsymbol{\theta}$ has the form

$$W_n\left(\boldsymbol{\theta}\right) = \frac{1}{n} \sum_{i=1}^{n} \boldsymbol{\Psi}_i\left(Y_i, \boldsymbol{\theta}\right) = \mathbf{0}, \qquad (1.15)$$

where $\boldsymbol{\Psi}_i\left(Y_i, \boldsymbol{\theta}\right)$ is a vector-valued function of Y_i and $\boldsymbol{\theta}$. The solution is called an *M-estimator* of $\boldsymbol{\theta}$ (Huber, 1964). The estimating equation (1.15) is *unbiased* if $E\left[\boldsymbol{\Psi}_i\left(Y_i, \boldsymbol{\theta}\right)\right] = \mathbf{0}$. The estimate obtained by solving a set of unbiased estimating equations is asymptotically consistent and follows a normal distribution. More precisely, let $\widehat{\boldsymbol{\theta}}_n$ denote the estimating equation estimate, the solution to (1.15). Then, $\widehat{\boldsymbol{\theta}}_n$ has the following asymptotic distribution:

$$\sqrt{n}\left(\widehat{\boldsymbol{\theta}}_n - \boldsymbol{\theta}\right) \to_d N\left(\mathbf{0}, \Sigma_{\boldsymbol{\theta}} = A\left(\boldsymbol{\theta}\right)^{-1} B\left(\boldsymbol{\theta}\right) A\left(\boldsymbol{\theta}\right)^{-\top}\right), \qquad (1.16)$$

where $A\left(\boldsymbol{\theta}\right) = E\left[\frac{\partial}{\partial \boldsymbol{\theta}} \boldsymbol{\Psi}_i\left(Y_i, \boldsymbol{\theta}\right)\right]$ and $B\left(\boldsymbol{\theta}\right) = Var\left(\boldsymbol{\Psi}_i\left(Y_i, \boldsymbol{\theta}\right)\right)$. The covariance matrix $\Sigma_{\boldsymbol{\theta}}$ can be estimated using their corresponding sample moments:

$$\widehat{\Sigma}_{\boldsymbol{\theta}} = \widehat{A}_n^{-1} \widehat{B}_n \widehat{A}_n^{-\top}, \quad \widehat{A}_n = \frac{1}{n} \sum_{i=1}^{n} \frac{\partial}{\partial \boldsymbol{\theta}} \boldsymbol{\Psi}_i\left(Y_i, \widehat{\boldsymbol{\theta}}_n\right), \quad \widehat{B}_n = \frac{1}{n} \boldsymbol{\Psi}_i\left(Y_i, \widehat{\boldsymbol{\theta}}_n\right) \boldsymbol{\Psi}_i^\top\left(Y_i, \widehat{\boldsymbol{\theta}}_n\right). \qquad (1.17)$$

This estimate is known as the *sandwich variance estimate* of the variance of $\widehat{\boldsymbol{\theta}}_n$.

The estimating equation approach also applies to regression models. Consider the following semiparametric model:

$$E\left(Y_i \mid \mathbf{X}_i\right) = f\left(\mathbf{X}_i; \boldsymbol{\theta}\right), \quad 1 \le i \le n,$$

where f is a known function of the vector of parameters $\boldsymbol{\theta}$. We assume a parametric form for $E\left(Y_i \mid \mathbf{X}_i\right)$, but we do not assume any specific distribution for $Y_i \mid \mathbf{X}_i$. The estimating equations usually have the form

$$W_n\left(\boldsymbol{\theta}\right) = \frac{1}{n} \sum_{i=1}^{n} G(\mathbf{X}_i)\left[Y_i - f\left(\mathbf{X}_i; \boldsymbol{\theta}\right)\right] = 0, \qquad (1.18)$$

where $G(\mathbf{X}_i)$ is some known vector-valued function of $\mathbf{X}_i$ and $\boldsymbol{\theta}$. Usually we may choose $G(\mathbf{X}_i) = \frac{\partial}{\partial \boldsymbol{\theta}} f\left(\mathbf{X}_i; \boldsymbol{\theta}\right) Var\left(Y_i \mid \mathbf{X}_i; \boldsymbol{\theta}\right)^{-1}$, which is optimal in sense that the estimate has minimum variance among the class of estimating equations (EEs) (Tsiatis, 2006). Since $W_n(\boldsymbol{\theta})$ is unbiased, i.e., $E\left[G(\mathbf{X}_i)(Y_i - f(\mathbf{X}_i;\boldsymbol{\theta}))\right] = \mathbf{0}$, the estimate $\widehat{\boldsymbol{\theta}}_n$ as the solution to (1.17) is again consistent and asymptotically normal. The asymptotic variance and a consistent estimate are given by the same expressions in (1.16) and (1.17), except for substituting $G(\mathbf{X}_i)\left[Y_i - f(\mathbf{X}_i;\boldsymbol{\theta})\right]$ for $\boldsymbol{\Psi}_i(Y_i, \boldsymbol{\theta})$.

Example 1.8

Consider the linear regression model in (1.12). If the normal assumption for the error term ϵ_i is replaced by any distribution with mean 0 and variance σ^2, then the model becomes

$$Y_i = \mathbf{X}_i^\top \boldsymbol{\beta} + \epsilon_i, \quad \epsilon_i \sim \left(0, \sigma^2\right), \quad 1 \le i \le n, \tag{1.19}$$

where $\left(0, \sigma^2\right)$ denotes any distribution with mean 0 and variance σ^2. The revised model is distribution free since it does not impose any assumption on the distribution of ϵ_i other than a zero mean and finite variance. It is impossible to write down the likelihood, and inference about $\boldsymbol{\beta}$ cannot be based on maximum likelihood.

Under (1.19) and using the theorem of iterated conditional expectations, it is readily checked that

$$E\left[\mathbf{X}_i\left(Y_i - \mathbf{X}_i^\top \boldsymbol{\beta}\right)\right] = E\left\{E\left[\mathbf{X}_i\left(Y_i - \mathbf{X}_i^\top \boldsymbol{\beta}\right)\right] \mid \mathbf{X}_i\right\}$$
$$= E\left\{\mathbf{X}_i E\left[\left(Y_i - \mathbf{X}_i^\top \boldsymbol{\beta}\right)\right] \mid \mathbf{X}_i\right\} = E\left\{\mathbf{X}_i\left[E\left(Y_i \mid \mathbf{X}_i\right) - \mathbf{X}_i^\top \boldsymbol{\beta}\right]\right\} = \mathbf{0}.$$

Thus, the following EEs are unbiased:

$$W_n(\boldsymbol{\theta}) = \sum_{i=1}^n \boldsymbol{\Psi}_i\left(Y_i, \mathbf{X}_i, \boldsymbol{\theta}\right) = \sum_{i=1}^n \mathbf{X}_i\left(Y_i - \mathbf{X}_i^\top \boldsymbol{\beta}\right) = \mathbf{0}. \tag{1.20}$$

Solving the equations, we obtain $\widehat{\boldsymbol{\beta}} = \left(\sum_{i=1}^n \mathbf{X}_i \mathbf{X}_i^\top\right)^{-1}\left(\sum_{i=1}^n \mathbf{X}_i Y_i\right)$, which is the same as the MLE obtained by applying maximum likelihood to (1.12).

The two estimates differ in their asymptotic variances. As derived in Example 1.7, the asymptotic variance for the MLE is $\Sigma_{MLE} = \frac{1}{n}\sigma^2 E^{-1}\left(\mathbf{X}_i \mathbf{X}_i^\top\right)$ with a consistent estimate $\widehat{\Sigma}_{MLE} = \frac{\widehat{\sigma}^2}{n}\left(\frac{1}{n}\sum_{i=1}^n \mathbf{X}_i \mathbf{X}_i^\top\right)^{-1}$. The validity of the asymptotic MLE inference actually does not require the normality assumption, but it still depends on the assumption of homoscedasticity of the errors. In fact, without the normality assumption, the EE (1.20) can be obtained by minimizing $\sum_{i=1}^n\left(Y_i - \mathbf{X}_i^\top \boldsymbol{\beta}\right)^2$. This ordinary least square (OLS) method actually predates the MLE. The point estimate by OLS is the same, and it is the best among all unbiased linear estimators (BLUE) of $\boldsymbol{\beta}$; here "best" means the one with minimum variance. OLS inference estimates σ^2 by $\frac{1}{n-p}\sum_{i=1}^n\left(Y_i - \mathbf{X}_i^\top \widehat{\boldsymbol{\beta}}\right)^2$ where p is the number of parameters in $\boldsymbol{\beta}$; thus, it is asymptotically equivalent to the MLE. Both the MLE and OLS estimates are invalid in the presence of heteroscedasticity where $\text{Var}(\epsilon_i)$ varies with the subjects.

The asymptotic variance of the estimating equations estimate is

$$\Sigma_{EEE} = \frac{1}{n}E^{-1}\left(\mathbf{X}_i \mathbf{X}_i^\top\right) E\left[\mathbf{X}_i\left(Y_i - \mathbf{X}_i^\top \boldsymbol{\beta}\right)^2 \mathbf{X}_i^\top\right] E^{-1}\left(\mathbf{X}_i \mathbf{X}_i^\top\right). \tag{1.21}$$

In cases of homoscedasticity, $E\left[\mathbf{X}_i\left(Y_i - \mathbf{X}_i^\top \boldsymbol{\beta}\right)^2 \mathbf{X}_i^\top\right] = \sigma^2 E\left[\mathbf{X}_i \mathbf{X}_i^\top\right]$ and the variance is reduced to that of the MLE. The advantage of the EE inference is that it can also be applied to heteroscedasticity situations, as the EE (1.20) is still unbiased. We can simply plug in $\widehat{\beta}$ for β in (1.21) to obtain a sandwich variance estimate

$$\widehat{\mathbf{\Sigma}}_{EEE} = \left(\sum_{i=1}^{n} \mathbf{X}_i \mathbf{X}_i^{\top}\right)^{-1} \left(\sum_{i=1}^{n} \mathbf{X}_i \left(Y_i - \mathbf{X}_i^{\top}\widehat{\beta}\right)^2 \mathbf{X}_i^{\top}\right) \left(\sum_{i=1}^{n} \mathbf{X}_i \mathbf{X}_i^{\top}\right)^{-1}.$$

However, the EE inference based on (1.20) may not be efficient. When $\sigma_i^2 = \mathrm{Var}(\epsilon_i)$ is known, we may use weighted EE to obtain optimal efficiency. See Problem 1.10. □

The estimating equation approach yields more robust estimates than MLEs. However, EE estimates are usually less efficient than MLEs. If parametric models are appropriate for the data at hand, it is best to use MLE for inference. In many real data applications, it is quite difficult to validate parametric assumptions especially for complex longitudinal models with missing data. Distribution-free models are particularly useful in such settings to provide valid inference.

1.4.5 U-statistics

The asymptotic results discussed in Sections 1.4.1–1.4.4 apply to most, but not all, of the models in this book when used to find the asymptotic distributions of parameter estimates. For example, a popular measure of association between two ordinal categorical outcomes X and Y is the Goodman-Kruskal γ. Consider an i.i.d. sample of bivariate ordinal outcomes $\mathbf{Z}_i = (X_i, Y_i)^{\top}$ $(1 \le i \le n)$. Suppose X_i (Y_i) has K (M) levels indexed by k (m). For a pair of subjects $\mathbf{Z}_i = (k, m)^{\top}$ and $\mathbf{Z}_j = (k', m')^{\top}$, concordance and discordance are defined as follows:

$$(\mathbf{Z}_i, \mathbf{Z}_j) \text{ is } \begin{cases} \text{concordant if } X_i < (>)X_j, \ Y_i < (>)Y_j \\ \text{discordant if } X_i > (<)X_j, \ Y_i < (>)Y_j \\ \text{neither} \quad \text{if otherwise.} \end{cases}$$

Let p_s and p_d denote the probability of concordant and discordant pairs. The Goodman-Kruskal γ $\left(= \frac{p_s - p_d}{p_s + p_d}\right)$ is one of the measures for the association between X and Y. Let $\theta = (p_s, p_d)^{\top}$, then γ is a function of θ, denoted by $f(\theta)$. Also, let C_2^n denote the set of all distinct combinations of two indices (i, j) from the integer set $\{1, 2, \ldots, n\}$ and $I_{\{A\}}$ be a binary indicator with $I_{\{A\}} = 1$ if the condition A is true and 0 if otherwise. We can estimate γ by $\widehat{\gamma} = f\left(\widehat{\theta}\right)$, where $\widehat{\theta}$ is the vector of sample proportions,

$$\widehat{\theta} = \binom{n}{2}^{-1} \sum_{(i,j) \in C_2^n} \mathbf{h}(\mathbf{Z}_i, \mathbf{Z}_j), \tag{1.22}$$

where $\mathbf{h}(\mathbf{Z}_i, \mathbf{Z}_j) = \left(I_{\{(X_i - X_j)(Y_i - Y_j) > 0\}}, I_{\{(X_i - X_j)(Y_i - Y_j) < 0\}}\right)^{\top}$. If $\widehat{\theta}$ has an asymptotic normal distribution, then so does $\widehat{\gamma}$ by the delta method. It is readily checked that $\widehat{\theta}$ above is an unbiased estimate of θ, i.e., $E\left(\widehat{\theta}\right) = \theta$ (see Problem 1.14). However, $\widehat{\theta}$ is not a sum of i.i.d. terms, although all the terms are identically distributed. For example, $\mathbf{h}(\mathbf{Z}_1, \mathbf{Z}_2)$ and $\mathbf{h}(\mathbf{Z}_1, \mathbf{Z}_3)$ are not independent since they share $\mathbf{Z}_1$ in common. As a result, the CLT in Section 1.4.1 cannot be applied to show the asymptotic normality of $\widehat{\theta}$.

Although the sum in (1.22) does not have i.i.d. terms, it has a particular structure. Such a structured sum has been extensively studied and is known as a *U-statistic*. Let $\mathbf{X}_i$ $(1 \le i \le n)$ be an i.i.d. sample of random vectors and $\mathbf{h}$ a vector-valued symmetric function with m arguments. Consider the parameter vector of interest $\boldsymbol{\theta} = E\left[\mathbf{h}\left(\mathbf{X}_1, \ldots, \mathbf{X}_m\right)\right]$. We estimate $\boldsymbol{\theta}$ by $\widehat{\boldsymbol{\theta}} = \binom{n}{m}^{-1} \sum_{(i_1, \ldots, i_m) \in C_m^n} \mathbf{h}\left(\mathbf{X}_{i_1}, \ldots, \mathbf{X}_{i_m}\right)$, which can be shown to be an unbiased estimate of $\boldsymbol{\theta}$ (see Problem 1.14). Further, $\widehat{\boldsymbol{\theta}}$ is consistent and asymptotically normal. Let

$$\mathbf{h}_1\left(\mathbf{X}_1\right) = E\left[\mathbf{h}\left(\mathbf{X}_1, \ldots, \mathbf{X}_m\right) \mid \mathbf{X}_1\right] \quad \text{and} \quad \widetilde{\mathbf{h}}_1\left(\mathbf{X}_1\right) = \mathbf{h}_1\left(\mathbf{X}_1\right) - \boldsymbol{\theta}, \tag{1.23}$$

then $\widehat{\boldsymbol{\theta}} \sim_a N\left(\boldsymbol{\theta}, \frac{1}{n} m^2 Var\left[\widetilde{\mathbf{h}}_1\left(\mathbf{X}_1\right)\right]\right)$. For a proof of the theorem and more applications, check Kowalski and Tu (2008).

Example 1.9
Consider an i.i.d. sample X_i with a finite mean $\mu = E(X)$ and variance σ^2 ($1 \le i \le n$). Let $h(x) = x$. Then, $\mu = E\left[h\left(X_i\right)\right] = E\left(X_i\right)$. The U-statistic estimate of μ is $\widehat{\mu} = \binom{n}{1}^{-1} \sum_{i \in C_1^n} h\left(X_i\right) = \frac{1}{n} \sum_{i=1}^{n} X_i$, which is just the sample mean of μ.

In this case $m = 1$, $h_1\left(X_1\right) = E\left[h\left(X_1\right) \mid X_1\right] = X_1$, $\widetilde{h}\left(X_1\right) = X_1 - \mu$, and $Var\left[\widetilde{h}\left(X_1\right)\right] = Var\left(X_1\right) = \sigma^2$. Thus, $\widehat{\mu} \sim_a N(\mu, \frac{1}{n}\sigma^2)$. Note that the asymptotic normality of $\widehat{\mu}$ can also be obtained by applying the CLT since $\widehat{\mu}$ is a sum of i.i.d. terms in this special case. $\quad$ ▯

Example 1.10
Let us use U-statistics to estimate the variance in Example 1.9.

Let $h\left(x_1, x_2\right) = \frac{1}{2}\left(x_1 - x_2\right)^2$. Since $E\left[h(X_1, X_2)\right] = \frac{1}{2} E\left(X_1^2 - 2X_1 X_2 + X_2^2\right) = Var(X)$ estimate of 2, the U-statistic

$$\widehat{\sigma}_n^2 = \binom{n}{2}^{-1} \sum_{(i,j) \in C_2^n} h\left(X_i, X_j\right) = \frac{2}{n(n-1)} \sum_{1 \le i < j \le n} \frac{1}{2}\left(X_i - X_j\right)^2 \tag{1.24}$$

is an unbiased estimate of σ^2. In fact, it is readily checked that $\widehat{\sigma}_n^2$ above is actually the sample variance s_n^2 in Example 1.5 (see Problem 1.15). Further, since

$$\widetilde{h}\left(X_1\right) = E\left[h\left(X_1, X_2\right) \mid X_1\right] - \sigma^2 = \frac{1}{2}\left[\left(X_1 - \mu\right)^2 - \sigma^2\right],$$

it follows that

$$Var\left(\widetilde{h}_1\left(X_1\right)\right) = \frac{1}{4} Var\left(\left(X_1 - \mu\right)^2\right) = \frac{1}{4}\left(\mu_4 - \sigma^4\right).$$

Thus, $\widehat{\sigma}_n^2 - \sigma^2 \sim_a N\left(0, \frac{1}{n}\left(\mu_4 - \sigma^4\right)\right)$. This is the same asymptotic distribution of the sample variance s_n^2 we derived in Example 1.5 using a different approach. $\quad$ ▯

It is not possible to express the asymptotic variance for the U-statistic vector in (1.22) in closed form. However, it can be shown that (see Problem 1.16)

$$Var\left[E\left[\mathbf{h}\left(\mathbf{Z}_1, \mathbf{Z}_2\right) \mid \mathbf{Z}_1\right]\right] = E\left(\mathbf{h}\left(\mathbf{Z}_1, \mathbf{Z}_2\right) \mathbf{h}^\top\left(\mathbf{Z}_1, \mathbf{Z}_3\right)\right) - \boldsymbol{\theta}\boldsymbol{\theta}^\top. \tag{1.25}$$

We can estimate $\boldsymbol{\theta}\boldsymbol{\theta}^\top$ by $\widehat{\boldsymbol{\theta}}\widehat{\boldsymbol{\theta}}^\top$. To estimate $\Psi = E\left[\mathbf{h}\left(\mathbf{Z}_1, \mathbf{Z}_2\right) \mathbf{h}^\top\left(\mathbf{Z}_1, \mathbf{Z}_3\right)\right]$ in (1.25), we can construct another U-statistic. Let

$$\mathbf{g}\left(\mathbf{Z}_1, \mathbf{Z}_2, \mathbf{Z}_3\right) = \mathbf{h}\left(\mathbf{Z}_1, \mathbf{Z}_2\right) \mathbf{h}^\top\left(\mathbf{Z}_1, \mathbf{Z}_3\right),$$

$$\widetilde{\mathbf{g}}\left(\mathbf{Z}_1, \mathbf{Z}_2, \mathbf{Z}_3\right) = \frac{1}{3}\left(\mathbf{g}\left(\mathbf{Z}_1, \mathbf{Z}_2, \mathbf{Z}_3\right) + \mathbf{g}\left(\mathbf{Z}_2, \mathbf{Z}_1, \mathbf{Z}_3\right) + \mathbf{g}\left(\mathbf{Z}_3, \mathbf{Z}_2, \mathbf{Z}_1\right)\right).$$

Then, $\widetilde{\mathbf{g}}(\mathbf{Z}_1, \mathbf{Z}_2, \mathbf{Z}_3)$ is a symmetric with respect to the permutations of $\mathbf{Z}_1$, $\mathbf{Z}_2$, and $\mathbf{Z}_3$. The matrix $\widehat{\Psi} = \binom{n}{3}^{-1} \sum_{(i,j,k) \in C_3^n} \widetilde{\mathbf{g}}(\mathbf{Z}_i, \mathbf{Z}_j, \mathbf{Z}_k)$ is a U-statistic and thus is a consistent estimate of Ψ.

1.4.5.1 Multigroup U-statistics

The above technique of U-statistics can also be applied to estimate parameters that involve multiple independent groups. We give a brief review for the two-sample case as this is what we will need in this book, and it would be sufficient to suggest how to generalize it to general situations. Let X_i $(1 \leq i \leq n)$ and Y_j $(1 \leq j \leq m)$ be two independent i.i.d. samples of random vectors from two different populations. Let h be a vector-valued symmetric function with $s + t$ arguments. Consider the parameter vector of interest $\theta = E[\mathbf{h}(\mathbf{X}_1, \ldots, \mathbf{X}_s, \mathbf{Y}_1, \ldots, \mathbf{Y}_t)]$. We can estimate θ by

$$\widehat{\theta} = \binom{n}{s}^{-1} \binom{m}{t}^{-1} \sum_{(i_1,\ldots,i_s) \in C_s^n} \sum_{(j_1,\ldots,j_t) \in C_t^m} \mathbf{h}(\mathbf{X}_{i_1}, \ldots, \mathbf{X}_{i_s}, \mathbf{Y}_{j_1}, \ldots, \mathbf{Y}_{j_t}).$$

It can be shown that the estimate is an unbiased estimate of θ (see Problem 1.17). Further, $\widehat{\theta}$ is consistent and asymptotically normal. Let

$$\mathbf{h}_1(\mathbf{X}_1) = E[\mathbf{h}(\mathbf{X}_1, \ldots, \mathbf{X}_s, \mathbf{Y}_1, \ldots, \mathbf{Y}_t) \mid \mathbf{X}_1]$$
$$\mathbf{h}_2(\mathbf{Y}_1) = E[\mathbf{h}(\mathbf{X}_1, \ldots, \mathbf{X}_s, \mathbf{Y}_1, \ldots, \mathbf{Y}_t) \mid \mathbf{Y}_1]. \tag{1.26}$$

Let $N = n + m$; assume that $\lim_{N \to \infty} \frac{N}{n} = \rho_1 (> 1)$ and $\lim_{N \to \infty} \frac{N}{m} = \rho_2 (> 1)$. Then, we have

$$\frac{1}{\sqrt{N}}\left(\widehat{\theta} - \theta\right) \to_d N\left(\mathbf{0}, \rho_1^2 s^2 Var[\mathbf{h}_1(\mathbf{X}_1)] + \rho_2^2 t^2 Var[\mathbf{h}_2(\mathbf{Y}_1)]\right). \tag{1.27}$$

For a proof of the result and more applications, check Kowalski and Tu (2008).

Example 1.11
The area under the ROC curve (AUC) is a commonly used summary index for the accuracy of a continuous test for a binary diagnosis of a disease. Let X and Y be the outcomes of a test for disease from diseased and non diseased subjects, then $AUC = \Pr(X > Y)$, assuming that the higher the test outcome the more likely to be diseased. See Chapter 10 for a detailed discussion on ROC curves.

Let $h(X, Y) = I(X > Y)$, where $I(\cdot)$ is the indicator function, so $I(X > Y) = 1$ if $X > Y$ and $= 0$ if otherwise. Then, AUC $= E(h(X, Y))$, and we can estimate the AUC by $\frac{1}{mn} \sum_{i=1}^{n} \sum_{j=1}^{m} I(X_i > Y_j)$. When there are ties between diseased and non diseased subjects, we may use the ties adjusted estimate $\frac{1}{mn} \sum_{i=1}^{n} \sum_{j=1}^{m} \left[I(X_i > Y_j) + \frac{1}{2} I(X_i = Y_j)\right]$.
☐

1.5 Software

As study designs get more complex and sizes for modern clinical trials become large, it is impossible to perform data analysis without the help of statistical software. There are many excellent statistical software packages available for use when studying the materials in the book. However, the audience is encouraged to choose from R, SAS, SPSS, and

Stata, as the sample code written for many examples in the book is available online for free download from the publisher's website. All of these four packages are powerful, but each has its own special features. R is most convenient for coding new statistical methods and performing simulation studies. This package is especially popular among academic researchers, especially those engaged in methodological research, and hence many new statistical methods first become available in R. However, for the very same reason, some R procedures, especially those for new statistical methods, may not be as reliable as their commercial counterparts such as SAS because of the limited resources committed to testing and documenting them. In comparison, the commercial packages SAS, SPSS, and Stata offer more rigorously tested procedures, with a much wider usership. In addition, these packages provide better formatted output, more intuitive user interface, simpler programming, and more detailed documentation. SPSS and Stata even offer a menu-driven system for commonly used procedures so users can point and click on pop-up menus to select the desired models and test statistics.

If you are a practitioner interested in applying statistical models for data analysis, you may choose one of the three commercial packages when working out the examples in the book. If you are primarily interested in methodological research with the goal of developing new models and adapting existing statistical models for research purposes, you may also consider R. The latter package can be downloaded for free from www.r-project.org, which also contains some useful tutorials. The official websites for SAS, SPSS, and Stata are www.sas.com, www.spss.com, and www.stata.com. Finally, we would like to point out that it may be worthwhile to get to know all the four packages since it is often necessary to use multiple packages to efficiently and effectively address statistical problems in practice.

Exercises

1.1 If a fair die is thrown, then each number from 1 to 6 has the same chance of being the outcome. Let X be the random variable to indicate whether the outcome is 5, i.e.,

$$X = \begin{cases} 1 \text{ if the outcome is 5,} \\ 0 \text{ if the outcome is not 5.} \end{cases}$$

(a) Describe the distribution of X and find the mean and variance of X.

(b) For a sequence of i.i.d. variables that follow the distribution, find the limit of the sample mean.

1.2 (Theorem of iterated conditional expectations) For random variables X and Y, show that $E\left[E\left(X \mid Y\right)\right] = E\left(X\right)$ and $Var(X) = Var(E(X \mid Y)) + E(Var(X \mid Y))$.

1.3 The sequence $\left\{\frac{1}{n}\right\}_{n=1}^{\infty}$ converges to 0. If we treat each constant, $\frac{1}{n}$, as a constant random variable, then the corresponding CDF is

$$F_n(x) = \begin{cases} 0 \text{ if } x < \frac{1}{n}, \\ 1 \text{ if } x \geq \frac{1}{n}. \end{cases}$$

Show that the sequence also converges to 0 in distribution and probability. Note, however, that $F_n(0) = 0$ for all n, but $F(0) = 1$, where F is the CDF of constant 0.

1.4 Suppose $X_n \sim \chi_n^2$, the chi-square with n degrees of freedom. Show that

$$\frac{1}{\sqrt{2n}}\left(X_n - n\right) \to_d N(0, 1).$$

1.5 Let $X_1, ..., X_n$, be a sequence of i.i.d. random variables that follow the Poisson distribution with mean μ. Then the sample average $\overline{X}_n = \frac{X_1+\cdots+X_n}{n}$ is a consistent estimate of μ and $\overline{X}_n^2$ is a consistent estimate of μ^2.

   (a) Determine the asymptotic distribution of $\overline{X}_n$.

   (b) Determine the asymptotic distribution of $\overline{X}_n^2$.

1.6 Follow the steps below to prove the LLN without using CLT.

   (a) (Chebyshev's inequality) Let X be a random variable with mean μ and variance σ^2. Then for any real number $\alpha > 0$, $\Pr(|X - \mu| \geq \alpha) \leq \frac{\sigma^2}{\alpha^2}$.

   (b) Apply Chebyshev's inequality to prove the LLN.

1.7 Prove Slutsky's theorem.

1.8 Prove that under some regularity conditions such as the exchangeability of the integral and differentiation, we have

   (a) $E\left[\frac{1}{f(X_i, \theta)} \frac{\partial}{\partial \theta} f(X_i, \theta)\right] = 0$. This shows that the score equation of the MLE is unbiased.

   (b) $Var\left[\frac{1}{f(X_i, \theta)} \frac{\partial}{\partial \theta} f(X_i, \theta)\right] = -E\left[\frac{\partial^2}{\partial \theta \partial \theta^\top} l(\theta)\right]$.

1.9 A random variable X follows an exponential distribution with parameter λ that takes positive values and $\Pr(X < t) = 1 - \exp(\lambda t)$. Suppose that X_i $(i = 1, \ldots, n)$ is a random sample following an exponential distribution with parameter λ. Find the MLE of λ as well as its asymptotic distribution.

1.10 For an independent sample of Y_i and X_i $(1 \leq i \leq n)$, suppose that

$$Y_i = \mathbf{X}_i^\top \boldsymbol{\beta} + \epsilon_i, \quad \epsilon_i \sim \left(0, \sigma_i^2\right), \quad 1 \leq i \leq n, \tag{1.28}$$

where $\sigma_i^2 = Var(\epsilon_i)$ is known. Show that the weighted EE

$$\sum_{i=1}^{n} \frac{1}{\sigma_i^2} \mathbf{X}_i \left(Y_i - \mathbf{X}_i^\top \boldsymbol{\beta}\right) = \mathbf{0} \tag{1.29}$$

will produce the best linear unbiased estimate (BLUE) for $\boldsymbol{\beta}$.

1.11 Let $\mathbf{f}(\boldsymbol{\theta})$ be a $n \times 1$ and $\mathbf{g}(\boldsymbol{\theta})$ a $1 \times m$ vector-valued function of $\boldsymbol{\theta} = (\theta_1, \ldots, \theta_q)$. The derivatives of $\frac{\partial}{\partial \boldsymbol{\theta}} \mathbf{f}$ and $\frac{\partial}{\partial \boldsymbol{\theta}} \mathbf{g}$ are defined as follows:

$$\frac{\partial}{\partial \boldsymbol{\theta}} \mathbf{f} = \begin{pmatrix} \frac{\partial f_1}{\partial \theta_1} & \cdots & \frac{\partial f_n}{\partial \theta_1} \\ \vdots & \ddots & \vdots \\ \frac{\partial f_1}{\partial \theta_q} & \cdots & \frac{\partial f_n}{\partial \theta_q} \end{pmatrix}_{q \times n}, \quad \frac{\partial}{\partial \boldsymbol{\theta}} \mathbf{g} = \begin{pmatrix} \frac{\partial g_1}{\partial \theta_1} & \cdots & \frac{\partial g_1}{\partial \theta_q} \\ \vdots & \ddots & \vdots \\ \frac{\partial g_m}{\partial \theta_1} & \cdots & \frac{\partial g_m}{\partial \theta_q} \end{pmatrix}_{m \times q} \tag{1.30}$$

Thus, $\frac{\partial}{\partial \boldsymbol{\theta}} \mathbf{g} = \left(\frac{\partial}{\partial \boldsymbol{\theta}} \mathbf{g}^\top\right)^\top$. As a special case, if $f(\boldsymbol{\theta})$ is a scalar function, it follows from (1.30) that $\frac{\partial f}{\partial \boldsymbol{\theta}} = \left(\frac{\partial}{\partial \theta_1} f, \ldots, \frac{\partial}{\partial \theta_q} f\right)^\top$ is a $q \times 1$ column vector. Let A be a $m \times n$ matrix of constants, $\mathbf{g}(\boldsymbol{\theta})$ a $m \times 1$ vector-valued function of $\boldsymbol{\theta}$, and $h(\boldsymbol{\theta})$ a function of $\boldsymbol{\theta}$. Then, we have the following:

(a) $\frac{\partial}{\partial \theta} (A\mathbf{f}) = \left(\frac{\partial}{\partial \theta}\mathbf{f}\right) A^{\top}$,

(b) $\frac{\partial}{\partial \theta} (h\mathbf{f}) = \left(\frac{\partial}{\partial \theta}h\right) \mathbf{f}^{\top} + h\frac{\partial}{\partial \theta}\mathbf{f}$,

(c) $\frac{\partial}{\partial \theta} \left(\mathbf{g}^{\top} A\mathbf{f}\right) = \left(\frac{\partial}{\partial \theta}\mathbf{g}\right) A\mathbf{f} + \left(\frac{\partial}{\partial \theta}\mathbf{f}\right) A^{\top}\mathbf{g}$.

1.12 Use the properties of differentiation in Problem 1.11 to prove (1.13) and (1.14).

1.13 Prove (1.21).

1.14 Let $\mathbf{X}_i$ $(1 \le i \le n)$ be an i.i.d. sample of random vectors and let $\mathbf{h}$ be a vector-valued symmetric function m arguments. Then,

$$\widehat{\theta} = \binom{n}{m}^{-1} \sum_{(i_1,\ldots,i_m) \in C_m^n} \mathbf{h}\left(\mathbf{X}_{i_1}, \ldots, \mathbf{X}_{i_m}\right)$$

is an unbiased estimate of θ. This shows that $\widehat{\theta}$ in (1.22) is an unbiased estimate of $\theta = E\left[\mathbf{h}\left(\mathbf{X}_1, \ldots, \mathbf{X}_m\right)\right]$.

1.15 Show that the U-statistic $\widehat{\sigma}^2$ in (1.24) is the sample variance of σ^2, i.e., $\widehat{\sigma}^2$ can be expressed as $\widehat{\sigma}^2 = \frac{1}{n-1} \sum_{i=1}^{n} \left(X_i - \overline{X}_n\right)^2$.

1.16 Consider the function $\mathbf{h}\left(\mathbf{Z}_1, \mathbf{Z}_2\right)$ for the U-statistic in (1.22).

(a) Show

$$Var\left[\widetilde{\mathbf{h}}_1\left(\mathbf{Z}_1\right)\right] = E\left(\mathbf{h}_1\left(\mathbf{Z}_1\right)\mathbf{h}_1^{\top}\left(\mathbf{Z}_1\right)\right) - \theta\theta^{\top}.$$

(b) Use the iterated conditional expectation to show

$$E\left(\mathbf{h}_1\left(\mathbf{Z}_1\right)\mathbf{h}_1^{\top}\left(\mathbf{Z}_1\right)\right) = E\left(\mathbf{h}\left(\mathbf{Z}_1, \mathbf{Z}_2\right)\mathbf{h}^{\top}\left(\mathbf{Z}_1, \mathbf{Z}_3\right)\right).$$

1.17 Let $\mathbf{X}_i$ $(1 \le i \le n)$ and $\mathbf{Y}_i$ $(1 \le i \le n)$ be two independent i.i.d. samples of random vectors from two different population. Let $\mathbf{h}$ be a vector-valued symmetric function with $s + t$ arguments. Show that

$$\widehat{\theta} = \binom{m}{t}^{-1} \binom{n}{s}^{-1} \sum_{(i_1,\ldots,i_s) \in C_s^n} \sum_{(j_1,\ldots,j_t) \in C_t^m} \mathbf{h}\left(\mathbf{X}_{i_1}, \ldots, \mathbf{X}_{i_s}, \mathbf{Y}_{j_1}, \ldots, \mathbf{Y}_{j_t}\right).$$

is an unbiased estimate of $\theta = E\left[\mathbf{h}\left(\mathbf{X}_1, \ldots, \mathbf{X}_s, \mathbf{Y}_1, \ldots, \mathbf{Y}_t\right)\right]$.

1.18 Install the statistical software packages that you will use for the book in your computer. Read the DOS data set using your statistical software and find out the number of observations and the number of variables in the data sets.

Chapter 2

Contingency Tables

In this chapter, we discuss statistical inference for frequency or contingency tables. As noted in Chapter 1, such tables arise when the underlying discrete variables have a finite range, which include binary, ordinal, and nominal outcomes. For convenience, we refer to all such variables as categorical outcomes.

Suppose that we are interested in several such variables simultaneously. Since each variable has only a finite number of possible values, there are finitely many combinations of outcomes formed from these variables. A frequency or contingency table records the frequency for each combination.

If there is only one categorical variable, we call it a one-way frequency table. If there are two categorical variables, we refer to it as a two-way frequency (or contingency) table, etc. If we want to emphasize the range of each categorical variable, e.g., if there are two categorical variables, one with s possible values and the other with r possible values, then we call it a two-way $s \times r$ contingency table, or simply an $s \times r$ contingency table. For example, for the Metabolic Syndrome study, using gender as the row variable and the metabolic syndrome (MS) status as the column variable, the data were summarized in a 2×2 table (Table 1.1).

It is easy to input aggregated frequency data into a contingency table by hand. Alternatively, if data are presented as individual responses, then such a table can be easily generated using statistical software such as SAS, SPSS, Stata, and R.

We will discuss one-way frequency tables in Section 2.1. The remainder of the chapter will be devoted to inference of two-way contingency tables. The simplest cases, 2×2 tables, are studied in Section 2.2, followed by studies of $2 \times s$ tables in Section 2.3. General two-way contingency tables are discussed in Section 2.4. Measures of association are introduced in the last section.

2.1 Inference for One-Way Frequency Table

We start with the simplest case in which we only have one categorical variable. The random variable is known to follow a *multinomial* distribution. We are interested in inference about this distribution. We first discuss in detail the special *binary* case where the variable has only two categories and then briefly go over the general multinomial case as the latter is a straightforward extension of the former.

For completeness, we also discuss Poisson distributions for count variables. A count variable differs from a categorical variable in that it has an infinite range, thus giving rise to fundamentally different models and methods for inference. Nonetheless, observed data from such a variable in practice are finite in range and can be displayed in a one-way table.

DOI: 10.1201/9781003109815-2

2.1.1 Binary case

A binary variable x has two potential outcomes, which are often denoted by 0 and 1. Thus, the random nature of x is completely determined by the probabilities with which x takes on the two values. Since the two probabilities add up to 1, only one of them is needed to characterize this *Bernoulli* distribution. By convention, the probability that x assumes the value 1, $p = \Pr(x = 1)$, is used as the parameter for the Bernoulli distribution, *Bernoulli*(p). Thus, for a binary variable, we are interested in estimating p (point estimate), assessing the accuracy of the estimate (confidence interval), and confirming our knowledge about p (hypothesis testing).

2.1.1.1 Point estimate

Let $x_i \sim$ i.i.d. *Bernoulli*(p), where i.i.d. denotes an *independently* and *identically distributed* sample. It follows from the theory of maximum likelihood in Chapter 1 that the sample mean, $\widehat{p} = \frac{1}{n}(x_1 + \cdots + x_n)$, is a consistent estimate of p and follows an asymptotically normal distribution, i.e.,

$$\widehat{p} \to_p p, \quad \widehat{p} \sim_a N\left(p, \frac{1}{n}p(1-p)\right), \quad \text{as } n \to \infty, \tag{2.1}$$

where p is the mean and $p(1-p)$ is the variance of x_i. The symbols ' $\to_p$' and ' $\sim_a$' above denote convergence in probability (or consistency) and asymptotic distribution as defined in Chapter 1. In layman's terms, consistency means that the probability of observing a difference between the estimate $\widehat{p}$ and parameter p becomes increasingly small as n grows. Similarly, asymptotic normality implies that the error in approximating the sampling distribution of $\widehat{p}$ by the normal in (2.1) dissipates as n gets large. Note that $\widehat{p}$ is called an *estimator* to emphasis that it represents a rule for estimating the parameter p for all samples with n i.i.d. Bernoulli outcomes. Being a function of the random variables, the estimator $\widehat{p}$ is also a random variable with the mean and variance. This is different from an estimate based on a specific sample, which is simply a value. Since it is generally clear from the context of discussion, we simply use estimate for both situations throughout the book.

By Slutsky's theorem (see Chapter 1), the asymptotic variance can be estimated by $\frac{\widehat{p}(1-\widehat{p})}{n} = \frac{k(n-k)}{n^3}$, where $k = x_1 + \cdots + x_n$. Based on $\widehat{p}$ and this estimated asymptotic variance, we may make inference about p, e.g., by constructing confidence intervals of p. However, we must be mindful about the behavior of the asymptotic distribution of $\widehat{p}$, as a larger sample size is generally required for the asymptotic normal distribution to provide a good approximation to the distribution of $\widehat{p}$. In particular, the required sample size depends on the value of p; the closer p is to the extreme values 0 and 1, the larger the sample size. A frequently cited convention is that $np > 5$ and $n(1-p) > 5$, where n is the sample size.

Another problem with small sample size and extreme values of the probability p is that the sample observed may consist of observations with the same outcome such as all 0's. In such cases, the point estimate would be 0 or 1, yielding 0 for the asymptotic variance and making inference impossible. To overcome this problem, a popular approach is to add one half to both counts of the observations of 0 and 1, resulting in a revised estimate, $\widetilde{p} = \frac{1}{n+1}(x_1 + \cdots + x_n + 0.5)$. The variance can then be estimated using $\frac{\widetilde{p}(1-\widetilde{p})}{n+1}$. A justification for this approach is provided in Chapter 4.

2.1.1.2 Confidence interval

In addition to a point estimate $\widehat{p}$, we may also compute *confidence intervals* (CIs) to provide some indication of accuracy of the estimate. Confidence intervals are random intervals derived to cover the true value of p with a certain probability, the confidence level.

In other words, if the sampling procedure is repeated independently, the CIs will change from sample to sample, with the percentage of times that the CIs contain the true value of p approximately equal to the confidence level. Note that we have adopted the "frequentist" interpretation by viewing the parameter p as fixed. Thus, the random CIs move around the true value of p from sample to sample.

For large samples, *Wald* CIs based on the asymptotic distribution of $\widehat{p}$ are often used. The $100(1-\alpha)\%$ Wald confidence interval for p has the limits, $\widehat{p} \pm z_\alpha \sqrt{\frac{\widehat{p}(1-\widehat{p})}{n}}$, where α is typically 0.01, 0.05, and 0.10 and z_α is the $100(1-\alpha/2)$th percentile of the standard normal distribution. For example, if $\alpha = 0.05$, $100(1-\alpha/2) = 0.975$ and $z_\alpha = 1.96$. A 95% Wald confidence interval has the limits, $\widehat{p} \pm 1.96 \sqrt{\frac{\widehat{p}(1-\widehat{p})}{n}}$. Like the point estimate, the Wald interval does not work when $\widehat{p} = 0$ or 1. Wald confidence intervals may be used if n is large, and the coverage of the Wald interval will be close to the nominal $100(1-\alpha)\%$ level. The coverage of the Wald interval, however, can be erratic for extreme values of p even for large samples due to the discrete nature of the underlying sampling distribution (see, e.g., Brown et al. (2001)). In such cases, the actual covering probabilities may deviate considerably from the nominal ones even when some commonly suggested requirements such as $np(1-p) \geq 5$, or both $np \geq 5$ and $n(1-p) \geq 5$, are met. Thus, for extremely small p, alternative methods may be considered. For example, by taking into consideration the discrete nature of binary variable, the following CIs generally provide better coverage for samples of small to moderate sizes.

Note that it is not the case that CIs with higher confidence levels are better. Although CIs with higher confidence levels are more likely to cover the true values of p than those with lower levels, they may become too wide to be practically useful. In most applications, 95% CIs are used.

Because Wald CIs are based on the asymptotic distribution of $\widehat{p}$, a 95% Wald CI generally does not cover the true value of p exactly 95% of the times, especially with small and moderate sample sizes. Various alternatives have been proposed to improve the coverage accuracy, and some popular ones are described below.

Let $\widetilde{x} = x_1 + \cdots + x_n + z_\alpha^2/2$, $\widetilde{n} = n + z_\alpha^2$, and $\widetilde{p} = \widetilde{x} \,/\, \widetilde{n}$. Agresti and Coull (1998) suggested to estimate the $100(1-\alpha)\%$ CI for p by $\left(\widetilde{p} - z_\alpha \sqrt{\frac{\widetilde{p}(1-\widetilde{p})}{\widetilde{n}}}, \widetilde{p} + z_\alpha \sqrt{\frac{\widetilde{p}(1-\widetilde{p})}{\widetilde{n}}} \right)$. Wilson (1927) proposed another estimate: $\left(\widetilde{p} - \frac{z_\alpha \sqrt{n}}{\widetilde{n}} \sqrt{\widehat{p}(1-\widehat{p}) + \frac{z_\alpha^2}{4n}}, \widetilde{p} + \frac{z_\alpha \sqrt{n}}{\widetilde{n}} \sqrt{\widehat{p}(1-\widehat{p}) + \frac{z_\alpha^2}{4n}} \right)$. Rubin and Schenker (1987) suggested yet another estimate of the confidence interval. This approach first computes the confidence interval for $\theta = \mathrm{logit}(p) = \log\left(\frac{p}{1-p}\right)$ and then transforms the estimate back to the original scale to obtain the limits of the $100(1-\alpha)\%$ CI for p: $\mathrm{logit}^{-1}\left(\mathrm{logit}(\widetilde{p}) \pm \frac{z_\alpha}{\sqrt{(n+1)\widetilde{p}(1-\widetilde{p})}} \right)$, where $\widetilde{p} = \frac{x_1 + \cdots + x_n + 0.5}{n+1}$.

The above methods estimate the confidence intervals by trying various adjustments to improve the approximations of the asymptotic normal distribution to the sampling distribution. By taking a different approach, Clopper and Pearson (1934) suggested to construct confidence intervals based on the actual binary sampling distribution. Let $k = \sum_{i=1}^{n} x_i$. If $k = 0$, then the lower limit is 0. On the other hand, if $k = n$, then the upper limit is 1. In general, the upper (p_u) and lower (p_l) limits satisfy

$$\sum_{x=0}^{k} \binom{n}{x} p_u^x (1-p_u)^{n-x} = \alpha/2, \quad \sum_{x=k}^{n} \binom{n}{x} p_l^x (1-p_l)^{n-x} = \alpha/2. \qquad (2.2)$$

The solutions to (2.2) are unique (see Problem 2.5). As the interval is based on the true sampling distribution, it is also known as the *exact confidence interval*. This interval estimate is guaranteed to have coverage probability of at least $1 - \alpha$; thus, it is conservative because of the discrete nature of the binomial distribution.

Some studies have compared the performance of these different interval estimates. For example, Agresti and Coull (1998) recommended approximate intervals over exact ones. Brown et al. (2001) recommended that the Wilson CIs be used for $n \leq 40$. They also preferred the Agresti–Coull interval over the Wald in this case. For larger n, the Wilson and the Agresti–Coull intervals are comparable to each other, but the Agresti–Coull interval was recommended because of its simpler form.

2.1.1.3   Hypothesis testing

Sometimes, we may want to test whether p equals some a priori known value. For example, if we want to know whether the outcomes $x = 0$ and $x = 1$ are equally likely, we can test the null hypothesis, $H_0 : p = 0.5$. By means of the estimate $\widehat{p}$ and its asymptotic distribution, we can readily carry out the test by computing the probability of the occurrence of outcomes that are as or more extreme than the value of $\widehat{p}$ based on the observed data. The meaning of "extreme" depends on the *alternative hypothesis H_a*, which the null hypothesis is against. For example, if we have no idea about which outcome of x will be more likely to occur, we may test the null hypothesis $H_0 : p = 0.5$ against the *two-sided* alternative $H_a : p \neq 0.5$. On the other hand, if we believe that the outcome $x = 1$ is more (less) likely to occur, then the *one-sided* alternative $H_a : p > 0.5$ $(p < 0.5)$ may be considered.

More precisely, suppose the null hypothesis is $H_0 : p = p_0$, and let $\widehat{p}_0$ denote the value of the statistic $\widehat{p}$ computed based on the observed data. If the alternative is $H_a : p \neq p_0$, then the values of $\widehat{p}$ based on data from all potential samples drawn from a population with $p \neq p_0$ will generally deviate more from p_0 than $\widehat{p}_0$ does. Thus, the probability $\Pr\left(|\widehat{p} - p_0| \geq |\widehat{p}_0 - p_0|\right)$ calculated based on the null hypothesis $H_0 : p = p_0$ indicates how unlikely such potential outcomes of $\widehat{p}$ will occur under H_0. Such a probability, called *p-value*, is widely used for making a decision regarding the null hypothesis. Since the p-value is the probability of observing a value of the statistic $\widehat{p}$ that is as unlikely or more unlikely than $\widehat{p}_0$ under the null hypothesis H_0, it indicates the level of error committed when making a decision to reject the null hypothesis (*type I error*) when it is true. In practice, we control the rate of type I error by determining a priori that we will reject the null hypothesis if the p-value falls below the nominal type I error probability α (also called the *significance level* α). Thus, our likelihood of making a type I error is controlled to be under α. Choices of the α may depend on the questions, but cannot be made arbitrarily small. This is because lower significance levels mean that it is harder to reject the null hypothesis and thus more likely fail to reject the null hypothesis when it is not true (*type II error*). In other words, for hypothesis testing, lower probabilities of type I errors are associated with higher probabilities of type II errors, or lower *power*, which is defined as 1 − probability of type II errors. The popular cut-points for the p-value, or significance level α, are 0.05 and 0.01.

When the sample size is large, the p-value can be computed based on the asymptotic distribution of $\widehat{p}$ in (2.1). More precisely, by standardizing $\widehat{p} - p_0$, we obtain the following Z-score, which follows a standard normal distribution asymptotically under the null hypothesis

$$Z = \frac{\widehat{p} - p_0}{\sqrt{p_0\left(1 - p_0\right)/n}} \sim_a N\left(0, 1\right).$$

The p-value for the two-sided test is the probability that $|Z|$ is at least $|z|$, where z is the Z-score based on observed data. Thus, the p-value equals $2 \times \left(1 - \Phi\left(|z|\right)\right) = 2\Phi\left(-|z|\right)$, where

$\Phi\left(\cdot\right)$ is the cumulative distribution function (CDF) of a standard normal. Similarly, for one-sided alternative $H_a : p > p_0$, the p-value is the probability that $Z \geq z$, $1 - \Phi\left(z\right) = \Phi\left(-z\right)$. The p-value for the other alternative $p < p_0$ is the probability of $Z \leq z$, $\Phi\left(z\right)$. Note that when calculated based on the normal asymptotic distribution, the p-value for the two-sided test is twice that of the smaller p-value for the two one-sided test, because of the symmetry of the normal distribution.

2.1.1.4 Exact inference

If the sample size is small or if the value of p is close to 1 or 0, the asymptotic distribution may not provide accurate approximations to the sampling distribution of the estimate. In such cases, inference based on the exact distribution of $\widehat{p}$ under the null hypothesis may be preferred. Since it is equivalent to calculating the p-value using the distribution of either the sample mean or the sum of the observations $\sum_{i=1}^{n} x_i$, we discuss exact inference based on the latter statistic for convenience.

The random variable $K = \sum_{i=1}^{n} x_i$ follows a binomial distribution with probability p and size n, $BI\left(n, p\right)$. Let k denote the value of K based on a particular sample. For testing the null hypothesis $H_0 : p = p_0$ against the one-sided alternative $H_a : p > p_0$, the p-value or the probability of observing K as or more extreme than k under the null hypothesis is

$$\tau_u = \Pr\left(K \geq k \mid H_0\right) = \sum_{i=k}^{n} \binom{n}{i} p_0^i \left(1 - p_0\right)^{n-i}. \tag{2.3}$$

For the other one-sided alternative $H_a : p < p_0$, the p-value is defined as the probability of observing K as small or smaller than k under H_0:

$$\tau_l = \Pr\left(K \leq k \mid H_0\right) = \sum_{i=0}^{k} \binom{n}{i} p_0^i \left(1 - p_0\right)^{n-i}. \tag{2.4}$$

If $\widehat{p} > p_0$, then in general $\tau_u < \tau_l$, and in this case it is not necessary to test the one-sided alternative $H_a : p < p_0$. Another way to think about this logic is that if the null hypothesis is unlikely, then the alternative will be even more so. Thus, we only need to consider the following one-sided test:

$$H_0 : p = p_0 \quad \text{vs.} \quad H_a : p > p_0.$$

Similarly, if $\widehat{p} < p_0$, we only consider the one-sided alternative: $H_a : p < p_0$.

Following the definition above for one-sided tests in (2.3) and (2.4), the p-value for the two-sided alternative $H_a : p \neq p_0$ is defined as

$$p = \sum_{i} \binom{n}{i} p_0^i (1 - p_0)^{n-i}, \tag{2.5}$$

where the sum is taken over all i such that $\binom{n}{i} p_0^i (1 - p_0)^{n-i} \leq \binom{n}{k} p_0^k (1 - p_0)^{n-k}$.

Sometimes, the two-sided p-value is also defined by $p = 2\min(\tau_l, \tau_u)$. These two definitions are consistent for large sample size. For small samples, there may be some difference between the p-values obtained from the two methods.

Note that because of the discrete nature of the distribution of K, the test is conservative. For example, let $n = 20$ and the null hypothesis $p = 0.5$. For a one-sided test with the alternative $p < 0.5$ with type I error 0.05, the null hypothesis will be rejected if and only if $k \leq 5$. This is because $\Pr(K \leq 5) = 0.021$ and $\Pr(K \leq 6) = 0.057$. In this case, the actual type I error of the test is 0.021, smaller than the nominal type I error level 0.05. For more discussion on this issue, see Hutson (2006).

Example 2.1

In the Metabolic Syndrome study, we are interested in the prevalence of MS among people taking clozapine and want to test whether the prevalence is 40% in this study population.

Since there are 48 patients with MS among the 93 patients, the maximum likelihood estimate (MLE) of the prevalence is $\hat{p} = \frac{48}{93} = 51.61\%$. The 95% Wald CI is $(0.4146, 0.6177)$, and the 95% Wilson and Agresti–Coull CIs are both equal to $(0.4160, 0.6150)$. The 95% exact CI is $(0.4101, 0.6211)$.

Both asymptotic and exact tests are applied to the null hypothesis $H_0 : p = 0.4$. The Z-score is $\frac{48/93 - 0.4}{\sqrt{0.4 \times 0.6/93}} = 2.2860$. Thus, the p-value under the (two-sided) asymptotic test is $2 \times \Phi(-2.286) = 0.0223$, where $\Phi(\cdot)$ is the CDF of a standard normal. For the exact test, the p-value for the one-sided test $H_a : p > 0.4$ is $\sum_{i=48}^{93} \binom{93}{i} \times 0.4^i \times 0.6^{93-i} = 0.0153$. When doubled, the number, 0.0307, can also be viewed as the p-value for the two-sided exact test. Alternatively, by evaluating the tail probabilities of the binomial distribution with $i \leq 26$ and $i \geq 48$, the p-value defined by (2.5) is $\sum_{i=0}^{26} \binom{93}{i} \times 0.4^i \times 0.6^{93-i} + \sum_{i=48}^{93} \binom{93}{i} \times 0.4^i \times 0.6^{93-i} = 0.0259$. Based on all these tests, we reject the null hypothesis at the 5% significance level, i.e., with nominal probability 0.05 of type I error.   ⬚

2.1.2   Inference for multinomial variable

A categorical variable x with more than two levels is said to follow the *multinomial* distribution. Let j index the possible levels of x $(j = 1, 2, \ldots, k)$. Then, the multinomial distribution is defined by the probabilities that x takes on each of these values, i.e.,

$$\Pr(x = j) = p_j, \quad p_j \geq 0 \quad j = 1, 2, \ldots, k, \quad \sum_{j=1}^{k} p_j = 1. \tag{2.6}$$

Note that since $\sum_{j=1}^{k} p_j = 1$, only $(k-1)$ of the p_j's are free parameters. For n independent trials according to (2.6), the joint distribution of the outcome is given by $\Pr(X_j = m_j, j = 1, \ldots, k) = \frac{n!}{m_1! \cdots m_k!} p_1^{m_1} \cdots p_k^{m_k}$, where X_j is the number of occurrences of the event j, if $\sum_{j=1}^{k} m_j = n$ and 0 otherwise. We denote the multinomial model with probabilities $\mathbf{p} = (p_1, \ldots, p_{k-1})^\top$ and size n by $\mathrm{MN}(n, \mathbf{p})$. If $k = 2$, the multinomial distribution reduces to the binomial model with $p_1 + p_2 = 1$. As noted earlier, we often denote the two levels of x in this special case as 0 and 1 and the binomial model by $BI(n, p)$ with $p = \Pr(x = 1)$.

Information about x in a sample is summarized in a one-way frequency table, and the parameters vector $\mathbf{p}$ can be estimated from the table. For example, the depression status in the Depression of Seniors Study (DOS) has three levels: major, minor, and no depression. The information can be compactly recorded in the following one-way frequency table (Table 2.1).

In general, if the sample has n subjects and the number of subjects with $x = j$ is n_j, then the MLE of $\mathbf{p}$ is $\hat{\mathbf{p}} = (\hat{p}_1, \ldots, \hat{p}_{k-1})$, with $\hat{p}_j = \frac{n_j}{n}$ $(1 \leq j \leq k-1)$ and the asymptotic variance of $\hat{p}_j$ is $\frac{p_j(1-p_j)}{n}$. The asymptotic covariance between $\hat{p}_j$ and $\hat{p}_{j'}$ is $-\frac{1}{n} p_j p_{j'}$ if

TABLE 2.1: Depression diagnosis in the DOS

Major Dep	Minor Dep	No Dep	Total
128	136	481	745

$k > 2$ (see Problem 2.6). We may also use chi-square statistics to test whether x follows a particular multinomial distribution, i.e.,

$$H_0 : \ p_j = c_j, \quad \text{for } j = 1, 2, \dots, k,$$

where c_j are the prespecified constants satisfying $c_j \geq 0$ and $\sum_{j=1}^{k} c_j = 1$. Under the null hypothesis above, the expected count for $x = j$ is $w_j = nc_j$ $(1 \leq j \leq k)$, and the statistic,

$$\sum_{j=1}^{k} \frac{(n_j - w_j)^2}{w_j} \sim_a \chi_{k-1}^2,\tag{2.7}$$

where χ_{k-1}^2 denotes a *chi-square distribution with $k-1$ degrees of freedom*. If $x_1, \dots, x_k$ are independent standard normal random variables, then the sum of their squares, $\sum_{j=1}^{k} x_j^2 \sim \chi_k^2$. The important result (2.7) can be proved using the central limit theorem (CLT). However, since $\sum_{j=1}^{k} (n_j - w_j) = 0$, the CLT cannot be applied directly to the vector statistic $\mathbf{u}_n = \frac{1}{n} (n_1 - w_1, \dots, n_k - w_k)^\top$. By selecting a subvector of $\mathbf{u}_n$ with $k - 1$ entries, say the first $k - 1$ components, $\mathbf{v}_n = \frac{1}{n} (n_1 - w_1, \dots, n_{k-1} - w_{k-1})^\top$, and applying the CLT, it is readily shown that $\mathbf{v}_n$ has an asymptotic normal and thus $\mathbf{v}_n^\top \Sigma^{-1} \mathbf{v}_n$ has an asymptotic chi-square distribution with $k - 1$ degrees of freedom, where Σ is the asymptotic variance of $\mathbf{v}_n$. Further, it can be shown that $\mathbf{v}_n^\top \Sigma^{-1} \mathbf{v}_n$ has the same asymptotic distribution as the statistic in (2.7). Interested readers may consult Cramér (1946) for details. Please note that when $k = 2$ the statistic in (2.7) is the same as the one based on the asymptotic normal distribution of $\widehat{p}$ in (2.1) (see Problem 2.3).

Exact inference may be considered if the sample size is small. In the binary case, the frequency table is determined by either of the two cell counts. For a multinomial with k levels, the table is determined by any $k - 1$ cell counts. So, it is impossible to define extremeness by comparing cell counts. The idea is that if x follows a multinomial distribution, the cell counts of the frequency table can be predicted by the cell probabilities specified and the sample size n. Thus, extremeness is measured by the probability of the occurrence of a frequency table with the same sample size n; the occurrence of the table is less likely if this probability is small, and tables with probabilities smaller than that of the observed one are considered to be more extreme. Under the null hypothesis, the exact p-value is defined as the sum of the probabilities of all the potential tables with probabilities equal to or smaller than that of the observed one.

Example 2.2

In the DOS, we are interested in testing the following hypothesis concerning the distribution of depression diagnosis for the entire sample:

$$\Pr(\text{No Depression}) = 0.65, \quad \Pr(\text{Minor Depression}) = 0.2,$$
$$\Pr(\text{Major Depression}) = 0.15.$$

Given the sample size 745, the expected counts for no, minor, and major depression are 484.25, 149, and 111.75, respectively. Hence, the chi-square statistic based on Table 2.1

is $\frac{(481-484.25)^2}{484.25} + \frac{(136-149)^2}{149} + \frac{(128-111.75)^2}{111.75} = 3.519$. The p-value based on the asymptotic distribution χ_2^2 is 0.1721. For the exact test, we need to first compute the probability of the occurrence of the observed table, which is $p_o = \binom{745}{481}\binom{745-481}{136} \times 0.65^{481} \times 0.2^{136} \times 0.15^{128} = 2.6757 \times 10^{-4}$ based on the multinomial distribution. Next, we compute the distribution of occurrences of all potential tables and add all the probabilities that are less than or equal to p_o to find the p-value. Because of the large sample size, it is not practical to compute these probabilities by hand. However, using a software package such as one of those mentioned in Chapter 1, it is easy to obtain the exact p-value, 0.1726. The two p-values are very close, and the null hypothesis should not be rejected at the 5% significance level. The same conclusion reached by both tests are expected because of the large sample size of the study. ☐

Instead of a specific distribution, sometimes we may be interested in learning if the multinomial distribution follows some specific patterns. For example, we often group outcomes above some threshold into one group, either at data collection or data management, for count variables. The resulting multinomial distribution will be described a single parameter if the underlying count variable follows a Poisson distribution which is described by a single parameter.

2.1.3   Inference for Count variable

One-way frequency tables may also be used to display distributions for count variables, even without grouping. Although in theory the potential outcome of a count variable ranges between 0 and infinity, the observed numbers from a real study sample are always finite because of the limited sample size; the range of the observed data is at most the sample size. However, since the observed range of a count variable varies from sample to sample and generally increases with the sample size, its distribution can no longer be described by the multinomial distribution. In addition, even if the range of the observed outcomes is limited in practical studies, it may be too large to be meaningfully modeled by the multinomial distribution. A commonly applied model for such a variable is the Poisson distribution.

In statistics, the *Poisson distribution* plays similar important role in modeling count responses as the normal distribution does in modeling continuous outcomes. If a variable y follows the Poisson distribution with parameter λ, Poisson(λ), it has the following distribution function:

$$f_P(k \mid \lambda) = \Pr(y = k) = \frac{\lambda^k \exp(-\lambda)}{k!}, \quad \lambda > 0, k = 0, 1, \ldots. \tag{2.8}$$

It is easy to show that λ is the mean and variance of y (see Problem 2.7). Since the Poisson distribution is determined by λ, inference concerns the parameter λ. Like the normal distribution, the Poisson model has many nice properties. For example, if two variables y_1 and y_2 are independent and $y_j \sim$ Poisson(λ_j) $(j = 1, 2)$, the sum $y_1 + y_2 \sim$ Poisson($\lambda_1 + \lambda_2$) (see Problem 2.7).

Consider a random sample of n subjects, with each y_i following the Poisson in (2.8). Then the likelihood and log-likelihood are given by

$$L = \prod_{i=1}^{n} \frac{\lambda^{y_i} \exp(-\lambda)}{y_i!}, \quad l = \log(L) = \sum_{i=1}^{n} [y_i \log \lambda - \lambda - \log(y_i!)].$$

By solving the score equation,

$$\frac{\partial}{\partial \lambda} l = \sum_{i=1}^{n} \left(\frac{y_i}{\lambda} - 1\right) = 0,$$

we obtain the MLE $\widehat{\lambda} = \frac{1}{n} \sum_{i=1}^{n} y_i$. The asymptotic distribution of $\widehat{\lambda}$ is also readily obtained from the theory of maximum likelihood for inference about λ (see Problem 2.8).

If the responses larger than some threshold are combined into a single category, we obtain a multinomial variable with the multinomial distributions determined by the single parameter λ. More precisely, let $\{y_i; 1 \leq i \leq n\}$ denote count observations from a sample of n subjects. Let m denote the cut-point for grouping all responses $y_i \geq m$ and the observed cell count n_j for the resulting multinomial model as follows:

$$n_j = \begin{cases} \text{number of } \{i : y_i = j\} \text{ if } 0 \leq j \leq m - 1 \\ \text{number of } \{i : y_i \geq j\} \text{ if } j = m. \end{cases}$$

Under the null hypothesis that the distribution is Poisson, we can determine the cell probabilities of the induced multinomial:

$$p_j = \begin{cases} \frac{\lambda^j \exp(-\lambda)}{j!} & \text{if } 0 \leq j \leq m - 1 \\ \sum_{k \geq m} \frac{\lambda^k \exp(-\lambda)}{k!} & \text{if } j = m, \end{cases} \tag{2.9}$$

where $f_P(\cdot \mid \lambda)$ is the Poisson distribution for the underlying count variable before grouping. The parameter λ can then be estimated by maximum likelihood method based on these grouped data.

Unlike the multinomial case, there is in general no a priori guarantee that a count variable will follow the Poisson distribution. In fact, non-Poisson variables arise quite often in practical applications. Thus, we are often interested in testing whether a count variable follows the Poisson law. For grouped data, we may apply a Pearson-type chi-square statistic to test whether the multinomial distribution follows equation (2.9). If $\widehat{p}_j$ is an estimate of p_j, we can define a Pearson-type chi-square statistic, $P = \sum_{j=0}^{m} \frac{(n_j - n\widehat{p}_j)^2}{n\widehat{p}_j}$, to test the null hypothesis that the count variable follows a Poisson distribution. However, the distribution of this statistic can be quite complex depending on how the $\widehat{p}_j$ are estimated. If we estimate λ by the multinomial likelihood based on the grouped data with p_j modeled by $f_P(j \mid \lambda)$ in (2.9), then the statistic P has asymptotically a chi-square distribution with $m-1$ degrees of freedom. Note that the chi-square reference distribution has a degree of freedom that is one less than m, due to estimating the mean λ. This type of loss of degrees of freedom will be discussed in more detail in Chapter 4 when we consider more complex regression analyses for categorical variables.

If the original count outcomes are available ungrouped, we may still combine the responses larger than some threshold into a single category and apply the above approach. In practice, it is more convenient to estimate λ by the MLE $\widehat{\lambda} = \frac{1}{n} \sum_{i=1}^{n} y_i$ based on the null hypothesis that the distribution is Poisson and substitute $\widehat{\lambda}$ in place of λ in (2.9) to estimate p_j. Since this estimate uses information in the original individual responses y_i, rather than the grouped data, it incurs no loss of power. With such estimates of p_j, the corresponding Pearson statistic P becomes a *Chernoff–Lehmann-type* statistic and its asymptotic distribution is no longer a chi-square, but rather $\chi^2_{m-1} + c\chi^2_1$, where c is a constant between 0 and 1. However, as mentioned in Lancaster (1969), χ^2_{m-1} is still a good approximation to the asymptotic distribution of P.

Example 2.3

In the Sexual Health pilot study, the number of vaginal sex encounters during the past three months was collected at the first visit. Presented in Table 2.2 are the counts, with all counts ≥ 6 combined.

If a Poisson distribution is fit, the MLE of λ based on the raw data (not the grouped data in the table) is $\widehat{\lambda} = \overline{y} = 9.1$. Based on the estimated $\widehat{\lambda}$, the cell probabilities are 0.0001,

TABLE 2.2:   Frequency of protected vaginal sex

Sex encounters	0	1	2	3	4	5	≥ 6
# of subjects	32	4	5	5	5	6	41

0.0010, 0.0046, 0.0140, 0.0319, 0.0580, and 0.8904. By computing w_i based on these fitted cell probabilities and comparing the statistic $\sum_{i=0}^{k} \frac{(n_i - w_i)^2}{w_i} = 93934.89$ against a chi-square distribution with $7 - 1 - 1 = 5$ degrees of freedom, we obtain a p-value that is less than 0.0001. However, as mentioned in the last section, the chi-square distribution is not the true asymptotic distribution of this statistic, which is a more complex linear combination of chi-squares. More seriously, the expected cell counts for $y = 0$ and $y = 1$ are extremely small, and thus the test is not appropriate. An alternative is to estimate λ using a multinomial for the grouped data and compute w_i using the following cell probabilities p_j:

$$p_j = \begin{cases} f_P\left(j \mid \lambda\right) & \text{if} \quad 0 \leq j < 6 \\ 1 - \sum_{j=0}^{5} f_P\left(j \mid \lambda\right) & \text{if} \quad j = 6. \end{cases}$$

By maximizing the likelihood of the multinomial with the above p_j based on the grouped data, we obtain the MLE $\widetilde{\lambda} = 3.6489$. The chi-square statistic based on such an estimate is 417.37, which gives the p-value < 0.0001. Note that it is more involved to compute the estimate of λ under this approach. Further, since only the grouped data, not the raw data, are used, there may be some loss of power.   ⬚

If the distribution of a count response is deemed not to follow the Poisson model, more complex models may be used to fit the data. For example, in many applications, there is an excessive number of zeros above and beyond what is expected by the Poisson law. By considering the data as a mixture of a degenerate distribution centered at 0 and a Poisson, we may apply the *zero-inflated Poisson* (ZIP) to fit the data. Again, we will discuss this and other models to address this and other similar issues within a more general regression context for count responses in Chapters 6 and 7.

2.2   Inference for 2×2 Table

Contingency tables are often used to summarize the relationship between two categorical variables x and y; one is designated as the row and the other as the column variable. If both variables are binary, there are four combinations of possible values from these two variables, and thus their occurrence in a sample can be displayed in a 2×2 contingency table.

Two-by-two contingency tables arise from a variety of contexts and sampling schemes. Listed below are some common examples that give rise to such a table.

- A single random sample from a target population. The row and column variables may represent two different characteristics of the subject or they may be repeated measures of the same characteristics collected at a time before and after an intervention as in the pre-post study design. Table 1.1 (Chapter 1) follows such a sampling scheme with gender and MS representing two characteristics of interest.

- A stratified random sample from two independent groups. In this case, one of the variables is the group indicator, and subjects are randomly sampled from each of the groups in the population. Randomized clinical trials with two treatment conditions and case/control studies are such examples.

- Two judges prescribe ratings for each subject in a sample based on a binary scale.

The origin of the contingency table is important to make valid inference. For example, it may look like the probabilities $\Pr(x = 1)$, $\Pr(x = 1, y = 1)$, or $\Pr(x = 1 \mid y = 1)$ can be easily estimated from Table 2.3 using the cell counts. But we must be mindful about how the table is obtained in order for the estimates to be interpretable. For example, in a diagnostic test study, x may represent the status of a disease D, and y the result of a test T in detecting the presence of the disease. Commonly used indices for the accuracy of binary tests include the true positive fraction (TPF), or sensitivity, $\Pr(T = 1 \mid D = 1)$; true negative fraction (TNF), or specificity, $\Pr(T = 0 \mid D = 0)$; positive predictive value (PPV), $\Pr(D = 1 \mid T = 1)$; and negative predictive value (NPV), $\Pr(D = 0 \mid T = 0)$. If subjects are randomly selected from the target population (single sample), then all these indices can be easily estimated using sample proportions. However, if they are independently sampled from diseased and nondiseased (case-control study), TPF and TNF can be directly estimated, but without further information, PPV and NPV cannot be estimated from the table. Similarly, if subjects are sampled based on the test results $T = 1$ or 0, only PPV and NPV can be estimated from the table. If disease prevalence is available, all the indices may be computed; however, more complex approaches are needed than simple sample proportions (see Problem 2.21).

We are generally interested in studying the relationship between the row and column variables. In the remaining sections of this chapter, we discuss how to test independence between the two variables and introduce various measures of association to describe the strengths of the relationships when they are not independent. We also discuss how to test whether the variables have the same marginal distribution, which has important applications in assessing intervention effects for pre-post study designs. If the contingency table is obtained as ratings on a sample of subjects from two raters, then an assessment of their agreement is typically of interest.

Consider two binary variables, x and y, with outcomes displayed in the following 2×2 table:

TABLE 2.3: A typical 2×2 contingency table

x	y 1	0	Total
1	n_{11}	n_{12}	n_{1+}
0	n_{21}	n_{22}	n_{2+}
Total	n_{+1}	n_{+2}	n

Let $p_{ij} = \Pr(x = i, y = j)$, $p_{i+} = \Pr(x = i)$, and $p_{+i} = \Pr(y = i)$ $(i, j = 0, 1)$.

2.2.1 Testing association

If the row and column variables are two measures on the subjects from a single sample, the first question about the 2×2 contingency table is whether the two variables are independent. In case-control and randomized clinical trial studies, the 2×2 table contains only one outcome, say the column variable y, as the other variable is used to denote study type (case vs. control) or treatment condition (e.g., intervention vs. control). In this case, we can compare the case and control or two treatment groups by testing whether the distributions of y are the same between the two groups defined by x.

Consider first the one-sample case where we are interested in assessing independence between the two variables represented by the row and columns. By definition, stochastic independence means that

$$\Pr(x = i, y = j) = \Pr(x = i)\Pr(y = j), \quad i, j = 0, 1.$$

Thus, the condition for independence can be expressed as the following null hypothesis:

$$H_0 : p_{11} - p_{1+}p_{+1} = 0. \tag{2.10}$$

Intuitively, if an estimate of $p_{11} - p_{1+}p_{+1}$ is far from zero, then we may reject this null hypothesis. Test statistics may be constructed based on the following estimate:

$$\widehat{p}_{11} - \widehat{p}_{1+}\widehat{p}_{+1} = \frac{n_{11}}{n} - \frac{n_{1+}}{n}\frac{n_{+1}}{n} = \frac{n_{11}n - n_{1+}n_{+1}}{n^2} = \frac{n_{11}n_{22} - n_{12}n_{21}}{n^2}.$$

It can be shown that $\widehat{p}_{11} - \widehat{p}_{1+}\widehat{p}_{+1}$ has an asymptotic normal distribution under the independence of the row and column variables:

$$\widehat{p}_{11} - \widehat{p}_{1+}\widehat{p}_{+1} \sim_a N\left(p_{11} - p_{1+}p_{+1}, \frac{1}{n}\left[p_{1+}p_{+1}(1 - p_{1+})(1 - p_{+1})\right]\right). \tag{2.11}$$

If we standardize $\widehat{p}_{11} - \widehat{p}_{1+}\widehat{p}_{+1}$ and replace p_{1+} and p_{+1} by their respective sample proportions, we obtain the following Z score:

$$Z = \sqrt{n}\frac{\frac{n_{11}n_{22} - n_{12}n_{21}}{n^2}}{\sqrt{\frac{n_{1+}}{n}\frac{n_{+1}}{n}\frac{n_{2+}}{n}\frac{n_{+2}}{n}}} = \sqrt{n}\frac{n_{11}n_{22} - n_{12}n_{21}}{\sqrt{n_{1+}n_{2+}n_{+1}n_{+2}}} \sim_a N(0, 1), \tag{2.12}$$

or equivalently the chi-square statistic:

$$Q = \left(\sqrt{n}\frac{n_{11}n_{22} - n_{12}n_{21}}{\sqrt{n_{1+}n_{2+}n_{+1}n_{+2}}}\right)^2 = n\frac{(n_{11}n_{22} - n_{12}n_{21})^2}{n_{1+}n_{2+}n_{+1}n_{+2}} \sim_a \chi_1^2. \tag{2.13}$$

A closely related quantity, called the *Pearson* chi-square statistic, is $P = \frac{n}{n-1}Q$. Since $\frac{n}{n-1} \to 1$ as $n \to \infty$, it follows from Slutsky's theorem that $P \sim_a \chi_1^2$. Thus, the two test statistics are equivalent in terms of having the same asymptotic distribution.

If the data are derived from two independent samples such as in a two-treatment or case-control study, interest becomes centered on testing whether the outcome y has the same distributions between the two samples defined by the two levels of x, i.e., the following null hypothesis:

$$H_0 : \Pr(y = 1 \mid x = 1) = \Pr(y = 1 \mid x = 0).$$

Let $p_1 = \Pr(y = 1 \mid x = 1)$ and $p_2 = \Pr(y = 1 \mid x = 0)$. We can estimate p_1 by $\widehat{p}_1 = \frac{n_{11}}{n_{1+}}$ and p_2 by $\widehat{p}_2 = \frac{n_{21}}{n_{2+}}$. If the difference between the two proportions $\widehat{p}_1$ and $\widehat{p}_2$ is far from zero, it provides evidence for rejecting the null hypothesis. Thus, the following difference can be used to construct a test statistic:

$$\widehat{p}_1 - \widehat{p}_2 = \frac{n_{11}}{n_{1+}} - \frac{n_{21}}{n_{2+}} = \frac{n_{11}n_{2+} - n_{1+}n_{21}}{n_{1+}n_{2+}} = \frac{n_{11}n_{22} - n_{12}n_{21}}{n_{1+}n_{2+}}. \tag{2.14}$$

This statistic is asymptotically normal (see Problem 2.11). By estimating p_1 and p_2 using $\widehat{p}_1 = \widehat{p}_2 = \frac{n_{+1}}{n}$ under the null hypothesis $H_0 : p_1 = p_2$ and normalizing the statistic using its asymptotic variance estimate, we obtain

$$\frac{\sqrt{n}\,(\widehat{p}_1 - \widehat{p}_2)}{\sqrt{Var_a\left(\sqrt{n}\,(\widehat{p}_1 - \widehat{p}_2)\right)}} = \frac{\sqrt{n}\left(\frac{n_{11}n_{22}-n_{12}n_{21}}{n_{1+}n_{2+}}\right)}{\sqrt{\left(\frac{n}{n_{1+}} + \frac{n}{n_{2+}}\right)\frac{n_{+1}}{n}\frac{n_{+2}}{n}}} \sim_a N(0,1). \qquad (2.15)$$

By simple algebra, we can simplify the above statistic to $\sqrt{n}\frac{n_{11}n_{22}-n_{12}n_{21}}{\sqrt{n_{1+}n_{2+}n_{+1}n_{+2}}}$, which is exactly the same Z score test statistic in (2.12) derived from the one-sample case considered earlier. Similarly, by squaring the term, we obtain the same chi-square statistic for the one-sample case.

Thus, we can use the same statistics regardless of whether we test for row and column independence as in the one-sample case or equal proportion of the response $y = 1$ across the levels of x for the two-sample case. Although identical in statistics, we must be mindful about the difference in interpretation between the two sampling cases.

Note that since the Z and chi-square statistics are discrete, the normal and chi-square distributions may not provide good approximations to the sampling distributions of these statistics. Yates suggested a version of the chi-square statistic corrected for continuity to improve accuracy. The Yates' statistic is $n\frac{\left(|n_{11}n_{22}-n_{12}n_{21}|-\frac{n}{2}\right)^2}{n_{1+}n_{2+}n_{+1}n_{+2}}$.

Example 2.4

In a study on treatment for schizophrenic patients, 32 schizophrenic patients were treated by some antipsychotic drugs for a brief period of time, and their recidivism status was collected 12 months before and after the treatment. Table 2.4 shows the recidivism rate for each of the 12-month period. Let us test whether the post treatment recidivism depends on pre-treatment recidivism.

TABLE 2.4: Recidivism before and after treatment

	12 month post treatment		
12 month pre-treatment	**No (0)**	**Yes (1)**	**Total**
No (0)	12	3	15
Yes (1)	9	8	17
Total	21	11	32

Based on the table, it is straightforward to compute the statistic in (2.12) $Z = 1.6082$. Thus, the p-value for the test of association is $2 \times \Phi(-1.6082) = 0.108$, where as before Φ stands for the CDF of a standard normal. ⬚

2.2.1.1 Exact tests

When sample size is small (e.g., when the cell mean in any of the cells of the table is below 10), the chi-square test based on the asymptotic theory may not be appropriate (or sufficiently accurate), and hence exact inference should be considered. The idea is that if there is no association between x and y, the cell counts can be predicted by the marginal counts of the variables, i.e., the distribution of the cell count conditioning on the marginal counts can be exactly computed.

In Table 2.3, n_{1+}, n_{2+}, n_{+1}, and n_{+2} are marginal counts. When they are held fixed, the contingency table is determined by the value of any one of the four cells. Under the null hypothesis of no x and y association, each cell count follows a *hypergeometric distribution*. For example, n_{11} satisfies a hypergeometric distribution with parameters n, n_{1+}, and n_{+1}. Based on this exact distribution, the p-value can be computed as the total probability of observing values of n_{11} that are as or more extreme than the observed cell counts under the null hypothesis. Next we describe the hypergeometric distribution and explain why the conditional distribution of n_{11} follows such a parametric model.

2.2.1.2 Hypergeometric distribution with parameter

Before introducing the hypergeometric distribution, it is important to distinguish between two popular sampling processes. Suppose a bag contains n balls, n_{1+} of which are white, and the remaining n_{2+} ones being black. Now, draw n_{+1} balls from the bag. Let K be the number of white balls among the n_{+1} balls sampled. If the balls are drawn one at a time with the color of the ball recorded and then put back to the bag before the next draw, this process is called sampling *with replacement*. Under this sampling procedure, the draws in general can be considered independent and identically distributed. Thus, we can calculate the probability of observing $K = k$ number of white balls using the binomial distribution.

If the n_{+1} balls are drawn simultaneously or sequentially without putting the one drawn back in the bag before the next draw, the balls are sampled *without replacement*. In this case, we cannot compute the probability of observing $K = k$, the number of white balls, based on the binomial distribution since the number of balls decreases as the sampling process continues and the proportion of white balls dynamically changes from draw to draw. Thus, under this sampling procedure, the draws are *dependent*. In most real studies, subjects are sampled without replacement. However, in data analysis, most methods are based on the i.i.d. assumption, which means that sampling with replacement is assumed. If the target population is very large compared to the sample size, the difference will be small and this assumption is still reasonable.

Now let us compute the distribution of K under sampling without replacement. Clearly, K cannot exceed the number of balls sampled n_{+1} and the total number of white balls in the bag n_{1+}, i.e., $K \leq \min\{n_{+1}, n_{1+}\}$. Similarly, if the number of balls sampled n_{+1} exceeds the number of black balls n_{2+} in the bag, we will draw at least $n_{+1} - n_{2+}$ white balls and hence $K \geq \max\{0, n_{1+} + n_{+1} - n\}$.

Since each ball has the same chance to be drawn, regardless of its color, each combination is equally likely to be the outcome, i.e., each of the $\binom{n}{n_{+1}}$ possible combinations has the same probability $\binom{n}{n_{+1}}^{-1}$ to be the outcome. Out of the $\binom{n}{n_{+1}}$ combinations, $\binom{n_{1+}}{K}\binom{n - n_{1+}}{n_{+1} - K}$ of them have the same configuration with K white and $n_{+1} - K$ black balls. Thus, the probability of having K white balls is

$$f_{HG}\left(K = k \mid n, n_{1+}, n_{+1}\right) = \frac{\dbinom{n_{1+}}{k}\dbinom{n - n_{1+}}{n_{+1} - k}}{\dbinom{n}{n_{+1}}},$$

$$\max\{0, n_{1+} + n_{+1} - n\} \leq k \leq \min\{n_{+1}, n_{1+}\}.$$

The above is a hypergeometric distribution $HG(k; n, n_{1+}, n_{+1})$ with parameters n, n_{1+}, and n_{+1}. It can be shown that the mean and variance of k are $\frac{n_{1+}n_{+1}}{n}$ and $\frac{n_{1+}n_{+1}n_{+2}n_{2+}}{n^2(n-1)}$, respectively (see Problem 2.15).

By mapping the variables x and y to the context of the above sampling process involving the different-colored balls, we can immediately see that the distribution of the cell count n_{11} conditional on the marginal counts follows the hypergeometric distribution $HG(k; n, n_{1+}, n_{+1})$:

Whole sample: n subjects $\leftrightarrow$ n balls in the bag
Two levels of x $\leftrightarrow$ White $(x = 1)$ and black $(x = 0)$
Two levels of y $\leftrightarrow$ Drawn $y = 1$ and not $y = 0$
x and y not associated $\leftrightarrow$ All balls are equally likely to be drawn.

2.2.1.3 Fisher's exact test

Knowing the exact distribution of the cell count under the null hypothesis enables us to find p-values similar to the exact test for the single proportion case discussed in Section 2.1.1. For a two-tailed test, the p-value for testing the null hypothesis of no row by column association is the sum of the probabilities of $K = k$ that are as or more extreme than the probability of the observed cell count n_{11}. This is called the *Fisher's exact* test.

There are also one-sided Fisher's exact tests. While such tests may not be important or intuitively clear for assessing independence between the row and column variables in a one-sample case, they are natural to consider when we compare two conditional probabilities, $p_1 = \Pr(y = 1 \mid x = 1)$ and $p_2 = \Pr(y = 1 \mid x = 0)$, in a two-sample setting. If the alternative is $H_a : p_1 > p_2$ (right sided), the p-value is the sum of the probabilities over $K \geq n_{11}$, and if $H_a : p_1 < p_2$ (left sided), the p-value is the sum of the probabilities over $K \leq n_{11}$.

Note that Fisher's exact test is based on the conditional distribution of a cell count such as n_{11} with fixed marginal totals. Exact p-values may also be calculated for other test statistics such as the chi-square statistic by using the exact rather than asymptotic distribution of the statistic. For example, exact p-values for the chi-square statistic are computed in essentially the same way as for Fisher's exact test. The p-value is the probability of observing a chi-square statistic not smaller than the one of the observed table. Note also that although most studies may not satisfy the requirement that both marginal counts be fixed, Fisher's exact test still provides valid inference. For example, a single random sample will create random marginals for both row and column variables. We will provide a theoretical justification in a more broad regression context when discussing the *conditional logistic regression* in Section 4.2.3.

Example 2.5

Some of the cell counts in Table 2.4 are small, so let us test the association using the exact method. Since $n = 32$, $n_{1+} = 15$, and $n_{+1} = 21$, it follows that $\max\{0, n_{1+} + n_{+1} - n\} = 4$ and $\min\{n_{+1}, n_{1+}\} = 15$, and the permissible range of k is $\{4, 5, \ldots, 15\}$. The probabilities are shown in Table 2.5 for each k in the permissible range.

The probability of the observed table or $n_{11} = 12$ is $\Pr(K = 12) = 0.086$. The p-value for the two-sided is the sum of the probabilities less than or equal to 0.086. Since the probabilities for $K = 4, 5, 6, 7, 12, 13, 14$, and 15 are all less than or equal to this threshold, summing up these probabilities yields the p-value $p = 0.1475$. If the alternative is $p_1 > p_2$ (right-sided), the probabilities for $K = 12, 13, 14$, and 15 are added up to yield the p-value $p_r = 0.108$. If the significance level is set at 0.05, then we will not reject the null hypothesis. Note that the problem associated with the discrete nature of the data for proportions also exists here:

TABLE 2.5: Table probabilities

k	4	5	6	7	8	9
p	0.000	0.000	0.005	0.034	0.119	0.240
k	10	11	12	13	14	15
p	0.288	0.206	0.086	0.020	0.002	0.000

At the nominal 0.05 level, the true type one error level is actually 0.022. If the alternative is $p_1 < p_2$ (left-sided), the probabilities for $K = 4$ to 12 are added up to yield the p-value $p_l = 0.978$.

Note that as in the case of a single proportion, $p_r + p_l = 1 + \Pr(K = 12) > 1$, and thus one of p_r and p_l will be large ($> \frac{1}{2}$). So, only one of the one-sided tests is worth considering. In this example, $\widehat{p}_1 = \frac{12}{15} > \widehat{p}_2 = \frac{9}{17}$. Without further computation, we know that it is senseless to consider the one-sided alternative $H_a : p_1 < p_2$, since if the null hypothesis $H_0 : p_1 = p_2$ is unlikely, then H_a will be even more so.

2.2.2 Measures of association

When two variables (or row and column) are actually associated, we may want to know the nature of the association. There are many indices that have been developed to characterize the association between two variables. If one of the two variables is an outcome and the other a group indicator as in a case-control or randomized trial study with two treatment conditions, one may use the difference between the proportions as a measure of association between the two variables. Other common measures for assessing association between the outcome and group indicator include the odds ratio and relative risk. If both variables are outcomes measuring some characteristics of interest for subjects in a single sample, correlations are used as measures of association. We start with the two-sample case with one variable being the group indicator.

2.2.2.1 Difference between proportions

Assume two groups defined by x and let

$$p_1 = \Pr(y = 1 \mid x = 1), \quad p_2 = \Pr(y = 1 \mid x = 0).$$

As described earlier, p_1 and p_2 are estimated by the sample proportions: $\widehat{p}_1 = \frac{n_{11}}{n_{1+}}$ and $\widehat{p}_2 = \frac{n_{21}}{n_{2+}}$. When there is no difference between the population proportions, i.e., $p_1 = p_2$, their difference $p_1 - p_2$ is zero and the sample analogue $\widehat{p}_1 - \widehat{p}_2$ should be close to 0. To formally account for sampling variability in $\widehat{p}_1 - \widehat{p}_2$, we need to know its asymptotic variance. Since

$$Var(\widehat{p}_1 - \widehat{p}_2) = \frac{1}{n_{1+}}p_1(1 - p_1) + \frac{1}{n_{2+}}p_2(1 - p_2), \tag{2.16}$$

by substituting $\widehat{p}_k$ in place of p_k in (2.16), we immediately obtain an estimate of the asymptotic variance and use it to construct confidence intervals for $p_1 - p_2$.

If y is the indicator of a disease with $y = 1$ (0) indicating disease (nondisease), then the difference between two proportions is commonly called *attributable risk* (AR) (Rothman, 1998). In theory, the range of AR is $[-1, 1]$; however, if x is the indicator of a risk factor (e.g., subjects with $x = 1$ have a higher risk of the disease than those with $x = 0$), AR is positive. AR is an important concept in epidemiology, especially in disease prevention/intervention, as it measures the excess risk of disease for the subjects exposed to the risk over those not exposed to the risk. Thus, it is an index for how effective the prevention/intervention can be if we try to block the effect of the risk factor on the exposed subjects. For example, if we can successfully treat n exposed subjects, the expected number of diseased subjects will change from np_1 to np_2, or a reduction by nAR number of subjects in the diseased population. The ratio $\frac{n}{nAR} = \frac{1}{AR}$, called the *number needed to treat* (NNT), indicates the number of the exposed subjects we need to treat in order to reduce the number of diseased subjects by 1. Thus, cost-effective intervention strategies may target risk factors with a low NNT or high AR.

Here are some additional frequently used and closely related concepts within the current context. The *attributable risk fraction* (ARF) is defined as

$$\text{ARF} \equiv \frac{p_1 - p_2}{p_2} = \text{RR} - 1, \tag{2.17}$$

where $\text{RR} = \frac{p_1}{p_2}$ is the *relative risk,* another measure of association to be discussed shortly. The *population attributable risk* (PAR) is defined as

$$\Pr(y = 1) - \Pr(y = 1 \mid x = 0) = \Pr(x = 1) \cdot AR,$$

which measures the excess risk of disease in the entire population because of the risk factor. In other words, if the risk effect could be completely eliminated for the exposed subjects, then PAR would be the reduction of disease prevalence in the population. Thus, if only a small portion of the population is exposed to the risk, PAR will be small even when AR is large. The proportion of the reduction of PAR among all the diseased subjects, $\text{PAR}/\Pr(y = 1)$, is called the *population attributable risk fraction* (PARF). All these indices may also be viewed as measures of association between x and y and will equal 0 if x and y are independent. Hence, departures from zero of ARF, PAR, or PARF imply association.

Note that AR and ARF can be estimated if the data are from either a single random sample or a stratified random sample from two independent groups ($x = 0$ and $x = 1$). Since the disease prevalence, $\Pr(y = 1)$, is not estimable directly from the sample in the latter case, PAR and PARF are not estimable. However, if the prevalence is known, PAR and PARF can still be estimated (see Problem 2.21).

2.2.2.2 Odds ratio

The odds ratio is the most popular index for association between two binary outcomes. The odds of response $y = 1$ for each group is defined as

$$\frac{\Pr(y = 1 \mid x = 1)}{\Pr(y = 0 \mid x = 1)} = \frac{p_1}{1 - p_1}, \quad \frac{\Pr(y = 1 \mid x = 0)}{\Pr(y = 0 \mid x = 0)} = \frac{p_2}{1 - p_2}.$$

The *odds ratio* of group 1 to group 2 or $x = 1$ to $x = 0$ is defined by

$$OR = \frac{\Pr(y = 1 \mid x = 1)}{\Pr(y = 0 \mid x = 1)} \Big/ \frac{\Pr(y = 1 \mid x = 0)}{\Pr(y = 0 \mid x = 0)} = \frac{p_1(1 - p_2)}{p_2(1 - p_1)}.$$

When the two population proportions are equal to each other, $OR = 1$. An odds ratio greater than 1 indicates that the odds of response in group 1 is higher than that for group 2.

If the data are from a simple random sample or a stratified sample with subjects independently sampled from $x = 1$ and $x = 0$, then $\Pr(y = 1 \mid x = 1)$ and $\Pr(y = 0 \mid x = 1)$ can be estimated by $\frac{n_{11}}{n_{1+}}$ and $\frac{n_{12}}{n_{1+}}$, and the odds of response $\frac{\Pr(y=1 \mid x=1)}{\Pr(y=0 \mid x=1)}$ can be estimated by $\frac{n_{11}}{n_{12}}$. Similarly, we estimate the odds $\frac{\Pr(y=1 \mid x=0)}{\Pr(y=0 \mid x=0)}$ by $\frac{n_{21}}{n_{22}}$. Thus, the odds ratio is estimated by

$$\widehat{OR} = \frac{n_{11} n_{22}}{n_{12} n_{21}}. \tag{2.18}$$

To obtain the asymptotic variance of $\widehat{OR}$ in (2.18) for inference, we first find the asymptotic variance of $\log\left(\widehat{OR}\right)$ by the delta method:

$$\widehat{\sigma}^2_{\log(OR)} = \widehat{Var}_a\left(\log\widehat{OR}\right) = \frac{1}{n_{11}} + \frac{1}{n_{12}} + \frac{1}{n_{21}} + \frac{1}{n_{22}}.$$

Note that this estimate is contingent upon the assumption that none of the four cell counts is 0; if any one of them is 0, the estimate will be infinite. By again applying the delta method to the above, we immediately obtain the asymptotic variance of $\widehat{OR}$:

$$\widehat{\sigma}^2_{OR} = \left(\exp\left(\log\widehat{OR}\right)\right)^2 \widehat{Var}_a\left(\log\widehat{OR}\right) = \widehat{OR}^2\left(\frac{1}{n_{11}} + \frac{1}{n_{12}} + \frac{1}{n_{21}} + \frac{1}{n_{22}}\right). \qquad (2.19)$$

This asymptotic variance is used to calculate standard errors of $\widehat{OR}$ as well as p-values for inference about OR.

Note that confidence intervals for $\widehat{OR}$ are usually obtained by first constructing such intervals for $\log(OR)$ and then transforming them to the scale of OR. For example, a $(1 - \alpha)$ confidence interval for $\log(OR)$ is given by

$$\left(\log\widehat{OR} - z_{1-\frac{\alpha}{2}}\widehat{\sigma}_{\log(OR)}, \ \log\widehat{OR} + z_{1-\frac{\alpha}{2}}\widehat{\sigma}_{\log(OR)}\right), \qquad (2.20)$$

where $z_{1-\frac{\alpha}{2}}$ is the percentile of the standard normal distribution. By exponentiating the limits of the interval in (2.20), we can obtain a $(1 - \alpha)$ confidence interval for OR:

$$\left(\widehat{OR}\exp\left(-z_{1-\frac{\alpha}{2}}\widehat{\sigma}_{\log(OR)}\right), \ \widehat{OR}\exp\left(z_{1-\frac{\alpha}{2}}\widehat{\sigma}_{\log(OR)}\right)\right).$$

Given two binary variables x and y, eight versions of odds ratios can be defined by switching either the roles of x and y or the order of the groups in x and y. For example, the odds ratio of response $x = 0$ of $y = 1$ to $y = 0$ can be similarly defined as $\frac{\Pr(x=0|y=1)}{\Pr(x=1|y=1)}$ / $\frac{\Pr(x=0|y=0)}{\Pr(x=1|y=0)}$. These odds ratios satisfy simple relations, and one can be computed from any other using such relationships (see Problem 2.13). Because of the symmetry of the variance in (2.19), odds ratios can be computed if the data are from a simple random sample or a stratified sample with two independent groups either defined by $x = 1$ and $x = 0$, or by $y = 0$ and $y = 1$.

2.2.2.3 Relative risk

The *risk ratio* or *relative risk* (RR) of response $y = 1$ of the population $x = 1$ to population $x = 0$ is the ratio of the two population proportions:

$$RR = \frac{\Pr(y = 1 \mid x = 1)}{\Pr(y = 1 \mid x = 0)} = \frac{p_1}{p_2}.$$

A relative risk greater (less) than 1 indicates that the probability of response is larger (smaller) in group $x = 1$ than in group $x = 0$. The relative risk is estimated by $\widehat{RR} = \frac{n_{11}}{n_{1+}} / \frac{n_{21}}{n_{2+}}$. The relative risk is also often referred to as the *incidence rate ratio* (IRR).

As in the case of odds ratio, we first estimate the asymptotic variance of $\log\left(\widehat{RR}\right)$ by the delta method:

$$\widehat{\sigma}^2_{\log(RR)} = \widehat{Var}_a(\log\widehat{RR}) = \frac{1 - \widehat{p}_1}{n_{11}} + \frac{1 - \widehat{p}_2}{n_{22}}.$$

Then, by invoking the delta method again, we obtain the asymptotic variance $\sigma_{\widehat{RR}}^2$ of $\widehat{RR}$:

$$\widehat{\sigma}_{RR}^2 = \left(\exp \left(\log \widehat{RR} \right) \right)^2 \widehat{Var}_a \left(\log \widehat{RR} \right) = \widehat{RR}^2 \left(\frac{1 - \widehat{p}_1}{n_{11}} + \frac{1 - \widehat{p}_2}{n_{22}} \right).$$

Similarly, we can obtain a $(1 - \alpha)$ confidence interval for RR by transforming the following interval for $\log (RR)$,

$$\left(\log \widehat{RR} - z_{1 - \frac{\alpha}{2}} \widehat{\sigma}_{\log(RR)}, \ \log \widehat{RR} + z_{1 - \frac{\alpha}{2}} \widehat{\sigma}_{\log(RR)} \right),$$

to the scale of RR,

$$\left(\widehat{RR} \exp \left(-z_{1 - \frac{\alpha}{2}} \widehat{\sigma}_{\log(RR)} \right), \ \widehat{RR} \exp \left(z_{1 - \frac{\alpha}{2}} \widehat{\sigma}_{\log(RR)} \right) \right).$$

Note that the estimates of RR do not share the same symmetry as that of OR, and thus RR will be inestimable if the data are independently sampled from $y = 0$ and $y = 1$ such as in case-control studies. Again, there are eight versions of RR. However, their relations are much more complex, and in particular, unlike OR, one may not be computed from another without further information (see Problem 2.14). However, for rare diseases or outcomes where $\Pr(y = 1 \mid x = 1)$ and $\Pr(y = 1 \mid x = 0)$ are very small, the OR approximates the RR.

2.2.2.4 Phi coefficient

When x also becomes an outcome of interest rather than a group indicator, we may also use correlation to define measures of association between the two variables. Denote the two levels of x and y as 0 and 1. The product-moment correlation between variables x and y is defined as $\rho = \frac{E(xy) - E(x)E(y)}{\sqrt{Var(x)Var(y)}}$. If x and y are independent, $E(xy) - E(x)E(y) = 0$ and thus $\rho = 0$. If the two variables are identical (opposite), i.e., $y = x$ ($y = 1 - x$), then $\rho = 1$ (-1). Thus, for interval or ordinal outcomes x and y, ρ provides a measure of association between x and y, with the sign and magnitude of ρ indicating, respectively, the direction (direct or inverse) and strength of the relationship.

Note that although the computation of ρ requires some numerical coding of the levels, the value of ρ does not, as long as the order of the two levels is kept the same; i.e., the higher level is coded with a higher value. If the order of one of the variables is reversed, the correlation coefficient will change signs. We will discuss this invariance property about ρ in detail in Section 2.5.1.1.

The Pearson correlation is an estimate of the product-moment correlation by substituting the sample moments in place of the respective parameters. In particular, when both x and y are binary as in the current context, these moments are given by

$$\widehat{E(xy)} = \frac{n_{11}}{n}, \quad \widehat{E(x)} = \frac{n_{11} + n_{12}}{n}, \quad \widehat{E(y)} = \frac{n_{11} + n_{21}}{n},$$

$$\widehat{Var(x)} = \frac{(n_{11} + n_{12})(n_{21} + n_{22})}{n^2}, \quad \widehat{Var(y)} = \frac{(n_{11} + n_{21})(n_{12} + n_{22})}{n^2},$$

and the Pearson correlation is

$$\widehat{\rho} = \frac{n_{11}n_{22} - n_{12}n_{21}}{\sqrt{(n_{11} + n_{12})(n_{21} + n_{22})(n_{11} + n_{21})(n_{12} + n_{22})}}. \tag{2.21}$$

This version of the Pearson correlation in (2.21) is known as the *Phi coefficient*.

Example 2.6

For Table 2.4, it is straightforward to compute the measures of association between the recidivisms before and after treatment.

Since $\widehat{\Pr}(y=1|x=1) = \frac{8}{17}$ and $\widehat{\Pr}(y=1|x=0) = \frac{3}{15}$, the difference in probabilities is $\frac{8}{17} - \frac{3}{15} = 0.2706$, with asymptotic variance $\frac{1}{17} \times \frac{8}{17} \times \left(1 - \frac{8}{17}\right) + \frac{1}{15} \times \frac{3}{15} \times \left(1 - \frac{3}{15}\right) = 0.0$ 253. The odds ratio of $y=1$ to $y=0$ of $x=1$ over $x=0$ is estimated by $\frac{8}{9}/\frac{3}{12} = 3.5556$, and the relative risk is estimated by $\frac{8}{17}/\frac{3}{15} = 2.3529$. If the roles of $y=1$ and $y=0$ are switched, the estimates for odds ratio and RR are $\frac{9}{8}/\frac{12}{3} = 0.28125$ and $\frac{9}{17}/\frac{12}{15} = 0.6618$, respectively. It can be verified that the two odds ratios are reciprocal to each other, but the RRs do not have such a simple relation. The Phi coefficient is 0.2843. ▯

2.2.3   Test for marginal homogeneity

For 2×2 tables, our interest thus far has been in the independency between x and y or the difference in response rates between two groups. Although this is the primary interest in most contingency table analyses, it is not always the case. A notable exception is when comparing dependent proportions in a matched pair or pre-post treatment study design.

Table 2.6 contains information of depression diagnosis of those patients who completed the one-year follow-up in the DOS at the baseline (year 0) and 1 year after the study.

TABLE 2.6:   Depression of patients at years 0 and 1 (DOS)

Year 0	Year 1		Total
	No	Dep	
No	276	41	317
Dep	9	155	164
Total	285	196	481

We can check the prevalence of depression at the two time points to assess the effect of the treatment, i.e., test $H_0 : p_{1+} = p_{+1}$, where $p_{1+} = \Pr(x=1)$ and $p_{+1} = \Pr(y=1)$. Since p_{1+} and p_{+1} are readily estimated by

$$\widehat{p}_{1+} = \frac{n_{1+}}{n}, \quad \widehat{p}_{+1} = \frac{n_{+1}}{n},$$

we can again use their difference $\widehat{p}_{1+} - \widehat{p}_{+1}$ as a test statistic. However, as $\widehat{p}_{1+}$ and $\widehat{p}_{+1}$ are dependent, the methods discussed in Section 2.2.2 for comparing p_1 and p_2 do not apply; since $\widehat{p}_{1+}$ and $\widehat{p}_{+1}$ are dependent, the variance of their difference is not the sum of the variances of each individual proportions $\widehat{p}_{1+}$ and $\widehat{p}_{+1}$, i.e., $Var_a(\widehat{p}_{1+} - \widehat{p}_{+1}) \neq Var_a(\widehat{p}_{1+}) + Var_a(\widehat{p}_{+1})$.

By some simple algebra, $\widehat{p}_{1+} - \widehat{p}_{+1} = \frac{n_{12} - n_{21}}{n}$. So, the difference between the two proportions is essentially a function of $n_{12} - n_{21}$, the difference between the two off-diagonal cell counts. By conditioning on the total $n_{21} + n_{12}$, we can use the proportion $\frac{n_{21}}{n_{21}+n_{12}}$ as a test statistic, since $\frac{n_{12}}{n_{21}+n_{12}} = 1 - \frac{n_{21}}{n_{21}+n_{12}}$ is a function of this proportion. This statistic has an asymptotic normal distribution. By standardizing it using the asymptotic variance and squaring the resulting Z statistic, we obtain the following chi-square test statistic:

$$\frac{(n_{12} - n_{21})^2}{n_{12} + n_{21}} \sim_a \chi_1^2. \tag{2.22}$$

The above is known as *McNemar's* test (McNemar, 1947). A version of McNemar's test after correction for continuity is

$$\frac{(|n_{12} - n_{21}| - 1)^2}{n_{12} + n_{21}} \sim_a \chi_1^2. \tag{2.23}$$

2.2.3.1 Exact inference

If the sample size is small, or more specifically, if $n_{12} + n_{21}$ is small, then the McNemar test in (2.22) or (2.23) may not be appropriate and exact inference may be considered. Since the two proportions p_{2+} and p_{+2} agree under the null hypothesis, a subject with discordant x and y is equally likely to be in either the $(1, 2)$ or $(2, 1)$ cell. So, conditional on such subjects, we are essentially testing whether $p = 0.5$ in a Bernoulli(p). This is known as the *sign* test.

Example 2.7

For Table 2.6, the McNemar statistic is $\frac{(9-41)^2}{9+41} = 20.48$, and hence the asymptotic p-value is < 0.0001. The p-value based on exact method is also very small. Hence, we reject the null hypothesis of marginal homogeneity. □

2.2.4 Agreement

When both the row and column variables are measures on the same scale, we may be interested in the agreement between the two variables. For example, if the two variables are the ratings of n subjects by two independent raters such as radiologists when reading X-ray images, we may want to know to what degree the two raters agree with each other on their ratings. The most popular measures for such rater or observer agreement are Cohen's kappa coefficients.

Intuitively, we may use the cell probabilities on the diagonal of the table to measure the agreement, which represent the likelihood of having the same ratings by the two raters with respect to each of the rating categories. However, even if the two raters give their evaluations randomly and independent of each other, these probabilities will not be zero, since just by chance they may agree on their ratings for some subjects. For example, consider two people who are asked to guess the outcome of a coin when tossed. Suppose both believe that the coin is fair. Then, half of the times their guesses will be head (or tail). If they guess independently from each other, then their guesses will be close to the proportions in the following table after a larger number of guesses:

	Face	Tail	Total
Face	0.25	0.25	0.5
Tail	0.25	0.25	0.5
Total	0.5	0.5	1

The total proportion on the diagonal is $0.25 + 0.25 = 0.5$. Thus, even though the subjects are randomly guessing the outcomes, half of the times they appear to agree with each other's guesses. Since the agreement here is purely by chance, we must find and remove the chance factor if we want to determine the raters' true agreement.

Suppose the two raters are given ratings independently according to their marginal distributions. Then, the probability of a subject being rated as 0 by chance by both raters is $p_{2+}p_{+2}$, and similarly, the probability of a subject being rated as 1 by chance by both raters is $p_{1+}p_{+1}$. Thus, the agreement rate by chance is the sum of the products of the marginal probabilities, $p_{1+}p_{+1} + p_{2+}p_{+2}$. By excluding this term from the agreement probability $p_{11} + p_{22}$, we obtain the probability of agreement: $p_{11} + p_{22} - (p_{1+}p_{+1} + p_{2+}p_{+2})$ corrected for the chance factor. Further, by normalizing it, Cohen (1960) suggested the following coefficient:

$$\kappa = \frac{p_{11} + p_{22} - (p_{1+}p_{+1} + p_{2+}p_{+2})}{1 - (p_{1+}p_{+1} + p_{2+}p_{+2})}. \tag{2.24}$$

This simple kappa coefficient varies between -1 and 1, depending on the marginal probabilities. If the two raters completely agree with each other, then $p_{11} + p_{22} = 1$ and thus $\kappa = 1$, and the converse is also true. On the other hand, if the judges rate the subjects at random, then the observer agreement is completely by chance and as a result, $p_{ii} = p_{i+}p_{+i}$ for $i = 0, 1$ and the kappa coefficient in (2.24) equals 0. In general, when the observer agreement exceeds the agreement by chance, kappa is positive, and when the raters really disagree on their ratings, kappa is negative. The magnitude of kappa indicates the degree of agreement or disagreement depending on whether κ is positive (negative).

By plugging in the sample proportions as the estimates of the corresponding probabilities in (2.24), the kappa index can be estimated by

$$\widehat{\kappa} = \frac{\widehat{p}_{11} + \widehat{p}_{22} - (\widehat{p}_{1+}\widehat{p}_{+1} + \widehat{p}_{2+}\widehat{p}_{+2})}{1 - (\widehat{p}_{1+}\widehat{p}_{+1} + \widehat{p}_{2+}\widehat{p}_{+2})}. \tag{2.25}$$

The estimates of the various parameters in (2.25) follow a multivariate normal distribution. Since $\widehat{\kappa}$ is a function of consistent estimates of these parameters, it is also consistent. The delta method can be used to obtain the asymptotic distribution of $\widehat{\kappa}$, upon which we can make inference about κ if the sample size is large. For small sample sizes, exact methods may be applied.

Example 2.8
In the DDPC, informants were recruited for 200 subjects (probands). Table 2.7 displays the probands' depression diagnoses based on the probands (row) and informants' (column) ratings:

TABLE 2.7: Depression diagnoses based on the probands and informants

Proband	Informant		Total
	No	Dep	
No	66	19	85
Dep	50	65	115
Total	116	84	200

The estimate of the kappa coefficient is

$$\widehat{\kappa} = \frac{\frac{66}{200} + \frac{65}{200} - \left(\frac{116}{200}\frac{85}{200} + \frac{84}{200}\frac{115}{200}\right)}{1 - \left(\frac{116}{200}\frac{85}{200} + \frac{84}{200}\frac{115}{200}\right)} = 0.3262,$$

with an asymptotic standard error of 0.0630. This gives a 95% CI (0.2026, 0.4497). The positive kappa indicates some degree of agreement between the probands and informants. However, the agreement is not high, as a value of kappa larger than 0.8 is generally considered as high agreement. ∎

Note that although widely used in assessing the agreement between two raters, the kappa index is not perfect. For example, consider the following table representing proportions of agreement about some disease condition from two judges in a hypothetical example:

	No	**Yes**	**Total**
No	0.8	0.1	0.9
Yes	0.1	0	0.1
Total	0.9	0.1	1

By assuming a very large sample, we can ignore the sampling variability and interpret the following estimate as the kappa coefficient:

$$\kappa = \frac{0.8 + 0 - 0.9 \times 0.9 - 0.1 \times 0.1}{1 - 0.9 \times 0.9 - 0.1 \times 0.1} = -0.1111.$$

The small magnitude and negative sign of κ suggest there is some slight disagreement between the two raters. However, if we look at the proportions in the agreement table, it seems that the two raters agree to a high degree, since they agree on 80% of the subjects. This paradox was discussed in Feinstein and Cicchetti (1990). A primary cause of the problem is imbalances in the distribution of marginal totals. Sometimes, the marginal distributions are also informative when considering agreement between raters, but Cohen's kappa does not take into account such information. For example, the low prevalence assumed by both raters in the above table may not be coincidental, but rather reflecting the raters' prior knowledge about the disease in this particular population.

2.3 Inference for $2 \times r$ Tables

In this section, we examine the special case of the general $s \times r$ table when one of the row and column variables is binary and the other is ordinal. For such a table, we may be interested in learning if the binary proportion increases (decreases) as the ordinal variable increases, i.e., whether there is any trend in the binary response as the ordinal variable changes. Alternatively, if the binary variable represents two independent groups, we may want to know whether the distribution of the ordinal response is the same between the groups. The Cochran–Armitage trend test is designed to examine the former, while the Mann–Whitney–Wilcoxon test can be used to address the latter. Note that both tests assess the association between the row and column variables, and the null hypothesis is the same (the row and column variables are independent). However, they emphasize different features under the alternative hypothesis and thus may have different power in different alternative scenarios. For convenience, we assume the row variable x is binary and data are displayed in a $2 \times r$ table as below:

	y				
x	1	2	$\cdots$	r	**Total**
1	n_{11}	n_{12}	$\cdots$	n_{1r}	n_{1+}
0	n_{21}	n_{22}	$\cdots$	n_{2r}	n_{2+}
Total	n_{+1}	n_{+2}	$\cdots$	n_{+r}	n

Let $p_{ij} = \Pr(x = i \text{ and } = j)$, $p_{i+} = \Pr(x = i)$, and $p_{+j} = \Pr(y = j)$; $i = 0, 1$; $j = 1, \ldots, r$.

2.3.1   Cochran–Armitage trend test

For the binary row variable x and ordered column variable y, we are interested in whether the proportions of $x = 1$ follow some patterns as a function of the levels of y. For example, if the levels of y are indexed by the integers j ($1 \leq j \leq r$), the proportions of $x = 1$ may have a linear relationship with j. More generally, let R_j denote the ordinal values of y ($1 \leq j \leq r$) and $p_{1|j} = \Pr(x = 1 \mid y = j)$. The Cochran–Armitage trend test is designed to test whether $p_{1|j}$ and R_j satisfy some linear relationship, i.e., $p_{1|j} = \alpha + \beta R_j$ (Armitage, 1955). The trend test is based on an estimate of $\sum_{j=1}^{r} p_{1j}(R_j - E(y))$, where $E(y) = \sum_{j=1}^{r} p_{+j} R_j$ is the total mean score of y. This statistic is motivated by testing whether $\beta = 0$. Since $E(x \mid y = j) = p_{1|j}$, the coefficient β can be estimated by the linear regression:

$$E(x \mid y) = \alpha + \beta y, \tag{2.26}$$

which gives the estimate $\widehat{\beta} = \frac{\overline{xy} - \overline{x}\,\overline{y}}{\overline{y^2} - \overline{y}^2}$ (see Problem 2.25). Obviously, $\overline{xy} - \overline{x} \cdot \overline{y}$ is an estimate of $E(xy) - E(x)E(y) = \sum_{j=1}^{r} p_{1j}(R_j - E(y))$.

Under the null hypothesis,

$$H_0 : \frac{p_{11}}{p_{+1}} = \frac{p_{12}}{p_{+2}} = \cdots = \frac{p_{1r}}{p_{+r}} = p_{1+}. \tag{2.27}$$

It then follows from (2.27) that

$$\sum_{j=1}^{r} p_{1j}(R_j - E(y)) = \sum_{j=1}^{r} p_{1+} p_{+j}(R_j - E(y)) = 0.$$

Thus, if an estimate of $\sum_{j=1}^{r} p_{1j}(R_j - E(y))$ is far from 0, the null hypothesis is likely to be false. Furthermore, if the proportions p_{1j} of $x = 1$ increase or decrease as y increases, then

$$\sum_{j=1}^{r} p_{1j}(R_j - E(y)) > 0 \quad \text{or} \quad \sum_{j=1}^{r} p_{1j}(R_j - E(y)) < 0.$$

Thus, one-sided tests may also be carried out based on the estimate.

Substituting $\overline{y} = \frac{1}{n}\sum_{j=1}^{r} n_{+j} R_j$ for $E(y)$ and $\frac{n_{1j}}{n}$ for p_{1j} in $\sum_{j=1}^{r} p_{1j}(R_j - E(y))$, we obtain an estimate, $\sum_{j=1}^{r} \frac{n_{1j}}{n}(R_j - \overline{y})$, of this quantity. Under the null hypothesis, the above statistic converges to 0 in probability as $n \to \infty$. Thus, by standardizing it using its asymptotic variance, we have

$$Z_{CA} = \frac{\sum_{j=1}^{r} n_{1j}(R_j - \overline{y})}{\sqrt{\left[\sum_{j=1}^{r} n_{+j}(R_j - \overline{y})^2\right] \widehat{p}_{1+}(1 - \widehat{p}_{1+})}} \sim_a N(0,1).$$

The statistic Z_{CA} is called the *Cochran–Armitage* statistic, and the above asymptotic distribution allows us to use this statistic for inference about one-sided and two-sided tests.

Example 2.9

In the DOS, we are interested in testing whether the proportion of females increases or decreases as the depression level increases. In Table 2.8 MinD and MajD denote the group of patients diagnosed with minor and major depression, respectively.

If the three depression levels are assigned scores 1, 2, and 3, respectively, then $Z_{CA} = 4.2111$ and p-value $= 2 \times \Phi(-4.2111) < 0.0001$ for the two-sided Cochran–Armitage trend

TABLE 2.8:   Depression diagnoses by gender in the DOS

	No	**MinD**	**MajD**	**Total**
Female	274	105	93	472
Male	207	31	35	273
Total	481	136	128	745

test. As Z_{CA} is positive, we conclude that women had more severe depression than men in this study sample.

When the sample size is small, exact inference may be considered. The computational procedure is similar to Fisher's exact test. With the row and column marginal counts fixed, the exact distribution of the table then follows a generalized hypergeometric distribution. To use the statistic Z_{CA} for exact inference, we compute the probabilities for all the tables with different cell counts while holding the row and column margins fixed. The exact p-values are then computed as the sum of the probabilities of the tables to yield a Z_{CA} with its absolute value no smaller than that of the one based on the observed table.

For the example above, the p-value is also less than 0.0001 based on the exact test. Thus, both the asymptotic and exact tests yield the same conclusion. This may not be surprising given the relatively large sample size in this study.

2.3.2   Mann–Whitney–Wilcoxon test

If the row binary variable x indicates two groups, then we.can test whether the ordinal column variable y has the same distributions across the levels of x using the two-group Mann–Whitney–Wilcoxon (MWW) statistic (Wilcoxon, 1945, Mann and Whitney, 1947). This statistic is initially developed to provide a nonparametric alternative for comparing two continuous distributions as it does not assume any mathematical model for the distributions of the variables such as the normal as in the case of the popular two-sample t test. For this historical reason, we first introduce this statistic within the context of continuous variables and then discuss its adaptation to the current setting.

Consider two independent samples y_{1i} $(1 \leq i \leq n_1)$ and y_{2j} $(1 \leq j \leq n_2)$ of sizes n_1 and n_2. We are interested in testing whether the two samples are selected from the same study population, i.e., whether y_{1i} and y_{2j} follow the same distribution. Under this null hypothesis, y_{1i} and y_{2j} follow the same distribution, with the form of the distribution unspecified such as the normal as in the case of t test. Thus, under the null hypothesis, y_{1i} has an equal chance to be larger or smaller than y_{2j}, i.e., $\Pr(y_{1i} > y_{2j}) = \Pr(y_{1i} < y_{2j})$, or equivalent, $\Pr(y_{1i} > y_{2j}) + \frac{1}{2}\Pr(y_{1i} = y_{2j}) = \frac{1}{2}$. The MWW statistic is defined as

$$M = \text{number of } \{(i,j) \mid y_{1i} > y_{2j}, i = 1, \ldots, n_1, j = 1, \ldots, n_2\}$$
$$+ \frac{1}{2}\text{number of } \{(i,j) \mid y_{1i} = y_{2j}, i = 1, \ldots, n_1, j = 1, \ldots, n_2\}.$$

Since there are a total of $n_1 n_2$ pairs (y_{1i}, y_{2j}), we can estimate $\Pr(y_{1i} > y_{2j}) + \frac{1}{2}\Pr(y_{1i} = y_{2j})$ by $\frac{M}{n_1 n_2}$. If the estimate is far from $\frac{1}{2}$, then the null hypothesis is likely to be rejected.

When expressed in terms of contingency tables (assuming higher columns have higher order), we have

$$M = \sum_{j=2}^{r}\left[n_{1j}\left(\sum_{k=1}^{j-1} n_{2k}\right)\right] + \frac{1}{2}\sum_{j=1}^{r} n_{1j}n_{2j}. \tag{2.28}$$

This is an example of U-statistics for two independent groups, and the asymptotic variance can be computed using the theory of U-statistics. The asymptotic variance under the independence of the row and column variables is given by (see Problem 2.23)

$$\frac{n_1 n_2 (N+1)}{12} - \left[\frac{n_1 n_2}{12N(N-1)} \sum_{j=1}^{r} (n_{+j} - 1) n_{+j} (n_{+j} + 1) \right], \qquad (2.29)$$

where $N = n_1 + n_2$ is the total sample size.

The MWW statistic M above can also be expressed as a sum of ranks. We first order the pooled observations from y_{1i} and y_{2j} from the smallest to the largest; if there are ties among the observations, they are arbitrarily broken. The ordered observations are then assigned rank scores based on their rankings, with tied observations assigned midranks. For example, consider the following 2×3 table:

	$y = 1$	$y = 2$	$y = 3$
$x = 1$	1	2	2
$x = 2$	2	1	2

Then, all the observations from both groups (subscripts indicating group memberships) are

$$1_1, \; 2_1, \; 2_1, \; 3_1, \; 3_1, \; 1_2, \; 1_2, \; 2_2, \; 3_2, \; 3_2.$$

The ordered version is

$$1_1, \; 1_2, \; 1_2, \; 2_1, \; 2_1, \; 2_2, \; 3_1, \; 3_1, \; 3_2, \; 3_2,$$

with the assigned rank scores given by

$$\frac{1+2+3}{3}, \; \frac{1+2+3}{3}, \; \frac{1+2+3}{3}, \; \frac{4+5+6}{3}, \; \frac{4+5+6}{3}, \; \frac{4+5+6}{3},$$
$$\frac{7+8+9+10}{4}, \; \frac{7+8+9+10}{4}, \; \frac{7+8+9+10}{4}, \; \frac{7+8+9+10}{4}.$$

It can be shown that the sum of ranks for the first group is

$$R_1 = M + \frac{1}{2} n_{1+}(n_{1+} + 1), \qquad (2.30)$$

where M given in (2.28) contains the ranks contributed from the second group, while $\frac{1}{2} n_{1+}(n_{1+} + 1)$ is the sum of ranks from the first group. For example, for the 2×3 table above,

$$n_{1+} = 5, \; M = \frac{1}{2} \times 2 + 2 \times 2 + \frac{1}{2} \times 2 + 2 \times (2+1) + \frac{1}{2} \times 2 \times 2 = 14,$$
$$R_1 = \frac{1+2+3}{3} + \frac{4+5+6}{3} + \frac{4+5+6}{3} + \frac{7+8+9+10}{4} + \frac{7+8+9+10}{4} = 29.$$

Under the null hypothesis, each subject has an expected rank of $\frac{N+1}{2}$ and thus $E(R_1) = n_{1+} \frac{N+1}{2}$, where $N = n_1 + n_2$ is the total sample size. We can test the null hypothesis by checking how far R_1 is from the expected value. Since R_1 differs from M only by a constant (since for all the tests, we assume n_{1+} and n_{2+} are fixed), these two statistics in (2.28) and (2.30) are equivalent, and for this reason, MWW is also known as the *Wilcoxon rank sum test*.

Example 2.10

For Example 2.9, we can also use the MWW statistic to test whether females have systematic worse or better depression outcome. The MWW test gives a p-value <0.0001. Thus, we reject the null hypothesis that there is no difference. Further, by looking at the average ranks, we can see that in general females have a worse depression outcome. $\square$

Note that in the rank sum test approach, we test whether there is a difference between the average ranks of the two groups. The expected rank depends on the sample size of two groups. If the levels of y can be assigned scores and their mean scores are meaningful as is the case for interval variables, then under the null hypothesis of independence between x and y, the mean scores of y will be the same across the rows of x. Thus, we may compare the mean scores and reject the null hypothesis if they are significantly different.

Let a_j represent the score for the jth-ordered response of y $(1 \le j \le r)$. Then, the mean score for the ith row is $\overline{f}_i = \frac{1}{n_{i+}} \sum_{j=1}^{r} a_j n_{ij}$ $(i = 1, 2)$. By using the mean scores of y, we reduce the two-way $2 \times r$ table to a one-way 2×1 table and can then apply techniques for one-way table for inference. The statistic is

$$Q_S = \frac{(n-1) \left[n_{1+}(\overline{f}_1 - \widehat{\mu}_a)^2 + n_{2+}(\overline{f}_2 - \widehat{\mu}_a)^2 \right]}{n \widehat{v}_a},$$

where $\widehat{\mu}_a = \frac{1}{m} \sum_{j=1}^{r} a_j n_{+j}$ is the overall average score of the whole sample and $\widehat{v}_a = \sum_{j=1}^{r} (a_j - \widehat{\mu}_a)^2 \frac{n_{+j}}{n}$. It follows asymptotically a chi-square distribution with one degree of freedom if the row and column variables are independent. This *mean score test* is much simpler compared to MWW test since the scores are fixed.

Note that in the mean score approach, we frequently assume that the column levels represent the ordinal response of y. Sometimes, however, to make the mean scores more meaningful or interpretable for a given application, other numerical scores may be assigned to each level of y. For example, if y is a five-level Likert scale:

strongly disagree, disagree, neutral, agree, strongly agree,

we may use the column levels, 1, 2, 3, 4, 5, to compute the mean score of y. Alternatively, if we regard the change from "strongly disagree" ("strongly disagree") to disagree (agree) as representing a bigger jump than the change between disagree (agree) and neutral, we may want to change the equal spacing in 1, 2, 3, 4, 5 to something like 1, 4, 5, 6, 9 to emphasize the differential effect when moving across the response categories.

The MWW test could have been introduced as a special case for multiple groups within the context of general $s \times r$ tables (see the Kruskal–Wallis and Jonckheere–Terpstra tests in Section 2.4.1). However, because of the popularity of the two-sample test and simplified algebra in this special case, we think that it is worth presenting the two-group version separately. Note also that there is no obvious generalization of the Cochran–Armitage trend test to the general case. When one variable is binary and one is nominal, we can apply methods for general $s \times r$ tables to be discussed next.

2.4 Inference for $s \times r$ Table

Consider again two categorical outcomes x and y. Now suppose that x (y) has s (r) levels. The outcomes of the pairs (x, y) can be displayed in an $s \times r$ contingency table:

			y			
x	1	2	$\cdots$	r	Total	
1	n_{11}	n_{12}	$\cdots$	n_{1r}	n_{1+}	
2	n_{21}	n_{22}	$\cdots$	n_{2r}	n_{2+}	
$\vdots$	$\vdots$	$\vdots$	$\ddots$	$\vdots$	$\vdots$	
s	n_{s1}	n_{s2}	$\cdots$	n_{sr}	n_{s+}	
Total	n_{+1}	n_{+2}	$\cdots$	n_{+r}	n	

Let $p_{ij} = \Pr(x = i, y = j)$, $p_{i+} = \Pr(x = i)$, and $p_{+j} = \Pr(y = i)$ for $i = 1, \ldots, s$ and $j = 1, \ldots, r$. As in the 2×2 table case, we are again interested in whether x and y are independent, and the nature and strength of association if they are not. Depending on the nature of the row and column variables and targeting alternatives, different tests may be applied. We first discuss tests of general association when both variables are considered nominal, and then follow with methods for situations when one or both of the row and column variables are ordinal.

2.4.1 Tests of association

2.4.1.1 Pearson test for general Association

Pearson chi-square statistic is defined for the general $s \times r$ table and used to test the general association between the row and column variables. In this test, both the row x and column y variables are treated as a nominal outcome, even if one or both may be ordinal. Under the null hypothesis, we assume that there is no association between the row and column variables, i.e., they are independent, $H_0 : p_{ij} = p_{i+}p_{+j}$ for all i and j. As in the 2×2 table case, we may consider $\frac{n_{ij}}{n} - \frac{n_{i+}n_{+j}}{n^2}$ as an estimate of $p_{ij} - p_{i+}p_{+j}$ and use it to test the null hypothesis. Unlike the 2×2 table, however, there are $(i - 1) \times (j - 1)$ independent equations. Thus, we need to combine them in some fashion to form a single statistic.

Pearson chi-square statistic is defined based on such a strategy. By squaring these terms and summing them up, this statistic transforms all the difference terms into positive values to avoid the cancellation of terms because of the different directions of the differences:

$$Q_P = \sum_{i=1}^{s} \sum_{j=1}^{r} \frac{(n_{ij} - m_{ij})^2}{m_{ij}},$$

where $m_{ij} = \frac{n_{i+}n_{+j}}{n}$ is the expected cell count in the cell (i, j). Under the null hypothesis of no row by column association, the statistic Q_P has an asymptotic chi-square distribution with $(s - 1)(r - 1)$ degrees of freedom. The degree of freedom is based on the fact that with fixed marginal counts, the cell counts in any $(s - 1)$ by $(r - 1)$ submatrix will determine the entire table. As a special case for the 2×2 table, $(s - 1)(r - 1) = 1$, and as we have seen earlier, a single cell count in any of the four cells such as n_{11} identifies the table in this special case.

Note that if we test a multinomial distribution with rs levels, the degree of freedom of the associated chi-square statistic is $rs - 1$, rather than $(r - 1)(s - 1)$ as in the above. Thus, it seems that we have lost $sr - 1 - (s - 1)(r - 1) = s + r - 2$ degrees of freedom. This is because under the null hypothesis of row and column independence, the distribution of the table is determined by $s + r - 2$ parameters, with $s - 1$ for the row and $r - 1$ for the column marginals. The loss of degrees of freedom is due to estimating these parameters. This is a general phenomenon for this type of statistics. We will see more such statistics in this and subsequent chapters.

Another way to look at the degrees of freedom is to fix the marginal counts as in deriving Fisher's exact test procedure for 2×2 tables. Under the null hypothesis, no parameters need to be estimated and the cell counts in any $(s-1) \times (r-1)$ submatrix determine the table. The vector $\mathbf{v}$ formed by the entries in such a submatrix follows an asymptotic normal distribution, with a nonsingular asymptotic variance Σ. Thus, $\mathbf{v}^\top \Sigma^{-1} \mathbf{v}$ follows an asymptotic chi-square distribution with $(s-1)(r-1)$ degrees of freedom. Similar to the discussion of the statistic (2.7) in Section 2.1.2, the asymptotic distribution of $\mathbf{v}^\top \Sigma^{-1} \mathbf{v}$ is equal to Q_P regardless of the choice of the $(s-1) \times (r-1)$ submatrix.

2.4.1.2   Kruskal–Wallis and Jonckheere–Terpstra tests

Assume that the data are from several independent groups defined by the row variable x. If the column variable y is ordinal, then the MWW statistic can be generalized to test whether the column variable y follows the same distribution across the group levels of x. If we rank the subjects as described in Section 2.3, then the average rank of the ith group, W_i, will have an expected value $\frac{N+1}{2}$ under the null hypothesis. The sum of squares of the deviations from the means of the rank sums can be used as a test statistic for the null hypothesis.

More specifically, Kruskal (1952) and Kruskal and Wallis (1952) introduced the statistic

$$Q_{KW} = \frac{12}{N(N+1)} \sum_{i=1}^{s} n_{i+} \left(W_i - \frac{N+1}{2} \right)^2$$

for continuous variables. It follows an asymptotic chi-square distribution with $s-1$ degrees of freedom, if x and y are independent, and all the groups have comparable large sizes. Note that since we test differences among more than two groups with no specific direction, the test is two-sided.

To apply the Kruskal–Wallis test to ordinal variables, we need to handle ties in the outcomes of y. First, as in the computation of MWW rank sum test, tied subjects are assigned the average ranks. Thus, all subjects in the first column have $\frac{1+n_{+1}}{2}$, the average of $1, 2, \ldots, n_{+1}$, as their ranks. In general, subjects in the jth column have the rank $\sum_{k=1}^{j-1} n_{+k} + \frac{1+n_{+j}}{2}$ for $j > 1$. Hence, the average rank for the ith group is $W_i = \frac{1}{n_{i+}} \sum_{j=1}^{r} n_{ij} \left(\sum_{k=1}^{j-1} n_{+k} + \frac{1+n_{+j}}{2} \right)$. Because of tied observation, the asymptotic variance will be smaller than that computed based on the formula for continuous variable. For contingency tables, we may use the following tie-corrected version of Kruskal–Wallis statistic:

$$Q_{KW} = \frac{12}{N(N+1)\left[1 - \sum_{j=1}^{r} \left(n_{+j}^3 - n_{+j} \right) / (N^3 - N) \right]} \sum_{i=1}^{s} n_{i+} \left(W_i - \frac{N+1}{2} \right)^2.$$

Under the null of row and column independence, the statistic asymptotically follows a chi-square distribution with $s-1$ degrees of freedom.

If the row variable x is also ordinal, with higher row levels indicating larger response categories, then more restricted alternative may be of interest. For example, we may want to know if the expected ranks for the groups change monotonically with group levels. Further, we may be interested in whether such changes follow some specific directions. In such cases, we may use the Jonckheere–Terpstra test. As it considers if the response y increases or decreases as x increases, this test generally yields more power than the Kruskal–Wallis test.

The *Jonckheere–Terpstra* statistic, introduced in Terpstra (1952) and Jonckheere (1954), is developed by considering all the possible pairs of groups. For any two levels i and i' of x

with $i < i'$, let $M_{i,i'}$ denote the tie-corrected Mann–Whitney–Wilcoxon statistic:

$$M_{i,i'} = \sum_{j=2}^{r} \left[n_{i'j} \left(\sum_{k=1}^{j} \right) n_{ik} \right] + \frac{1}{2} \sum_{j=1}^{r} n_{i'j} n_{ij}.$$

The Jonckheere–Terpstra statistic is defined as

$$J = \sum_{1 \le i < i' \le s} M_{i,i'}. \tag{2.31}$$

Note that the sum in (2.31) is over the terms $M_{i,i'}$ with $i < i'$. Under the alternative hypothesis, $M_{i,i'}$ are all very likely to lie on the same side of their means. Thus, the sum in J can accumulate the differences, leading to increased power for rejecting the null hypothesis. The statistic in (2.31) can be applied to both one- and two-sided tests. One-sided alternatives test whether the change of expected rank follows a specific direction.

If the sample size is small, an exact test may again be considered. The computational procedure is similar to that of the Cochran–Armitage test.

Example 2.11

In the Postpartum Depression Study (PPD), each mother was diagnosed as major, minor, or no depression, based on Structured Clinical Interview for DSM-IV-TR (SCID). They were also screened with Edinburgh Postnatal Depression Scale (EPDS) questionnaires.

Applying the Kruskal–Wallis test to the three groups consisting of major, minor, and no depression, the statistic is 89.0736 and the p-value is < 0.00001. Thus, we reject the null hypothesis that there is no difference among the three groups. The Kruskal–Wallis test ignores the order of the group levels. To take the order of major, minor, and no depression into consideration, we apply Jonckheere–Terpstra test. The statistic is 10408, and the p-value < 0.00001. This shows the trend of the EPDS scores among patients from the three groups; major depression patients have the highest scores, while patients without depression have the lowest.

Note that similar to the $2 \times s$ cases, we can compare the mean scores across the rows of x and reject the null hypothesis if they are significantly different. Let a_j represent the score for the jth-ordered response of y $(1 \le j \le r)$. Then, the mean score for the ith row is $\overline{f}_i = \frac{1}{n_{i+}} \sum_{j=1}^{r} a_j n_{ij}$ $(1 \le i \le s)$. By using the mean scores of y, we reduce the two-way $s \times r$ table to a one-way $s \times 1$ table and then apply techniques for one-way table for inference. The statistic is

$$Q_S = \frac{(n-1) \sum_{i=1}^{s} n_{i+} (\overline{f}_i - \widehat{\mu}_a)^2}{n \widehat{v}_a},$$

where $\widehat{\mu}_a = \sum_{j=1}^{r} \frac{a_j n_{+j}}{n}$ and $\widehat{v}_a = \sum_{j=1}^{r} (a_j - \widehat{\mu}_a)^2 \frac{n_{+j}}{n}$. It follows asymptotically a chi-square distribution with $s - 1$ degrees of freedom if the row and column variables are independent.

□

2.4.1.3   Correlation test for ordinal row and column variables

When both row and column variables are ordinal with each level assigned a score as if there were interval variables, we can also use Pearson correlation coefficients to test their association. Recall that for two continuous outcomes x and y, the product-moment correlation is defined by

$$Corr(x,y) = \frac{Cov(x,y)}{\sqrt{Var(x)Var(y)}}.$$

With data from a sample of n subjects, we can estimate $Var(x)$, $Var(y)$, and $Cov(x,y)$ by their respective sample moments,

$$\frac{1}{n-1}\sum_{i=1}^{n}(x_i - \overline{x})^2, \quad \frac{1}{n-1}\sum_{i=1}^{n}(y_i - \overline{y})^2, \quad \frac{1}{n-1}\sum_{i=1}^{n}(x_i - \overline{x})(y_i - \overline{y}),$$

where $\overline{x} = \frac{1}{n}\sum_{i=1}^{n}x_i$ and $\overline{y} = \frac{1}{n}\sum_{i=1}^{n}y_i$. Substituting these sample moments for their corresponding variance and covariance components, we obtain the following Pearson correlation coefficient:

$$P = \frac{\frac{1}{n-1}\sum_{i=1}^{n}(x_i - \overline{x})(y_i - \overline{y})}{\sqrt{\frac{1}{n-1}\sum_{i=1}^{n}(x_i - \overline{x})^2 \frac{1}{n-1}\sum_{i=1}^{n}(y_i - \overline{y})^2}}.$$

For two continuous outcomes (x, y), the Pearson correlation coefficient P has the interpretation that if (x, y) follows a linear relationship, then a positive (negative) P implies a positive (negative) relation between the two variables.

Within our context, if both the row x and column y variables are ordinal and some assignment of scores to the levels of each variable is meaningful, then we can apply the Pearson correlation coefficient P to the assigned row and column scores and use the resulting statistic as a measure of association between the variables.

Let R_i (C_j) denote the score assigned to the ith row of x (jth column of y), and $\overline{R} = \frac{1}{n}\sum_{i=1}^{s}n_{i+}R_i$ ($\overline{C} = \frac{1}{n}\sum_{j=1}^{r}n_{+j}C_j$), the mean of R_i (C_j). Then, writing in terms of these scores and cell counts in the contingency table, we can express the Pearson correlation coefficient when applied to the assigned row and column scores as

$$Q = \frac{\sum_{i=1}^{s}\sum_{j=1}^{r}n_{ij}(R_i - \overline{R})(C_j - \overline{C})}{\sqrt{\sum_{i=1}^{s}n_{i+}(R_i - \overline{R})^2 \sum_{j=1}^{r}n_{+j}(C_j - \overline{C})^2}}. \tag{2.32}$$

The interpretation of this correlation Q is similar to that for two continuous outcomes. For example, a positive correlation implies that as the score of one variable, say x increases, and the score of the other variable y also increases and vice versa, while a negative correlation indicates the reversal of this relationship (an increase in the score of x is associated with a decrease in the score of y). We can use the asymptotic distribution of Q to test the null hypothesis of no association between the row and column variables. Note that since the Pearson correlation coefficient only tests linear association between the row and column variables, the coefficient depends on the scores assigned to the levels. Spearman (1904) suggested to compute the correlation coefficient based on ranks, eliminating such a dependence.

Spearman's rank correlation, originally developed for two continuous outcomes, is widely used as an alternative to Pearson correlation when the outcomes are not linearly related. For example, if the scatterplot of x vs. y shows a curved relationship between x and y, the Pearson correlation may be low even though x and y are closely related, giving rise to incorrect indication of the association between them. Spearman's rank correlation addresses this limitation.

Spearman's rank correlation is defined similarly to (2.32) with scores replaced by ranks. Suppose that the row (column) levels represent the ordering of the outcome x (y), i.e., observations in the ith (jth) row have a lower order than observations in the i'th (j'th) row if $i < i'$ ($j < j'$). Then, the rank scores for the row and column levels can be computed from the table, using the following formula:

$$R_i = \sum_{k=1}^{i-1}n_{k+} + \frac{1+n_{i+}}{2}, \quad C_j = \sum_{k=1}^{j-1}n_{+k} + \frac{1+n_{+j}}{2}.$$

Let $\overline{R}$ and $\overline{C}$ be the mean rank scores of R_i and C_j. The Spearman rank correlation coefficient expressed in terms of data in the frequency table is

$$Q_S = \frac{\sum_{i=1}^{s} \sum_{j=1}^{r} n_{ij} \left(R_i - \overline{R}\right)\left(C_i - \overline{C}\right)}{\sqrt{\sum_{i=1}^{s} n_{i+} \left(R_i - \overline{R}\right)^2 \sum_{j=1}^{r} n_{+j} \left(C_i - \overline{C}\right)^2}}. \tag{2.33}$$

Example 2.12

Consider the association between gender and depression in the DOS. Although gender is nominal in nature, its binary representation in that study allows us to treat it as an ordinal variable so that we can apply the Q statistic to measure the direction and strength of association between gender and severity of depression. Assign scores to the row and column levels as follows:

$$R_1 = 0, \ R_2 = 1, \quad C_1 = 0, \ C_2 = 1, \ C_3 = 3.$$

Then, based on Table 2.8, we have

$$\overline{R} = \frac{472 \times 0 + 273 \times 1}{745} = 0.3664$$

$$\overline{C} = \frac{481 \times 0 + 136 \times 1 + 128 \times 3}{745} = 0.6980.$$

It follows that $Q = -0.1364$. The p-value for a null hypothesis of zero correlation is < 0.0001. The negative correlation Q confirms our prior belief that female patients in this study group were at an increased risk for depression.

For the Spearman correlation, the row and column rank scores are

$$R_1 = \frac{1 + 472}{2} = 236.5, \quad R_2 = 472 + \frac{1 + 273}{2} = 609,$$

$$C_1 = \frac{1 + 481}{2} = 241 \ , \quad C_2 = 481 + \frac{1 + 136}{2} = 549.5,$$

$$C_3 = 481 + 136 + \frac{1 + 128}{2} = 681.5.$$

The mean row and column rank scores are $\overline{R} = \overline{C} = \frac{1+745}{2} = 373$. By (2.33), the Spearman rank correlation coefficient is $\rho = -0.1688$.

If the third column is assigned the score 2 ($C_3 = 2$), the Pearson correlation coefficient will change to -0.1543, but the Spearman correlation coefficient remains unchanged. ⬚

In appearance, at least from the formulas in (2.32) and (2.33), the Spearman coefficient looks like a special case of the Pearson correlation. But in principle, they are quite different. The Pearson correlation coefficient measures the strength of association under a linear relationship between the scales of two outcomes, and the value of the coefficient in general depends on the scoring systems used for the row and column variables. In contrast, the Spearman coefficient measures the strength of association by examining the association between the rank scores of the variables without assuming any shape or form for the relationship between the two variables. The use of rank score removes the artifact introduced by scoring the variables and thus provides a more robust measure of association. For 2×2 tables, the Pearson and Spearman correlation coefficients are identical. This follows from our later discussion on the invariance property of the Pearson correlation coefficient under orientation-preserving, affine linear transformations (see Section 2.5.1.1).

Note also that the rank score for each observation depends on other observations in the sample. The computation of the asymptotic variance of the Spearman correlation coefficient is much more complicated than for the Pearson correlation coefficient and is typically facilitated by employing the theory of U-statistics (Kowalski and Tu, 2008).

2.4.1.4   Exact test

Fisher's exact test has been extended to the general $s \times r$ table by Freeman and Halton (1951), and hence the exact test for the $s \times r$ table is also known as the *Freeman–Halton* test. The basic principle of the exact test of Freeman and Halton for the $s \times r$ table is essentially the same as Fisher's original test for the 2×2 table. Conditional on the fixed marginal counts, the distribution of tables with varying cell counts follows a multivariate generalization of the hypergeometric probability function under the null hypothesis. The p-value is defined as the sum of the probabilities of all tables whose probabilities of occurrence are less than or equal to the probability of the observed table.

Unlike the 2×2 table, however, the distribution of the tables for the $s \times r$ case ($\max(s, r) > 2$) is not determined by just one cell count. As a result, we cannot use just a single cell count to order all potential tables as in the 2×2 case. For this reason, the test is inherently two-sided.

Similar to the 2×2 case, we may also compute exact p-values for the asymptotic tests described above. Under such exact inference, we compute the value of a chi-square statistic under consideration for each potential table with the same fixed marginal counts as the observed table. The exact p-value is the sum of the probabilities of the occurrence of the tables that yield either as large or larger chi-square statistic values than the observed one. Again, all such tests are two-sided.

2.4.2   Marginal homogeneity and symmetry

There are situations in which both the row and column variables are measures on the same scale. For example, as discussed in Section 2.2.3, measurements from each pair of individuals in a matched-pair study or from the pre- and post-intervention assessments of each individual in a pre-post study follow the same scale. Since the row and column variables represent the same measurement scale, they are identical in terms of the possible categorical outcomes, creating perfectly square contingency tables. For such square tables, common questions are if the row and column variables have homogeneous marginal distribution ($p_{i+} = p_{+i}$) and whether the table is symmetric, i.e., whether $p_{ij} = p_{ji}$ for all i and j. For the special case of 2×2 tables, these two questions are identical, and McNemar's test in Section 2.2.3 can be applied. For general $r \times r$ tables, we can use generalized versions of McNemar's test developed by Stuart (1955) and Maxwell (1970) for marginal homogeneity and by Bowker (1948) for symmetry.

McNemar's statistic is motivated by considering the difference between corresponding marginal counts. The Stuart–Maxwell test focuses on a similar vector statistic consisting of differences between corresponding marginal counts $\mathbf{d} = (d_1, \ldots, d_r)^\top$, where $d_i = n_{i+} - n_{+i}$ and $i = 1, \ldots, r$. Under the null hypothesis of marginal homogeneity, the true values of all the elements of the vector (the population mean) are zero. Using the theory of multinomial distribution, we can find the components of the variance $Var(\mathbf{d})$ as follows:

$$Cov(d_i, d_j) = \begin{cases} n \left[p_{i+} + p_{+i} - 2p_{ii} - (p_{i+} - p_{+i})^2 \right] & \text{if } i = j \\ -n \left[p_{ji} + p_{ij} + (p_{i+} - p_{+i})(p_{j+} - p_{+j}) \right] & \text{if } i \neq j. \end{cases}$$

Since $\sum_{i=1}^r d_i = 0$, $Var(\mathbf{d})$ is not of full rank. By removing one component from $\mathbf{d}$, say d_r, the reduced vector statistic $\widetilde{\mathbf{d}} = (d_1, \ldots, d_{r-1})^\top$ has a full-rank variance matrix $\Sigma_{(r-1) \times (r-1)}$, with the (i, j)th entry equal to $n_{i+} + n_{+i} - 2n_{ij}$ if $i = j$ and $-(n_{ji} + n_{ij})$ if $i \neq j$. The Stuart–Maxwell statistic is defined as

$$Q_{SM} = \widetilde{\mathbf{d}}^\top \Sigma_{(r-1) \times (r-1)}^{-1} \widetilde{\mathbf{d}}. \tag{2.34}$$

This statistic follows a chi-square distribution of $r - 1$ degrees of freedom asymptotically under the null hypothesis.

The McNemar statistic is based on the comparison of the cell count in $(1, 2)$ with the one in its symmetric cell $(2, 1)$. Bowker's test similarly compares the cell count in (i, j) with its symmetric counterpart in cell (j, i) for all $i \neq j$:

$$Q_B = \sum_{i<j} \frac{(n_{ij} - n_{ji})^2}{n_{ij} + n_{ji}}. \tag{2.35}$$

Similar to McNemar's test, each term in (2.35) is asymptotically chi-square distributed. Further, since they are asymptotically independent, Q_B approximately follows a chi-square distribution with $\frac{r(r-1)}{2}$ degrees of freedom under the null hypothesis of symmetry for large samples. For 2×2 tables, both Q_{SM} and Q_B reduce to the McNemar statistic.

Example 2.13

For the DDPC, the diagnosis of both probands and informants are actually available in three levels. The information is summarized in Table 2.9. Let us test its marginal homogeneity and symmetry.

TABLE 2.9: Depression diagnoses based on the probands and informants

Probands	Informants			Total
	No	MinD	MajD	
No	66	13	6	85
MinD	36	16	10	62
MajD	14	12	27	53
Total	116	41	43	200

The statistic for the Stuart–Maxwell test of marginal homogeneity is 13.96. Comparing it with a chi-square distribution with two degrees of freedom, we obtain p-value = 0.0009. The statistic for Bowker's test for symmetry is 14.1777. Comparing it with a chi-square distribution with three degrees of freedom, we obtain p-value = 0.0027. ▯

2.4.3 Agreement

Another common situation in which we may obtain square tables arises from rating data where there are two raters rating each subject in a sample. Since both raters use the same scale, the rating data can be presented in a square table. In such situations, we may be interested in the agreement between the two variables. When the rating is on a two-level scale, we have discussed how to use Cohen's kappa coefficient to assess the agreement in Section 2.2.4. In this section, we generalize the kappa coefficient to general square tables.

Consider a rating scale with k categories (nominal or ordinal). The agreement rate by chance is the sum of the products of the marginal probabilities, $\sum_{i=1}^{k} p_{i+}p_{+i}$. By excluding this term from the agreement probability $\sum_{i=1}^{k} p_{ii}$, we obtain the probability of agreement corrected for the chance factor: $\sum_{i=1}^{k} p_{ii} - \sum_{i=1}^{k} p_{i+}p_{+i}$. By normalization, Cohen's kappa coefficient for a general $r \times r$ square table is defined as

$$\kappa = \frac{\sum_{i=1}^{k} p_{ii} - \sum_{i=1}^{k} p_{i+} p_{+i}}{1 - \sum_{i=1}^{k} p_{i+} p_{+i}}. \tag{2.36}$$

The coefficient κ varies between -1 and 1, depending on the marginal probabilities. If the two raters completely agree with each other, $\sum_{i=1}^{k} p_{ii} = 1$ and $\kappa = 1$, and the converse is also true. On the other hand, if the judges rate the subjects at random, then the observer agreement is completely by chance and, as a result, $p_{ii} = p_{i+} p_{+i}$ for all i, and the kappa coefficients become 0. In general, when the observer agreement exceeds the agreement by chance, kappa is positive, and when the raters really disagree on their ratings, kappa is negative. The magnitude of kappa indicates the degree of agreement (disagreement) when kappa is positive (negative).

The kappa index in (2.36) is estimated by

$$\widehat{\kappa} = \frac{\sum_{i=1}^{k} \widehat{p}_{ii} - \sum_{i=1}^{k} \widehat{p}_{i+} \widehat{p}_{+i}}{1 - \sum_{i=1}^{k} \widehat{p}_{i+} \widehat{p}_{+i}},$$

where $\widehat{p}_{ii}$, $\widehat{p}_{i+}$, and $\widehat{p}_{+i}$ denote the moment estimates of the respective parameters.

The definition in (2.36) assumes that the rating categories are treated equally. If the rating categories are ordered, say, for example, by the Likert scale:

strongly disagree, disagree, neutral, agree, strongly agree,

then the disagreement between *strongly disagree* and *strongly agree* represents a larger difference than the disagreement between *agree* and *strongly agree*. The *simple* kappa coefficient in (2.36) does not reflect such a varying degree of disagreement (agreement) across the categories, and the weighted kappa can be used to address this limitation.

Let w_{ij} be a set of known numbers defined for the ith row and jth column satisfying

$$0 \le w_{ij} < 1, \quad w_{ij} = w_{ji}, \quad \text{for all } i, j, \quad w_{ii} = 1, \quad \text{for all } i.$$

By assigning these weights to the cell probabilities and following the same philosophy as in developing the kappa coefficient in (2.36), we can account for the varying degree of disagreement (agreement) by the following weighted kappa:

$$\kappa_w = \frac{\sum_{i=1}^{k} \sum_{j=1}^{k} w_{ij} p_{ij} - \sum_{i=1}^{k} \sum_{j=1}^{k} w_{ij} p_{i+} p_{+j}}{1 - \sum_{i=1}^{k} \sum_{j=1}^{k} w_{ij} p_{i+} p_{+j}}.$$

If $w_{ij} = 0$ for all $i \ne j$, the weighted kappa coefficient κ_w reduces to the simple kappa. We can plug in the moment estimates $\widehat{p}_{ii}$, $\widehat{p}_{i+}$, and $\widehat{p}_{+i}$ to obtain estimate $\widehat{\kappa}_w$ for κ_w.

Similar to the 2×2 case, the delta method can be applied to develop the asymptotic distribution of the estimate $\widehat{\kappa}$ and $\widehat{\kappa}_w$. Also, for 2×2 tables, there is no weighted version of kappa, since $\widehat{\kappa}$ remains the same and equals the simple kappa no matter what weights are used (see Problem 2.22).

In theory, any weight system satisfying the defining condition may be used. In practice, however, additional constraints are often imposed to make the weights more interpretable and meaningful. For example, since the degree of disagreement (agreement) is often a function of the difference between the ith and jth rating categories, we assume that $w_{ij} = f(i - j)$, where f is some decreasing function satisfying

$$0 \le f(x) < 1, \quad f(x) = f(-x), \quad f(0) = 1.$$

Two such types of weighting systems based on column scores are commonly used. Suppose the column scores are ordered, say $C_1 \leq C_2 \ldots \leq C_r$. Then, the Cicchetti–Allison weight type defines weights w_{ij} according to the following criteria:

$$w_{ij} = 1 - \frac{|C_i - C_j|}{|C_1 - C_r|}.$$

The other, called the Fleiss–Cohen weight type, is defined by

$$w_{ij} = 1 - \frac{(C_i - C_j)^2}{(C_1 - C_r)^2}.$$

Example 2.14

The estimate of the unweighted kappa coefficient for Table 2.9 is

$$\widehat{\kappa} = \frac{\frac{66}{200} + \frac{16}{200} + \frac{27}{200} - \left(\frac{116}{200}\frac{85}{200} + \frac{41}{200}\frac{62}{200} + \frac{43}{200}\frac{53}{200}\right)}{1 - \left(\frac{116}{200}\frac{85}{200} + \frac{41}{200}\frac{62}{200} + \frac{43}{200}\frac{53}{200}\right)} = 0.2812.$$

The weighted kappa estimates are 0.3679 if the Cicchetti–Allison weight is used and 0.4482 if the Fleiss–Cohen weight is applied. Thus, agreement between probands and informants is not high in this example. It is seen from the table that the low agreement is due to the fact that a larger number of subjects with minor depression have been missed by the informants.
▯

In practice, one or both raters may not use all the rating categories in the ratings of subjects due either to rater bias or to small samples, yielding nonsquare tables. Since agreement data conceptually create square tables, most software packages will not compute and output kappa coefficients. In some cases, the rater-endorsed rating categories may still produce a square table, although the resulting table may completely change the meaning of the original rating scale. For example, suppose a scale for rater agreement has three categories, A, B, and C. If one rater only used B and C and the other only A and B in their ratings, we may obtain a table that looks like the following:

	B	C	Total
A	10	3	13
B	5	11	16
Total	15	14	29

Some software packages may compute the kappa coefficients treating it as a 2×2 table. Of course, this is not what we mean to compute, and the kappa coefficients based on the 2×2 table do not provide the correct information about the agreement of interest. To obtain the correct kappa coefficients for the original three-categorical rating scale in this situation, we must add observations with zero counts for the rating categories not endorsed by the raters.

	A	B	C	Total
A	0	10	3	13
B	0	5	11	16
C	0	0	0	0
Total	0	15	14	29

2.5 Measures of Association

As in the case of the 2×2 table, we like to know both the direction and strength of association when two variables (or row and column) are associated. Again, various indices have been developed to characterize the association between the two variables in the general $s \times r$ table case.

2.5.1 Measures of association for ordinal outcome

2.5.1.1 Pearson correlation coefficient

If both the row and column variables are ordinal and some assignment of scores to the levels of each variable is meaningful, then the Pearson correlation coefficient Q discussed in last section can be used as a measure of association between the row and column variables. However, the Pearson correlation coefficients depend on the scores assigned to the levels.

Example 2.15
In Example 2.12, we considered the scores:

$$R_1 = 0, \ R_2 = 1, \quad C_1 = 0, \ C_2 = 1, \ C_3 = 3. \tag{2.37}$$

Now consider two different scoring methods for the columns:

$$C_1' = 0, \ C_2' = 2, \ C_3' = 6; \ C_1'' = 3, \ C_2'' = 2, \ C_3'' = 0. \tag{2.38}$$

By straightforward calculations, Pearson correlation coefficient estimates are -0.1364 and 0.1364. ⬜

In the example, the Pearson correlation for the first remains the same, but the second differs by flipping the sign. This is not a coincidence, as the Pearson correlation coefficient is invariant under affine linear transformations that preserve the orientation of the original score. An *affine linear* transformation relates the original scores R_i and C_j to a new set of scores R_i' and C_j' as follows:

$$R_i' = aR_i + b, \quad C_j' = cC_j + d,$$

where a, b, c, and d are known constants. The transformation is orientation preserving if and only if $ac > 0$. For example, consider

$$\text{Transformation 1: } R_i' = 2R_i, \quad C_j' = C_j,$$
$$\text{Transformation 2: } R_i' = -2R_i, \quad C_j' = C_j,$$
$$\text{Transformation 3: } R_i' = -2R_i, \quad C_j' = -C_j.$$

The first linear transformation stretches the original R_i in the same direction, thus retaining the original orientation of (R, C). The second also changes the direction of R_i, thus altering the original orientation of this variable. The third transformation changes the directions of both R_i and C_j. However, it does not alter the original orientation of the scores. To see the invariance property of Q under orientation-preserving, affine linear transformations, let $\overline{R'}$ ($\overline{C'}$) denote the mean of R_i' (C_j'). Then, since

$$R_i' - \overline{R}' = a(R_i - \overline{R}), \quad C_j' - \overline{C}' = c(C_j - \overline{C}),$$

it is readily checked that the numerator of Q in (2.32) based on (R', C') changes by a factor of ac and the denominator by a factor of $|ac|$. Thus, the Pearson correlation coefficient Q will be unaffected if $ac > 0$ and change to $-Q$ if $ac < 0$.

The first scoring method in Example 2.15 is an orientation-preserving transformation, while the second is an orientation-reserving transformation, of the scores in (2.37). So, it is not surprising that one retains the same value, but the other reverses the sign.

Note that not all orientation-preserving transformations are affine linear transformations. For example, in the DOS sample, consider a new scoring method for the column:

$$C_1' = 0, \ C_2' = 1, \ C_3' = 2.$$

It is readily checked that C_j' retains the original orientation, but it is not possible to express C_j' through an affine linear transformation of C_j. However, for a 2×2 table, any transformation (R', C') is an affine transformation. To see this, let (R, C) and (R', C') denote the two scoring methods. Then, since both the row x and column y variables have only two levels, we can always find four values a, b, c, and d such that

$$R_i' = aR_i + b, \quad C_i' = cC_i + d.$$

Thus, for 2×2 tables, the Pearson correlation coefficient Q is invariant under orientation-preserving transformations.

If no meaningful ordinal scores can be assigned to the variables, the Pearson correlation usually does not apply, and other measures of association may be considered.

2.5.1.2 Goodman–Kruskal Gamma, Kendall's tau-b, Stuart's tau-c, and Somers' D

All these measures consider whether the column variable y tends to increase as the row variable x increases and vice versa by exploiting the notion of concordant and discordant pairs. For a pair of two subjects, (x_1, y_1) and (x_2, y_2), they are concordant if

$$x_1 < x_2 \quad \text{and} \quad y_1 < y_2, \quad \text{or} \quad x_1 > x_2 \quad \text{and} \quad y_1 > y_2,$$

and discordant if

$$x_1 < x_2 \quad \text{and} \quad y_1 > y_2, \quad \text{or} \quad x_1 > x_2 \quad \text{and} \quad y_1 < y_2,$$

If a pair of subjects share the same value in either x or y or both, i.e., the two subjects have a tie in either x or y or both, they form neither a concordant nor a discordant pair.

Let p_s (p_d) denote the probability of a concordant (discordant) pair. To estimate p_s and p_d, consider a sample of n subjects and sample of two subjects independently with replacement from the sample. Since two subjects (x_i, y_i) and (x_j, y_j) in a pair are sampled individually with replacement, the same two subjects are considered to form two pairs,

$$\{(x_i, y_i), (x_j, y_j)\}, \quad \{(x_j, y_j), (x_i, y_i)\},$$

which differ in their ordering but are otherwise identical. Altogether, there are a total of n^2 such distinct pairs. Let

$$C = \text{Number of concordant pairs}, \quad D = \text{Number of discordant pairs}.$$

Then, we can estimate the concordance and discordance probabilities by $\widehat{p}_s = \frac{C}{n^2}$ and $\widehat{p}_d = \frac{D}{n^2}$. Below, we describe how to compute C and D through a simple example.

	$y(1)$	$y(2)$	$y(3)$	Total
$x(1)$	2	1	5	8
$x(2)$	4	2	3	9
$x(3)$	3	2	2	7
Total	9	5	10	24

Consider the 3×3 table above from a hypothetical study. Suppose x and y are ordered by the row and column levels as shown in the table. Consider two subjects, with the first subject from the cell (i, j) and the second from the cell (i', j'). By changing the order of the two subjects if necessary, we may assume without the loss of generality that $j < j'$. Since they form a concordant pair, we must have $i < i'$ and $j < j'$.

For example, consider a subject in the cell $(1, 1)$. Then, the subjects that form concordant pairs with this subject are found in the cells with higher row and column levels, $(2, 2)$, $(2, 3)$, $(3, 2)$, and $(3, 3)$. By adding all subjects from these cells, we obtain the total number of concordant pairs with this subject: $2 + 3 + 2 + 2 = 9$. Since there are two subjects in $(1, 1)$, the total concordant pairs formed by the subjects in the cell $(1, 1)$ are $2 \times 9 = 18$. This is the number of concordant pairs if we are sampling without replacement. For sampling with replacement as in our case, the order of two subjects in a pair also counts, and thus the total number of concordant pairs will be twice as large: $C_{11} = 2 \times 18 = 36$. Similarly, we find $C_{12} = 2 \times (3 + 2) = 10$, $C_{13} = 0$, $C_{21} = 2 \times 4 \times (2 + 2) = 32$, etc.

Now, consider the discordant pairs and assume the first and second subjects are from cell (i, j) and (i', j'), respectively, with a smaller i', but larger j', i.e., $i > i'$ and $j < j'$. For example, for cell $(1, 1)$, there are no discordant pairs and so $D_{11} = 0$. Similarly, $D_{12} = D_{13} = 0$. For cell $(2, 1)$, the subjects forming discordant pairs with the subjects in this cell are found in cells $(1, 2)$ and $(1, 3)$. Thus, $D_{21} = 2 \times 4 \times (1 + 5) = 48$. For cell $(2, 2)$, the subjects forming discordant pairs with those in this cell are found in cells $(1, 3)$, and thus $D_{22} = 2 \times 2 \times 5 = 20$. Note that the subjects in $(2, 2)$ may also form discordant pairs with those in cell $(3, 1)$. But these discordant pairs will be counted when computing D_{31}. So, their exclusion when computing D_{22} is to avoid double counting. Similarly, $D_{23} = 0$, $D_{31} = 2 \times 3 \times (1 + 5 + 2 + 3) = 66$, $D_{32} = 2 \times 2 \times (5 + 3) = 32$, and $D_{33} = 0$.

By repeating this process across all the cells and summing the number of concordant C_{ij} (discordant, D_{ij}) pairs over all the cells, we have

$$C = \sum_{i,j} \left(2n_{ij} \sum_{i'>i, j'>j} n_{i'j'} \right), \quad D = \sum_{i,j} 2n_{ij} \sum_{i>i', j<j'} n_{i'j'}. \tag{2.39}$$

For example, consider Table 2.8 from the DOS. If we ordered the female before the male subjects, then a pair consisting of a depressed female and a nondepressed male is a discordant pair and a pair consisting of a depressed male and a nondepressed female is a concordant pair. Hence,

$$C = 2 \times (274 \times (31 + 35) + 105 \times 35) = 43518,$$
$$D = 2 \times (207 \times (105 + 93) + 31 \times 93) = 87738.$$

The total possible pairs is $n^2 = 745^2 = 555\,025$. Thus, the estimated probabilities of concordant and discordant pairs are

$$\widehat{p}_s = \frac{C}{n^2} = \frac{43518}{555025} = 0.0784, \quad \widehat{p}_d = \frac{D}{n^2} = \frac{87738}{555025} = 0.1581.$$

Because C and D, or the normalized p_s and p_d, can be expressed as U-statistics, the theory of U-statistics can be used to compute the variance of the measures we will introduce next.

Interested readers may check Brown and Benedetti (1977) and Gibbons and Chakraborti (2003) for details.

Goodman–Kruskal Gamma

The *Goodman–Kruskal gamma* is defined as the difference between the probability of concordant pair and that of the discordant pair, conditional on all such concordant and nonconcordant pairs, i.e.,

$$\gamma = \frac{p_s - p_d}{p_s + p_d}. \tag{2.40}$$

As noted earlier, we estimate p_s and p_d by $\widehat{p}_s = \frac{C}{n^2}$ and $\widehat{p}_d = \frac{D}{n^2}$, and thus we can estimate γ by $\widehat{\gamma} = \frac{C-D}{C+D}$.

The Goodman–Kruskal γ ranges between -1 and 1. Under independence between x and y, $p_s - p_d = 0$ and thus $\gamma = 0$. A positive γ means that we are more likely to obtain concordant pairs than discordant ones; i.e., subjects with larger x are more likely to have larger y, yielding a positive association. Likewise, a negative γ implies that subjects with larger x are more likely to be associated with smaller y, giving rise to a negative association. The extreme value $\gamma = 1$ occurs if and only if $p_d = 0$, in which case there is no discordant pair and the ordering of x is *almost* in perfect agreement with the ordering of y, i.e., larger x corresponds to larger y and vice versa. Similarly, at the other end of the spectrum, $\gamma = -1$ if and only if $p_s = 0$. There is no concordant pair in this extreme case, and the ordering of x is *almost* in perfect agreement with the reverse of the ordering of y, i.e., larger x leads to smaller y and vice versa. We use the word "almost" to indicate the fact that the agreement or disagreement in the ordering between x and y may still include ties.

For example, consider the left 3×4 table below:

	$y(1)$	$y(2)$	$y(3)$	$y(4)$
$x(1)$	*	0	0	0
$x(2)$	0	0	*	*
$x(3)$	0	0	0	*

	$y(1)$	$y(2)$	$y(3)$	$y(4)$
$x(1)$	0	0	0	*
$x(2)$	0	0	*	*
$x(3)$	*	*	0	0

where $*$ denotes some nonzero counts. For this table, $\gamma = 1$ since $D = 0$. However, the subjects in cell $(2,3)$ and $(2,4)$ do not form concordant pairs because of the tied x values. Similarly, for the right table above, $\gamma = -1$, but we again have subjects that are either tied in x or y levels.

Example 2.16
For Table 2.8, $\widehat{\gamma} = \frac{43518 - 87738}{43518 + 87738} = -0.3369$.   ⌷

Kendall's tau-b

Kendall's tau-b is defined similarly as Goodman–Kruskal gamma, except with some adjustment for ties. This index is the ratio of $p_s - p_d$ over the geometric mean of the probability of no tie in both x and y. Since pairs of subjects are individually sampled with replacement from the sample, the probability of having a tie in x, P_x, and the probability of having a tie in y, P_y, are given by $P_x = \sum_{i=1}^{s} p_{i+}^2$ and $P_y = \sum_{j=1}^{r} p_{+j}^2$. Kendall's tau-b is defined as

$$\tau_b = \frac{p_s - p_d}{\sqrt{(1 - P_x)(1 - P_y)}} = \frac{p_s - p_d}{\sqrt{(1 - \sum_i p_{i+}^2)(1 - \sum_j p_{+j}^2)}}.$$

By estimating the parameters with data from the table, we obtain an estimate of τ_b:

$$\widehat{\tau}_b = \frac{C - D}{\sqrt{(n^2 - \sum_i n_{i+}^2)(n^2 - \sum_j n_{+j}^2)}},$$

where $n^2 - \sum_i n_{i+}^2$ and $n^2 - \sum_j n_{+j}^2$ are the number of pairs with no ties in x and in y, respectively.

Kendall's tau-b also ranges from -1 for perfect discordance to 0 for no association and to 1 for perfect concordance. As no ties are allowed when $\tau_b = \pm 1$, subjects from different cells form strict concordant and discordant pairs in these two extreme cases. Consequently, $\tau_b = \pm 1$ are generally stronger than $\gamma = \pm 1$. Thus, $\tau_b = 1$ ($\tau_b = -1$) corresponds to a diagonal (skewed diagonal) table.

For example, consider the two 4×5 tables below.

*	0	0	0	0
0	0	*	0	0
0	0	0	0	0
0	0	0	*	0

*	0	0	0	0
0	0	*	*	0
0	0	0	0	0
0	0	0	*	0

The second and fifth columns and the third row of the left table above all have zeros. After deleting these columns and rows, we obtain the 3×3 table on the left below. As only the diagonal cells have nonzero counts, $\tau_b = 1$ and $\gamma = 1$.

*	0	0
0	*	0
0	0	*

*	0	0
0	*	*
0	0	*

The second 4×5 table also reduces to a 3×3 table after the second and fifth columns, and the third row are removed (shown on the right above). As this is not a diagonal table, ties are present and hence $\tau_b < 1$. However, $\gamma = 1$ since there is no discordant pair and $p_d = 0$.

Example 2.17

For Table 2.8, we have

$$\widehat{\tau}_b = \frac{43518 - 87738}{\sqrt{(745^2 - (472^2 + 273^2))(745^2 - (481^2 + 136^2 + 128^2))}} = -0.1621.$$

Thus, $|\widehat{\tau}_b| = 0.1621 < 0.3369 = |\widehat{\gamma}|$. □

Note that like other popular association measures such as Pearson and Spearman correlations, Kendall's tau was initially developed for continuous outcomes. The index τ_b is a variation of the original version for tabulated data.

Stuart's tau-c

Stuart's tau-c is closely related to Goodman–Kruskal γ. It is also premised on the difference $p_s - p_d$, but adjustment is made according to the table size.

Let $m = \min(r, s)$. Then, it can be shown that the maximum of p_s and that of p_d both equal to $\frac{m-1}{m}$ (see Problem 2.30). Stuart's tau-c is defined as

$$\tau_c = \frac{p_s - p_d}{(m-1)/m}.$$

With this normalizing factor $\frac{m}{m-1}$ for $p_s - p_d$, τ_c has a range from -1 to 1. The Stuart tau-c is estimated by $\widehat{\tau}_c = \frac{m(C-D)}{(m-1)n^2}$.

Example 2.18

For Table 2.8, $m = \min(2, 3) = 2$. Thus, $\widehat{\tau}_c = \frac{2 \times (43518 - 87738)}{745^2} = -0.1593$, which is closer to $\widehat{\tau}_b = -0.1621$ than to $\widehat{\gamma} = -0.3369$. □

Somers' D

Somers' D is again defined based on the difference $p_s - p_d$ and is closely related to Kendall's τ_b. However, unlike τ_b, Somer's D adjusts for ties in the row x and column y variables individually by creating two indices:

$$D(y \mid x) = \frac{p_s - p_d}{(1 - P_x)} = \frac{p_s - p_d}{1 - \sum_{i=1}^{s} p_{i+}^2},$$

$$D(x \mid y) = \frac{p_s - p_d}{(1 - P_y)} = \frac{p_s - p_d}{1 - \sum_{j=1}^{r} p_{+j}^2},$$

where P_x, $\sum_{i=1}^{s} p_{i+}^2$, P_x, and $P_y = \sum_{j=1}^{r} p_{+j}^2$ have the same interpretation as in Kendall's τ_b. Thus, $D(y \mid x)$ $(D(x \mid y))$ is the difference between the probabilities p_s and p_d given that there is no tie in the row x (column y) variable. Somers' Ds each range from -1 to 1.

Somer's Ds are estimated from data in the table by

$$\widehat{D}(y \mid x) = \frac{C - D}{n^2 - \sum_{i=1}^{s} n_{i+}^2}, \quad \widehat{D}(x \mid y) = \frac{C - D}{n^2 - \sum_{j=1}^{r} n_{+j}^2}.$$

Somers' $D(C \mid R)$ and $D(C \mid R)$ are asymmetric modifications of τ_b, but they differ from τ_b in that they adjust for ties for the row and column variables separately.

Example 2.19

For Table 2.8, we have

$$\widehat{D}(y \mid x) = \frac{43518 - 87738}{745^2 - (472^2 + 273^2)} = -0.1716,$$

$$\widehat{D}(x \mid y) = \frac{43518 - 87738}{745^2 - (481^2 + 136^2 + 128^2)} = -0.1531.$$

Both are close to Kendall's $\widehat{\tau}_b = -0.1621$. ⬜

2.5.2 Measures of association for nominal outcome

When both row and column variables are nominal, the measures described above relying on the orders of the levels of the variables do not apply. We introduce two indices for describing association of such nominal variables (Goodman and Kruskal, 1954).

Uncertainty Coefficient

The *uncertainty coefficients* are defined based on the concept of *entropy*, first introduced by Shannon (1948). Entropy is frequently used to measure uncertainty in information theory. For a discrete random variable x, with distribution function $p_i = \Pr(x = i)$, the entropy of x is defined as

$$H(x) = -\sum_i \Pr(x = i) \log(\Pr(x = i)) = -\sum_i p_i \log p_i. \tag{2.41}$$

If x is a constant, $p_i = 1$ and $H(x) = 0$. Thus, $H(x) \geq 0$, with larger $H(x)$ indicating more uncertainty. For example, if x is binary, we are most uncertain about the outcome of x when $p = \frac{1}{2}$. It is readily checked that $H(x)$ has its maximum at $p = \frac{1}{2}$ (see Problem 2.29).

Within our context, it follows from (2.41) that the entropy of the column variable y, without any information about x, is $H(y) = -\sum_j p_{+j} \log p_{+j}$. Given $x = i$, the distribution of y conditional on this information about x is $\Pr(y = j \mid x = i) = \frac{p_{ij}}{p_{i+}}$. Substituting this conditional distribution of y into the definition, we obtain the entropy of y given the particular value i of x:

$$H(y \mid x = i) = -\sum_{j=1}^{r} \frac{p_{ij}}{p_{i+}} \log \frac{p_{ij}}{p_{i+}}.$$

Hence, the entropy of y given x is given by

$$H(y \mid x) = -\sum_{i=1}^{s} p_{i+} \sum_{j=1}^{r} \frac{p_{ij}}{p_{i+}} \log \frac{p_{ij}}{p_{i+}} = -\sum_{i,j} p_{ij} \log p_{ij} + \sum_{i=1}^{s} p_{i+} \log p_{i+},$$

or, in other words,

$$H(y \mid x) = H(xy) - H(x). \tag{2.42}$$

In particular, if x and y are independent, $H(y \mid x) = H(y)$ and (2.42) reduces to $H(xy) = H(x) + H(y)$. In general, $H(y \mid x) \leq H(y)$.

The uncertainty coefficient for y (x), $U(y \mid x)$ ($U(x \mid y)$), is the proportion of entropy in the variable y (x) explained by x (y):

$$U(y \mid x) = \frac{H(y) - H(y \mid x)}{H(y)} = \frac{H(x) + H(y) - H(xy)}{H(y)},$$

$$U(x \mid y) = \frac{H(x) + H(y) - H(xy)}{H(x)}.$$

The overall uncertainty coefficient for both the row and column variables is

$$U = \frac{2[H(x) + H(y) - H(xy)]}{H(x) + H(y)}.$$

All the three versions of the uncertainty coefficient are readily estimated from data by substituting estimates for the respective parameters:

$$\widehat{H}(x) = -\sum_{i=1}^{s} \frac{n_{i+}}{n} \log \frac{n_{i+}}{n}, \quad \widehat{H}(y) = -\sum_{j=1}^{r} \frac{n_{+j}}{n} \log \frac{n_{+j}}{n},$$

$$\widehat{H}(xy) = -\sum_{i=1}^{s} \sum_{j=1}^{r} \frac{n_{ij}}{n} \log \frac{n_{ij}}{n}.$$

Example 2.20

We again use the DOS data as an example to illustrate the calculations, although those measures introduced earlier for ordinal outcomes such as γ and τ_b are more appropriate as the depression outcome is an ordinal variable.

Based on Table 2.8, we have

$$\widehat{H}(x) = -\sum_{i=1}^{s} \frac{n_{i+}}{n} \log \frac{n_{i+}}{n} = -\left(\frac{472}{745} \log \frac{472}{745} + \frac{273}{745} \log \frac{273}{745}\right) = 0.6570$$

$$\widehat{H}(y) = -\sum_{j=1}^{r} \frac{n_{+j}}{n} \log \frac{n_{+j}}{n} = -\left(\frac{481}{745} \log \frac{481}{745} + \frac{136}{745} \log \frac{136}{745} + \frac{128}{745} \log \frac{128}{745}\right)$$

$$= 0.8956$$

$$\widehat{H}(xy) = -\sum_{i=1}^{s}\sum_{j=1}^{r} \frac{n_{ij}}{n} \log \frac{n_{ij}}{n} = -\left(\frac{274}{745} \log \frac{274}{745} + \frac{105}{745} \log \frac{105}{745} + \frac{93}{745} \log \frac{93}{745}\right.$$

$$\left. + \frac{207}{745} \log \frac{207}{745} + \frac{31}{745} \log \frac{31}{745} + \frac{35}{745} \log \frac{35}{745}\right) = 1.5356.$$

Thus,

$$\widehat{U}(y \mid x) = \frac{0.6570 + 0.8956 - 1.5356}{0.8956} = 0.0190,$$

$$\widehat{U}(x \mid y) = \frac{0.6570 + 0.8956 - 1.5356}{0.6570} = 0.0259,$$

and $\widehat{U} = \frac{2(0.6570+0.8956-1.5356)}{0.6570+0.8956} = 0.0219.$ ▯

Lambda Coefficient

The asymmetric *lambda coefficient* $\lambda(y \mid x)$ measures the improvement in percentage of the predictability of the column variable y, given the row variable x. If we are asked to give a guess for the column level of a subject without any information, then the optimal guess would be the level of y with maximum marginal probability, $p_{+M} = \max_{\{1 \le j \le r\}} p_{+j}$. The probability of incorrect prediction for such a guess is $1 - p_{+M}$.

Now suppose that we know the row level and want to predict the column level. The optimal guess would be the column level with maximum conditional probability given the row level. Hence, for a subject in the ith row level, the probability of incorrect prediction for such a guess is $1 - \frac{p_{iM_i}}{p_{i+}}$, where $p_{iM_i} = \max_{\{1 \le j \le r\}} p_{ij}$. The overall false prediction rate for the entire sample is the weighted average

$$\sum_{i=1}^{s} p_{i+}\left(1 - \frac{p_{iM_i}}{p_{i+}}\right) = 1 - \sum_{i=1}^{s} p_{iM_i}.$$

The lambda coefficient $\lambda(y \mid x)$ is defined as the improvement in percentage of prediction by utilizing the information in the two variables x:

$$\lambda(y \mid x) = \frac{1 - p_{+M} - \sum_{i=1}^{s} p_{i+}\left(1 - \frac{p_{iM_i}}{p_{i+}}\right)}{1 - p_{+M}} = \frac{\sum_{i=1}^{s} p_{iM_i} - p_{+M}}{1 - p_{+M}}. \tag{2.43}$$

Similarly, the lambda coefficient for predicting the row given the column information is defined by

$$\lambda(x \mid y) = \frac{\sum_{j=1}^{r} p_{M_j j} - p_{M+}}{1 - p_{M+}}, \tag{2.44}$$

where $p_{M_j j} = \max_{\{1 \le i \le s\}} p_{ij}$ and $p_{M+} = \max_{\{1 \le i \le s\}} p_{i+}$.

The symmetric lambda coefficient indicates improvement in percentage of the predictability if half the time we are asked to guess the row levels and half the time to guess the column levels. If no additional information is given, the optimal guess would be the row (column) level with maximum marginal probability for predicting the row (column) level. Thus, the probability of overall error in predicting the row and column levels is

$$\frac{1}{2}\left(1 - p_{+M}\right) + \frac{1}{2}\left(1 - p_{M+}\right) = 1 - \frac{1}{2}\left(p_{M+} + p_{+M}\right). \qquad (2.45)$$

As before, if we know the column (or row) level, we would select the row (column) level with maximum conditional probability for our prediction. The probability of overall error incurred in predicting both row and column levels is

$$\frac{1}{2}\left(1 - \sum_{i=1}^{s} p_{iM_i}\right) + \frac{1}{2}\left(1 - \sum_{j=1}^{r} p_{M_j j}\right) = 1 - \frac{1}{2}\left(\sum_{i=1}^{s} p_{iM_i} + \sum_{j=1}^{r} p_{M_j j}\right). \qquad (2.46)$$

It follows from (2.45) and (2.46) that the symmetric lambda λ is given by

$$\lambda = \frac{\left(\sum_{i=1}^{s} p_{iM_i} + \sum_{j=1}^{r} p_{M_j j}\right) - \left(p_{M+} + p_{+M}\right)}{2 - \left(p_{M+} + p_{+M}\right)}. \qquad (2.47)$$

All three versions of the lambda coefficient have the range $[0, 1]$, with 0 implying no improvement or association. The lambda coefficients are estimated by substituting estimates for the respective parameters.

Example 2.21

For Table 2.8, it is easy to verify that all the lambda coefficients are 0. In this particular example, "Female" has the maximum conditional probabilities across all levels of depression diagnosis and "No" has the maximum conditional probabilities across both gender categories. For example, the estimated conditional probabilities of each gender level given the "No" level of depression diagnosis are given by

$$\text{Pr (Female | No depression)} = \frac{274}{481}, \quad \text{Pr (Male | No depression)} = \frac{207}{481}.$$

Since the denominator is the same for both conditional probabilities, finding the maximum conditional probabilities is equivalent to locating the maximum cell counts between the rows. Given the large cell count in the Female by No cell, the conditional probability of Female is larger than the conditional probability of Male given the "No" level of depression diagnosis. Similarly, the conditional probabilities of Female are larger than the corresponding conditional probabilities of Male for the "MinD" and "MajD" levels of y. From the table, it is also readily checked that the "No" category of y has the maximum conditional probability across both gender levels.

Thus, the prediction for the row variable x is always the "Female" level regardless of the levels of y. Likewise, the prediction for y is always the "No" depression category irrespective of the levels of x. Thus, all three lambda coefficients are zeros. As we have demonstrated before, there are associations between gender and depression diagnosis when considering the ordered levels of the depression diagnosis outcome. Thus, if measures of nominal variables are applied to ordinal outcomes, some associations may be missed. For those variables, it is more appropriate to use measures of ordinal variables. $\square$

Exercises

2.1 A random sample of 16 subjects was taken from a target population to study the prevalence of a disease p. It turned out that six of them were diseased.

   (a) Estimate the disease prevalence p.
   (b) Use the asymptotic procedure to test

$$H_0 : p = 0.3 \quad \text{vs.} \quad H_a : p > 0.3. \tag{2.48}$$

   (c) Change $H_a : p > 0.3$ in (2.48) to $H_a : p < 0.3$ and repeat (b).
   (d) Change $H_a : p > 0.3$ in (2.48) to $H_a : p \neq 0.3$ and repeat (b).

2.2 Since the sample size in Problem 2.1 is not very large, it is better to use exact tests.

   (a) Apply exact tests to test the hypothesis in (2.48) for the data in Problem 2.1 and compare your results with those derived from asymptotic tests.
   (b) Change $H_a : p > 0.3$ in (2.48) to $H_a : p < 0.3$ and repeat (a).
   (c) Change $H_a : p > 0.3$ in (2.48) to $H_a : p \neq 0.3$ and repeat (a).
   Provide the steps that lead to your answers in (a)–(c).

2.3 Check that in the binary case ($k = 2$), the statistic in (2.7) is equivalent to the one in (2.1).

2.4 In the DOS, we are interested in testing the following hypothesis concerning the distribution of depression diagnosis for the entire sample:

$$\Pr(\text{No depression}) = 0.5$$
$$\Pr(\text{Minor depression}) = 0.3$$
$$\Pr(\text{Major depression}) = 0.2$$

   (a) Use the DOS data to test this hypothesis. First, use the chi-square test and then follow with the exact test.
   (b) Compare the results from the chi-square and exact test.
   (c) Describe your findings and conclusions based on the test results.

2.5 Suppose $x \sim BI(n, p)$ follows a binomial distribution of size n and probability p. Let k be an integer between 0 and n. Show that $\Pr(x \geq k)$, looking as a function of p with n and k fixed, is an increasing function of p.

2.6 Suppose $\mathbf{x} \sim MN(n, \boldsymbol{\pi})$ follows a multinomial distribution of size n and probability $\boldsymbol{\pi}$. Derive the variance matrix of $\mathbf{x}$.

2.7 Prove that

   (a) If $y \sim \text{Poisson}(\lambda)$, then both the mean and variance of y are λ.
   (b) If y_1 and y_2 are independent and $y_j \sim \text{Poisson}(\lambda_j)$ ($j = 1, 2$), then the sum $y_1 + y_2 \sim \text{Poisson}(\lambda_1 + \lambda_2)$.

2.8 Following the MLE method, the information matrix is closely related with the asymptotic variance of MLE (see Chapter 1). For the MLE of Poisson distribution,

(a) First compute the Fisher information matrix then plug in the MLE $\widehat{\lambda}$ to estimate the variance of $\widehat{\lambda}$.

(b) Plug in $\widehat{\lambda}$ in the observed Fisher information matrix to estimate the variance of $\widehat{\lambda}$.

(c) Compare part (a) and (b).

2.9 Derive the negative binomial (NB) distribution.

(a) Suppose y follows a Poisson(λ), where the parameter λ itself is a random variable following a gamma distribution Gamma(p, r). Derive the distribution of y. (Note that the density function of a Gamma(p, r) is $\frac{\lambda^{r-1} \exp(-\lambda p/(1-p))}{\Gamma(r)((1-p)/p)^r}$ for $\lambda > 0$ and 0 otherwise.)

(b) Derive the distribution of the number of trials needed to achieve r successes, where each trial is independent and has the probability of success p. Compare it with the distribution in part (a).

2.10 Prove (2.11).

2.11 Consider the statistic in (2.14).

(a) Show that this statistic is asymptotically normal with the asymptotic variance given by

$$Var_a \left(\sqrt{n} \left(\widehat{p}_1 - \widehat{p}_2 \right) \right) = \frac{n}{n_{1+}} p_1 \left(1 - p_1 \right) + \frac{n}{n_{2+}} p_2 \left(1 - p_2 \right).$$

(b) By estimating p_1 and p_2 using $\widehat{p}_1 = \widehat{p}_2 = \frac{n_{+1}}{n}$, confirm the asymptotic distribution in (2.15).

2.12 For the DOS, test whether education is associated with depression. To simplify the problem, we dichotomize both variables; use no and major/minor for depression diagnosis and at most and more than 12 years education for education.

2.13 Derive the relationships among the eight versions of odds ratios of Section 2.2.2.

2.14 Let $p_1 = \Pr \left(y = 1 \mid x = 1 \right) = 0.8$ and $p_2 = \Pr \left(y = 1 \mid x = 0 \right) = 0.4$.

(a) Compute the relative risk of response $y = 1$ of population $x = 1$ to population $x = 0$, and the relative risk of response $y = 0$ of population $x = 1$ to population $x = 0$.

(b) Change the values of p_1 and p_2 so that one of RRs in part (a) remains unchanged (then the other RR will change, and this shows that one RR may not determine the other).

2.15 Show that the hypergeometric distribution $HG \left(k; n, n_{1+}, n_{+1} \right)$ has mean $\frac{n_{1+}n_{+1}}{n}$ and variance $\frac{n_{1+}n_{+1}n_{+2}n_{2+}}{n^2(n-1)}$.

2.16 In the PPD, each subject was diagnosed for depression using SCID along with several screening tests including EPDS. By repeatedly dichotomizing the EPDS outcome, answer the following questions:

(a) For all possible EPDS cut-points observed in the data, compute the kappa coefficients between SCID and dichotomized EPDS.

(b) Which cut-point gives the highest kappa?

2.17 The data set "intake" contains baseline information of the Catheter Study. Use the two binary outcomes on whether urinary tract infections (UTIs) and catheter blockages occurred during the last two months to assess

(a) whether catheter blockage is associated with UTI;

(b) the relative risk and odds ratio for UTI with the blockage as a risk factor;

(c) whether catheter blockage and UTI have the same marginal distribution;

(d) agreement as measured by the kappa coefficient.

2.18 Group the count responses on UTIs and catheter blockages in the data set "intake" into three levels: no occurrence, only once, and more than once. Use these three-level outcomes to assess

(a) whether catheter blockage is associated with UTI;

(b) whether catheter blockage and UTI have the same marginal distribution;

(c) whether the joint distribution of catheter blockage and UTI is symmetric;

(d) agreement as measured by the kappa coefficient.

2.19 For the DOS, use the three-level depression diagnosis and dichotomized education (more than 12 years education or not) to check the association between education and depression.

(a) Test whether education and depression are associated;

(b) Compare the results of part (a) with that from Problem 2.12.

2.20 The data set "DosPrepost" contains depression diagnosis of patients at baseline (pre-treatment) and one year after treatment (posttreatment) in the DOS. We are interested in whether there is any change in depression rates between pre- and post treatment.

(a) Carry out the two-sided asymptotic and exact tests and summarize your results.

(b) Carry out the two one-sided exact tests and summarize your results. Please write down the alternatives and the procedure you use to obtain the p-value.

2.21 Let p denote the prevalence of a disease of interest. Express PPV and NPV as a function of p, Se, and Sp.

2.22 Prove that the weighted kappa for 2×2 tables will reduce to the simple kappa, no matter which weights are assigned to the two levels.

2.23 Verify the variance formula for the MWW statistics (2.29).

2.24 Use the three-level depression diagnosis and dichotomized education (more than 12 years education or not) in the DOS data to test the association between education and depression.

(a) Use the Pearson chi-square statistic for the test;

(b) Use the row mean score test;

(c) Use the Pearson correlation test;

(d) Compare results from (a), (b), and (c) and describe your findings.

2.25 For the $2 \times r$ table with scores as in Section 2.3.1,

(a) verify that the MLE of β in the linear regression model in (2.26) is $\widehat{\beta} = \frac{\overline{xy} - \overline{x}\,\overline{y}}{\overline{y^2} - \overline{y}^2}$;

(b) prove $E\,(xy) - E\,(x)\,E\,(y) = \sum_{j=1}^{r} p_{1j}\,(R_j - E\,(y))$.

2.26 For the DOS, compute the indices, Pearson correlation, Spearman correlation, Goodman–Kruskal γ, Kendall's τ_b, Stuart's τ_c, Somers' D, lambda coefficients, and uncertainty coefficients, for assessing association between education (dichotomized with cut-point 12) and depression using two different score systems for the depression outcome specified in (a) and (b) below.

(a) Scores for depression are 0 for no, 1 for minor, and 2 for major depression.

(b) Scores for depression are 0 for no, 1 for minor, and 3 for major depression.

(c) Compare results from (a) and (b) and state which indices are unchanged under the different scoring methods.

(d) Is the invariance true under all score systems that preserve orientation?

2.27 Many measures of association for two-way frequency tables consisting of two ordinal variables are based on the numbers of concordant and discordant pairs. To compute such indices, it is important to count each concordant (discordant) pair exactly once with no misses and repetitions. In Section 2.5.1, we discussed one such counting procedure. Now, let us consider two other alternatives.

(a) For subjects in the cell in the ith row and jth column, they will form concordant (discordant) pairs with the subjects in the cells lying to the left of and above (left and below) that cell. The total number of concordant (discordant) pairs is obtained by summing such concordant (discordant) pairs over all the cells in the table. The formula following this approach is given by

$$C = \sum_{i,j} C_{ij}, \quad D = \sum_{i,j} D_{ij},$$

where

$$C_{ij} = 2n_{ij} \sum_{i' < i, j' < j} n_{i'j'}, \quad D_{ij} = 2n_{ij} \sum_{i < i', j > j'} n_{i'j'}.$$

Compute the concordant and discordant pairs for Table 2.8 using this alternative method.

(b) For subjects in the cell in the ith row and jth column, they will form concordant (discordant) pairs with the subjects in the cells to the left of and above (left of and below) that cell and to the right of and below (right and above) that cell. The formula following this alternative approach is given by

$$C = \sum_{i,j} C_{ij}, \quad D = \sum_{i,j} D_{ij},$$

where

$$C_{ij} = n_{ij} \left(\sum_{i'>i,j'>j} n_{i'j'} + \sum_{i'<i,j'<j} n_{i'j'} \right),$$

$$D_{ij} = n_{ij} \left(\sum_{i>i',j<j'} n_{i'j'} + \sum_{i<i',j>j'} n_{i'j'} \right).$$

Compute the concordant and discordant pairs for Table 2.8 using this alternative method.

(c) Compare the results.

2.28 Suppose x is a random variable with m levels such that $\Pr(x = i) = p_i$ for $i = 1, 2, \ldots, m$ with $\sum_{i=1}^{m} p_i = 1$. In other words, $x \sim MN(1, \mathbf{p})$. Let x_1 and x_2 be two independent random variables following the distribution of x.

(a) Compute $\Pr(x_1 = x_2)$.

(b) Among all the possible distributions of x, i.e., among all different p_i with $\sum_{i=1}^{m} p_i = 1$, when will $\Pr(x_1 = x_2)$ have the minimum value? Please determine the minimum value of $\Pr(x_1 = x_2)$ and the distribution of x in this *optimal* case.

2.29 Let x be a binary variable with outcomes 0 and 1. Let $p = \Pr(x = 1)$. Show that entropy has the maximum at $p = 0.5$.

2.30 For an $r \times s$ table, the probability of concordant (discordant) pair p_s $(p_d) \leq \frac{m-1}{m}$, where $m = \min(r, s)$.

2.31 EPDS is an instrument (questionnaire) for depression in postpartum women. This instrument is designed so that a person with a higher EPDS score has a higher chance to be depressed. Use the PPD data to confirm this defining property of the instrument. More specifically,

(a) Use the Cochran–Armitage test to examine whether the proportion of depression increases as the EPDS becomes bigger.

(b) Use the Jonckheere–Terpstra test to test whether the depressed subgroup has larger EPDS.

(c) Compare (a) and (b) and summarize your finding.

2.32 Suppose x is a random variable with at least two levels, with $\Pr(x = x_i) = p_i$, for $i = 1, 2$. Let x' be the new random variable based on x with the two levels x_1 and x_2 combined, i.e.,

$$x' = \begin{cases} x_1 & \text{if } x = x_1 \text{ or } x_2, \\ x & \text{if otherwise,} \end{cases}$$

then $H(x) = H(x') + (p_1 + p_2) H(z)$, where $H(z)$ is the entropy conditional on $x = x_1$ or x_2, or, in other words, $z \sim Bernoulli(\frac{p_1}{p_1 + p_2})$.

Chapter 3

Sets of Contingency Tables

Sometimes, we may have a set of similar contingency tables. For example, for large-scale clinical trials where a large number of patients are required, it is common to involve multiple medical centers to help with study recruitments so that the trials can be completed in a timely fashion. For example, one of the largest studies for treating alcohol dependence, COMBINE Combined Pharmacotherapies and Behavioral Interventions (COMBINE), randomized 1,383 recently alcohol-abstinent subjects into nine pharmacological and/or psychosocial treatment conditions from 11 academic sites in the United States (COMBINE Study Research Group, 2006). A study of this scale would have been much more difficult to conduct for a single medical facility. For rare diseases, such an approach is normally required because it is almost impossible to enroll enough patients at one site. Stratified studies also improve power; through stratification, subjects within the same stratum are more homogenous, and the reduced between-subject variability helps increase power. Since patients from different sites may be different in terms of their health conditions and varying levels of quality of healthcare services they receive from the different hospitals, treatments are likely to have varying effects across the sites. To account for such differences in the analysis, we cannot pull all patients' data into one contingency table and apply the methods in Chapter 2.

The controlling variables used for stratification might themselves be of no interest, but they may affect the relationship among variables of interest. Such variables are called *confounding variables,* or *confounders*. The study site in COMBINE is an example of a confounder. Since they affect the relationship among the variables of interest, it is not valid to ignore them. In Section 3.1, we illustrate how confounding effects can seriously alter analysis results when ignored. In the subsequent sections, we address the effect of confounding on two-way contingency tables by a categorical variable. We start with methods for sets of 2×2 contingency tables in Section 3.2 and then discuss sets of general two-way tables in Section 3.3.

3.1 Confounding Effects

It is important to consider a set of contingency tables to address confounding caused by a categorical variable. As mentioned above, a common confounding factor in multisite studies is study site because of differences in patients and healthcare systems across multiple sites. Other common examples of confounders include gender, race, disease severity, and education. Although we are often primarily interested in making inference about the association between the row and column variables for the overall population, we cannot simply apply the methods in Chapter 2 to one contingency table based on the pulled data because potentially different relationships may exist across the levels of the confounder.

DOI: 10.1201/9781003109815-3

For example, suppose that a new treatment is being tested against an existing alternative (a control condition) at several hospitals across the different cites in a multi-site randomized trial. We are interested in learning if the treatment has (superior) effects over its control counterpart. In other words, we want to test the null hypothesis that the odds ratio (OR) is 1. Since patients from the different sites may be different in terms of their health conditions and levels of quality of care received, the new treatment is likely to yield differential treatment effects across the sites. If we pull all patients' data into one contingency table regardless of the differences among patients and hospitals, we may miss the opportunity to study treatment variability across the sites and causes of such variability. Further, aggregating data across different tables stratified by the levels of the confounding variable has far more serious ramifications, as the next example illustrates.

Example 3.1

Suppose that there are two hospitals serving residents in a community. Hospital A is staffed with better surgeons than hospital B for some hypothetical surgery. Table 3.1 compares the success rates of surgery between the two hospitals over a period of time.

TABLE 3.1:   Success rates of two hospitals

Hospital	Outcome		Total
	Success	**Fail**	
A (Good)	50	50	100
B (Bad)	68	32	100
Total	118	82	200

The data seem to suggest that the bad hospital (Hospital B) had a higher success rate (0.68 vs. 0.5). The OR (or relative risk) in comparing the success rate is less than 1 in favor of the bad hospital. So, we may conclude based on the evidence in the data that the bad hospital actually performed better!   ⬚

What is going on? Are the data lying here?

Actually, there is nothing wrong with the data. The problem is that the aggregated data in the above table does not tell the whole story about surgeries performed by the two hospitals. If we stratify the data by disease severity before surgery, we obtain Table 3.2:

TABLE 3.2:   Success rates of two hospitals stratified by disease severity

Severity	Hospital	Outcome		Total
		Success	**Fail**	
Less severe	A (Good)	18	2	20
	B (Bad)	64	16	80
More severe	A (Good)	32	48	80
	B (Bad)	4	16	20

If we now compare success rates within each level of severity before surgery, we can see that Hospital A (good) always performed better than Hospital B (bad). However, in comparison with hospital B (bad), hospital A (good) received far more patients with a more severe disease before surgery. This selection of sicker patients to Hospital A is what caused Hospital A to appear to have a lower overall success rate when disease severity is ignored. In statistics, this phenomenon is called Simpson's paradox. In this example, more severe patients selected (either self-select or through referrals) the good hospital, and this disproportionality lowered the overall success rate for this hospital. In other words, disease severity simultaneously affects the hospital selection and the surgery outcome, leading to biased conclusions when ignored. This is commonly referred to as confounding effect, and variables such as the disease severity in this example are called confounding variables, confounders, or covariates in statistical lingo. Note that in this example, if the hospitals have control on patient selection and select patients based on their disease severity, the confounding effect is often referred to as selection bias (Haneuse, 2016). We do not make such distinctions and simply refer the bias as confounding in both cases. Confounding bias in estimates of effects is one of the most important issues in the field of statistics. In fact, most cutting-edge topics in statistical research in recent years such as causal inference and longitudinal data analysis all attempt to address this key issue.

The above example shows that we cannot simply ignore confounding variables and collapse multiple tables into a single one. A correct approach is to account for differences in the individual tables when making inference about the association between the row and column variables. For example, for stratified 2×2 tables, the Cochran–Mantel–Haenszel (CMH) test is the most popular method to derive inference about the association between the row and column variables while taking into account the differences across the tables.

Sometimes, the source of bias is difficult to detect. For example, as we discussed in Chapter 1, disease surveillance systems may underreport caseloads if the disease of interest has a long latency time such as HIV/AIDS (see Figure 1.1). Here, bias is the result of our limited observation time and is less obvious than other confounding factors such as disease severity and demographic differences across patients in most treatment and cohort studies.

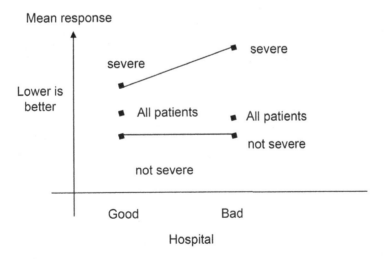

FIGURE 3.1: Mean responses of a continuous outcome for two hospitals.

Note that Simpson's paradox illustrates the effect of bias by a confounder for categorical outcomes. The same phenomenon occurs for continuous responses as well. The diagram above (Figure 3.1) illustrates the effect of confounding bias within the same setting of the example, but involving a continuous response, with lower values indicating better outcomes. If we ignore disease severity before treatment, we may again conclude that hospital B (bad) performed better than hospital A (good). By accounting for this confounding factor in the analysis, we will be able to tease out the effect of hospital from that of disease severity, leading to correct conclusions. Such a procedure is called "control for the effect of covariates," "control for covariates," "covariance analysis," or "covariate analysis." When comparing the mean response of a continuous outcome across two or more groups, we can apply the *analysis of covariance* (ANCOVA) model (Neter et al., 1990).

Another related type of analysis, which is becoming increasingly popular, particularly in research in the behavioral and social sciences in recent years, is *moderation analysis*. We will not discuss the concept of moderation in this chapter, but would like to point out that moderation analysis investigates a different phenomenon, which is both conceptually and technically different from covariance analysis.

3.2   Sets of 2×2 Tables

Suppose that there are q 2×2 tables, each having the same row and column variables. We use a superscript to index the tables, as shown in Table 3.3. Again, we are interested in testing whether there is an association between the row and column variables.

TABLE 3.3:   A set of q 2×2 tables

x	y 1	0	Total			x	y 1	0	Total
1	$n_{11}^{(1)}$	$n_{12}^{(1)}$	$n_{1+}^{(1)}$	, $\ldots$,		1	$n_{11}^{(q)}$	$n_{12}^{(q)}$	$n_{1+}^{(q)}$
0	$n_{21}^{(1)}$	$n_{22}^{(1)}$	$n_{2+}^{(1)}$			0	$n_{21}^{(q)}$	$n_{22}^{(q)}$	$n_{2+}^{(q)}$
Total	$n_{+1}^{(1)}$	$n_{+2}^{(1)}$	$n^{(1)}$			Total	$n_{+1}^{(q)}$	$n_{+2}^{(q)}$	$n^{(q)}$

The null hypothesis is that there is no row by column association in any of the tables. Thus, for each table, we may use the difference between the observed and expect counts in one of the cells to test the row by column independence, as described in Chapter 2. For the hth table, the observed counts in the (1,1) cell are $n_{11}^{(h)}$ ($1 \leq h \leq q$). Let $m_{11}^{(h)}$ be the expected cell counts of this cell. Then, as described in Section 2.2, the difference $n_{11}^{(h)} - m_{11}^{(h)}$ measures how likely the row and column variables are dependent within the hth table, and we may use it to test whether the row and column variables are independent in the hth table. By repeating this procedure for each of the tables, we may reject the null hypothesis if at least one of the tables shows a significant association between the row and column variables. Although tempting, this is not a valid approach for testing independence between the row and column variables in a set of tables. There are two major problems.

First, type I error will be inflated under such a one-table-at-a-time approach because of increased false significance rates. This is a common problem when multiple tests are conducted and is known as the *multiple testing* issue in the statistical literature. For example, randomized trials are widely used to avoid bias in treatment assignment. Since demographic variables such as age, gender, race, and other variables for characterizing patients' differences such as comorbid medical and mental health conditions often have impact on treatment effects, it is important to compare them across different treatment groups to see whether randomization is successful. Ideally, subjects in the different treatment groups will have similar characteristics, and thus the distributions of these covariates should be comparable between the groups. However, if a large number of covariates are compared, some of them are likely to be significantly different across the different groups, especially under simple randomization, i.e., patients are randomized into treatment groups without considering their differences with respect to these covariates. For example, if ten covariates are compared with a type I error rate of 5% for each, then the rejection rate for the null hypothesis of no difference in all ten variables, assuming the variables are independent, will be about 40%, considerably elevated beyond 5% (see Problem 3.2). To address a large number of covariates, many studies employ stratified randomization methods to achieve balance of treatment assignment within each covariate.

Another reason relates to power. If the number of tables is large, the sample size for each table is likely small, limiting the power to detect associations between the row and column variables. For example, if a sample of 100 subjects is distributed across 5 tables, there will be about 20 subjects within each table. Since binary outcomes have much lower power than continuous ones, a sample size of 20 in a two-way table does not provide much power to detect associations.

To address these flaws, one would prefer methods that derive a single statistic by combining information across all tables. In general, the statistics for stratified data are developed as weighted combination of those for individual tables. The weights are generally chosen according to the variance or precision of corresponding estimates for individual tables. In this section, we deal with the relatively simpler 2×2 case, deferring considerations for general two-way tables to the next section.

Note that there is no need to discuss a set of one-way frequency tables, since such tables can be easily presented using a two-way contingency table. To see this, let the column denote the variable of interest and the row indicate the different strata. Then, all data in a series of one-way tables can be displayed through a two-way contingency table, and associated questions can be addressed using methods for two-way contingency tables discussed in the last chapter. For example, consider a simple binary outcome indicating the presence/absence of a disease of interest. We are interested in whether the distribution of this outcome is homogeneous across the levels of a categorical variable such as gender. By representing the association using a two-way table with the column designating the disease status and the row identifying the association variable, the question becomes testing the row-column independence of the two-way table.

Although we can similarly use a three-way contingency table to present a set of two-way tables, analysis of such tables is generally quite complicated. As a result, we will not discuss contingency table methods for three-way tables in this chapter. Instead, we will discuss a unified approach for the analysis of general contingency tables with the help of log-linear models in Chapter 7.

3.2.1 Cochran–Mantel–Haenszel test for independence

In Chapter 2, we introduced the chi-square test for 2×2 tables. To generalize this test to a set of q such tables, consider the hth table ($1 \leq h \leq q$). The chi-square test for this table is based on $n_{11}^{(h)} - m_{11}^{(h)}$, where $m_{11}^{(h)} = \frac{1}{n^{(h)}} n_{1+}^{(h)} n_{+1}^{(h)}$ is the expected cell count of the $(1,1)$ cell of the hth table under the null hypothesis of row and column independence with given marginal counts. Under the null hypothesis of independence, $n_{11}^{(h)} - m_{11}^{(h)}$ are equally likely to be positive or negative, making the sum $\sum_{h=1}^{q}\left(n_{11}^{(h)}-m_{11}^{(h)}\right)$ more likely to be close to 0. On the other hand, if there are row by column associations, the observed cell counts will differ from their expected counterparts in certain directions, creating more positive or negative terms. The accumulated differences will shift the sum $\sum_{h=1}^{q}\left(n_{11}^{(h)}-m_{11}^{(h)}\right)$ away from 0, making it a good candidate for an overall test for the set of tables.

We can easily obtain the variance of $\sum_{h=1}^{q}\left(n_{11}^{(h)}-m_{11}^{(h)}\right)$ based on the independence among the strata:

$$Var\left(\sum_{h=1}^{q}\left(n_{11}^{(h)}-m_{11}^{(h)}\right)\right)=\sum_{h=1}^{q}Var\left(n_{11}^{(h)}-m_{11}^{(h)}\right).$$

Further, since $n_{11}^{(h)}$ follows a hypergeometric distribution with parameters $\left(n^{(h)},n_{1+}^{(h)},n_{+1}^{(h)}\right)$, we have $Var\left(n_{11}^{(h)}-m_{11}^{(h)}\right)=\frac{n_{1+}^{(h)}n_{+1}^{(h)}n_{2+}^{(h)}n_{+2}^{(h)}}{(n^{(h)})^2(n^{(h)}-1)}=v_{11}^{(h)}$. Thus, normalizing it, we obtain the following statistic:

$$Q_{CMH}=\frac{\left[\sum_{h=1}^{q}\left(n_{11}^{(h)}-m_{11}^{(h)}\right)\right]^2}{\sum_{h=1}^{q}v_{11}^{(h)}}. \tag{3.1}$$

The test statistic was first suggested by Mantel and Haenszel (1959), so it is called the *Mantel–Haenszel test.* Since Cochran (1954) derived and proposed a similar statistic using a different approach, the statistic in (3.1) is often referred to as the CMH *statistic.*

Under the null hypothesis, this statistic follows approximately a chi-square distribution with one degree of freedom, if the total sample size is large (see Problem 3.3). In other words, reliable inference is still obtained, even if some of the tables have small sample sizes as long as the total size is large. The following version, with $\frac{1}{2}$ subtracted from the total difference count to improve the approximation to the distribution of this discrete statistic by a chi-square variate, is also commonly used:

$$\frac{\left[\left|\sum_{h=1}^{q}\left(n_{11}^{(h)}-m_{11}^{(h)}\right)\right|-\frac{1}{2}\right]^2}{\sum_{h=1}^{q}v_{11}^{(h)}}.$$

Note that tables with larger sizes are assigned larger weights in the statistic. This can be made clearer by writing the sum as a linear combination of quantities in the same scale $\sum_{h=1}^{q}\left(n_{11}^{(h)}-m_{11}^{(h)}\right)=\sum_{h=1}^{q}n^{(h)}\left(\widehat{p}_{11}^{(h)}-p_{11}^{(h)}\right)$, where $p_{11}^{(h)}$ is the expected and $\widehat{p}_{11}^{(h)}$ is the observed proportion of the counts in $(1,1)$ cell of the hth table. Also, under the null hypothesis, the row and column variables are independent in each table, but the marginal probabilities can still be different. So the marginal probabilities, expectations, and variances of the cell counts should be estimated separately for each table. Finally, although $n_{11}^{(h)}$ has been used to derive the test, the same statistic is obtained if a different cell count is used. For example, if cell $(1,2)$ is used instead, then because $n_{12}^{(h)}-m_{12}^{(h)}=-\left(n_{11}^{(h)}-m_{11}^{(h)}\right)$ and $\left(n_{11}^{(h)}-m_{11}^{(h)}\right)$ and $\left(n_{12}^{(h)}-m_{12}^{(h)}\right)$ have the same variance, the statistic based on the cell $(1,2)$ is identical to the one given in (3.1).

Example 3.2

For Table 3.2 in the hypothetical example above, with fixed marginal counts, the means of success for Hospital A are $\frac{20\times82}{100}=16.4$ for less severe patients and $\frac{80\times36}{100}=28.8$ for more severe patients. The corresponding variances are $\frac{82\times18\times20\times80}{100^2\times(100-1)} = 2.3855$ and $\frac{36\times64\times80\times20}{100^2\times(100-1)} = 3.7236$. Hence, the CMH statistic is $Q_{CMH} = \frac{(18-16.4+32-28.8)^2}{2.3855+3.7236} = 3.7714$, and the corresponding p-value is 0.052.

Thus, based on the stratified tables, there is some evidence for a difference in the success rates between the two hospitals, although the p-value is still slightly larger than 0.05. □

Example 3.3

In depression studies, levels of education are often found to be associated with depression outcomes. In the DOS, consider the association between gender and depression, stratified by education levels according to whether subjects had completed 12 years of education, as shown in Table 3.4, where Dep = "Yes" for major/minor depression and Dep = "No" for no depression.

TABLE 3.4: Depression by gender, stratified by education

Gender	Dep Yes	No	Total	Gender	Dep Yes	No	Total
Female	85	105	190	Female	112	169	281
Male	24	48	72	Male	42	159	201
Total	109	153	262	Total	154	328	482
Education ≤ 12				Education > 12			

Consider the (female, No) cells. The expected cell counts for the two tables are $\frac{190\times153}{262} = 110.95$ and $\frac{281\times328}{482} = 191.22$, and their corresponding variances are $\frac{190\times153\times72\times109}{262^2\times(262-1)} = 12.734$ and $\frac{281\times328\times201\times154}{482^2\times(482-1)} = 25.53$. The CMH statistic is $Q_{CMH} = \frac{[105-110.95+169-191.22]^2}{12.734+25.53} = 20.739$, with a p-value < 0.0001. Hence, after controlling for the levels of education, gender and depression are still highly associated. □

3.2.2 Estimates and tests of common odds ratios

If the row and column variables in Table 3.3 are found to be associated (e.g., if the null hypothesis of independence is rejected by the CMH test), then we would like to know if the associations between the row and column variables are similar across the tables. In this section, we discuss how to test the homogeneity of ORs and how to estimate the common OR. We discuss estimates first, since they will be used in the test statistic.

By weighting the tables according to their sample sizes, Woolf (1955) proposed an estimate by combining such weighted averages with a logarithm transformation. More specifically, the Woolf estimate is defined by the exponential of the weighted average of the logarithm of the ORs of the separate tables:

$$OR_W = \exp\left(\frac{\sum_h w_h \log \widehat{OR}_h}{\sum_h w_h}\right), \tag{3.2}$$

where $\widehat{OR}_h = \frac{n_{11}^{(h)} n_{22}^{(h)}}{n_{12}^{(h)} n_{21}^{(h)}}$ is the estimate of the ORs for the hth stratum, and the weight

function given by $w_h = \left[\frac{1}{n_{11}^{(h)}} + \frac{1}{n_{12}^{(h)}} + \frac{1}{n_{21}^{(h)}} + \frac{1}{n_{22}^{(h)}}\right]^{-1}$ is the variance of $\log \widehat{OR}_h$ given in

Section 2.2. The variance of the statistic in (3.2) is estimated by $(\sum_h w_h)^{-1}$. All the tables are required to have large sample sizes for OR_W to behave well; if a large number of tables have small sizes, the estimate may be seriously biased (Gart, 1970). For this reason, people in general prefer the Mantel–Haenszel estimate.

The Mantel–Haenszel estimate, defined as

$$OR_{MH} = \frac{\sum_h \frac{1}{n^{(h)}} n_{11}^{(h)} n_{22}^{(h)}}{\sum_h \frac{1}{n^{(h)}} n_{12}^{(h)} n_{21}^{(h)}}, \tag{3.3}$$

was first suggested by Mantel and Haenszel (1959). This estimate behaves well if the total sample size is large. This includes the situations where there are a large number of small tables. As noted earlier, Woolf's estimate may not work well in such situations, but the Mantel and Haenszel estimate does. There are different versions of the variances of the estimate OR_{MH}, depending on the different limiting conditions. Most have quite complex formulas, and interested readers may consult Kuritz et al. (1988) for details.

The common OR may also be estimated using the maximum likelihood method. For example, conditioning on the marginal counts, the distributions of all the tables are determined by the common OR, and hence the (conditional) likelihood can be computed and the theory of maximum likelihood can be applied. Birch (1964) showed that the conditional maximum likelihood estimate (MLE) is the solution to the equation

$$\sum_{h=1}^{q} n_{11}^{(h)} = E\left[\sum_{h=1}^{q} n_{11}^{(h)} \mid \text{marginal counts}, \psi\right]$$

where ψ is the parameter of common OR. As a solution to high-degree polynomial equations, the conditional MLE has no simple expression. There is also an unconditional version of MLE where the marginal counts are themselves treated as random. Gart (1970, 1971) showed that the conditional and unconditional MLEs are asymptotically equivalent. The maximum likelihood and Mantel–Haenszel test statistics are also asymptotically equivalent when each individual table is large. Like the Mantel–Haenszel estimate, the conditional MLE is consistent if the total sample size is large. But the unconditional MLE requires a large sample size in each stratum.

Compared to the MLEs, the Mantel–Haenszel estimate (3.3) is much simpler. Thus, the Mantel–Haenszel estimate, rather than its MLE counterpart, has been widely used in practice. Note also that unlike the MLEs, which are based on the assumption of a common OR, the Mantel–Haenszel estimate, as a weighted average of the individual ORs, may still be meaningful if this assumption does not hold, in which case it may be viewed as an *overall* OR.

3.2.2.1 Confidence intervals

Confidence intervals for the common OR may be constructed based on the asymptotic distribution of the estimates for large samples. As in the case of a single table, a logarithmic transformation is typically applied to obtain better confidence intervals. More precisely, the confidence interval of $\log(OR)$ is first computed based on the asymptotic distribution of $\log(OR)$, which is then exponentiated to yield the confidence interval of the OR.

3.2.2.2 Tests of a common odds ratio

The common OR estimates discussed above are based on the assumption that the ORs are the same across the tables. Thus, before using such an estimate for a given application, we must test to see whether such an assumption holds. The null hypothesis is that the ORs from the tables are all equal to each other. This is different from the null hypothesis of no row by column association considered in Section 3.2.1, which posits not only the equality of the ORs, but also the value 1 of the common ORs. Thus, the *Breslow–Day test* (Breslow et al., 1982) in the current setting investigates whether the strength of relationship between the row and column variables is moderated by the different levels of the stratification covariate.

The idea of the Breslow–Day test is again to compare the observed cell counts with those expected under the null hypothesis. However, we cannot compute the expected counts directly based on the marginal distributions of the row and column variables. Indeed, the OR is a monotone function of p_{11} if the marginal distributions are held fixed (see Problem 3.7). Thus, the expected counts also depend on the unknown common odds ratio. We need to estimate the common odds ratio to obtain the expected $(1,1)$ cell count $m_{11}^{(h)'}$ and the variance $v_{11}^{(h)'}$ of the difference $n_{11}^{(h)} - m_{11}^{(h)'}$. Note that since $m_{11}^{(h)'}$ and $v_{11}^{(h)'}$ are calculated based on the null hypothesis of equal odds ratios, $m_{11}^{(h)'}$ and $v_{11}^{(h)'}$ are generally different from their respective counterparts computed under the null hypothesis of row by column independence, which assumes not only the equality of all the odds ratios but also the specific value 1 of the common odds ratios. Thus, in general, $m_{11}^{(h)'} \neq m_{11}^{(h)} = n^{(h)} p_{1+}^{(h)} p_{+1}^{(h)}$ and $v_{11}^{(h)'} \neq v_{11}^{(h)} = \frac{n_{1+}^{(h)} n_{+1}^{(h)} n_{2+}^{(h)} n_{+2}^{(h)}}{(n^{(h)})^2 (n^{(h)} - 1)}$.

To compute the mean and variance of cell counts with a given odds ratio $(\neq 1)$, first consider the distribution of the cell counts for a single table. With fixed row and column marginals and a given odds ratio, the cell counts follow a noncentral hypergeometric distribution given by

$$\Pr(n_{11} = k; \psi) = \frac{\binom{n_{1+}}{k} \binom{n_{2+}}{n_{+1} - k} \psi^k}{P_0(\psi)}, \tag{3.4}$$

where $P_0(\psi) = \sum_k \binom{n_{1+}}{k} \binom{n_{2+}}{n_{+1} - k} \psi^k$ with the summation over the range of all possible integers, i.e., $\max(0, n_{+1} - n_{2+}) \leq k \leq \min(n_{1+}, n_{+1})$. The distribution can be derived by considering two independent sampling groups with $x = 1$ and $x = 0$, with Bernoulli parameter p_1 and p_2 (hence $\psi = \left(\frac{p_1}{1 - p_1}\right) / \left(\frac{p_2}{1 - p_2}\right)$). If a total of n_{1+} (n_{2+}) subjects are sampled from the first (second) group, then the probability that there are a total of k subjects with $y = 1$ in the first $(n_{11} = k)$ and l subjects with $y = 1$ in the second group $(n_{21} = l)$ is

$$\Pr(n_{11} = k, n_{21} = l) = \binom{n_{1+}}{k} \binom{n_{2+}}{l} p_1^k (1 - p_1)^{n_{1+} - k} p_2^l (1 - p_2)^{n_{2+} - l}$$

$$= \binom{n_{1+}}{k} \binom{n_{2+}}{l} \left(\frac{p_1}{1 - p_1}\right)^k (1 - p_1)^{n_{1+}} \left(\frac{p_2}{1 - p_2}\right)^l (1 - p_2)^{n_{2+}}$$

$$= \binom{n_{1+}}{k} \binom{n_{2+}}{l} \psi^k (1 - p_1)^{n_{1+}} \left(\frac{p_2}{1 - p_2}\right)^{n_{+1}} (1 - p_2)^{n_{2+}}. \tag{3.5}$$

If we fix the column marginal counts as well, i.e., $k + l = n_{+1}$ is fixed, then one cell count will determine the other three. By computing the conditional probabilities, it is straightforward to verify the distribution in (3.4) based on (3.5). When $\psi = 1$, the distribution reduces to the hypergeometric distribution discussed in Chapter 2. Given the complexity in computing the mean and variance of a noncentral hypergeometric distribution, McCullagh and Nelder (1989) suggested the following simple approximation procedure.

For a 2×2 table with odds ratio ψ, it can be shown that $E(n_{11}n_{22}) = \psi E(n_{12}n_{21})$ (see Problem 3.9), and hence

$$\psi = \frac{\mu_{11}\mu_{22} + v}{\mu_{12}\mu_{21} + v}, \tag{3.6}$$

where v is the variance. In addition, v can be approximated by

$$v = \frac{n}{n-1}\left(\frac{1}{\mu_{11}} + \frac{1}{\mu_{12}} + \frac{1}{\mu_{21}} + \frac{1}{\mu_{22}}\right)^{-1}. \tag{3.7}$$

Solving the equations in (3.6) and (3.7) simultaneously provides an approximate estimate of the mean and variance.

Under homogeneous odds ratios, the Breslow–Day statistic has the following form:

$$Q_{BD} = \sum_{h=1}^{q} \frac{\left(n_{11}^{(h)} - m_{11}^{(h)\prime}\right)^2}{v_{11}^{(h)\prime}} \tag{3.8}$$

where $m_{11}^{(h)\prime}$ and $v_{11}^{(h)\prime}$ are estimates of the mean and variance of $n_{11}^{(h)}$. This statistic has approximately a chi-square distribution with $q-1$ degrees of freedom, if the sample sizes of the tables are all large. Note that the loss of one degree of freedom is due to estimation of the common odds ratio across the tables. Unlike the Mantel–Haenszel statistic, the Breslow–Day test requires a relatively large sample size within each table for reliable inference. This is because the terms corresponding to each of the tables are normalized before they are combined so that each approximately follows a chi-square distribution. If the sample size is small for a table, its approximation to the chi-square may be poor. In comparison, the Mantel–Haenszel statistic in (3.3) normalizes the sum rather than each term within the sum. As a result, the statistic is not sensitive to small sample sizes within some tables so long as the total sample size is large.

We used the word "approximately" to indicate that the Breslow–Day test statistic Q_{BD} in (3.8) does not have an asymptotic chi-square distribution. Tarone (1985) first noticed this and suggested an adjustment to this statistic to make it follow an asymptotic chi-square distribution. This corrected Breslow–Day–Tarone statistic is given by

$$Q_{BDT} = Q_{BD} - \frac{\left[\sum_{h=1}^{q}\left(n_{11}^{(h)} - m_{11}^{(h)\prime}\right)\right]^2}{\sum_{h=1}^{q} v_{11}^{(h)\prime}}.$$

Tarone (1985) showed that Q_{BDT} has an asymptotically chi-square distribution with $q-1$ degrees of freedom. Since the difference between the two statistics is usually small in practice, the Breslow–Day statistic is still widely used.

Exact tests should be used if the sample size is small. Zelen (1971) generalized Fisher's exact test for a single 2×2 table to a set of stratified tables. Similar to Fisher's exact test, the exact p-value is computed conditional on fixed marginal counts. Since each 2×2 table follows a hypergeometric distribution, the distribution of q such tables conditional on the respective fixed margins is a product of hypergeometric probabilities. As usual, the p-value is the sum of the probabilities of all those potential sets of tables that are deemed as or more extreme than the one observed. Zelen's test uses the table probabilities themselves to define extremeness; the smaller the table probability, the more extreme. Hence, the p-value of Zelen's test is the sum of probabilities of all those potential sets of tables with probabilities not bigger than that of the observed one. We may also use statistics such as Q_{BD} and Q_{BDT} to define extremeness.

Example 3.4

In Example 3.3, we found an association between gender and depression, stratified by education levels. Let us now test whether the association is the same across the two education levels.

For the test, we need an overall estimate of the odds ratio. The Mantel and Haenszel estimate is $\left(\frac{105\times24}{262} + \frac{169\times42}{482}\right) / \left(\frac{85\times48}{262} + \frac{112\times159}{482}\right) = 0.46354$. Based on this estimate of the common odds ratio, we can estimate the mean and variance of the count in the (female, no depression) cell for the first table by solving the following equations:

$$
\begin{cases}
\dfrac{x \times (262 - 190 - 153 + x) + v}{(153 - x)(190 - x) + v} = 0.46354 \\[2ex]
v = \dfrac{262/261}{\left(\dfrac{1}{x} + \dfrac{1}{153 - x} + \dfrac{1}{190 - x} + \dfrac{1}{262 - 190 - 153 + x}\right)}
\end{cases}
$$

Similarly, we can estimate the mean and variance of the count in the (female, no depression) cell for the second table by solving the following equations:

$$
\begin{cases}
\dfrac{x \times (482 - 281 - 328 + x) + v}{(281 - x)(328 - x) + v} = 0.46354 \\[2ex]
v = \dfrac{482/481}{\left(\dfrac{1}{x} + \dfrac{1}{281 - x} + \dfrac{1}{328 - x} + \dfrac{1}{482 - 281 - 328 + x}\right)}
\end{cases}
$$

The estimated means and variances of counts in the (female, no depression) cells are 101.64 and 11.269 for the first table and 172.38 and 23.053 for the second table. Thus, the Breslow-Day statistic is $\frac{(105-101.64)^2}{11.269} + \frac{(169-172.38)^2}{23.053} = 1.4974$, and the p-value $= 0.2211$. It is easy to check that the Tarone corrected version yields quite a similar value to the Breslow-Day statistic. $\square$

Note that although we introduced several indices for the single 2×2 table, we only discussed inference about homogeneity of odds ratios for sets of such tables. A primary reason is the wide popularity of the odds ratio index when assessing the strength of association for binary outcomes. In addition, as odds ratios do not depend on how the levels of stratification are selected, the common odds ratio is well defined. Other indices do not have such a nice invariance property. For example, the relative risk (RR) depends on the levels of stratification considered. Although there is some work on estimating the common RR, we find it difficult to interpret such estimates and thus will not discuss them in this book.

If one variable is considered as response and the other as predictor within the context of regression, the common odds ratio is equivalent to the fact that there is no interaction between the strata and the other variable in the regression model. We will discuss this in more detail in Chapter 4.

3.3   Sets of $s \times r$ Tables

As in the case of 2×2 tables, if we stratify the association between the row and column variables by the levels of a third categorical variable, we get a set of $s \times r$ tables. As illustrated by the analysis of sets of 2×2 tables, it is generally not appropriate to collapse all tables into one and apply the methods for single $s \times r$ tables to the resulting table. In this section, we extend the tests for a single $s \times r$ table to a set of such tables. As in the single table case, we will first discuss the test of row and column independence, in different situations according to whether the variables are treated as nominal, ordinal, or interval variables. For sets of square tables where the row and column variables represent the same scale, we will discuss how to generalize the kappa coefficients to assess the agreement between two raters.

3.3.1   Tests of general association

By ignoring the orders of row and column variables, if any, and treating both as nominal outcomes, we can combine the ideas of the Pearson chi-square test for general association introduced in Section 2.4 and the CMH statistic for stratified 2×2 tables to develop tests for row–column independence in stratified $s \times r$ tables. These multivariate extensions of the CMH test to sets of $s \times r$ tables are derived based on the multivariate hypergeometric distribution for each table with fixed marginal counts. For ease of notation and understanding, we start with a relatively simple case involving a set of q 3×2 tables.

Consider such a set of tables below:

x	y 1	2	Total	
1	$n_{11}^{(h)}$	$n_{12}^{(h)}$	$n_{1+}^{(h)}$	
2	$n_{21}^{(h)}$	$n_{22}^{(h)}$	$n_{2+}^{(h)}$	, $1 \le h \le q$.
3	$n_{31}^{(h)}$	$n_{32}^{(h)}$	$n_{3+}^{(h)}$	
Total	$n_{+1}^{(h)}$	$n_{+2}^{(h)}$	$n^{(h)}$	

How do we determine whether the row and column variables are independent? Recall that for a 2×2 table, if the marginal counts are fixed, then the count in any one of the four cells completely determines the table. In other words, a 2×2 table has only one degree of freedom if the marginals are fixed. For a 3×2 table, the situation is more complicated. In particular, a single cell count will not determine the table, and a 2×1 submatrix is required to identify the table. So, unlike the case of 2×2 table, we need to concentrate on a 2×1 submatrix, say the one formed by the first two rows and first column, i.e., the cell counts $n_{11}^{(h)}$ and $n_{21}^{(h)}$ in the $(1, 1)$ and $(2, 1)$ cells shown in the above tables.

The basic idea again is to combine the different observed and expected counts across the q tables. Let $\mathbf{u}^{(h)} = \left(n_{11}^{(h)}, n_{21}^{(h)} \right)^{\top}$. Under the null hypothesis of no row and column association across all tables, each table then follows a multivariate hypergeometric distribution, if the marginal counts are held fixed. Based on the properties of multivariate hypergeometric distribution, we find the expectation and variance of $\mathbf{u}^{(h)}$ as follows:

$$\mathbf{e}^{(h)} = E\left[\mathbf{u}^{(h)} \mid H_0 \right] = \left(n_{1+}^{(h)} n_{+1}^{(h)} / n^{(h)}, n_{2+}^{(h)} n_{+1}^{(h)} / n^{(h)} \right)^{\top},$$

$$V^{(h)} = Var\left[\mathbf{u}^{(h)} \mid H_0 \right]$$

$$= \begin{pmatrix} \dfrac{n_{1+}^{(h)} n_{+1}^{(h)} \left(n^{(h)} - n_{1+}^{(h)}\right) \left(n^{(h)} - n_{+1}^{(h)}\right)}{(n^{(h)})^2 (n^{(h)} - 1)} & \dfrac{n_{1+}^{(h)} n_{+1}^{(h)} n_{2+}^{(h)} \left(n^{(h)} - n_{+2}^{(h)}\right)}{(n^{(h)})^2 (n^{(h)} - 1)} \\[6mm] \dfrac{n_{1+}^{(h)} n_{+1}^{(h)} n_{2+}^{(h)} \left(n^{(h)} - n_{+2}^{(h)}\right)}{(n^{(h)})^2 (n^{(h)} - 1)} & \dfrac{n_{2+}^{(h)} n_{+1}^{(h)} \left(n^{(h)} - n_{2+}^{(h)}\right) \left(n^{(h)} - n_{+1}^{(h)}\right)}{(n^{(h)})^2 (n^{(h)} - 1)} \end{pmatrix}.$$

Since $\mathbf{u}^{(h)}$ are independent across the tables (this is so as different tables involve different sets of subjects), $Var\left(\sum_{h=1}^{q} \left(\mathbf{u}^{(h)} - \mathbf{e}^{(h)}\right)\right) = \sum_{h=1}^{q} V^{(h)}$. Thus, it follows from the central limit theorem that

$$\boldsymbol{\xi} = \left[\sum_{h=1}^{q} V^{(h)}\right]^{-1/2} \left[\sum_{h=1}^{q} \left(\mathbf{u}^{(h)} - \mathbf{e}^{(h)}\right)\right] \sim_d N\left(\mathbf{0}, \begin{pmatrix} 1 & 0 \\ 0 & 1 \end{pmatrix}\right). \tag{3.9}$$

By Slutsky's theorem, the following *generalized Cochran–Mantel–Haenszel (GCMH) statistic*,

$$Q_{GCMH} = \left[\sum_{h=1}^{q} \left(\mathbf{u}^{(h)} - \mathbf{e}^{(h)}\right)\right]^{\top} \left[\sum_{h=1}^{q} V^{(h)}\right]^{-1} \left[\sum_{h=1}^{q} \left(\mathbf{u}^{(h)} - \mathbf{e}^{(h)}\right)\right], \tag{3.10}$$

has an asymptotic chi-square distribution with two degrees of freedom under the null hypothesis. We may want to compare this statistic to the CMH statistic for a set of q 2×2 tables in (3.1). The two statistics have the same form except for the obvious difference in the dimensions of the difference statistics; while $n_{11}^{(h)} - m_{11}^{(h)}$ is a scalar in the CMH statistic, $\mathbf{u}^{(h)} - \mathbf{e}^{(h)}$ is a 2×1 vector in (3.10).

In the above, we chose to focus on the submatrix $\left(n_{11}^{(h)}, n_{21}^{(h)}\right)^{\top}$. However, one may select any other 2×1 submatrices such as $\left(n_{11}^{(h)}, n_{31}^{(h)}\right)^{\top}$, $\left(n_{21}^{(h)}, n_{31}^{(h)}\right)^{\top}$, and $\left(n_{12}^{(h)}, n_{22}^{(h)}\right)^{\top}$ and apply the above considerations to each of these. Since all different choices are linearly equivalent and the GCMH statistics, as quadratic forms of the differences, are invariant under linear transformation (see Problem 3.4), the statistics do not depend on choices of submatrices. One may also consider statistics like $\sum_{h=1}^{q} \left(\mathbf{u}^{(h)} - \mathbf{e}^{(h)}\right)^{\top} \left[V^{(h)}\right]^{-1} \left(\mathbf{u}^{(h)} - \mathbf{e}^{(h)}\right)$. However, as discussed in Section 3.2.2, this approach requires a large sample size for each table and may not work well when the requirement is not met. In contrast, the GCMH statistic is valid as long as the total sample size is large.

Now consider a set of q $s \times r$ tables:

		y			
x	1	$\cdots$	r	Total	
1	$n_{11}^{(h)}$	$\cdots$	$n_{1r}^{(h)}$	$n_{1+}^{(h)}$	
$\vdots$	$\vdots$	$\ddots$	$\vdots$	$\vdots$	, $1 \le h \le q$.
s	$n_{s1}^{(h)}$	$\cdots$	$n_{sr}^{(h)}$	$n_{s+}^{(h)}$	
Total	$n_{+1}^{(h)}$	$\cdots$	$n_{+r}^{(h)}$	$n^{(h)}$	

If the marginals are held fixed, then an $s \times r$ table is determined by any of its $(s-1) \times (r-1)$ submatrices. We may similarly compare the observed and expected cell counts from these submatrices by defining a test statistic based on such differences as in the case of a set of 3×2 tables just discussed. Without loss of generality, let us consider the upper left $(s-1) \times (r-1)$ submatrix and reexpress the entries using a vector:

$$\mathbf{n}^{(h)} = \left(n_{11}^{(h)}, \ldots, n_{1r-1}^{(h)}, \ldots, n_{s-11}^{(h)}, \ldots, n_{s-1r-1}^{(h)}\right)^{\top}, \quad 1 \le h \le q.$$

The above is a $(s-1)(r-1)$ column vector containing all the cell counts of the upper left $(s-1) \times (r-1)$ submatrix of the hth table. Conditional on the marginal counts, each of the tables follows a multivariate hypergeometric distribution. Let $\mathbf{e}^{(h)} = E\left(\mathbf{n}^{(h)} \mid H_0\right)$ be the expectation and $V^{(h)} = Var\left(\mathbf{n}^{(h)} \mid H_0\right)$ the variance of $\mathbf{n}^{(h)}$ under the null hypothesis. Then, it can be shown that

$$\mathbf{e}^{(h)} = \frac{1}{n^{(h)}}\left(n_{1+}^{(h)}n_{+1}^{(h)}, \ldots, n_{1+}^{(h)}n_{+(r-1)}^{(h)}, \ldots, n_{(s-1)+}^{(h)}n_{+1}^{(h)}, \ldots, n_{(s-1)+}^{(h)}n_{+(r-1)}^{(h)}\right)^{\top},$$

$$Cov\left(n_{ij}^{(h)}, n_{i'j'}^{(h)}\right) = \frac{n_{i+}^{(h)}\left(\delta_{ii'}n^{(h)} - n_{i'+}^{(h)}\right)n_{j+}^{(h)}\left(\delta_{jj'}n^{(h)} - n_{+j'}^{(h)}\right)}{(n^{(h)})^2(n^{(h)}-1)}, \tag{3.11}$$

where $\delta_{ii'} = 1$ if $i = i'$ and 0 otherwise. Similar to the 3×2 case, we obtain the GCMH statistic for a set of q $s \times r$ tables:

$$Q_{GCMH} = \left[\sum_{h=1}^{q}\left(\mathbf{n}^{(h)} - \mathbf{e}^{(h)}\right)\right]^{\top}\left[\sum_{h=1}^{q}V^{(h)}\right]^{-1}\left[\sum_{h=1}^{q}\left(\mathbf{n}^{(h)} - \mathbf{e}^{(h)}\right)\right]. \tag{3.12}$$

The above statistic again follows an asymptotic chi-square distribution with $(s-1)(r-1)$ degrees of freedom under the null hypothesis.

The considerations above can also be applied to other types of associations such as those involving the mean score statistics and correlations, which we consider next. As noted above for the 3×2 case, the statistic in general does not depend on the choice of submatrix.

3.3.2   Mean score statistic

If the column variable has ordinal levels, we may assign the levels some numerical scores and use the mean score to construct a dimensional scale for the column variable. As discussed in Chapter 2 for such ordinal levels, we may be interested in alternatives about mean scores such as H_a : The mean scores are not the same across different rows. Note that the null hypothesis is still that the row and column variables are independent. Note also that although the same column levels in different tables may be assigned different scores, they typically receive the same scores in practice.

First, consider a relatively simpler case involving a set of q 3×2 tables. Let $\mathbf{a}^{(h)} = \left(a_1^{(h)}, a_2^{(h)}\right)^{\top}$ be a column vector with $a_j^{(h)}$ denoting the assigned score for the jth level of the response y in the hth table ($1 \leq j \leq 2$, $1 \leq h \leq q$). Then, the observed total score for the ith row of the hth table is

$$t_i^{(h)} = a_1^{(h)}n_{i1}^{(h)} + a_2^{(h)}n_{i2}^{(h)}, \quad 1 \leq i \leq 3, \quad 1 \leq h \leq q.$$

As in the case of assessing association between nominal row and column variables, we compare the observed total scores with those expected under the null hypothesis.

If the mean scores are the same across all the rows of the hth table, then the expected total score for the ith row is

$$e_i^{(h)} = n_{i+}^{(h)}\left(a_1^{(h)}n_{+1}^{(h)} + a_2^{(h)}n_{+2}^{(h)}\right)/n^{(h)}, \quad 1 \leq i \leq 3, \quad 1 \leq h \leq q.$$

Since

$$\sum_{i=1}^{3}\left(t_i^{(h)} - e_i^{(h)}\right) = a_1^{(h)}\sum_{i=1}^{3}\left(n_{i1}^{(h)} - \frac{n_{i+}^{(h)}n_{+1}^{(h)}}{n^{(h)}}\right) + a_2^{(h)}\sum_{i=1}^{3}\left(n_{i2}^{(h)} - \frac{n_{i+}^{(h)}n_{+2}^{(h)}}{n^{(h)}}\right) = 0,$$

only two of the three $t_i^{(h)} - e_i^{(h)}$ are free to vary. For convenience, we consider the first two rows, but the same argument applies to any other pair of rows. Let $V^{(h)}$ be the variance of $\mathbf{t}^{(h)} - \mathbf{e}^{(h)} = \left(t_1^{(h)} - e_1^{(h)}, \bar{t}_2^{(h)} - e_2^{(h)}\right)$ under the null hypothesis that the row and column variables are independent. As $t_i^{(h)}$ is a linear combination of $n_{ij}^{(h)}$, its variance matrix $V^{(h)}$ can be computed based on (3.11). Following the discussion in Section 3.3.1, the GCMH statistic,

$$
Q_{GCMH} = \left[\sum_{h=1}^{q} \mathbf{t}^{(h)} - \mathbf{e}^{(h)}\right]^{\top} \left[\sum_{h=1}^{q} V^{(h)}\right]^{-1} \left[\sum_{h=1}^{q} \mathbf{t}^{(h)} - \mathbf{e}^{(h)}\right],
$$

follows an asymptotic chi-square distribution with two degrees of freedom under the null hypothesis.

In general, let $\mathbf{a}^{(h)} = \left(a_1^{(h)}, a_2^{(h)}, \ldots, a_r^{(h)}\right)^{\top}$ be a column vector with $a_j^{(h)}$ denoting the assigned score for the jth level of the response y in the hth table ($1 \leq j \leq r$, $1 \leq h \leq q$). Then, the observed total score for the ith row of the hth table is

$$
t_i^{(h)} = \sum_{j=1}^{r} a_j^{(h)} n_{ij}^{(h)}, \quad 1 \leq i \leq s, \quad 1 \leq h \leq q.
$$

Under the null hypothesis, the expected total score for the ith row of the hth table is

$$
e_i^{(h)} = \sum_{j=1}^{r} a_j^{(h)} n_{i+}^{(h)} n_{+j}^{(h)} / n^{(h)}, \quad 1 \leq i \leq s \quad 1 \leq h \leq q.
$$

Let $\mathbf{t}^{(h)} = \left(t_1^{(h)}, t_2^{(h)}, \ldots, t_{s-1}^{(h)}\right)^{\top}$, $\mathbf{e}^{(h)} = \left(e_1^{(h)}, e_2^{(h)}, \ldots, e_{s-1}^{(h)}\right)^{\top}$, and $V^{(h)} = Var\left(\mathbf{t}^{(h)} - \mathbf{e}^{(h)}\right)$ under the null hypothesis that the row and column variables are independent. As in the above, $V^{(h)}$ can be computed based on (3.11) and the GCMH statistic,

$$
Q_{GCMH} = \left[\sum_{h=1}^{q} \mathbf{t}^{(h)} - \mathbf{e}^{(h)}\right]^{\top} \left[\sum_{h=1}^{q} V^{(h)}\right]^{-1} \left[\sum_{h=1}^{q} \mathbf{t}^{(h)} - \mathbf{e}^{(h)}\right],
$$

follows an asymptotic chi-square distribution with $(s-1)$ degrees of freedom under the null hypothesis.

3.3.3 Correlation statistic

If both the row and column variables are ordinal with scores treated as interval variables, we may test the linear association between the row and column variables based on the Pearson correlation coefficient. For a set of q $s \times r$ tables, we can again develop a similar GCMH statistic Q_{GCMH} by combining information across the q tables.

Let $\mathbf{R}^{(h)} = \left(R_1^{(h)}, R_2^{(h)}, \ldots, R_s^{(h)}\right)^{\top}$ and $\mathbf{C}^{(h)} = \left(C_1^{(h)}, C_2^{(h)}, \ldots, C_r^{(h)}\right)^{\top}$ be the vectors corresponding to the row and column scores of the hth table. Consider the sample moment of XY for the hth table, $\overline{XY} = \sum_{ij} \frac{n_{ij}}{n} R_i^{(h)} C_j^{(h)}$. Its expectation under the null hypothesis of no linear association between X and Y is $E\left(\overline{XY}\right) = E(X) E(Y)$, which can be estimated by $\overline{X} \, \overline{Y}$.

Let $a^{(h)} = \overline{XY} - \overline{X} \, \overline{Y}$. Then $E\left(a^{(h)}\right) = 0$. Let $V^{(h)}$ be the variance of $a^{(h)}$ and consider the following GCMH statistic Q_{GCMH}:

$$
Q_{GCMH} = \frac{\left(\sum_{h=1}^{q} a^{(h)}\right)^2}{\sum_{h=1}^{q} V^{(h)}}.
$$

This statistic follows asymptotically a chi-square distribution with one degree of freedom. Since the statistic is based on a linear combination of correlations between the row and column variables across the tables, the statistic is also called the GCMH statistic for correlation. This correlation statistic has more power than either the general association statistic or the mean score statistic to detect linear association between the row and column variables.

Example 3.5

In Example 3.3, we tested the association between dichotomized depression diagnosis and gender stratified by the levels of education. Now, consider testing this association using a three-level depression diagnosis outcome, with 0, 1, and 2 representing no, minor, and major depression, respectively.

When applied to the resulting two 2×3 tables, all three statistics yield very small p-values (all p-values < 0.0001), where scores 1, 2, and 3 are assigned to no, minor, and major depression for the mean score test. Thus, we still reach the same conclusion regarding a strong relationship between gender and depression. When the two tables are analyzed separately, however, no association of gender and depression is found for the lower education group, but significant association is detected for the higher education group. It is interesting that the association between gender and depression is a function of the education level for the patients in this study.   ⬜

3.3.4 Kappa coefficients for stratified tables

When the row and column variables represent ratings on the same subjects by two raters, kappa coefficients have been discussed as a measure for observer agreement between the two raters. When subjects are stratified by a third categorical variable, we may have several tables and hence several kappa coefficients, one for each stratum. Similar to odds ratios, one would naturally ask the question whether agreement is unanimous across the strata and how to estimate the overall agreement if that is the case. As in the case of homogeneous odds ratios, we first need to estimate the overall agreement and then check whether estimates of agreement from the individual tables significantly differ from it. Thus, we start with the estimation of the overall agreement.

The overall kappa coefficient, which is a weighted average of the individual kappa coefficients, may be used as an overall agreement across all strata. Let $\widehat{\kappa}^{(h)}$ denote the estimate of the agreement between the two raters for the hth table, which may be simple or weighted kappa coefficient, and $var\left(\widehat{\kappa}^{(h)}\right)$ the variance of the kappa estimate $\widehat{\kappa}^{(h)}$ for the hth table. Note that if weighted kappas are used, the same weighting scheme is usually applied to all tables. The overall kappa coefficient is defined as

$$\widehat{\kappa}_{overall} = \frac{\sum \widehat{\kappa}^{(h)} \left[var\left(\widehat{\kappa}^{(h)}\right)\right]^{-1}}{\sum \left[var\left(\widehat{\kappa}^{(h)}\right)\right]^{-1}}. \tag{3.13}$$

The asymptotic variance of the overall kappa is estimated by $\frac{1}{\sum [var(\widehat{\kappa}^{(h)})]^{-1}}$ (see Problem 3.13). This variance estimate can be used for inference about the overall kappa; for example, the confidence interval can be constructed as

$$\left(\widehat{\kappa}_{overall} - z_{\alpha/2} \sqrt{\frac{1}{\sum \left[var\left(\widehat{\kappa}^{(h)}\right)\right]^{-1}}}, \ \widehat{\kappa}_{overall} + z_{\alpha/2} \sqrt{\frac{1}{\sum \left[var\left(\widehat{\kappa}^{(h)}\right)\right]^{-1}}} \right).$$

3.3.4.1 Tests for equal kappa coefficients

If individual kappa coefficients are the same, then the overall kappa coefficient, which is a weighted average of the individual ones, equals the common kappa coefficient. Because of sampling variability, this is unlikely to happen in practice. However, the individual kappa estimates from the different tables should be close to the estimated overall kappa coefficient, if the null hypothesis of homogeneous Kappa coefficients is true. Hence, we may use the following statistic to test the equality of kappa coefficients:

$$Q_\kappa = \sum_{h=1}^{q} \frac{\left(\widehat{\kappa}^{(h)} - \widehat{\kappa}_{overall} \right)^2}{Var\left(\widehat{\kappa}^{(h)} \right)}.$$

Under the null hypothesis, the statistic Q_κ follows an asymptotic chi-square distribution with $q - 1$ degrees of freedom, where q is the number of tables. Intuitively, if the overall kappa is known in the null hypothesis, then Q_κ will follow a chi-square distribution with q degrees of freedom. However, since the overall kappa is estimated from the data, the loss of one degree of freedom is due to estimation of that parameter.

Example 3.6
For the Detection of Depression in a Primary Care Setting Study (DDPC), consider testing whether there is a difference in agreement between probands and informants on depression diagnosis stratified by the gender of the informants. The stratified information is given in Table 3.5.

TABLE 3.5: Depression diagnosis, stratified by informant gender

Proband	Informant 0	1	2	Total	Proband	Informant 0	1	2	Total
0	44	8	5	57	0	22	5	1	28
1	24	12	10	46	1	12	4	0	16
2	9	7	19	35	2	5	5	8	18
Total	77	27	34	138	Total	39	14	9	62
Female informants					Male informants				

The estimated (unweighted) kappa coefficients for the two tables are 0.2887 and 0.2663, with the corresponding variances 0.00384 and 0.009 08. Thus, the estimated overall kappa is $\left(\frac{0.2887}{0.00384} + \frac{0.2663}{0.009\,08} \right) / \left(\frac{1}{0.00384} + \frac{1}{0.009\,08} \right) = 0.2820$ and the estimated asymptotic variance is $1/\left(\frac{1}{0.00384} + \frac{1}{0.009\,08} \right) = 0.002699$. A 95% confidence interval is $(0.1801, 0.3839)$. The test of the null hypothesis of equal kappa coefficients has a p-value 0.8435, indicating that agreement between probands and informants is similar for informant females and males. ☐

Exercises

3.1 Have you experienced Simpson's paradox in your professional and/or personal life? If so, please describe the context in which it occurred.

3.2 Suppose you test ten hypotheses and under the null hypothesis each hypothesis is to be rejected with type I error rate 0.05. Assume that the hypotheses (test statistics) are independent. Compute the probability that at least one hypothesis will be rejected under the null hypothesis.

3.3 Show that the asymptotic distribution for the CMH test for a set of q 2×2 tables is valid as long as the total size is large. More precisely, $Q_{CMH} \to_d N(0,1)$ if $\Sigma_{i=1}^{q} n^{(h)} \to \infty$.

3.4 Let $\mathbf{x}$ be a random vector and $\boldsymbol{V}$ its variance matrix. Show that $\mathbf{x}^\top \boldsymbol{V}^{-1} \mathbf{x}$ is invariant under linear transformation. More precisely, let $\boldsymbol{A}$ be some nonsingular square matrix, $\mathbf{x}' = \boldsymbol{A}\mathbf{x}$, and $\boldsymbol{V}'$ the variance of $\mathbf{x}'$. Then, $\mathbf{x}'^\top \left(\boldsymbol{V}'\right)^{-1} \mathbf{x}' = \mathbf{x}^\top \boldsymbol{V}^{-1} \mathbf{x}$.

3.5 Use a data set such as DOS to see how GCMH statistics change when different scoring systems are used for the levels of the row and column variables.

3.6 Use the DOS data to test whether there is gender and depression (dichotomized according to no and minor/major depression) association by stratifying medical burden and education levels, where medical burden has two levels (CIRS ≤ 6 and CIRS > 6) and education has two levels (edu ≤ 12 and edu >12). This is a problem of testing association for a set of four 2×2 tables.

3.7 Show that the odds ratio is a monotone function of p_{11} if marginal distributions are fixed.

3.8 Estimate the overall odds ratio of the set of tables in Problem 3.6, and test whether the odds ratios are the same across the tables.

3.9 Show that for a 2×2 table with odds ratio ψ, $E(n_{11}n_{22}) = \psi E(n_{12}n_{21})$ if marginals for both row and column variables are fixed. (See Mantel and Hankey (1975) for a complete proof.)

3.10 Verify (3.6).

3.11 In the Postpartum Depression Study (PPD), stratify the subjects according to the ages of the babies (0–6 months, 7–12 months, and 13–18 months) since it is known to affect postpartum depression. Apply methods for stratified tables to assess the association between Structured Clinical Interview for DSM-IV-TR (SCID) diagnosis and Edinburgh Postnatal Depression Scale (EPDS) screening tool.

3.12 Redo Problem 2.16 by stratifying the subjects according to baby ages as in Problem 3.11.

3.13 Prove that the asymptotic variance of $\dfrac{\sum \kappa^{(h)} \left[var\left(\kappa^{(h)}\right)\right]^{-1}}{\sum \left[var\left(\kappa^{(h)}\right)\right]^{-1}}$ is $\dfrac{1}{\sum \left[var\left(\kappa^{(h)}\right)\right]^{-1}}$.

3.14 Use statistic software to verify the given estimates of (unweighted) kappa coefficients and their variances in Example 3.6 for the two individual tables in Table 3.5.

3.15 Redo Problem 3.6, using the three-level depression diagnosis.

Chapter 4

Regression Models for Binary Response

In the last two chapters, we discussed how to make inference about association between two categorical variables with stratification (sets of contingency table) and without stratification (a single contingency table) by a third categorical variable. In such analyses, our primary interest lies in whether the two variables are associated as well as the direction of association, with stratification used to control for the effect of the categorical confounder. If there are many confounders or the confounding variable is continuous, the methods discussed in Chapter 3 for stratified tables may not work well or do not work at all. Furthermore, in many studies, we may want to know more about the relationship between the variables. Specifically, we want to know the amount of change in one variable per unit change of the other variable. The methods discussed in the last two chapters lack such specificity, and regression models, which are the focus of this chapter, come to rescue.

The logistic model is the most popular regression model to quantitatively characterize the relationship between a binary dependent variable (or response) and a set of independent variables (or predictors, explanatory, covariates, etc.). The logistic regression can be generalized to multi-level polytomous outcomes with more than two response levels. The multi-level response in the latter situation can be either nominal or ordinal. Alternative regression models for dichotomous and polytomous outcomes are also available. Among them, commonly used are probit and complementary log-log models. We will discuss them under the framework of generalized linear models later in this chapter.

In this chapter, we start with logistic regression for a binary response and discuss parameter interpretation in Section 4.1. In Section 4.2, we discuss estimation and inference for this relatively simple model case. In Section 4.3, we introduce the probit, complementary log-log, and linear probability models for binary responses after a brief introduction to the generalized linear model. We then discuss model diagnosis and evaluations in Section 4.4. We conclude this chapter by a discussion on analysis of aggregated binary responses in Section 4.5.

4.1 Logistic Regression for Binary Response

As discussed in Chapter 2, binary responses arise quite frequently in research and clinical studies. Further, as the treatment of such simplest discrete outcomes will provide a basic understanding of the regression approach and elucidate the development of more models for polytomous outcomes, we start with regression models for such a response.

4.1.1 Motivation of logistic regression

Consider a sample of n subjects. For each individual, let y_i denote a binary response of interest for the ith subject; it takes on one of two possible values, denoted for convenience by 0 and 1. Let $\mathbf{x}_i = (x_{i1}, \ldots, x_{ip})^\top$ denote a $p \times 1$ column vector of independent variables for the ith subject. Without using any information in the independent variables, we can make inference concerning the base response rate of $y_i = 1$, $\pi = \Pr(y_i = 1)$.

If we are interested in how this response rate changes as a function of $\mathbf{x}_i$, we can examine the conditional response rate given $\mathbf{x}_i$, $\pi(\mathbf{x}_i) = \Pr(y_i = 1 \mid \mathbf{x}_i)$, i.e., the probability that we obtain a response given the value of $\mathbf{x}_i$. For example, if $\mathbf{x}_i = x_i$ is a binary variable indicating two subgroups, say males ($x_i = 1$) and females ($x_i = 0$), then the response rates for each of the two subgroups are given by

$$\pi(1) = \Pr(y_i = 1 \mid x_i = 1), \quad \pi(0) = \Pr(y_i = 1 \mid x_i = 0).$$

If there is no association between y and x or if the response rate does not change as a function of x, then $\pi(1) = \pi(0)$. By displaying the data in a 2×2 table with x forming the rows and y forming the columns, testing such a row by column association has been discussed in Chapter 2. For example, we can use the chi-square statistic for a single 2×2 contingency table to test whether $\pi(1) = \pi(0)$ within our context. More generally, if x_i has more than two levels, say s levels indexed by j ($1 \leq j \leq s$), then the response rate for each level of j is given by

$$\pi(j) = \Pr(y_i = 1 \mid x_i = j), \quad 1 \leq j \leq s.$$

In this case, we may want to test whether the binomial proportions are constant across the row levels, i.e., the null hypothesis is

$$H_0 : \pi(1) = \cdots = \pi(s).$$

Pearson chi-square test may be applied in this case. If x_i is an ordinal variable, we can use the Cochran–Armitage test for trend alternatives.

In many real studies, variables arise in all shapes and forms, and it is quite common to have continuous independent variables. For example, in the DOS, we may also want to know whether the depression rate varies as a function of age. In this case, age is an independent variable. For such a continuous variable, it is generally not possible to display the data in a $s \times 2$ contingency table since in theory, age as a continuous outcome takes on infinitely many values in an interval, and as a result, none of the methods we have studied in Chapter 2 can be used to test the association between x_i and y_i. Regression models must be considered in order to be able to make inference about such a relationship involving a continuous row variable. More generally, regression models also make it possible to study the joint influence of multiple independent variables (continuous or otherwise) on the response rate.

Note that it is possible to tabulate data from a real study in an $s \times 2$ table even when x_i is a continuous independent variable, since the number of observed values of x_i is at most the same as the sample size and is thus finite. However, we cannot apply methods for $s \times 2$ tables for inference about the relationship between y_i and x_i. Inference is about the population from which the study subjects are sampled. For a genuine categorical variable with s levels, the row levels s in the $s \times 2$ are fixed and only the cell size changes when different samples of size n are drawn from the study population. Variations in the cell size across the cells in the table form the basis of sampling variability for inference in this case. For an intrinsic continuous variable such as age, however, the row levels will change from sample to sample, and this dynamic nature of the row levels invalidates any inference premised on treating the observed values of x_i as fixed row levels in an $s \times 2$ table.

If the two levels of y are coded by two different numbers such as 0 and 1, a linear regression model may be applied by treating the numerical representation as values of a continuous response. For example, the Cochran–Armitage test introduced in Chapter 2 was motivated by such a regression model. This is also equivalent to modeling the proportion of $y_i = 1$ using linear regression with x_i as a predictor. This approach, however, has a major flaw in that the fitted value, or the theoretical range of y_i, is $(-\infty, \infty)$, rather than the meaningful interval $(0, 1)$. Nowadays, the logistic model is the most popular regression model for binary responses.

4.1.2 Denfition of logistic models

The principal objective of a logistic model is to investigate the relationship between a binary response y and a set of independent variables $\mathbf{x} = (x_1, \ldots, x_p)^\top$. In many studies, a subset of $\mathbf{x}$ is often of primary interest, and the variables within such a subset are often called *explanatory* or *predictor variables*. The remaining variables in $\mathbf{x}$ are used to control for heterogeneity of the study sample such as differences in sociodemographic and clinical outcomes, and for this reason, they are called *covariates* or *confounding variables*. For example, in the hypothetical example used to illustrate Simpson's paradox when assessing success rates of some type of surgery of interest between two hospitals in Chapter 3, the type of hospital (good vs. bad) is of primary interest and as such is a predictor or explanatory variable. Patient's disease severity before surgery is used to stratify the sample so that comparisons of success rate can be made based on comparable subgroups of patients with similar disease severity prior to surgery. Thus, disease severity is of secondary interest and is a covariate or a confounding variable. However the variables in $\mathbf{x}$ are called, they are treated the same way when modeled in logistic regression. Thus, the differences in these variables only pertain to the interpretation of model parameters.

The logistic regression has the following general form:

$$y_i \mid \mathbf{x}_i \sim Bernoulli\,(\pi_i)\,, \quad \pi_i = \pi\,(\mathbf{x}_i) = \Pr\,(y_i = 1 \mid \mathbf{x}_i)$$

$$\mathrm{logit}(\pi_i) = \log\left(\frac{\pi_i}{1 - \pi_i}\right) = \beta_0 + \beta_1 x_{i1} + \ldots + \beta_p x_{ip} = \beta_0 + \boldsymbol{\beta}^\top \mathbf{x}_i, \qquad (4.1)$$

where $\boldsymbol{\beta} = (\beta_1, \ldots, \beta_p)^\top$ and $Bernoulli\,(\pi_i)$ denotes a Bernoulli random variable with success probability π_i. In the above, β_0 is the intercept and $\boldsymbol{\beta}$ is the vector of parameters for the independent variables. In the logistic model, we are modeling the effect of $\mathbf{x}$ on the response rate by relating $\mathrm{logit}(\pi)$ or log odds of response $\log\left(\frac{\pi}{1-\pi}\right)$ to a linear function of $\mathbf{x}$ of the form $\eta_i = \beta_0 + \boldsymbol{\beta}^\top \mathbf{x}$. Note that in this chapter all odds are defined as the probability of $y = 1$ to that of $y = 0$. This linear function η_i is also often called the *linear predictor*. Note that to compute $\mathbf{x}_i^\top \boldsymbol{\beta}$, it requires $\mathbf{x}_i$ to be a vector of numeric values, and thus it does not apply directly for categorical variables. For a binary covariate, we can represent the differential effect asserted by the two levels using a binary *indicator*, or *dummy variable*, taking the values 0 and 1. For a categorical covariate with k levels $(k > 2)$, we may designate one level as a reference and use $k-1$ binary indicators to represent the individual differences of each of the remaining $k - 1$ levels of the covariate relative to the selected reference. We will elaborate this approach as well as discuss other alternatives in Section 4.2.2.2.

Note that if you have studied linear regression models, it is interesting to compare them with logistic regression. For a continuous response y_i, the linear regression model has the following general form:

$$y_i \mid \mathbf{x}_i \sim N\left(\mu_i, \sigma^2\right), \quad \mu_i = \eta_i = \beta_0 + \mathbf{x}_i^\top \boldsymbol{\beta},$$

where $N\left(\mu_i, \sigma^2\right)$ denotes a normal distribution with mean μ_i and variance σ^2. The differences between the linear and logistic regression models lie in (a) the conditional distribution of y_i given $\mathbf{x}_i$ (random part) and (b) how the parameter of the conditional distribution, the mean of y_i given $\mathbf{x}_i$, is linked to the linear predictor η_i (deterministic part). In the linear model case, the random component is a normal distribution and the deterministic part is $\mu_i = E(y_i \mid \mathbf{x}_i)$, the mean of y_i given $\mathbf{x}_i$, which is linked to η_i by an identity function. Compared with the logistic model in (4.1), the random part is replaced by the Bernoulli distribution and the identity link in the deterministic component of the linear regression is replaced by the logit function. As we will discuss later in this chapter, these key steps are used to construct a wider class of models, known as generalized linear models, which in particular includes the linear and logistic regression models among their members.

As noted earlier, the deterministic component of the logistic regression models the effect of $\mathbf{x}$ on the mean of the response y through the following form:

$$\log\left(\frac{\pi}{1 - \pi}\right) = \beta_0 + \mathbf{x}^\top \boldsymbol{\beta}.$$

By exponentiating both sides, this part of the model can be equivalently written in terms of the odds of response as

$$\frac{\pi}{1 - \pi} = \exp(\beta_0 + \mathbf{x}^\top \boldsymbol{\beta}). \tag{4.2}$$

The response probability that $y = 1$ is then given by

$$\pi = \frac{\exp(\beta_0 + \mathbf{x}^\top \boldsymbol{\beta})}{1 + \exp(\beta_0 + \mathbf{x}^\top \boldsymbol{\beta})}.$$

Note that the expression for the response probability is responsible for the name of the logistic model because of its relationship to the following logistic curve or function:

$$F(x) = \frac{\exp(x)}{1 + \exp(x)}, \quad -\infty < x < \infty. \tag{4.3}$$

A (continuous) random variable with the above cumulative distribution function (CDF) is called the standard *logistic random variable*. Thus, the standard logistic variate has the probability density function (PDF):

$$f(x) = \frac{d}{dx}F(x) = \frac{\exp(x)}{(1 + \exp(x))^2}. \tag{4.4}$$

It is easy to check that $f(x)$ is symmetric about 0, i.e., $f(-x) = f(x)$, and $F(x)$ is S shaped and strictly increasing on $(-\infty, \infty)$ (see Problem 4.1).

Note that if all components of $\mathbf{x}_i$ are categorical, subjects with the same value of $\mathbf{x}_i$ are often pooled together so that each record in the data set indicates a unique value of $\mathbf{x}_i$, and the aggregated number of subjects with the event and size of the stratum corresponding to that value of $\mathbf{x}_i$. If we denote by y_i the number of subjects with the event and m_i the size of the stratum for the ith unique value of $\mathbf{x}_i$, then y_i follows a binomial model of size m_i with parameter π_i for the probability of success. Note that the independence assumption for the subjects within each stratum is critical to ensure the validity of modeling the aggregated data using the binomial model. For example, in the Combined Pharmacotherapies and Behavioral Interventions (COMBINE) study, one primary outcome for alcohol use, days of drinking for each subject over a certain time period such as a week or a month, is also presented in a binomial-like data format, with y_i indicating the number of events (days of drinking), m_i the total of number trials (number of days in the time period), and π_i the probability of success (probability of drinking). Since the events of drinking are aggregated over the same subject, they are no longer independent. Although binomial regression may still be a natural choice for modeling this outcome, the dependence in the occurrence of events will introduce extra-binomial variation, or *overdispersion* in y_i, rendering the binomial inappropriate for this outcome. We will discuss the notion of overdispersion and methods for addressing this issue in Section 4.4.2 and Chapter 6.

4.1.3   Parameter interpretation

The parameters in the logistic regression model are interpreted using odds ratios. Since $\beta_0 = \text{logit}(\Pr(y_i = 1 \mid \mathbf{x}_i = \mathbf{0}))$ or, equivalently, $\Pr(y_i = 1 \mid \mathbf{x}_i = \mathbf{0}) = \frac{\exp(\beta_0)}{1 + \exp(\beta_0)}$, the intercept represents the response rate. For example, $\beta_0 = 0$ implies that the response rate for subgroup $x_i = 0$, $\Pr(y_i = 1 \mid \mathbf{x}_i = \mathbf{0}) = \frac{1}{2}$. Usually this rate is unknown and needs to be estimated; an intercept should be included in those cases.

The parameters for the covariates indicate the association between the covariates and the response, controlling for other covariates in the model. Consider, for example, the covariate x_1 in the logistic model in (4.1), with β_1 denoting the coefficient of that covariate. Based on (4.2), the odds of response with $x_1 = a$ is $\exp\left(\beta_0 + \widetilde{\mathbf{x}}^\top \widetilde{\boldsymbol{\beta}} + \beta_1 a\right)$, where $\widetilde{\mathbf{x}}$ ($\widetilde{\boldsymbol{\beta}}$) denotes the vector $\mathbf{x}$ (the parameter vector $\boldsymbol{\beta}$) with the component of x_1 (β_1) removed. The odds of response for one unit increase in this covariate, i.e., $x_1 = a + 1$, with the remaining components of $\mathbf{x}$ held the same is $\exp\left(\beta_0 + \widetilde{\mathbf{x}}^\top \widetilde{\boldsymbol{\beta}} + \beta_1(a + 1)\right)$. Thus, the odds ratio of response per unit increase in x_1 is $\frac{\exp\left(\beta_0 + \mathbf{x}'^\top \boldsymbol{\beta}' + \beta_1(a+1)\right)}{\exp(\beta_0 + \mathbf{x}'^\top \boldsymbol{\beta}' + \beta_1 a)} = \exp(\beta_1)$.

Note that $\exp(\beta_1)$ is the odds ratio of the response $y = 1$ over $y = 0$ when modeling $\Pr(y_i = 1 \mid \mathbf{x}_i)$ as in (4.1). One can also model $\Pr(y_i = 0 \mid \mathbf{x}_i)$ using the logistic function. Because of the symmetry of the logistic function $f(x)$ in (4.4), it is easy to verify that this alternative model yields identical coefficients except for the reversal of the signs (see Problem 4.2).

Below, we illustrate with examples how the logistic model can be used to address hypotheses of interest concerning the relationship between y and x.

4.1.3.1   The 2×2 contingency table

To appreciate the odds ratio interpretation of the coefficient vector $\boldsymbol{\beta}$ for the independent variables in the logistic model, let us revisit our old friend, the 2×2 contingency table. In this relatively simple setting, the independent variable vector, $\mathbf{x}_i = x_i$, is binary. In the study of 2×2 contingency tables in Chapter 2, we discussed tests for association or independence between the row ($x = 0, 1$) and the column ($y = 0, 1$) variables and for equality of conditional response rates across the levels of x. We have shown that row and column independence is equivalent to the null hypothesis that the odds ratio equals 1. If a logistic regression model with a linear predictor $\eta = \beta_0 + \beta_1 x$ is applied to the 2×2 table, it follows from our earlier discussion that we have

$$\log(OR) = \beta_1 \quad \text{or} \quad OR = \exp(\beta_1),$$

where β_1 is the parameter of the row variable x. Thus, $OR = 1$ if and only if $\beta_1 = 0$. This shows that we can also state the above null hypothesis in terms of the parameter β_1 of the logistic model as follows:

$$H_0 : \beta_1 = 0.$$

We also discussed directions of association in the study of 2×2 tables. With the logistic model, we can easily assess the direction of association when x and y are correlated. More specifically, if the null of no association is rejected, we can then proceed to examine how the relationship changes as the level of x changes by the value of β_1, especially its sign.

Since $OR = \exp(\beta_1)$, it immediately follows that $OR > 1$ if and only if $\beta_1 > 0$, and $OR < 1$ if and only if $\beta_1 < 0$. Thus, $\beta_1 > 0$ ($\beta_1 < 0$) indicates that the group with $x = 1$ (exposed) has a higher (lower) response (disease) rate than the group with $x = 0$ (non-exposed). In addition, we have shown in Chapter 2 that when comparing the response rate of the $x = 1$ group to the $x = 0$ group, the relative risk $RR > 1$ if and only if $OR > 1$ and

$RR < 1$ if and only if $OR < 1$. It follows that $\beta_1 > 0$ ($\beta_1 < 0$) signifies that the group with $x = 1$ confers a higher (lower) risk for the response of y than the group with $x = 0$.

4.1.3.2   The $s \times 2$ contingency table

If x is a categorical covariate with k levels indexed by j, we can designate one level, say $j = 1$, as a reference and use $k - 1$ binary indicators to denote the difference from each of the remaining levels relative to this selected reference. Specifically, let $d_j = 1$ if $x = j$ and 0 if otherwise ($1 \leq j \leq k$). If a logistic regression model with a linear predictor $\eta = \beta_0 + \beta_1 d_1 + \cdots + \beta_{k-1} d_{k-1}$ is applied to the $s \times 2$ table, it follows from our earlier discussion that β_i gives the log odds ratio between $x = i$ and the reference level ($x = k$), $i = 1, \ldots, k - 1$. Thus, the null hypothesis of row (x) by column (y) independence can be equivalently expressed as

$$H_0 : \beta_1 = \cdots = \beta_{k-1} = 0. \tag{4.5}$$

Testing a composite null involving multiple equalities is more complex, and we will discuss this in Section 4.2.2.

In the above, all the levels of the categorical variable x are treated equally; i.e., x is treated as a nominal variable. However, if we would like to treat x as ordinal and test alternatives involving the ordinal scale of x such as some monotone trend of the proportions of y as a function of x, we may assign some scores to each of the levels of x, treat x as a continuous covariate, and apply the methods for testing hypotheses involving continuous covariates to be discussed next. Note that in this case, the null hypothesis reduces to a single parameter set to 0, and thus there is no need to use dummy variables.

4.1.3.3   Continuous covariate and invariance to linear transformation

For a continuous covariate x, the odds ratio interpretation of the parameter still applies. As discussed earlier, the coefficient β_1 in the logistic model with a linear predictor $\eta = \beta_0 + \beta_1 x$ measures the effect on the response expressed in terms of the odds ratio for every unit increase in x. However, since the value of a continuous variable is defined and meaningful with respect to a particular scale used, we must be careful when interpreting β_1, especially when different scales are involved. For example, body weight is measured in pounds in the United States, but in kilograms in many other countries in the world. Although the two weight scales are not identical, they are related to each other through a linear transformation as with many other different scales for measuring the same concept such as distance. Let us see how such a linear transformation will affect the parameter and its related odds ratio.

Suppose a new scale is applied to x, which not only shifts the original variable by an amount a but also scales it back by a factor $\frac{1}{k}$, i.e., $x' = a + \frac{1}{k}x$, or $x = kx' - ka$. In this case, $x' = 0$ corresponds to $x = -ka$, and one unit increase in x' results in a change of k in the original scale. The exponential of the coefficient of the new variable x' in the logistic model or the odds ratio per unit change in x' is

$$\exp\left(\beta_1'\right) = \frac{Odd\left(x' = 1\right)}{Odd(x' = 0)} = \frac{Odd\left(x = k - ka\right)}{Odd\left(x = -ka\right)} = \exp\left(k\beta_1\right).$$

Thus, $\beta_1' = k\beta_1$. This implies that $\beta_1' = 0$ if and only if $\beta_1 = 0$, as it should be since when two constructs are independent of each other, it does not matter how they are scaled. Also, if x has an effect on the response ($\beta_1 \neq 0$), the new scale will give rise to a coefficient k times as large.

For this reason, it is important to pay close attention to the scale used when interpreting the value of the odds ratio. However, the estimated variance of the estimate of a parameter will change along with the scale in a way that will not affect the distribution of the standardized estimate (see Section 4.4.1.5). This invariance property ensures that we obtain the same inference (same level of significance) regardless of the scales used.

Note that when interpreting the coefficient of one independent variable (x_1 in the above), we hold the others fixed ($\tilde{\mathbf{x}}$). Strictly speaking, this is only possible when all the independent variables are "independent." In most real studies, independent variables are usually correlated to some degree. However, as long as the correlation is not too high, such an interpretation is still sensible. Note also that the models above for the DOS data are for illustration purposes. If important covariates are missing, even if statistically nonsignificant themselves, these models may give biased estimates because of Simpson's paradox. We discuss techniques to build models in a systematic fashion in Chapter 7.

4.1.4   Invariance to study designs

Recall that for 2×2 tables, we illustrated with examples the importance of differentiating between perspective and retrospective case-control study designs to ensure meaningful interpretations of estimates constructed based on the cell count, though inference for association is independent of such design differences. This issue also arises when applying the logistic regression model. In this section, we discuss how interpretations of model parameters are different under the two types of study designs. For convenience and notational brevity, we illustrate the considerations for 2×2 tables.

4.1.4.1   Prospective study

Prospective study designs are the most popular in many areas of biomedical and psychosocial research, including randomized, controlled clinical trials (experimental) and cohort (observational) studies. For example, in a cohort study examining the effect of some type of exposure on certain diseases of interest, the levels of the exposure variable are fixed at the beginning and the disease status is ascertained at the end of the study. If we let $x \ (= 0, 1)$ denote the exposure and $y \ (= 0, 1)$ the disease status and display the data in a 2×2 contingency table (Table 4.1), then the row totals (number of exposed and nonexposed subjects) are fixed, while the column totals (number of disease status) are random.

TABLE 4.1:   A 2×2 contingency table for a prospective study

Exposure (x)	Response (y)		Total
	$y = 0$	$y = 1$	
$x = 0$	n_{00}	n_{01}	n_{0+} (fixed)
$x = 1$	n_{10}	n_{11}	n_{1+} (fixed)
Total	n_{+0}	n_{+1}	n

Let $\pi(x) = \Pr(y = 1 \mid x)$ denote the conditional probability of disease given the exposure status x; the odds ratio of disease by comparing the exposure to the nonexposure group (the odds of disease given exposure $x = 1$ to the odds of disease given no exposure $x = 0$) is given by

$$OR = \frac{\pi(1)/(1 - \pi(1))}{\pi(0)/(1 - \pi(0))}.$$

From the logistic regression model,

$$\text{logit} \left[\pi(x) \right] = \text{logit} \left[\Pr(y = 1 \mid x) \right] = \log \left(\frac{\pi(x)}{1 - \pi(x)} \right) = \beta_0 + \beta_1 x.$$

Thus, the log odds of disease for the nonexposed and exposed groups are $\text{logit}(\pi(0)) = \beta_0$ and $\text{logit}(\pi(1)) = \beta_0 + \beta_1$. Hence,

$$\log(OR) = \log \left(\frac{\pi(1)/(1 - \pi(1))}{\pi(0)/(1 - \pi(0))} \right) = \text{logit}(\pi(1)) - \text{logit}(\pi(0)) = (\beta_0 + \beta_1) - \beta_0 = \beta_1,$$

it follows that the parameter β_1 is the log odds ratio or, equivalently, $\exp(\beta_1)$ is the odds ratio for comparing the exposed and nonexposed groups.

4.1.4.2   Case-control retrospective study

In addition to prospective study designs, *case-control retrospective studies* are sometimes conducted especially for rare diseases and/or binary outcomes with extremely low frequency of events of interest. For such studies, controlled clinical trials or cohort studies may not yield a sufficiently large event rate to ensure adequate power within a reasonable study period. Although increasing sample size is an option, logistical considerations and prohibitive cost may argue against such a trial and, in some cases, may make it practically impossible to conduct such a trial.

In a case-control study on examining the relationship between some exposure variable and a disease of interest, we first select a random sample from a population of diseased subjects or cases. Such a population is usually retrospectively identified by chart reviews of patients' medical histories and records. We then randomly select a sample of disease-free individuals or controls from a nondiseased population based on similar sociodemographic and clinical variables. When compared to prospective study designs, the column totals (diseased and nondiseased subjects) of Table 4.1 above are fixed, but the row totals (exposed and nonexposed subjects) are random for such a study design.

Let z be a dummy variable denoting whether an individual is sampled or not from a population of interest. Then, the sampling probabilities for the diseased and nondiseased samples are

$$p_1 = \Pr(z = 1 \mid y = 1), \quad p_0 = \Pr(z = 1 \mid y = 0).$$

Let $a_i = \Pr(z = 1 \mid y = i, x) \Pr(y = i \mid x)$, $i = 0, 1$. By Bayes' theorem (see, for example, Rozanov (1977)), the disease probability among the sampled individuals with exposure status x is

$$\Pr[y = 1 \mid z = 1, x] = \frac{a_1}{a_1 + a_0} = \frac{p_1 \exp(\beta_0 + \beta_1 x)}{p_0 + p_1 \exp(\beta_0 + \beta_1 x)} = \frac{\exp(\beta_0^* + \beta_1 x)}{1 + \exp(\beta_0^* + \beta_1 x)},$$

where $\beta_0^* = \beta_0 + \log(p_1/p_0)$. We can see that the logistic model for the retrospective case-control study sample has the same coefficient β_1 as for the prospective study sample, albeit with a different intercept. For this reason, the logistic model is also widely used in case-control studies to assess the relationship between exposure and disease variables.

When applied to case-control study samples, the intercept of the logistic model is a function of the sampling probabilities of the selected cases and controls and is typically of no interest. Such a parameter is called a *nuisance parameter* in the nomenclature of statistics.

4.1.5 Simpson's paradox revisited

As we discussed in Chapter 3, the unexpected finding in the hypothetical example is the result of confounding bias in having more seriously ill patients in the good hospital (Hospital A) prior to surgery. Such an association reversal phenomenon or Simpson's paradox is of fundamental importance in epidemiological research since aggregated data are commonplace in such studies. Association reversal means that the direction of association between two variables is changed by aggregating data over a covariate. Samuels (1993) gave necessary and sufficient conditions for Simpson's paradox within a more general context. The logistic model can be used to address such bias.

For the hypothetical example, the bias is caused by having more seriously ill patients in the good hospital (Hospital A). Since sicker patients typically have a lower success rate, this imbalance in the patients' distribution between the two hospitals decreased the success rate for Hospital A. If we stratify patients by the level of severity and make comparison within each of the strata, we can control for this confounding effect. For the group of less severe patients, the odds ratio of success of Hospital A to Hospital B is

$$\widehat{OR}_{AB} = \frac{18 \times 16}{2 \times 64} = 2.25,$$

indicating a higher success rate for Hospital A than for Hospital B. Similarly, within the group of more severe patients, the odds ratio is

$$\widehat{OR}_{AB} = \frac{32 \times 16}{48 \times 4} = 2.667,$$

which again suggests that Hospital A has a higher success rate than Hospital B. Thus, after stratifying on patient's severity of illness, the odds ratios are all in the expected direction for both strata. We can also control for such confounding effects more elegantly using logistic regression.

For the unstratified data, let $x = 1$ ($x = 0$) for hospital A (B); then, the logistic model is given by

$$\text{logit}(\pi_i) = \log\left(\frac{\pi_i}{1 - \pi_i}\right) = \beta_0 + \beta_1 x_i.$$

Thus, the log odds of success for the good and bad hospitals based on this model are

$$\text{logit}(\pi(1)) = \log\left(\frac{\pi(1)}{1 - \pi(1)}\right) = \beta_0 + \beta_1, \ \text{logit}(\pi(0)) = \log\left(\frac{\pi(0)}{1 - \pi(0)}\right) = \beta_0.$$

As discussed, the imbalance in patients' distribution between the two hospitals with respect to disease severity prior to surgery led to a negative β_1 $\left(\hat{\beta}_1 = -0.753\right)$ or a lower odds of success for Hospital A than for Hospital B ($\text{logit}(\pi(1)) < \text{logit}(\pi(0))$), giving rise to the conclusion that the good hospital had a lower success rate.

To control for this bias in the logistic model, let $z = 1$ ($z = 0$) for patients with less (more) disease condition prior to surgery. By including z as a covariate in the original model, we have

$$\text{logit}(\pi_i) = \log\left(\frac{\pi_i}{1 - \pi_i}\right) = \beta_0 + \beta_1 x_i + \beta_2 z_i. \tag{4.6}$$

In this new model, the bias is controlled by the covariate z_i. More specifically, we can now compute the log odds of success separately for the less and more severe patients within each hospital. For less severe patients ($z_i = 1$):

$$\text{logit}(\pi(x = 1, z = 1)) = \beta_0 + \beta_1 + \beta_2, \ \text{logit}(\pi(x = 0, z = 1)) = \beta_0 + \beta_2.$$

For more severe patients $(z_i = 0)$,

$$\text{logit}\left(\pi\left(x = 1, z = 0\right)\right) = \beta_0 + \beta_1, \ \text{logit}\left(\pi\left(x = 0, z = 0\right)\right) = \beta_0.$$

Thus, the extra parameter β_2 allows us to model the difference in success rates between the two groups of different disease severity within each hospital, and as a result, it does not matter whether the distribution of disease severity is the same between the two hospitals.

Under this new model, we can compute two odds ratios (Hospital A to Hospital B) with one for each stratified patient group. For the less severe patients, $(z_i = 1)$: $\log\left(OR_1\right) = (\beta_0 + \beta_1 + \beta_2) - (\beta_0 + \beta_2) = \beta_1$, and for the more severe patients, $(z_i = 0)$: $\log\left(OR_0\right) = (\beta_0 + \beta_1) - \beta_0 = \beta_1$.

Thus, under this revised model, β_1 has the same interpretation as being the log odds ratio as in the original model. However, unlike the original model, β_1 is no longer subject to the confounding caused by imbalance in the distribution of patients' disease severity between the two hospitals since this is accounted for by the parameter β_2.

It is interesting to note that under the assumed logistic model, the log odds ratio β_1 is the same for both severity strata. Given the similar estimates of odds ratios for the two strata, 2.25 and 2.667, this assumption seems reasonable in this hypothetical example. We will discuss how to test the appropriateness of such a common odds ratio assumption and what to do if this assumption fails to describe the data, in the next section.

Note that we discussed the analogy of controlling for such confounding effect when modeling continuous responses using ANCOVA. Within the context of the hypothetical example, suppose that y is a continuous response. Then, the ANCOVA model has the following form:

$$y_i \mid \mathbf{x}_i \sim N\left(\mu_i, \sigma^2\right), \quad \mu_i = \beta_0 + \beta_1 x_i + \beta_2 z_i.$$

Within the context of ANCOVA, the mean responses μ_i of y_i for the two hospitals within each patient's stratum defined by z_i are given by

$$\text{for } z_i = 1, \quad \mu_i = \begin{cases} \mu_{1a} = \beta_0 + \beta_1 + \beta_2 \text{ if Hosp A} \\ \mu_{1b} = \beta_0 + \beta_2 \qquad \text{if Hosp B} \end{cases}$$

$$\text{for } z_i = 0, \quad \mu_i = \begin{cases} \mu_{0a} = \beta_0 + \beta_1 \ \text{if Hosp A} \\ \mu_{0b} = \beta_0 \qquad \text{if Hosp B.} \end{cases}$$

The mean responses for the two hospitals depend on the covariate z_i. As in the logistic model case, the parameter accounts for differences in mean response between the two patients' strata. Also, similar to the logistic model, the difference in mean responses between the two hospitals is the same across the patient's strata:

$$\mu_{1a} - \mu_{1b} = \mu_{0a} - \mu_{0b} = \beta_1.$$

The issue of model adequacy also arises in ANCOVA. We discuss how to address this issue for both logistic and ANCOVA models next.

4.1.6   Breslow–Day test and moderation analysis

In the hypothetical example, what if the difference in surgery success rates between the hospitals changes across patients' disease severity. For example, one of the reasons for Hospital A to have more seriously ill patients in the first place is the belief that such patients may have a higher success rate in a good hospital. In this case, the odds ratios OR_{AB} (Hospital A to Hospital B) may be different between the less and more severe groups of patients. In other words, disease severity prior to surgery will modify the effect of hospital type (good vs. bad) on the surgery success rate. Such a covariate is called an effect modifier, or a moderator, and the effect it exerts on the relationship between the predictor (hospital type) and response (surgery success) is called a *moderation effect*. Moderation effect and associated moderators are of great interest in many areas of research, especially in intervention studies.

In drug research and development, it is of great interest to know whether a drug will have varying efficacy for different patients. For example, many antidepressants only work for a selected patient population with varying degrees of efficacy, and it is quite common for a patient to be on two or more different antidepressants before the right medication can be found. In cancer intervention studies, disease staging (I, II, III, and IV) is often a moderator of treatment of cancer since it is more effective for patients in the early stages of cancer. In depression research, the effect of intervention may be modified by many factors such as age, race, gender, drug use, comorbid medical problems, etc.

In the study of sets of 2×2 tables in Chapter 2, we discussed the Breslow–Day statistic for testing homogeneity of odds ratios across the different tables. This test can be applied to the current context to assess moderation effects. Alternatively, we can also use the logistic model to examine such a hypothesis.

Consider again the hypothetical example on comparing surgery success rates between Hospital A (good) and Hospital B (bad). We applied the Breslow–Day test in Chapter 3 to Table 3.2 and found that there was no significant difference in odds ratio between the two severity strata. Thus, disease severity does not moderate (or modify) the effect of hospital type on surgery success. Alternatively, we can test for a moderation effect of severity by using the following logistic model:

$$\text{logit}(\pi_i) = \log \left(\frac{\pi_i}{1 - \pi_i} \right) = \beta_0 + \beta_1 x_i + \beta_2 z_i + \beta_3 x_i z_i. \tag{4.7}$$

If $\beta_3 = 0$, the above reduces to the logistic model in (4.6) we used to address the confounding due to imbalance in the distribution of disease severity between the two hospitals with respect to disease severity prior to surgery. In this reduced model case, we have shown that the (log) odds ratios for the two patient groups (less vs. more severe) are the same.

For the less severe patients $(z_i = 1)$, the log odds for the two hospitals are

$$\text{logit}\left(\pi\left(x = 1, z = 1\right)\right) = \beta_0 + \beta_1 + \beta_2 + \beta_3, \ \text{logit}\left(\pi\left(x = 0, z = 1\right)\right) = \beta_0 + \beta_2.$$

For the more severe patients $(z_i = 0)$, the log odds for the two hospitals are

$$\text{logit}\left(\pi\left(x = 1, z = 0\right)\right) = \beta_0 + \beta_1, \ \text{logit}\left(\pi\left(x = 0, z = 0\right)\right) = \beta_0.$$

Under this new model, we can compute the two log odds ratios of Hospital A to Hospital B for the two patient strata by

$$\log\left(OR_1\right) = (\beta_0 + \beta_1 + \beta_2 + \beta_3) - (\beta_0 + \beta_2) = \beta_1 + \beta_3,$$

for the less severe patients $(z_i = 1)$, and

$$\log\left(OR_0\right) = (\beta_0 + \beta_1) - \beta_0 = \beta_1$$

for the more severe patients $(z_i = 0)$. It is seen that unless $\beta_3 = 0$, the (log) odds ratios of Hospital A to Hospital B are assumed to be different between the two strata. Thus, under the logistic regression model, we can assess whether there is a moderation effect by testing the null $H_0 : \beta_3 = 0$. Using methods we will discuss in the next section, we obtain the test statistic 0.029 and p-value 0.865, indicating no moderation effect by disease severity.

In this example, the logistic model for moderation is created by adding a covariate (patient severity) by predictor (hospital type) interaction term. This is also the general strategy for testing for a moderation effect. For example, we discussed how to address confounding bias for the hypothetical example if y is a continuous response using the following ANCOVA:

$$y_i \mid \mathbf{x}_i \sim N\left(\mu_i, \sigma^2\right), \quad \mu_i = \beta_0 + \beta_1 x_i + \beta_2 z_i. \tag{4.8}$$

As in the logistic model case, under this ANCOVA, the difference in mean response between the two hospitals is the same across the patient's strata (see Chapter 3, Figure 3.1): $\mu_{1a} - \mu_{1b} = \mu_{0a} - \mu_{0b} = \beta_1$. When this difference differs between the two severity strata, ANCOVA will not apply. As noted earlier in the discussion of moderation, such a difference may not be so far fetched since it is reasonable to expect that patients with less severe disease may have the same outcomes between the two hospitals, but those with severe disease conditions may have better outcomes in the good hospital in the hypothetical study. The ANCOVA model above can be modified to account for such differential response rates, or a moderation effect by z_i. As in the logistic model case, this model is created by adding an interaction term between z_i and x_i to the ANCOVA model in (4.8):

$$y_i \mid \mathbf{x}_i \sim N\left(\mu_i, \sigma^2\right), \quad \mu_i = \beta_0 + \beta_1 x_i + \beta_2 z_i + \beta_3 x_i z_i.$$

The mean response μ_i of y_i for the two hospitals within each stratum of z_i is given by

$$z_i = 1 \quad \mu_i = \begin{cases} \mu_{1a} = \beta_0 + \beta_1 + \beta_2 + \beta_3 & \text{if Hosp A} \\ \mu_{1b} = \beta_0 + \beta_2 & \text{if Hosp B,} \end{cases}$$

$$z_i = 0 \quad \mu_i = \begin{cases} \mu_{0a} = \beta_0 + \beta_1 & \text{if Hosp A} \\ \mu_{0b} = \beta_0 & \text{if Hosp B.} \end{cases}$$

The differences in mean response between the two hospitals are given by

$$\mu_{1a} - \mu_{1b} = \beta_1 + \beta_3, \quad \mu_{0a} - \mu_{0b} = \beta_1.$$

Thus, unlike the original additive ANCOVA, the difference between the two hospitals now is a function of disease severity. As in the logistic model case, we can ascertain whether there is a moderation effect by testing the null $H_0 : \beta_3 = 0$.

Example 4.1

In the DOS, we found a significant relationship between gender and depression diagnosis. We are now interested in testing whether such a relationship is moderated by medical burdens (cumulative illnesses). Medical burdens are believed to be associated with depression, and we want to know whether this association is the same between males and females, i.e., whether medical burdens modify the risk of depression between males and females.

As before, let y_i denote the binary variable of depression diagnosis and x_i a binary variable for gender with $x_i = 1$ for male and $x_i = 0$ for female. We dichotomize the continuous medical burden measure (CIRS) using a cut-point 6 so that $z_i = 1$ if CIRS > 6 and $z_i = 0$ otherwise. Then, the logistic model is given by

$$\text{logit}(\pi_i) = \log\left(\frac{\pi_i}{1 - \pi_i}\right) = \beta_0 + \beta_1 x_i + \beta_2 z_i + \beta_3 x_i z_i.$$

We tested the null $H_0 : \beta_3 = 0$ for a potential moderation effect by CIRS. The p-value for the Wald test is 0.0019 (inference about H_0 will be discussed in the next section), indicating that CIRS has a moderating effect on the association between gender and depression.

In this example, we do not have to dichotomize CIRS to test for moderation. We can also use it as a continuous variable and the logistic model has the same form except that z_i is continuous. Again, we can test the null $H_0 : \beta_3 = 0$ to assess whether z_i is a moderator. The p-value of the Wald test is 0.0017. ▯

4.2 Inference About Model Parameters

In this section, we discuss how to estimate and make inference about the parameters. We start with the maximum likelihood estimate (MLE), which is the most popular type of estimate for almost all statistical models because of its nice asymptotic (large sample) properties. For small to moderate samples, asymptotic theory may be unreliable in terms of providing valid inference. More importantly, a phenomenon known as data separation may occur, in which case the MLE does not exist. Thus, we also introduce two alternative estimates, the conditional exact estimate and bias reduced estimate, to help address small samples as well as data separation issues.

4.2.1 Maximum likelihood estimate

As with most parametric statistical models, the method of maximum likelihood is the most popular to estimate and make inference about the parameters of logistic regression model. Consider a sample of n subjects, and let $y_i \, (= 0, 1)$ be a binary response of interest and $\mathbf{x}_i = (x_{i0}, x_{i1}, \ldots, x_{ip})^\top$ a vector of independent variables ($i = 1, \ldots, n$). For notational brevity, we assume $x_{i0} \equiv 1$, i.e., x_{i0} designates the intercept term so that the logistic model can be simply expressed as

$$y_i \mid \mathbf{x}_i \sim Bernoulli\,(\pi_i)\,, \ \pi_i = \Pr\,(y_i = 1 \mid \mathbf{x}_i)$$

$$\mathrm{logit}(\pi_i) = \log\left(\frac{\pi_i}{1 - \pi_i}\right) = \beta_0 x_{i0} + \beta_1 x_{i1} + \ldots + \beta_p x_{ip} = \mathbf{x}_i^\top \boldsymbol{\beta},$$

where $\boldsymbol{\beta} = (\beta_0, \beta_1, \ldots, \beta_p)^\top$ is the vector of parameters.

The likelihood function is

$$L(\boldsymbol{\beta}) = \prod_{i=1}^{n} \left[\pi_i^{y_i} (1 - \pi_i)^{1-y_i} \right] = \prod_{i=1}^{n} \left[\left(\frac{\pi_i}{1 - \pi_i}\right)^{y_i} (1 - \pi_i) \right]$$

$$= \exp\left(\sum_{i=1}^{n} y_i \mathbf{x}_i^\top \boldsymbol{\beta}\right) \prod_{i=1}^{n} \frac{1}{1 + \exp(\mathbf{x}_i^\top \boldsymbol{\beta})}.$$

The log-likelihood function is

$$l(\boldsymbol{\beta}) = \sum_{i=1}^{n} \left[y_i \log \pi_i + (1 - y_i) \log(1 - \pi_i) \right].$$

Note that $\pi_i = \Pr\,(y_i = 1 \mid \mathbf{x}_i)$ is a function of $\boldsymbol{\beta}$.

Since $\partial \pi_i / \partial \boldsymbol{\beta} = \pi_i (1 - \pi_i) \mathbf{x}_i$, the score equation is given by

$$S(\boldsymbol{\beta}) = \frac{\partial}{\partial \boldsymbol{\beta}} l(\boldsymbol{\beta}) = \sum_{i=1}^{n} (y_i - \pi_i) \mathbf{x}_i = \mathbf{0}.$$

As the second-order derivative of the log-likelihood given by

$$\frac{\partial^2}{\partial \boldsymbol{\beta} \partial \boldsymbol{\beta}^\top} l(\boldsymbol{\beta}) = -\sum_{i=1}^{n} \pi_i (1 - \pi_i) \mathbf{x}_i \mathbf{x}_i^\top$$

is negative definite, there is a unique solution $\widehat{\beta}$ to the score equation, which is the MLE of β (if exists). Although the score equation cannot be solved in closed form, $\widehat{\beta}$ can be numerically computed by the Newton–Raphson method.

By the theory of maximum likelihood, $\widehat{\beta}$ has asymptotically a normal distribution:

$$\widehat{\beta} \sim_a N\left(\beta, \mathbf{I}_n^{-1}(\beta)\right), \quad \mathbf{I}_n(\beta) = -\frac{\partial^2}{\partial\beta\partial\beta^\top} l(\beta),$$

where $\mathbf{I}_n(\beta)$ is the observed information matrix and is estimated by $\mathbf{I}_n\left(\widehat{\beta}\right)$. Thus, the asymptotic variance of $\widehat{\beta}$ is the inverse of the observed information matrix.

Example 4.2

For the hypothetical example given in Table 3.2 in Chapter 3, we model the surgery success rate with the logistic model:

$$\text{logit}(\pi_i) = \log\left(\frac{\pi_i}{1 - \pi_i}\right) = \beta_0 + \beta_1 x_i + \beta_2 z_i + \beta_3 x_i z_i, \tag{4.9}$$

where x and z are defined as in Section 4.1.5.

The estimates for the parameters are as follows:

Parameter	Estimate	Standard Error	P-value
β_0	−1.3863	0.5590	0.0131
β_1	0.9808	0.6038	0.1043
β_2	2.7726	0.6250	< 0.0001
β_3	−0.1699	0.9991	0.8650

Based on the estimated parameters, we can calculate the fitted probabilities. For example, when $x_i = z_i = 0$ the estimated response rate is

$$\Pr\left(y_i = 1 \mid x_i = 0, z_i = 0\right) = \frac{\exp\left(-1.3863\right)}{1 + \exp\left(-1.3863\right)} = 0.2,$$

and when $x_i = z_i = 1$ the estimated response rate is

$$\Pr\left(y_i = 1 \mid x_i = 1, z_i = 1\right) = \frac{\exp\left(-1.3863 + 0.9808 + 2.7726 - 0.1699\right)}{1 + \exp\left(-1.3863 + 0.9808 + 2.7726 - 0.1699\right)} = 0.9.$$

The log odds ratios for the two hospitals for each of the severity strata are $\log(OR_1) = \widehat{\beta}_1 + \widehat{\beta}_3 = 0.9808 - 0.1699 = 0.8109$ for less severe patients ($z_i = 1$) and $\log(OR_0) = \widehat{\beta}_1 = 0.9808$ for more severe patients ($z_i = 0$). Thus, the odds ratios for the two patients groups are 2.25 and 2.667, i.e., Hospital A has higher odds of surgical success in each severity group.

We can test the null $H_0 : \beta_3 = 0$, to assess potential moderation effect by z. Since the p-value = 0.865, the interaction is not significant, and thus there is no evidence for any moderating effect by z_i. Note that we obtained a similar conclusion based on the Breslow–Day test. Since the p-value indicates that there is no strong evidence against the null, we may remove the interaction term in the model in (4.9) by setting $\beta_3 = 0$. Under this *additive effect* model, the estimates become

Variable	Estimate	Standard Error	P-value
Hospital (x)	0.9206	0.4829	0.0566
Severity (z)	2.7081	0.4904	< 0.0001

The odds ratio of Hospital A to Hospital B is

$$OR_{AB} = \exp(\widehat{\beta}_1) = \exp(0.9206) = 2.51,$$

a constant across the strata of disease severity, in this model. One may compare this common odds ratio estimate with the Mantel–Haenszel estimate we obtained in Section 3.2.2. □

4.2.2 General linear hypotheses

So far, we have discussed inference concerning coefficients associated with a binary or continuous independent variable. Hypotheses concerning such coefficients are often expressed as

$$H_0 : \beta = 0 \quad \text{vs.} \quad H_a : \beta \neq 0, \tag{4.10}$$

where β is the coefficient associated with a binary or continuous independent variable in the model. Many hypotheses of interest, however, cannot be expressed in this simple form. For example, in the DOS, information is collected regarding the marital status of each subject, which is a risk factor for depression. This variable for marital status, MARITAL, has six nominal outcomes with 1 for single (never married), 2 for married and living with spouse, 3 for married and not living with a spouse, 4 for legally separated, 5 for divorced, and 6 for widowed. If we are interested in testing whether this variable is associated with a depression diagnosis, we cannot express the hypothesis in the form in (4.10), since we need to use more than one parameter to represent the different levels of this variable. Because of the small size for some levels, we combine some levels to obtain a simpler variable for marital status. MS is defined by grouping the six levels of MARITAL into three risk categories: MS = 1 for married and living with spouse (MARITAL = 2), MS = 2 for widowed (MARITAL = 6), and MS = 3 for Other (MARITAL = 1, 3, 4, 5). To represent the three levels of this new MS variable, let x_{ik} be an indicator for the group of subjects with MS = k, i.e.,

$$x_{i1} = \begin{cases} 1 \text{ if MS} = 1 \\ 0 \text{ if otherwise} \end{cases}, \quad x_{i2} = \begin{cases} 1 \text{ if MS} = 2 \\ 0 \text{ if otherwise} \end{cases}, \quad x_{i3} = \begin{cases} 1 \text{ if MS} = 3 \\ 0 \text{ if otherwise} \end{cases}.$$

By designating one level, say MS = 3, as the reference, we have the following logistic model:

$$\text{logit}(\pi_i) = \log\left(\frac{\pi_i}{1 - \pi_i}\right) = \beta_0 + \beta_1 x_{i1} + \beta_2 x_{i2}. \tag{4.11}$$

When dealing with a multi-level nominal variable, we first perform a global test to see whether there is an overall difference among the levels of the variable. Unfortunately, we cannot use the simple form in (4.10) to state such a hypothesis. For example, if the logistic model (4.11) is applied to assess the association between the three-level MS variable with depression in the DOS data, then the null can be expressed in a composite form as

$$H_0 : \beta_1 = 0, \quad \beta_2 = 0. \tag{4.12}$$

The null involves more than one parameter in the form of a linear combination.

4.2.2.1 Composite linear hypothesis

In general, a linear hypothesis can be expressed more elegantly in a matrix form as

$$H_0 : K\boldsymbol{\beta} = \mathbf{b}, \tag{4.13}$$

where K is a matrix of dimensions $l \times q$ and $\mathbf{b}$ an $l \times 1$ column vector of known constants with l indicating the number of equations and q the dimension of $\boldsymbol{\beta}$. If there are no redundant equations in the null hypothesis, then the rows of K are linear independent and K is of full rank. In the following, we assume this is the case. For example, for the three-level MS variable, we can express the null hypothesis of no MS and depression association as

$$K\boldsymbol{\beta} = \mathbf{0}, \quad K = \begin{pmatrix} 0 & 1 & 0 \\ 0 & 0 & 1 \end{pmatrix}, \quad \boldsymbol{\beta} = (\beta_0, \beta_1, \beta_2)^\top. \tag{4.14}$$

The most popular tests for linear hypotheses use the *Wald*, *score*, and *likelihood ratio* statistics. We briefly review these tests below.

Wald statistics. If $\widehat{\boldsymbol{\beta}} \sim_a N\left(\boldsymbol{\beta}, \frac{1}{n}\Sigma_{\boldsymbol{\beta}}\right)$, then it follows from properties of the multivariate normal distribution that

$$K\widehat{\boldsymbol{\beta}} \sim_a N\left(K\boldsymbol{\beta}, \frac{1}{n}K\Sigma_{\boldsymbol{\beta}}K^\top\right).$$

Under the null hypothesis that $K\boldsymbol{\beta} = \mathbf{b}$, $K\widehat{\boldsymbol{\beta}} \sim N\left(\mathbf{b}, \frac{1}{n}K\Sigma_{\boldsymbol{\beta}}K^\top\right)$. Because K is full rank, $K\Sigma_{\boldsymbol{\beta}}K^\top$ is invertible (see Problem 4.3). The statistic

$$Q_n^2 = n\left(K\widehat{\boldsymbol{\beta}} - \mathbf{b}\right)^\top \left(K\Sigma_{\boldsymbol{\beta}}K^\top\right)^{-1}\left(K\widehat{\boldsymbol{\beta}} - \mathbf{b}\right) \tag{4.15}$$

follows asymptotically a chi-square distribution with l degrees of freedom (χ_l^2), where l is the rank of K, if the null hypothesis is true. This statistic is known as a *Wald statistic*.

The Wald statistic does not depend on the specific forms of the linear hypothesis in (4.13), as long as they are equivalent (see Problem 4.4). For example, we may use K in (4.14) to test the null hypothesis (4.12). Because of this invariance property, we will obtain the same result if $K = \begin{pmatrix} 0 & -1 & 1 \\ 0 & 1 & 0 \end{pmatrix}$ (equations for no difference between MS = 1 and 2, and between MS =1 and 3) or $K' = \begin{pmatrix} 0 & -1 & 1 \\ 0 & 0 & 1 \end{pmatrix}$ (equations for no difference between MS = 1 and 2, and between MS =2 and 3) are used, as they are equivalent.

Score statistics. Let $l_i\left(\boldsymbol{\beta}\right) = l(Y_i; X_i, \boldsymbol{\beta})$ be the log-likelihood and $S(Y_i; X_i, \boldsymbol{\beta}) = \frac{\partial l_i}{\partial \boldsymbol{\beta}}\left(\boldsymbol{\beta}\right)$ the score associated with the ith subject. First, consider the case when $\text{rank}(K) = \dim(\boldsymbol{\beta})$. The null reduces to $H_0 : \boldsymbol{\beta} = \mathbf{c}$, where $\mathbf{c} = K^{-1}\mathbf{b}$. Since the score, $S(Y_i; X_i, \mathbf{c})$, has mean $\mathbf{0}$ and variance equal to the Fisher's information matrix $\mathbf{I}(\mathbf{c}) = E\left(S(Y_i; X_i, \mathbf{c})S^\top(Y_i; X_i, \mathbf{c})\right)$ has asymptotically a χ_p^2 distribution under the null hypothesis, where $p = \dim(\boldsymbol{\beta})$. The score statistics in such cases can be evaluated without actually fitting the model to data, since $\boldsymbol{\beta}$ is known under the null. Such an advantage may be important if the MLE does not exist. Moreover, score statistics usually have better performance than Wald statistics for small and moderate samples.

In general, $\text{rank}(K) < \dim(\boldsymbol{\beta})$, so the parameters are not completely determined by the null hypothesis. Let $\widetilde{\boldsymbol{\beta}}$ be the MLE of $\boldsymbol{\beta}$ under the null hypothesis, the score statistic can still be defined as $\frac{1}{n}\left(\sum_{i=1}^n S^\top(Y_i; X_i, \widetilde{\boldsymbol{\beta}})\right)\mathbf{I}^{-1}\left(\widetilde{\boldsymbol{\beta}}\right)\left(\sum_{i=1}^n S(Y_i; X_i, \widetilde{\boldsymbol{\beta}})\right)$, and it follows asymptotically a χ_p^2 distribution under the null hypothesis, where $p = \text{rank}(K)$. To better understand this, we can reparameterize $\boldsymbol{\beta}$ through a linear transformation such that $\boldsymbol{\beta}$ can be decomposed as $\boldsymbol{\beta}^\top = \left(\boldsymbol{\beta}_1^\top, \boldsymbol{\beta}_2^\top\right)$, with the dimension of $\boldsymbol{\beta}_2 = \text{rank}(K)$, and the null can be expressed as $\boldsymbol{\beta}_2 = \mathbf{c}$. The score equation can be decomposed as

$$\mathbf{w}_n^{(1)}\left(\boldsymbol{\beta}\right) = \frac{1}{n}\sum_{i=1}^n \frac{\partial l_i}{\partial \boldsymbol{\beta}_1}\left(\boldsymbol{\beta}\right) = \mathbf{0}, \quad \mathbf{w}_n^{(2)}\left(\boldsymbol{\beta}\right) = \frac{1}{n}\sum_{i=1}^n \frac{\partial l_i}{\partial \boldsymbol{\beta}_2}\left(\boldsymbol{\beta}\right) = \mathbf{0}. \tag{4.16}$$

Let $\widetilde{\boldsymbol{\beta}}_1$ be the MLE of $\boldsymbol{\beta}_1$ under the null hypothesis, obtained by solving $\mathbf{w}_n^{(1)}\left(\boldsymbol{\beta}_1, \mathbf{c}\right) = \mathbf{0}$. One may define the score statistic as

$$T_s\left(\widetilde{\boldsymbol{\beta}}_1, \mathbf{c}\right) = n\left(\mathbf{w}_n^{(2)}\left(\widetilde{\boldsymbol{\beta}}_1, \mathbf{c}\right)\right)^\top \widehat{\Sigma}_2^{-1}\mathbf{w}_n^{(2)}\left(\widetilde{\boldsymbol{\beta}}_1, \mathbf{c}\right),$$

where Σ_2 is the asymptotic variance of $\frac{\partial l_i}{\partial \beta_2}\left(\widetilde{\beta}_1, c\right)$ under the null hypothesis. Thus, it is clear that the degrees of freedom is rank(K). If we partition the Fisher information $\mathbf{I}$ under the null hypothesis according to $\beta^\top = \left(\beta_1^\top, \beta_2^\top\right)$, i.e., $\mathbf{I} = \begin{pmatrix} \mathbf{I}_{11} & \mathbf{I}_{12} \\ \mathbf{I}_{21} & \mathbf{I}_{22} \end{pmatrix}$ with $\mathbf{I}_{jk} = E\left(\left(\frac{\partial l_i}{\partial \beta_j}\right)\left(\frac{\partial l_i}{\partial \beta_k}\right)^\top\right)$ for $j, k = 1, 2$, then it can be proved that $\Sigma_2 = \mathbf{I}_{22} - \mathbf{I}_{21}\mathbf{I}_{11}^{-1}\mathbf{I}_{12}$. It is straightforward to check that the two definitions are equivalent. Since only β_1 is estimated based on $\beta_2 = c$, it does not require the existence of the MLE for the full model.

Likelihood ratio test. Let $L(\beta)$ denote the likelihood function. Let $\widehat{\beta}$ denote the MLE of β and $\widetilde{\beta}$ the MLE of the constrained model under the null hypothesis. Then, under the null hypothesis, the likelihood ratio statistic

$$2\log R\left(\widetilde{\beta}\right) = 2\left[\log L\left(\widehat{\beta}\right) - \log L\left(\widetilde{\beta}\right)\right] \sim_a \chi_l^2.$$

As the likelihood ratio test only depends on the height of the likelihood function rather than the curvature, it usually provides more accurate inference than the Wald statistic. Thus, it is in general preferred if available, although all three tests are asymptotically equivalent under the null hypothesis. Note that the likelihood ratio test only applies to parametric models as in the current context, while the Wald and score tests also apply to distribution-free models (see Chapters 6 and 9).

Example 4.3
We can test the null hypothesis (4.12) using all three tests. For example, since

$$\widehat{\beta} = \begin{pmatrix} -0.3915 \\ -0.4471 \\ 0.0638 \end{pmatrix}, \quad \mathbf{b} = \begin{pmatrix} 0 \\ 0 \end{pmatrix}, \quad \text{and } \frac{1}{n}\Sigma_\beta = \begin{pmatrix} 0.03492 & -0.03492 & -0.03492 \\ -0.03492 & 0.04737 & 0.03492 \\ -0.03492 & 0.03492 & 0.05247 \end{pmatrix},$$

straightforward calculations show that the Wald statistic is

$$n\left(K\widehat{\beta} - \mathbf{b}\right)^\top \left(K\Sigma_\beta K^\top\right)^{-1} \left(K\widehat{\beta} - \mathbf{b}\right) = 10.01.$$

The corresponding p-value is 0.0067; thus, we reject the null hypothesis. □

When more than one parameter is involved in the null, computations can be tedious. Fortunately, such complexity can be simplified by using statistical packages, provided that the coefficient matrix K and vector $\mathbf{b}$ are correctly specified.

4.2.2.2 Coding system

In the examples above, we used *indicator*, or *dummy*, variables to represent categorical variables. There are also other approaches to coding categorical variables. Since correct interpretation of the parameters depends on the coding systems used, we discuss some commonly used coding systems next. For example, to represent the three levels of the MS variable in DOS, let x_{ik} be an indicator for the group of subjects with MS $= k$, i.e.,

$$x_{i1} = \begin{cases} 1 \text{ if MS} = 1 \\ 0 \text{ if otherwise} \end{cases}, \quad x_{i2} = \begin{cases} 1 \text{ if MS} = 2 \\ 0 \text{ if otherwise} \end{cases}, \quad x_{i3} = \begin{cases} 1 \text{ if MS} = 3 \\ 0 \text{ if otherwise} \end{cases}.$$

Then, the logistic model for the depression outcome with MS as the only predictor may be given by

$$\text{logit}(\pi_i) = \log\left(\frac{\pi_i}{1 - \pi_i}\right) = \beta_0 + \beta_1 x_{i1} + \beta_2 x_{i2} + \beta_3 x_{i3}. \tag{4.17}$$

Since there are only three distinct response probabilities, $\pi_i(1)$, $\pi_i(2)$, and $\pi_i(3)$, one for each of the levels of MS, it is not possible to identify all the four β's without additional constraints imposed on them.

One popular approach is to set one of β_1, β_2, and β_3 to 0 or equivalently designate one of the three levels as a reference as we did in previous sections. For example, if we set $\beta_3 = 0$ or those in the "Other" category as the reference group, we can identify the three remaining parameters, β_0, β_1, and β_2. In this case, the logistic model reduces to (4.11). Thus, in terms of the indicator variables in the unconstrained model, this coding scheme retains only two of the three indicators (x_{i1} and x_{i2}). This approach is known as *reference coding*, or *dummy coding*, or *indicator coding*. We have used this method previously and will continue using it throughout the book except this section. The log odds for the three levels of the MS variable are given by

$$\log \frac{\pi_i(1)}{1 - \pi_i(1)} = \beta_0 + \beta_1, \quad \log \frac{\pi_i(2)}{1 - \pi_i(2)} = \beta_0 + \beta_2, \quad \log \frac{\pi_i(3)}{1 - \pi_i(3)} = \beta_0.$$

It follows that the log odds ratio and odds ratio of MS $= 1$ vs. MS $= 3$ and that of MS $= 2$ vs. MS $= 3$ are given by

$$\log(OR_{1 \text{ vs } 3}) = (\beta_0 + \beta_1) - \beta_0 = \beta_1, \quad OR_{1 \text{ vs } 3} = \exp(\beta_1),$$
$$\log(OR_{2 \text{ vs } 3}) = (\beta_0 + \beta_2) - \beta_0 = \beta_2, \quad OR_{2 \text{ vs } 3} = \exp(\beta_2).$$

Thus, β_0 is identified as the log odds for the reference group (MS $= 3$), and β_1 and β_2 represent the difference in log odds between each of the remaining groups to this MS $= 3$ reference group. Note that there is no particular reason to use MS $= 3$ as the reference level. We can designate any of the levels such as MS $= 1$ as the reference group when using this coding method.

One disadvantage of reference coding is that the interpretation of the intercept depends on the reference level selected. Another approach people often use is to impose the constraint $\beta_1 + \beta_2 + \beta_3 = 0$ and then solve for one of the β's. For example, if we solve for β_3, we have $\beta_3 = -\beta_1 - \beta_2$. The log odds for the three levels of the MS variable are given by

$$\log \frac{\pi_i(1)}{1 - \pi_i(1)} = \beta_0 + \beta_1, \quad \log \frac{\pi_i(2)}{1 - \pi_i(2)} = \beta_0 + \beta_2, \quad \log \frac{\pi_i(3)}{1 - \pi_i(3)} = \beta_0 - \beta_1 - \beta_2.$$

This is equivalent to the logistic model:

$$\text{logit}(\pi_i) = \log \left(\frac{\pi_i}{1 - \pi_i} \right) = \beta_0 + \beta_1 z_{i1} + \beta_2 z_{i2}. \tag{4.18}$$

with the following two variables defined by

$$z_{i1} = \begin{cases} 1 & \text{if MS} = 1 \\ 0 & \text{if MS} = 2 \\ -1 & \text{if MS} = 3 \end{cases}, \quad z_{i2} = \begin{cases} 0 & \text{if MS} = 1 \\ 1 & \text{if MS} = 2 \\ -1 & \text{if MS} = 3 \end{cases}.$$

The log odds ratio and odds ratio of MS $= 1$ vs. MS $= 3$ and that of MS $= 2$ vs. MS $= 3$ are given by

$$\log(OR_{1 \text{ vs } 3}) = 2\beta_1 + \beta_2, \quad OR_{1 \text{ vs } 3} = \exp(2\beta_1 + \beta_2),$$
$$\log(OR_{2 \text{ vs } 3}) = \beta_1 + 2\beta_2, \quad OR_{2 \text{ vs } 3} = \exp(\beta_1 + 2\beta_2). \tag{4.19}$$

Thus, under this *effect coding* method, β's have quite a different interpretation. In particular, β_0 cannot be interpreted as the log odds for a particular group as in the reference coding method. Rather, it can be interpreted simply as the average of log odds for all the MS levels.

Maybe the simplest approach is simply to set $\beta_0 = 0$, i.e., use the following logistic model without the intercept:

$$\text{logit}(\pi_i) = \log\left(\frac{\pi_i}{1 - \pi_i}\right) = \beta_1 x_{i1} + \beta_2 x_{i2} + \beta_3 x_{i3}. \tag{4.20}$$

Under this model, we have one parameter for each of response level, and the log odds for the three levels of the MS variable are given by

$$\log\frac{\pi_i(1)}{1 - \pi_i(1)} = \beta_1, \quad \log\frac{\pi_i(2)}{1 - \pi_i(2)} = \beta_2, \quad \log\frac{\pi_i(3)}{1 - \pi_i(3)} = \beta_3.$$

The log odds ratio and odds ratio of any two levels are given by the corresponding parameter difference and its exponential. For example, the log odds ratio and odds ratio of MS = 1 vs MS = 3 are given by

$$\log(OR_{1 \text{ vs } 3}) = \beta_1 - \beta_3, \quad OR_{1 \text{ vs } 3} = \exp(\beta_1 - \beta_3). \tag{4.21}$$

Under this *cell-means* coding method, all the levels of the categorical variable are treated the same way and all the β's in (4.20) are related to their corresponding specific levels. Note that this coding system still suffers the identifiability issue when applied to additive models with more than one categorical predictors such as model (4.8). In model (4.8), there are only three independent fitted values; thus, it is impossible to identify the four β's (two for x_i and two z_i) without additional constraints imposed on them. However, it would be fine if the interaction is included. In fact, the coding system may be convenient in special cases when all the interactions among all the categorical predictors are included in the model; see Problem 7.12.

All the coding schemes can be used to identify the levels of the categorical variable. There are also other coding systems available, and they are all equivalent as long as they can uniquely identify the parameters, although the interpretation of the parameters may be quite different, depending on the particular scheme. When using software packages, we usually do not have to create the required design variables to implement a coding method. For example, when using SAS, all we need to do is to declare such a variable using the CLASS statement and specify the coding system. To obtain correct interpretation, it is important to make sure which coding system is used.

Example 4.4

For the DOS data, consider the three-level variable, MS, for marital status. The estimates for (4.11) using reference coding are given in the table below.

Variable	Estimate	Standard Error	P-value
Intercept (β_0)	−0.3915	0.1869	0.0362
MS =1 (β_1)	−0.4471	0.2177	0.0400
MS =2 (β_2)	0.0638	0.2291	0.7806

The estimate of β_0 is the log odds for the MS = 3 group, while those of β_1 and β_2 represent the log odds ratios when comparing the MS =1 and MS = 2 to the reference MS = 3 group. For example, the odds ratio of MS = 1 to MS = 3 is $\exp(-0.4471) = 0.639$.

The estimates for the same model using the effect coding scheme, (4.18), are given in the following table.

Variable	Estimate	Standard Error	P-value
Intercept (β_0)	-0.5192	0.0849	< 0.0001
MS $=1$ (β_1)	-0.3193	0.1066	0.0027
MS $=2$ (β_2)	0.1916	0.1143	0.0938

As noted earlier, $\widehat{\beta}_0$ does not have the interpretation as the log odds for the MS $= 3$ group. In addition, we must use (4.19) to compute odds ratios. For example, the odds ratio of MS $= 1$ to MS $= 3$ is not $\exp(-0.3193) = 0.727$. The correct estimate of the odds ratio is

$$OR_{1 \text{ vs } 3} = \exp(2\widehat{\beta}_1 + \widehat{\beta}_2) = \exp(-0.3193 \times 2 + 0.1916) = 0.639.$$

The estimates for the same model using the cell-means coding scheme, (4.20), are given in the following table:

Variable	Estimate	Standard Error	P-value
MS $=1$ (β_1)	-0.8386	0.1116	$< .0001$
MS $=2$ (β_2)	-0.3277	0.1325	0.0134
MS $=3$ (β_3)	-0.3915	0.1869	0.0362

Note that there is no intercept in the model. Each of the parameter β_i represent the log odds for subjects in the corresponding MS level. For example, the log odds for the MS $= 3$ group is -0.3915, consistent with the numbers above. The log of odds ratios between two MS groups will be the difference of the corresponding β's. For example, the odds ratio of MS $= 1$ to MS $= 3$ is

$$OR_{1 \text{ vs } 3} = \exp(\widehat{\beta}_1 - \widehat{\beta}_3) = \exp(-0.8386 - (-0.3915)) = 0.639,$$

also consistent with the estimates we obtained above using the reference and effect coding systems.

Thus, the estimated ORs do not depend on the different coding methods, but to interpret the results correctly it is critical to make clear the coding method upon which the estimates are derived.   □

4.2.2.3 Offset

The examples above all involve testing linear hypotheses with $\mathbf{b} = \mathbf{0}$. We may easily test such hypotheses using software packages by simply specifying the matrix K. For example, the contrast statement in many SAS procedures allows one to specify such a K. What if $\mathbf{b} \neq \mathbf{0}$ in the linear hypothesis? Although there is no essential difference in the theory, this general linear hypothesis cannot be directly specified in a software package, and one may reparameterize the model so that the null hypothesis can be reexpressed as linear hypothesis with $\mathbf{b} = \mathbf{0}$.

Example 4.5
Consider again the MS variable in the DOS data. Suppose that we want to test the null

$$H_0 : \beta_1 - \beta_2 = 2,$$

under the model (4.11).

We first express the null as

$$H_0 : \beta_1 - \beta_2 - 2 = \beta_1 - (\beta_2 + 2) = 0. \tag{4.22}$$

Let

$$\alpha_1 = \beta_1, \quad \alpha_2 = \beta_2 + 2.$$

Under the new parameter vector $\mathbf{a} = (a_1, a_2)^\top$, we can express the original linear hypothesis using a linear contrast:

$$H_0 : \alpha_1 - \alpha_2 = 0.$$

When reparameterized, the logistic model is given by

$$\text{logit}(\pi_i) = \log \left(\frac{\pi_i}{1 - \pi_i} \right) = \beta_0 + a_1 x_{i1} + (a_2 - 2) x_{i2} = \beta_0 + (-2x_{i2}) + a_1 x_{i1} + a_2 x_{i2}.$$

The above has the same form as the original model except for the extra *offset* term $-2x_{i2}$. Offset terms are those items in the model with fixed coefficients. In prospective studies, offsets are commonly used to adjust the observation time if subjects are followed up for varying periods of time. $\square$

4.2.3 Exact inference for logistic regression

When sample size is small, the asymptotic theory may not be appropriate, and conclusions based on such a theory may not be valid. Furthermore, in some situations the maximum likelihood procedure may fail to produce any estimate. For example, consider the logistic regression

$$\text{logit}\left[\Pr\left(y = 1 \mid x\right)\right] = \log\left(\frac{\Pr\left(y = 1 \mid x\right)}{1 - \Pr\left(y = 1 \mid x\right)}\right) = \beta_0 + \beta_1 x \tag{4.23}$$

for Table 2.3. If one cell count equals 0, say $n_{12} = 0$, then the likelihood is

$$\left[\frac{1}{1 + \exp(\beta_0 + \beta_1)}\right]^{n_{11}} \left[\frac{1}{1 + \exp(\beta_0)}\right]^{n_{21}} \left[\frac{\exp(\beta_0)}{1 + \exp(\beta_0)}\right]^{n_{22}}.$$

Since the smaller the β_1, the bigger the likelihood, the MLE does not exist in this case. If a statistical software package is applied, a warning message usually appears, along with a negative (very large in magnitude) estimate of β_1. The estimate varies with the algorithm setting such as number of iterations and convergence criteria. The MLE does not exist in this situation because of a phenomenon known as *data separation*; all subjects with $y = 0$ have $x \leq 0$ (actually $x = 0$), and all subjects with $y = 1$ have $x \geq 0$ (Albert and Anderson, 1984, Santner and Duffy, 1986).

Exact conditional logistic regression can be used to deal with small sample sizes. In this approach, inference about the regression parameters of interest is made based on the distribution conditional on holding the sufficient statistics of the remaining parameters at their observed values. To better explain this approach, we first discuss the notation of sufficient statistics.

4.2.3.1 Sufficient statistics

A statistic is called *sufficient* if no other statistic can provide any additional information to the value of the parameter to be estimated (Fisher, 1922). Thus, a sufficient statistic summarizes all the information in the data necessary to estimate the parameter of interest. In other words, once we have a sufficient statistic, we can completely ignore the original data without losing any information as far as inference about the underlying parameter is concerned.

To technically characterize such a statistic, let y denote the observations for a variable following a parametric distribution parameterized by θ. A statistic $T(y)$ is sufficient for θ if $f(y \mid T(y) = t, \theta) = f(y \mid T(y) = t)$, where $f(y \mid z)$ is the conditional distribution of y given z. In other words, a statistic $T(y)$ is sufficient for the parameter θ if the conditional probability distribution of the data y given the statistic $T(y)$ is independent of the parameter θ. To verify such a condition in practice, we often use Fisher's factorization theorem (see, for example, Casella et al. (2002)). Under this theorem, $T(y)$ is sufficient for θ if and only if the probability distribution function of the data y, $f(y, \theta)$, can be written as a product of two functions $g(T(y), \theta)$ and $h(y)$, where $h(\cdot)$ does not depend on θ and $g(\cdot)$ depends on y only through $T(y)$.

For example, suppose we can decompose the covariates into two parts, and consider a logistic regression model given by $\mathrm{logit}(\pi_i) = \log\left(\frac{\pi_i}{1 - \pi_i}\right) = \mathbf{z}_i^\top \boldsymbol{\gamma} + \mathbf{x}_i^\top \boldsymbol{\beta}$, where $\pi_i = \Pr(y_i = 1 \mid \mathbf{z}_i, \mathbf{x}_i)$. Conditional on covariates $\mathbf{z}_i$ and $\mathbf{x}_i$ $(1 \le i \le n)$, the likelihood function is

$$
\begin{aligned}
L(y, \boldsymbol{\gamma}, \boldsymbol{\beta}) &= \prod_{i=1}^{n} \frac{\exp\left[y_i\left(\mathbf{z}_i^\top \boldsymbol{\gamma} + \mathbf{x}_i^\top \boldsymbol{\beta}\right)\right]}{1 + \exp\left(\mathbf{z}_i^\top \boldsymbol{\gamma} + \mathbf{x}_i^\top \boldsymbol{\beta}\right)} = \frac{\exp\left[\sum_{i=1}^{n} y_i(\mathbf{z}_i^\top \boldsymbol{\gamma} + \mathbf{x}_i^\top \boldsymbol{\beta})\right]}{\prod_{i=1}^{n}\left[1 + \exp\left(\mathbf{z}_i^\top \boldsymbol{\gamma} + \mathbf{x}_i^\top \boldsymbol{\beta}\right)\right]} \\
&= \frac{\exp\left(\sum_{i=1}^{n} y_i \mathbf{z}_i^\top \boldsymbol{\gamma}\right) \cdot \exp\left(\sum_{i=1}^{n} y_i \mathbf{x}_i^\top \boldsymbol{\beta}\right)}{\prod_{i=1}^{n}\left[1 + \exp\left(\mathbf{z}_i^\top \boldsymbol{\gamma} + \mathbf{x}_i^\top \boldsymbol{\beta}\right)\right]}.
\end{aligned} \tag{4.24}
$$

In (4.24), since we condition on $\mathbf{z}_i$ and $\mathbf{x}_i$, the term in the denominator is treated as a constant. Thus, it follows from the factorization theorem that $T(x) = \sum_{i=1}^{n} y_i \mathbf{x}_i$ is a sufficient statistic for $\boldsymbol{\beta}$, while $T(z) = \sum_{i=1}^{n} y_i \mathbf{z}_i$ is sufficient for $\boldsymbol{\gamma}$.

The importance of using sufficient statistics for inference about a parameter of interest is that by conditioning on sufficient statistics for all other parameters, the distribution of data is completely determined by that parameter. By applying this principle to our context, we can derive exact inference for logistic regression.

4.2.3.2 Conditional logistic regression

Let y_i be a binary response and $\mathbf{x}_i = (x_{i0}, \ldots, x_{ip})^\top$ be a vector of independent variables from the ith subject $(1 \le i \le n)$. Consider the logistic regression model

$$
\mathrm{logit}(\pi_i) = \log\left(\frac{\pi_i}{1 - \pi_i}\right) = \beta_0 x_{i0} + \cdots + \beta_p x_{ip} = \boldsymbol{\beta}^\top \mathbf{x}. \tag{4.25}
$$

We set $x_{i0} \equiv 1$ so that β_0 is the intercept. By an argument similar to (4.24), a sufficient statistic for the parameter β_j is given by $T_j(x) = \sum_{i=1}^{n} y_i x_{ij}$ $(1 \le j \le p)$ (see Problem 4.14). When making inference about a parameter such as β_j, we treat all other parameters as nuisance parameters. Thus, inference about β_j is derived based on the conditional distribution of $T_j(x)$ given $\{T_k(x); 0 \le k \le p, k \ne j\}$.

The conditional distribution of $T_j(x)$ given $\{T_k(x) : 0 \le k \le p, k \ne j\}$ is given by

$$
\begin{aligned}
f(t_j; \beta_j) &= \Pr\left(T_j(x) = t_j \mid T_k(x) = t_k, k \ne j\right) \\
&= \frac{c(T_k = t_k, k \ne j, T_j = t_j)\exp(\beta_j t_j)}{\sum_u c(T_k = t_k, k \ne j, T_j = u)\exp(\beta_j u)},
\end{aligned} \tag{4.26}
$$

where $c(T_k = t_k, k \ne j, T_j = u) = \#\{(y_1, \ldots, y_n) : T_k = t_k, k \ne j, T_j = u\}$, i.e., the number of observations such that $T_k = t_k, k \ne j$, and $T_j = u$. The above conditional distribution provides the premise for testing the hypothesis, $H_0 : \beta_j = 0$, since the conditional distribution is totally determined by β_j. For example, by using (4.26) we can immediately compute the p-value as the sum of all the probabilities that T_j are as or more extreme than the observed ones t_j. In addition to this *conditional probability test*, we may use some other statistics to define the extremeness of the samples. For example, the conditional score test computes the p-value as the sum of all the probabilities of T_j whose conditional scores equal or exceed the observed value of the test statistic.

As an application, let us apply the conditional logistic regression to (4.23). The sufficient statistic for β_0 is $T_0 = \sum_{i=1}^n y_i$, while the one for β_1 is $T_1 = \sum_{i=1}^n y_i x_i$. To assess the relationship between x and y, we need to estimate and make inference about β_1. These tasks can be completed based on the conditional distribution of T_1 given T_0.

Conditional on $T_0 = \sum_{i=1}^n y_i = 17$, as in Example 2.4, the possible values of $T_1 = \sum_{i=1}^n y_i x_i$ are $\{0, 1, 2, \ldots, 11\}$. For each $t_1 \in \{0, 1, 2, \ldots, 11\}$, the conditional distribution is given by

$$\Pr(T_1 = t_1 \mid T_0 = 17) = \frac{\binom{11}{t_1}\binom{21}{t_0 - t_1} \exp(t_1 \beta_1)}{\sum_{c=0}^{11} \binom{11}{c}\binom{21}{t_0 - c} \exp(c \beta_1)} \tag{4.27}$$

The above conditional distribution of T_1 (given T_0) is completely determined by β_1 and can be used for inference about β_1. For example, if we are interested in testing whether y is associated with x, i.e., the null $H_0 : \beta_1 = 0$, (4.27) reduces to

$$\Pr(T_1 = t_1 \mid T_0 = 17) = \frac{\binom{11}{t_1}\binom{21}{17 - t_1}}{\sum_{c=0}^{11} \binom{11}{c}\binom{21}{17 - c}}, \quad 0 \le t_1 \le 11. \tag{4.28}$$

It is seen that the conditional distribution in (4.28) is actually a hypergeometric and exact inference based on this distribution is identical to the exact methods we discussed for 2×2 tables in Chapter 2. This also gives a theoretical justification for Fisher's exact test.

For illustrative purposes, let us also test $H_0 : \beta_0 = 0$. In most studies, we would not be interested in such a test since we are primarily concerned about the association between x and y. Within the context of this particular example, $\beta_0 = 0$ is equivalent to the null $H_0 :$ logit$(\Pr(y = 1 \mid x = 0)) = 0$ or, equivalently, $H_0 : \Pr(y = 1 \mid x = 0) = \frac{1}{2}$. In other words, for subjects with $x = 0$, it is equally likely to have $y = 1$ and $y = 0$. Thus, we may use a test for proportions for such a null (see Chapter 2).

If we apply exact logistic regression theory, we make inference based on the conditional distribution of T_0 given T_1. This conditional distribution is given by $\Pr(T_0 = t_0 \mid T_1 = t_1) = \frac{c(t_0) \exp(t_0 \beta_0)}{\sum c(T_0) \exp(T_0 \beta_0)}$, where $c(T_0) =$ number of different combination of y_i $(i = 1, \ldots, n)$ such that $T_0 = \sum_{i=1}^n y_i$ and $\sum_{i=1}^n x_i y_i = t_1$ and the sum in the denominator is taken over the range of T_0. Under the null $H_0 : \beta_0 = 0$, it reduces to $\Pr(T_0 = t_0 \mid T_1 = t_1) = \frac{c(t_0)}{\sum c(T_0)}$. It is easy to verify that $c(T_0) = \binom{\sum_{i=1}^n (1 - x_i)}{T_0 - t_1}\binom{\sum_{i=1}^n x_i}{t_1}$, and the conditional distribution is actually a binomial $BI(m, p)$ with parameter $m = \sum_{i=1}^n (1 - x_i)$ and $p = \frac{1}{2}$ (see Problem 4.15). Thus, inference based on the exact conditional logistic regression is the same as testing the null hypothesis of a proportion of 0.5 on the restricted subset.

For simplicity, we have thus far only considered inference about one parameter. However, the same considerations can be applied to testing composite null hypotheses involving multiple parameters. Inference in this more general case is based on the *joint* distribution of the corresponding sufficient statistics conditional on those for the remaining parameters.

We can also estimate $\boldsymbol{\beta}$ based on the exact conditional distribution by maximizing the conditional likelihood. Since the general theory of maximum likelihood estimation can be applied, we will not discuss the details here. However, if the data are separated, then the conditional likelihood still cannot be maximized. Another *median unbiased estimate* (MUE) based on the exact conditional distribution may be used in such cases.

If T is a statistic such that the values of $\hat{\beta}$ vary monotonically with T, then for any observed value of $T = t$, the MUE is defined as the value of β for which $\Pr\left(T \le t\right) \ge 0.5$ and $\Pr\left(T \ge t\right) \ge 0.5$. For discrete distributions, this definition would generally be satisfied by a range of values. In this case, one may take the midpoint of this range as the point estimate. More explicitly, let β_- be defined by $\sum_{T \ge t} f\left(T \mid \beta_-\right) = 0.5$ and β_+ be defined by the equality $\sum_{T \le t} f\left(T \mid \beta_+\right) = 0.5$. Then, the MUE is defined as $\frac{\beta_+ + \beta_-}{2}$. In the extreme cases where the observed sufficient statistic is the maximum or minimum, the MUE is defined as $f\left(t \mid \beta_{MUE}\right) = 0.5$. Following the definition, MUE always exists.

Confidence intervals can also be computed based on the exact conditional distribution. Specially, a $100\left(1 - \alpha\right)\%$ confidence interval $\left(\beta_l, \beta_u\right)$ is defined as

$$\sum_{T \ge t} f\left(T \mid \beta_l\right) = \frac{\alpha}{2} \text{ and } \sum_{T \le t} f\left(T \mid \beta_u\right) = \frac{\alpha}{2}.$$

If the observed sufficient statistic is the maximum (minimum), then $\beta_u = \infty$ ($\beta_l = -\infty$).

Example 4.6

Consider the recidivism study presented in Example 2.4, using the posttreatment outcome as the response and pre-treatment measure as the predictor.

The conditional distribution, based on Table 2.4, was given in (4.27). We obtain $\beta_+ = 1.5214$ and $\beta_- = 0.9048$ by solving the equations

$$\frac{\sum_{c=0}^{8} \binom{11}{c}\binom{21}{17-c} \exp\left(c\beta_+\right)}{\sum_{c=0}^{11} \binom{11}{c}\binom{21}{17-c} \exp\left(c\beta_+\right)} = 0.5, \quad \frac{\sum_{c=8}^{11} \binom{11}{c}\binom{21}{17-c} \exp\left(c\beta_-\right)}{\sum_{c=0}^{11} \binom{11}{c}\binom{21}{17-c} \exp\left(c\beta_-\right)} = 0.5.$$

Thus, the MUE $\hat{\beta}_1 = \left(1.5214 + 0.9048\right)/2 = 1.2131$. Likewise, by solving the equations

$$\frac{\sum_{c=0}^{8} \binom{11}{c}\binom{21}{17-c} \exp\left(c\beta_1\right)}{\sum_{c=0}^{11} \binom{11}{c}\binom{21}{17-c} \exp\left(c\beta_1\right)} = 0.025, \text{ and } \frac{\sum_{c=8}^{11} \binom{11}{c}\binom{21}{17-c} \exp\left(c\beta_1\right)}{\sum_{c=0}^{11} \binom{11}{c}\binom{21}{17-c} \exp\left(c\beta_1\right)} = 0.025,$$

we obtain a 95% CI $(-0.5191, 3.2514)$. Since the CI contains 0, β_1 is not significantly different from 0.

If the cell count for $(x = 1, y = 0)$ is 0 and all other cell counts remain the same, then the MLE does not exist, as the data are separated. However, we can still compute the MUE. Since the conditional distribution is given by

$$\Pr\left(T_1 = t_1 \mid T_0 = t_0\right) = \frac{\binom{8}{t_1}\binom{21}{17-t_1} \exp\left(T_1\beta_1\right)}{\sum_{c=0}^{8} \binom{8}{c}\binom{21}{17-c} \exp\left(c\beta_1\right)}.$$

By solving the equation

$$\frac{\binom{8}{8}\binom{21}{17-8} \exp\left(8\beta_1\right)}{\sum_{c=0}^{8} \binom{8}{c}\binom{21}{17-c} \exp\left(c\beta_1\right)} = 0.5,$$

we obtain the MUE $\widehat{\beta}_1 = 2.5365$. A 95% confidence interval is obtained as $(0.5032, \infty)$, obtained by solving the equation

$$\frac{\binom{8}{8}\binom{21}{17-8}\exp(8\beta_1)}{\sum\limits_{c=0}^{8}\binom{8}{c}\binom{21}{17-c}\exp(c\beta_1)} = 0.025.$$

□

Hirji et al. (1989) studied the behaviors of the MLE and MUE for some small sample sizes and covariate structures. They found that the MUE was in general more accurate than the MLE. Exact inference is generally computationally intensive and time-consuming. This alternative approach became feasible and practical only recently after some major development of efficient algorithms and advent of modern computing power. Here are some important references in the development of computational algorithms. Tritchler (1984) gave an efficient algorithm for computing the distribution of the sufficient statistic using the fast Fourier transform (for a single explanatory variable), and Mehta et al. (1985) developed a network algorithm that can be used to compute this distribution for the special case when the logistic model can be depicted in terms of analysis of several 2×2 contingency tables. Algorithms for dealing with general logistic models for matched and unmatched designs were given by Hirji et al. (1987, 1988).

Note also that for exact logistic regression to be applicable and reliable, we need a large number of possible outcomes that produce the same conditional sufficient statistic. If one of the conditioning variables is continuous, it is very likely that only a few possible outcomes will result in the same sufficient statistic. For example, consider the situation of a continuous covariate x_i. The sufficient statistic for β_1 is $T_1 = \sum_{i=1}^{n} x_i y_i$ as above. However, since x_i is continuous, it is likely that only a few possible y_i's, or in the extreme case only the observed y_i's, will produce the sufficient statistic T_1 (see Problem 4.16). The conditional distribution in such cases will be very coarse, or even degenerate (in the extreme case), and inference based on such a distribution will be rather problematic. Thus, exact logistic regression may not work well when conditional on continuous variables.

Application to matched study. In Section 4.1.4, we showed that logistic regression can be applied to both prospective studies and retrospective case-control studies. The only difference is the interpretation of the intercept in the model; this term is well interpreted in the case-control study case. Following the same principle, we can also apply logistic regression models to matched study designs by treating the intercept as a nuisance parameter. However, ML inference may not be efficient if there are many matched groups, since each matched group creates an intercept and the number of such nuisance parameters increases with the sample size. By conditioning on the sufficient statistics of the intercepts, exact logistic regression excludes these parameters in the conditional likelihood, providing more efficient inference.

Note that in the special case of matched pair design where each matched group consists of a single case and control, the conditional likelihood is given by (see Holford et al. (1978))

$$\prod_{i=1}^{k}\frac{1}{1 + \exp(\boldsymbol{\beta}^{\top}\mathbf{d}_i)}, \tag{4.29}$$

where k is the number of matched pairs and $\mathbf{d}_i = \mathbf{x}_{i1} - \mathbf{x}_{i0}$ is the difference of the covariates between the case and control subjects in the ith pair $(1 \le i \le k)$. Inference can be carried out based on the conditional likelihood (4.29). This conditional likelihood is the same as the likelihood function of a logistic model with no intercept, based on a sample of k subjects, where the ith subject has covariates $\mathbf{d}_i$ and response 0. However, since all the responses are the same, the conditional logistic technique discussed above must be used to derive inference about $\boldsymbol{\beta}$ (Holford et al., 1978). Note that the unit of analysis in such analysis is the pair rather than each subject, and the pairs with the same covariates do not contribute to the inference. See Example 7.8 for an application

4.2.4 Bias-reduced logistic regression

The asymptotic bias of the MLE is of order n^{-1}, where n is the sample size. More precisely, the bias, $E(\widehat{\boldsymbol{\beta}} - \boldsymbol{\beta})$, can be expressed as

$$\mathbf{b}(\boldsymbol{\beta}) = \frac{\mathbf{b}_1(\boldsymbol{\beta})}{n} + \frac{\mathbf{b}_2(\boldsymbol{\beta})}{n^2} + \cdots,$$

where $\mathbf{b}_j(\boldsymbol{\beta})$ are functions of the parameter $\boldsymbol{\beta}$. Methods are available to reduce the bias to the order n^{-2} by attempting to remove the $\frac{\mathbf{b}_1(\boldsymbol{\beta})}{n}$ term above. One approach is to correct the bias using the MLE $\widehat{\boldsymbol{\beta}}$ of $\boldsymbol{\beta}$, with the bias corrected estimate $\widehat{\boldsymbol{\beta}}_{BC} = \widehat{\boldsymbol{\beta}} - \frac{\mathbf{b}_1(\widehat{\boldsymbol{\beta}})}{n}$. Alternatively, we may use a jackknife approach proposed by Quenouille (1949, 1956). Although easier to implement, Quenouille's approach is much more computationally intensive since it has to compute the MLE for each of the n subsamples created by deleting one observation at a time. Both approaches require that the MLE exists. Thus, to implement the latter *Jackknife* approach, the MLE must exist for each of the subsamples. However, within our context, complete or quasi-complete separation may occur and the MLE may not exist, especially for jackknife with small- to medium-sized samples (Albert and Anderson, 1984). To overcome this shortcoming, Firth (1993) suggested to correct bias without requiring the existence of the MLE.

Instead of working directly with the estimate, Firth's approach indirectly corrects the bias of the estimate by purposefully biasing the (score) estimating equations. Firth noticed two sources for bias: unbiasedness and curvature of the score function. Recall that, for i.i.d. observations, the MLE $\widehat{\boldsymbol{\beta}}$ is obtained by solving the score equation

$$\mathbf{U}\left(\boldsymbol{\beta}\right) = \sum_{i=1}^{n} \mathbf{U}_i\left(\boldsymbol{\beta}\right) = \mathbf{0}, \tag{4.30}$$

where $\mathbf{U}_i(\boldsymbol{\beta}) = \frac{\partial}{\partial \boldsymbol{\beta}} l_i(\boldsymbol{\beta})$ and $l_i(\boldsymbol{\beta})$ is the log-likelihood function based on the ith observation. As reviewed in Chapter 1, $E\left[\mathbf{U}\left(\boldsymbol{\beta}\right)\right] = \mathbf{0}$, i.e., the estimating equation (4.30) is unbiased. For notational brevity, we consider the case where β is a scalar. When $\boldsymbol{\beta}$ is a vector, the entire argument below can be similarly applied.

Based on a Taylor series expansion, we have

$$0 = \overline{U}(\widehat{\beta}) \approx \overline{U}(\beta) + \frac{d}{d\beta}\overline{U}(\beta)\left(\widehat{\beta} - \beta\right) + \frac{1}{2}\frac{d^2}{d\beta^2}\overline{U}\left(\beta\right)\left(\widehat{\beta} - \beta\right)^2, \tag{4.31}$$

where $\overline{U} = \frac{1}{n}U = \frac{1}{n}\sum_{i=1}^{n} U_i$ is the average of the individual score functions across the observations, U_i. Note that $E\left(\frac{d}{d\beta}\overline{U}(\beta)\right) = -I\left(\beta\right)$, where $I\left(\beta\right)$ is the Fisher information; thus it is always negative. It follows that the bias of the MLE $\widehat{\beta}$ is

$$b(\beta) \approx I\left(\beta\right)^{-1} E\left[\overline{U}(\beta) + \frac{1}{2}\frac{d^2}{d\beta^2}\overline{U}\left(\beta\right)\left(\widehat{\beta} - \beta\right)^2\right] \tag{4.32}$$

$$= \frac{1}{2}I\left(\beta\right)^{-1} E\left[\frac{d^2}{d\beta^2}\overline{U}\left(\beta\right)\left(\widehat{\beta} - \beta\right)^2\right].$$

Thus, if the score function has some curvature, i.e., $E\left[\frac{d^2}{d\beta^2}U_i(\beta)\right]\neq 0$, then $\widehat{\beta}$ is biased with the direction of bias depending on the sign of $E\left[\frac{d^2}{d\beta^2}U_i(\beta)\right]$; if $E\left[\frac{d^2}{d\beta^2}U_i(\beta)\right]$ is positive, then $\widehat{\beta}$ is biased upward and vice versa. Based on (4.32), it is clear that the bias of MLE is due to the curvature of the score function and the unbiasedness of the score equation so that the bias caused by the curvature cannot be cancelled. Firth's idea is to introduce an appropriate bias term into the score, or estimating equation, to counteract the effect of $\frac{d^2}{d\beta^2}U(\beta)$ so that the revised estimating equation will yield unbiased or less biased estimates of β.

If the score function is modified to $U^*(\beta) = U(\beta) - I(\beta)b(\beta)$, then the estimate based on the modified score equation will have the effect of $\frac{d^2}{d\beta^2}U(\beta)$ cancelled. Thus, although the modified score equation itself is biased, its solution as the estimate of β has a reduced bias in the order of $1/n^2$. The modified score equations can be obtained by maximizing the penalized likelihood function: $L^*(\beta) = L(\beta)\sqrt{I_n(\beta)}$. The factor $\sqrt{I_n(\beta)}$ introduces $\frac{d}{d\beta}\log\left(\sqrt{I_n(\beta)}\right) = \frac{1}{2}I_n^{-1}(\beta)\frac{d^2}{d\beta^2}U(\beta)$ into the modified score equation. This cancels the term that causes the bias in (4.31), $\frac{1}{2}\frac{d^2}{d\beta^2}\overline{U}(\beta)\left(\widehat{\beta} - \beta\right)^2$, because $E\left(\widehat{\beta} - \beta\right)^2 \approx I_n^{-1}(\beta)$.

For logistic regressions, the modified score equations above can also be obtained by maximizing the penalized likelihood function: $L^*(\boldsymbol{\beta}) = L(\boldsymbol{\beta})\det^{1/2}(I_n(\boldsymbol{\beta}))$, where $I_n(\boldsymbol{\beta})$ is the observed Fisher information. Firth's bias reduction method always produces valid point estimates and standard errors. Through the example below, we can see that it also provides a theoretical justification for the continuity correction method we discussed for contingency tables in Chapter 2.

Example 4.7

By designating one of the row and column variables as response and the other as predictor, we can apply logistic regression models to 2×2 tables. More precisely, consider the following logistic regression model

$$\text{logit}(\pi_i) = \beta_0 + \beta_1 x_i,$$

for an i.i.d. sample (x_i, y_i) $(i = 1, 2, \ldots, n)$, where x_i and y_i are binary variables taking values 0 and 1 and $\pi_i = \Pr(y_i = 1 \mid x_i)$. Then it is straightforward to check that

$$I_n(\boldsymbol{\beta}) = \begin{pmatrix} \sum_{i=1}^n \pi_i(1-\pi_i) & \sum_{i=1}^n \pi_i(1-\pi_i)X_i \\ \sum_{i=1}^n \pi_i(1-\pi_i)X_i & \sum_{i=1}^n \pi_i(1-\pi_i)X_i^2 \end{pmatrix}$$
$$= \begin{pmatrix} n_0\pi(0)(1-\pi(0)) + n_1\pi(1)(1-\pi(1)) & n_1\pi(1)(1-\pi(1)) \\ n_1\pi(1)(1-\pi(1)) & n_1\pi(1)(1-\pi(1)) \end{pmatrix}$$

where $\pi(j) = \Pr(y_i = 1 \mid x_i = j)$ and $n_j = $ number of subjects with $x_i = j$ $(j = 0, 1)$. Thus, the penalized likelihood is

$$\det^{1/2}(I_n(\boldsymbol{\beta}))\prod_{i=1}^n\left[\pi_i^{y_i}(1-\pi_i)^{1-y_i}\right]$$
$$= \pi(0)^{n_{01}}(1-\pi(0))^{n_{00}}\pi(1)^{n_{11}}(1-\pi(1))^{n_{10}}\left[n_0\pi_0(1-\pi_0)n_1\pi_1(1-\pi_1)\right]^{1/2},$$

where $n_{jk} = $ number of subjects with $x_i = j$ and $y_i = k$ $(j, k = 0, 1)$.

It is easy to check that this is equivalent to adding 0.5 to each of the four cells. If Firth's method is applied to Table 2.4, the odds ratio is estimated to be $\frac{8.5/9.5}{3.5/12.5} = 3.1955$. Hence, Firth's estimate of β_1 is $\log(3.1955) = 1.1617$. If the cell count for $(x = 1, y = 0)$ is 0 and others remain the same as in Example 4.6, the methods in Chapter 2 cannot be applied to estimate the odds ratio. But Firth's method is readily applied to give $\frac{8.5/9.5}{0.5/12.5} = 22.368$, yielding the estimate for $\beta_1 = \log(22.368) = 3.1076$. One may want to compare the Firth estimates with their exact counterparts given in Example 4.6. A similar argument applies to the estimation of proportions when there is no covariate. We would obtain the same estimate as the discreteness-corrected estimate of proportions discussed in Chapter 2: Add 0.5 to the number of successes as well as to the number of failures and then take the proportion. $\quad\square$

4.3   Generalized Linear Models for Binary Responses

The term "generalized linear model" (GLM) was first introduced in a landmark paper by Nelder and Wedderburn (1972), in which a wide range of seemingly disparate problems of statistical modeling and inference were framed in an elegant, unifying framework with high power and flexibility. This new class of models extends linear regressions for a continuous response to models for other types of responses such as binary and categorical outcomes. Examples of GLMs include linear regression, logistic regression, and Poisson regression. In this section, we introduce this new class of models and discuss its applications to modeling binary outcomes.

4.3.1   Generalized linear models

Recall that the multiple linear regression model has the form

$$y_i \mid \mathbf{x}_i \sim i.d. \ N\left(\mu_i, \sigma^2\right), \ \mu_i = \eta_i = \beta_0 + \beta_1 x_{i1} + \ldots + \beta_p x_{ip} = \beta_0 + \mathbf{x}_i^\top \boldsymbol{\beta}, \qquad (4.33)$$

where *i.d.* means independently distributed. The response y_i conditional on the covariates $\mathbf{x}_i$ is assumed to have a normal distribution with mean μ_i and common variance σ^2. In addition, μ_i is a linear function of the covariates $\mathbf{x}_i$. Since the right side of the model, $\eta_i = \beta_0 + \mathbf{x}_i^\top \boldsymbol{\beta}$, has a range in the real line R, which concurs with the range of μ_i on the left side, the linear model is not appropriate for modeling other types of noncontinuous responses. For example, if y_i is binary, the conditional mean of $y_i \mid \mathbf{x}_i$ is

$$\mu_i = E\left(y_i \mid \mathbf{x}_i\right) = \Pr\left(y_i = 1 \mid \mathbf{x}_i\right). \qquad (4.34)$$

Since μ_i is a value between 0 and 1, it may not be appropriate to model μ_i directly as a linear function of $\mathbf{x}_i$ as in (4.33). In addition, the normal distribution assumption does not apply to binary response.

To generalize the classic linear model to accommodate other types of response, we must modify (1) the normal distribution assumption and (2) the relationship between the conditional mean μ_i in (4.34) and the linear predictor η_i in (4.33). More specifically, the GLM is defined by the following two components:

1. *Random component.* This part specifies the conditional distribution of the response y_i given the independent variables $\mathbf{x}_i$.

2. *Deterministic component.* This part links the conditional mean of y_i given $\mathbf{x}_i$ to the linear predictor $\mathbf{x}_i$ by a one-to-one *link* function g:

$$g\left(\mu_i\right) = \eta = \beta_0 + \beta_1 x_{i1} + \ldots + \beta_p x_{ip}.$$

Thus, the linear regression is obtained as a special case if y given $\mathbf{x}$ follows a normal distribution, and $g\left(\mu\right)$ is the identity function, $g\left(\mu_i\right) = \mu_i$. By varying the distribution function for the random part and the link function $g(\cdot)$ in the deterministic part, we can use the family of GLMs to model a variety of response types with different distributions.

For example, for binary responses, we have thus far discussed the logistic regression in detail. This popular model is also a member of the family of GLMs with the random and deterministic components specified as follows:

1. The response y given $\mathbf{x}$ follows a Bernoulli distribution $Bernoulli\,(\pi)$ with the probability of success given by $E\,(y \mid \mathbf{x}) = \pi\,(\mathbf{x})$.

2. The conditional mean π is linked to the linear predictor η by the logit function, $\eta = g(\pi) = \log\,(\pi/\,(1-\pi))$.

We can change the link functions in the deterministic component to obtain different regression models for binary responses. We may also change the random component to accommodate further data types. For example, for count responses, we may define the random component as the response y given $\mathbf{x}$ follows a Poisson distribution or other distribution with mean μ. The link function should be chosen accordingly. Since the mean of a Poisson distribution has a theoretical range of all positive numbers, the logit link function is not appropriate. The logarithmic link is the most commonly used link function. Thus, we may have the following GLM for count response, combining both random and deterministic components together

$$y_i \mid \mathbf{x}_i \sim \text{Poisson}\,(\mu_i), \quad \log(\mu_i) = \mathbf{x}_i^\top \boldsymbol{\beta} = \beta_1 x_{i1} + \ldots + \beta_p x_{ip}. \qquad (4.35)$$

In general, the MLE method can be applied for the inference. However, note that "quasi-likelihood" estimation may be applied when a dispersion parameter is added to take care of the overdispersion issue. We defer the details until we discuss overdispersion for count outcomes in Chapter 6.

4.3.1.1   Regression models for binary response

Within the context of binary response, we have thus far discussed logistic regression. This popular model is also a GLM with the random and deterministic components specified as follows:

1. The response y given $\mathbf{x}$ follows a Bernoulli distribution $Bernoulli\,(\mu)$ with the probability of success given by $E\,(y \mid \mathbf{x}) = \pi\,(\mathbf{x}) = \pi$.

2. The conditional mean μ is linked to the linear predictor η by the logit function, $\eta = g(\pi) = \log\,(\pi/\,(1-\pi))$.

For binary response, the Bernoulli distribution is the only one available. However, we can change the link function to develop different models. Since the conditional mean $E\,(y \mid \mathbf{x})$ is a probability, it takes values in $(0,1)$. Thus, the link function should map $(0,1)$ onto R. Recall that onto means every real number is an image, so that the inverse of the link function will map all the real numbers into $(0,1)$, which can serve as probabilities. In theory, we can use any function that maps $(0,1)$ onto R as a link for modeling the binary response. However, only several link functions are commonly used in practice. In addition to the logit link, other popular functions used in practice for modeling the binary response include the *probit* link, $g(\pi) = \Phi^{-1}(\pi)$, where $\Phi(\cdot)$ denotes the CDF of standard normal, and *complementary log-log* link, $g(\pi) = \log(-\log(1-\pi))$. Note that although the identity and log link functions may not be appropriate link functions as they do not map $(0,1)$ onto R, they are still applied in practice because of some attractive features that we will mention later.

4.3.2 Probit model

The probit (probability unit) model, which has a long history in the analysis of binary outcomes (see Bliss (1935)), has the following general form:

$$\Pr\left(y_i = 1 \mid \mathbf{x}_i\right) = \Phi\left(\beta_0 + \mathbf{x}_i^\top \boldsymbol{\beta}\right). \tag{4.36}$$

This model is a natural choice if the binary response y is the result of dichotomizing a normally distributed latent continuous variable. More precisely, suppose there is an unobservable continuous variable $y_i^* = \beta_0 + \mathbf{x}_i^\top \boldsymbol{\beta} + \varepsilon_i$, where $\varepsilon_i \sim N(0,1)$ (a standard normal with mean 0 and variance 1) and y_i is determined by y_i^* as an indicator for whether this latent variable is positive:

$$y_i = \begin{cases} 1 & \text{if } y_i^* > 0, \text{ i.e., } -\varepsilon_i < \beta_0 + \mathbf{x}_i^\top \boldsymbol{\beta}, \\ 0 & \text{if otherwise.} \end{cases}$$

It is straightforward to verify that such assumptions imply the model in (4.36) (see Problem 4.22).

Compared to the logistic model, the interpretation of coefficients for the probit model is more complicated. If a predictor x_{ig} $(1 \leq g \leq p)$ increases by one unit with all others held fixed, the Z-score $\beta_0 + \mathbf{x}_i^\top \boldsymbol{\beta}$ increases by β_g. The sign of the coefficient β_g indicates the direction of association, with a negative (positive) sign defining higher likelihood for $y = 0$ $(y = 1)$. Such an interpretation is not as exquisite as the odds ratio for logistic models, which explains in part why logistic models are more popular. However, because of its connection to a normal latent continuous variable, the probit regression is still commonly used in models involving latent variables such as mixed-effects models for longitudinal and clustered data. Note that as in logistic models, the intercept β_0 represents the response rate: $\Pr\left(y_i = 1 \mid \mathbf{x}_i = \mathbf{0}\right) = \Phi\left(\beta_0\right)$, and a zero intercept indicates that the response rate for the reference group $x_i = 0$ is $\frac{1}{2}$.

The probit link is symmetric because $\Phi^{-1}\left(1 - \pi\right) = -\Phi^{-1}\left(\pi\right)$. This implies that there is no essential difference between modeling $\Pr\left(y_i = 1 \mid \mathbf{x}_i\right)$ and $\Pr\left(y_i = 0 \mid \mathbf{x}_i\right)$. Indeed, it is straightforward to check that the only difference between the two is the sign of the coefficients (see Problem 4.23).

4.3.3 Complementary log-log model

Another link function that people sometimes use, especially for rare events, is the complementary log-log function, $g(\pi) = \log(-\log(1 - \pi))$, which is the inverse function of the CDF of the extreme value distribution (Aranda-Ordaz, 1981). This model assumes $\log(-\log(1 - \pi)) = \beta_0 + \boldsymbol{\beta}^\top \mathbf{x}_i$ or, equivalently,

$$\Pr\left(y_i = 1 \mid \mathbf{x}_i\right) = 1 - \exp\left[-\exp\left(\beta_0 + \mathbf{x}_i^\top \boldsymbol{\beta}\right)\right]. \tag{4.37}$$

The sign of the coefficient also indicates the directions of association. If a coefficient is positive (negative), then the higher the values of the corresponding covariate (with others held fixed), the higher (lower) the probabilities for $y = 1$. To quantitatively characterize the association, rewrite the model as

$$\log\left(1 - \Pr\left(y_i = 1 \mid \mathbf{x}_i\right)\right) = -\exp\left(\beta_0 + \mathbf{x}_i^\top \boldsymbol{\beta}\right).$$

By taking the ratio of the above with $x_1 = a + 1$ to that with $x_1 = a$ (all other covariates are the same), we obtain

$$\frac{\log\left(1 - \Pr\left(y_i = 1 \mid x_1 = a + 1, \widetilde{\mathbf{x}}\right)\right)}{\log\left(1 - \Pr\left(y_i = 1 \mid x_1 = a, \widetilde{\mathbf{x}}\right)\right)} = \frac{-\exp\left(\beta_0 + \widetilde{\mathbf{x}}^\top \widetilde{\boldsymbol{\beta}} + \beta_1\left(a + 1\right)\right)}{-\exp\left(\beta_0 + \widetilde{\mathbf{x}}^\top \widetilde{\boldsymbol{\beta}} + \beta_1 a\right)} = \exp\left(\beta_1\right).$$

where $\widetilde{\mathbf{x}}$ ($\widetilde{\boldsymbol{\beta}}$) denotes the vector $\mathbf{x}$ (the parameter vector $\boldsymbol{\beta}$) with the component of x_1 (β_1) removed. So each unit increase in x_1 elevates the probability of $y = 0$ to the power of $\exp(\beta_1)$. Thus, a positive β_1 indicates that when x_1 increases, $\Pr(y = 0)$ decreases, and hence $\Pr(y = 1)$ increases. The intercept β_0 again represents the response rate: $\Pr(y_i = 1 \mid \mathbf{x}_i = \mathbf{0}) = 1 - \exp[-\exp(\beta_0)]$. However, a zero intercept indicates a $1 - \exp(-1) \approx 0.632$, not $\frac{1}{2}$, response rate for the reference group $\mathbf{x}_i = \mathbf{0}$. This also indicates that the complementary log-log link is asymmetric.

Note that unlike logistic and probit links, the complementary log-log function is not symmetric, i.e., $\log(-\log(\pi)) \neq -\log(-\log(1 - \pi))$. In fact, the link function behaves quite differently at the two ends, near 0 and 1. The function approaches negative infinity with order of $\log \pi$ as $\pi \to 0$ ($\log(-\log(1 - \pi)) \sim \log \pi$), while it approaches positive infinity at a much slower order of $\log(-\log(1 - \pi))$ as $1 - \pi \to 0$. Thus, modeling $\Pr(y_i = 1 \mid \mathbf{x}_i)$ and $\Pr(y_i = 0 \mid \mathbf{x}_i)$ using this link generally yields two different models. For this reason, it is important to make it clear which level of y is being modeled in a given application so that the findings can be replicated.

4.3.4 Linear probability model

The identity link function is also applied to binary responses in practice because the parameters can be easily interpreted in terms of the risks, the probabilities of the outcomes. A *linear probability model* assumes that

$$\Pr(y_i = 1 \mid \mathbf{x}_i) = \beta_0 + \boldsymbol{\beta}^\top \mathbf{x}_i. \tag{4.38}$$

For parameter interpretation, we take the difference of the fitted values when $x_1 = a + 1$ and $x_1 = a$ (all other covariates, $\mathbf{x}'$, are the same), and we obtain

$$\Pr(y = 1 \mid x_1 = a + 1, \mathbf{x}') - \Pr(y = 1 \mid x_1 = a, \mathbf{x}')$$
$$= \beta_0 + \mathbf{x}'^\top \boldsymbol{\beta}' + \beta_1(a + 1) - (\beta_0 + \mathbf{x}'^\top \boldsymbol{\beta}' + \beta_1 a) = \beta_1.$$

So each unit increase in x_1 elevates the probability of $y = 0$ by β_1. Thus, a positive β_1 indicates that when x_1 increases, $\Pr(y = 1)$ also increases. The intercept β_0 is simply the response rate or risk for the reference group $\mathbf{x} = \mathbf{0}$.

The linear probability model also does not depend on which level of the response is modeled, i.e., modeling $\Pr(y_i = 1 \mid \mathbf{x}_i)$ and $\Pr(y_i = 0 \mid \mathbf{x}_i)$ is equivalent. It is straightforward to check that the only difference for the coefficients $\boldsymbol{\beta}$ between the two is the sign and that the two intercepts β_0 add up to 1 (see Problem 4.25).

4.3.4.1 Inference

Maximum likelihood is typically used to provide inferences for GLMs. We have discussed how to compute and use such estimates for inference in logistic regression. The same procedures apply to the other link functions. For a GLM with the link function g and a linear η_i,

$$y_i \mid \mathbf{x}_i \sim Bernoulli(\pi_i), \quad g(\pi_i) = \eta_i = \mathbf{x}_i^\top \boldsymbol{\beta} = \beta_0 + \beta_1 x_{i1} + \ldots + \beta_p x_{ip},$$

for $1 \leq i \leq n$, the likelihood function is

$$L(\boldsymbol{\beta}) = \prod_{i=1}^{n} \left[\pi_i^{y_i} (1 - \pi_i)^{1 - y_i} \right] = \prod_{i=1}^{n} \left[\left(g^{-1}(\mathbf{x}_i^\top \boldsymbol{\beta}) \right)^{y_i} \left(1 - g^{-1}(\mathbf{x}_i^\top \boldsymbol{\beta}) \right)^{1 - y_i} \right].$$

The log-likelihood is

$$l(\boldsymbol{\beta}) = \sum_{i=1}^{n} \left[y_i \log g^{-1} \left(\mathbf{x}_i^\top \boldsymbol{\beta} \right) + (1 - y_i) \log(1 - g^{-1} \left(\mathbf{x}_i^\top \boldsymbol{\beta} \right)) \right].$$

The score equation is

$$S(\boldsymbol{\beta}) = \frac{\partial}{\partial \boldsymbol{\beta}} l(\boldsymbol{\beta}) = \sum_{i=1}^{n} \left[\frac{y_i h \left(\mathbf{x}_i^\top \boldsymbol{\beta} \right)}{g^{-1} \left(\mathbf{x}_i^\top \boldsymbol{\beta} \right)} \mathbf{x}_i^\top + \frac{(1 - y_i) h \left(\mathbf{x}_i^\top \boldsymbol{\beta} \right)}{1 - g^{-1} \left(\mathbf{x}_i^\top \boldsymbol{\beta} \right)} \mathbf{x}_i^\top \right]$$

$$= \sum_{i=1}^{n} \frac{y_i - \pi_i}{\pi_i (1 - \pi_i)} h \left(\mathbf{x}_i^\top \boldsymbol{\beta} \right) \mathbf{x}_i^\top = \mathbf{0}, \tag{4.39}$$

where $h = \left(g^{-1} \right)'$, the derivative function of g^{-1}. By solving the score equation (4.39), we obtain the MLE $\widehat{\boldsymbol{\beta}}$. As in the case of logistic regression, MLE usually can only be obtained using numerical methods such as the Newton–Raphson algorithm.

We can make inference about $\boldsymbol{\beta}$ using its asymptotic distribution, $\widehat{\boldsymbol{\beta}} \sim_a N \left(\boldsymbol{\beta}, \mathbf{I}_n^{-1} (\boldsymbol{\beta}) \right)$, with $\mathbf{I}_n (\boldsymbol{\beta})$ denoting the observed information matrix for i.i.d. observations. We can also readily construct the Wald and likelihood ratio statistics to test hypotheses concerning linear functions of $\boldsymbol{\beta}$.

Although the linear probability model bears the same appearance as an ordinary linear model: $E (y_i \mid \mathbf{x}_i) = \mathbf{x}_i^\top \boldsymbol{\beta}$, we cannot make inference by treating it as a linear regression model. As explained in Example 1.8, the normality-based MLE and ordinary least square (OLS) inference are not valid because of the so-called heteroskedasticity issue, as $\text{Var}(y_i \mid \mathbf{x}_i) = \mathbf{x}_i^\top \boldsymbol{\beta} \left(1 - \mathbf{x}_i^\top \boldsymbol{\beta} \right)$ varies with the covariates $\mathbf{x}_i$. We may apply robust inference to the estimating equations obtained by OLS, as the estimating equations (EE) inference is robust as long as the mean response is correctly specified. However, it may not be efficient because of the heteroskedasticity issue. The exact binomial-based MLE approach (4.39) actually weights the observations by the inverse of the variances of the responses and thus may be more efficient. However, for it to perform well, we need fitted probabilities to be well inside the region (far from endpoints of 0 and 1) to attain stable weights. If the probabilities can be very close to 0 or 1, then those subjects will have very large weights, making the maximum likelihood inference unreliable. In such cases, one may use estimating equation based robust inference to deal with the issue. More precisely, since $E (y_i \mid \mathbf{x}_i) = \mathbf{x}_i^\top \boldsymbol{\beta}$, it follows that

$$W_n (\boldsymbol{\theta}) = \frac{1}{n} \sum_{i=1}^{n} w_i (y_i - \mathbf{x}_i^\top \boldsymbol{\beta}) \mathbf{x}_i^\top = 0 \tag{4.40}$$

is unbiased for any weighting system w_i. One may put a ceiling on the weights to obtain bounded weights. For example, one may use the weight $w_i = \left(\mathbf{x}_i^\top \boldsymbol{\beta} (1 - \mathbf{x}_i^\top \boldsymbol{\beta}) \right)^{-1}$ if $\min(|\mathbf{x}_i^\top \boldsymbol{\beta}|, |1 - \mathbf{x}_i^\top \boldsymbol{\beta}|) < c$ and $= \frac{1}{c(1-c)}$ for some constant $c \in (0, 1)$; then the weights are bounded by $\frac{1}{c(1-c)}$. One may also simply use $w_i = 1$ (this is the EE obtained from OLS). However, use of this model should be avoided in such cases, since the fitted values often fall outside the permitted region.

Note that other link functions may still be applied. For example, the logarithmic link function may be applied and the resulting log-binomial model has the advantage of being directly related with the relative risk. For space considerations, we will not delve into the details. Interested readers may consult textbooks such as McNutt et al. (2003).

4.3.5   Comparison of the link functions

In theory, we can use any 1-1 function that maps $(0, 1)$ onto R as a link for modeling a binary response. However, the logit function has become the "standard" link and the resulting logistic regression the de facto method for modeling such a response. This model is available from all major statistical software packages such as R, SAS, SPSS, and Stata. One major reason for its popularity is the simple interpretation of parameters as the odds ratios of the response. In addition, the logistic regression is the only model for the binary response that can be applied to both prospective study designs and retrospective case-control study designs without altering the interpretation of model parameters (see Section 4.1.4).

Except for these key differences, however, the three link functions (logistic, probit, and complementary log-log) are very similar. For example, the CDFs for the three link functions all have the S shape. Plots of the cumulative distribution functions of logistic, normal, and extreme value distributions, with median 0 and variance 1, are plotted in Figure 4.1. Based on the plots, all three distributions are very similar, and the logit and probit functions are almost identical to each other over the interval $0.1 \leq \pi \leq 0.9$. For these reasons, it is usually difficult to discriminate between the models using goodness-of-fit tests. Although the estimates of parameters may not be comparable directly, the three models usually produce similar results, with similar interpretations. For example, as they are all monotone increasing functions, the corresponding model parameters indicate the same direction of association (signs) even though their values are generally different across the different models.

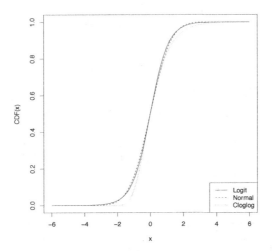

FIGURE 4.1: Cumulative distribution functions of logistic, normal, and extreme value distributions (median = 0, variance = 1).

The identity link function is more different; but, still the patterns are similar for the range not close to 0 and 1, say 0.2 to 0.8. Hence, as long as expected probabilities are not very close to 0 or 1, we would expect that the conclusions based on all these different link functions will be similar. Thus, choices of the link functions are often based on the convenience of interpretation and convention in the subject field. For example, the identity link function may be applied if we want to interpret the treatment effect in terms of a risk difference. However, as the identity link function is not a proper link function, it should in general be avoided, especially when there are subjects with expected probabilities close to 0 or 1.

It is interesting to point out that the logit function is also known as the *canonical link* for modeling binary responses (see Problem 4.13 (d)). The term "canonical" is derived from the *exponential family of distributions*. If y has a distribution from the one-parameter exponential family of distributions, then the density function of y can be expressed in the following form:

$$f(y \mid \theta) = \exp\left[(y\theta - b(\theta))/a(\phi) + c(y,\phi)\right], \qquad (4.41)$$

where θ is the canonical parameter (if ϕ is known) and ϕ is the dispersion parameter. The exponential family includes many distributions including normal, Bernoulli, binomial, exponential, etc. For example, for the normal distribution case, θ is the mean and ϕ the variance of the distribution. However, for the Bernoulli case, $\phi = 1$, but θ is not the mean (or probability of success) of the distribution. A canonical link is defined as a one-to-one function $g(\cdot)$ that maps the mean μ of the distribution to the canonical parameter θ, i.e., $g(\mu) = \eta = \theta$. It can be shown that for the exponential family defined above $\mu = \frac{d}{d\theta}b(\theta)$. Therefore, the canonical link function $g^{-1}(\cdot) = \frac{d}{d\theta}b(\cdot)$. Similarly, it is readily shown that the canonical link for a continuous outcome following a normal distribution is the identity function (see Problem 4.13).

Example 4.8

For the DOS, consider modeling depression diagnosis as a function of age, gender, education, marital status (MS), medical burdens (CIRS), and race. We treat age (centered at 75), education (Edu), and CIRS as continuous and gender, MS, and race as nominal variables. We use the MS = 3 as the reference, so the indicator variables corresponding to MS = 1 (MS1) and 2 (MS2) are included in the model. For race, we use the indicator variable for white race. So, we are fitting the following models

$$g(\pi) = \beta_0 + \beta_1 Age + \beta_2 Gender + \beta_3 Edu + \beta_4 CIRS + \beta_5 MS1 + \beta_6 MS2 + \beta_7 Race,$$

where $\pi = \Pr(Depd = 1)$ is the probability of being depressed and g is the link function. We fit the model with the four link functions (identity, logit, probit, and complementary log-log).

The linear probability model has convergence problems. Its minimum fitted probability when the estimating iteration stops is virtually zero ($< 2 \times 10^{-16}$). In fact, if a simple linear regression is applied, the fitted values may be negative. Thus, this model is not appropriate for the data. However, we still include it for comparison purpose. The parameter estimates for all the four models are provided in the following table.

Parameter	Identity Estimate	SE	p-value	Logistic Estimate	SE	p-value
Intercept	0.1086	0.1267	0.3911	−2.1711	0.6003	0.0003
Age	−0.0077	0.0029	0.0074	−0.0436	0.0132	0.0009
Gender	0.1405	0.0396	0.0004	0.7958	0.1843	< .0001
Edu	−0.0080	0.0071	0.2622	−0.0447	0.0334	0.1805
CIRS	0.0313	0.0063	< .0001	0.1707	0.0287	< .0001
MS1	−0.0442	0.0514	0.3899	−0.2554	0.2306	0.2680
MS2	0.0139	0.0566	0.8061	0.0681	0.2483	0.7839
Race	0.0791	0.0714	0.2678	0.5024	0.3313	0.1294
	Probit			Complementary Log-log		
Intercept	−1.3390	0.3620	0.0002	−2.0927	0.4653	< .0001
Age	−0.0264	0.00792	0.0009	−0.0340	0.0102	0.0008
Gender	0.4913	0.1100	< .0001	0.6134	0.1488	< .0001
Edu	−0.0269	0.0203	0.1839	−0.0339	0.0256	0.1853
CIRS	0.1054	0.0172	< .0001	0.1293	0.0207	< .0001
MS1	−0.1574	0.1403	0.2620	−0.1795	0.1791	0.3161
MS2	0.0422	0.1518	0.7811	0.0576	0.1880	0.7594
Race	0.3004	0.1985	0.1302	0.4165	0.2551	0.1025

It is seen from the analysis results that all the link functions give different estimates for the coefficients. However, the results are actually quite similar. First of all, all estimates retain the same direction of association (signs) across the different link functions. Age, gender, and CIRS are all very significant with the same direction, while Edu and race both have p-value over 0.1 under all the four models. For MS, there is no significant difference between MS = 2 and MS = 3, while p-values for the differences between MS = 1 and MS = 3 range from 0.0481 to 0.0908. The probit and logistic results look more closer, while the identity link is comparatively more different. Since the intercept being 0 has different meanings in the four models, we should not compare their estimates and p-values directly. In fact, the four intercepts correspond to response rates of 0.1086, 0.1024, 0.0903, and 0.1160 for the reference group (all covariates are 0), respectively, for these four models.

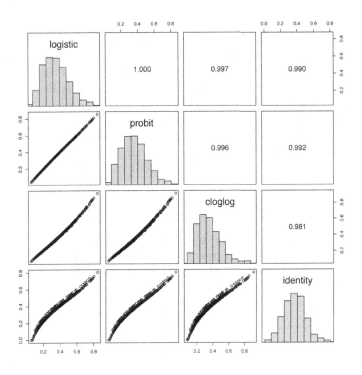

FIGURE 4.2: Fitted values under the four link functions.

We can also compare the fitted values. In Figure 4.2, the diagonal shows the histograms of the fitted values under the four link functions, and the lower triangle is their pairwise scatter plots. Based on the plots, it is clear that the fitted values are very similar, especially those of the logistic and probit models. This is confirmed by the estimated Pearson correlation coefficients, which are included above the diagonals. The correlation coefficient between the fitted values under the logistic and probit models is rounded to 1 in the plot. All the correlation coefficients are very high, with the one based on the identity link function comparatively more different from others. Thus, it is not surprising that it is hard to distinguish these models by goodness of fit. In fact, the Hosmer–Lemeshow goodness-of-fit tests, which will be discussed in the next section, show that all models fit the data well. □

4.4   Model Evaluation and Diagnosis

We discussed statistical inference about regression models based on the assumption that the models are correct. However, usually we don't have the "correct" models, or data-generating mechanisms; rather, we will be satisfied with models that well approximate the data at hand. Thus the question is how to obtain and determine an appropriate model. Models are usually evaluated by how well they fit the data, or goodness of fit. Models that do not fit the data well usually need to be modified. Model diagnosis may help find the issues and provide direction for improvement. However, we should not act solely based on goodness of fit when we select models. For space consideration, we defer the discussion of the model selection process to Chapter 7.

4.4.1   Goodness of fit

An important component of statistical modeling is assessment of model fit and evaluation of how well model-based predictions are in line with the observed data. Such a goodness-of-fit procedure includes detection of whether any important covariates are omitted, whether the link function is appropriate, or whether the functional forms of modeled predictors and covariates are correct.

Within the binary regression setting, the commonly used residual-based goodness-of-fit tests are the Pearson chi-square test, deviance statistic, and the Hosmer–Lemeshow test. All these tests are based upon comparing the observed vs. expected responses from various combinations of the independent variables.

4.4.1.1   Residuals for binary regression models

The differences between the observed outcomes and fitted values are the *raw residuals* for a regression model. For a binary regression with response y, let $\widehat{p}_i = \widehat{\Pr}\,(y = 1 \mid \mathbf{x}_i)$; the raw residual is $r_i = y_i - \widehat{p}_i$. The residuals can be standardized by the estimated variances. The Pearson residuals are defined as

$$r_i^P = \frac{y_i - \widehat{p}_i}{\sqrt{\widehat{p}_i\,(1 - \widehat{p}_i)}}. \tag{4.42}$$

Since the binary response can only take values 0 and 1, there are only two possible values for each $\widehat{p}_i$, depending on the fitted linear predictor $\mathbf{x}_i^\top \widehat{\boldsymbol{\beta}}$. Thus, a plot of the estimated linear predictor $\mathbf{x}_i^\top \widehat{\boldsymbol{\beta}}$ vs. the residual will have all the points on two curves. For example, for Pearson residuals, all the points fall on the curve $\left(t, -\frac{\exp(t)}{1+\exp(t)} \Big/ \sqrt{\frac{\exp(t)}{1+\exp(t)} \frac{1}{1+\exp(t)}}\right)$ or $\left(t, \frac{1}{1+\exp(t)} \Big/ \sqrt{\frac{\exp(t)}{1+\exp(t)} \frac{1}{1+\exp(t)}}\right)$. Thus, unlike in regular linear regression cases, visual assessment of the individual residuals in model assessment is usually of very limited use. Instead, statistics are usually derived from integrated residuals. For example, if the predictors are discrete, many subjects may have the same predicted values $\widehat{p}_i$ and the deviation of the observed proportion of these subjects with response $= 1$ from the fitted value $\widehat{p}_i$ may indicate lack of fit. In other words, the aggregated residuals can be very useful in the assessment of goodness of fit. In fact, the Pearson chi-square statistics we will introduce next are simply the sum of the squared aggregated Pearson residuals.

4.4.1.2 The Pearson chi-Square statistic

The Pearson chi-square statistics are mainly used for categorical independent variables. Given a sample of n subjects with a binary response and a number of categorical covariates, we can fit a logistic model and construct an $I \times 2$ table for the observed counts with the rows consisting of all I possible patterns of the categorical variables and the columns representing the categories of the binary response. With estimates from the fitted logistic model, we can then construct model-based expected counts for each pattern of the independent variables. The differences between the expected and observed counts will provide an indication as to how well the model fits the data.

Let $\widehat{p}_j = \Pr(y = 1 \mid X_j)$ be the fitted probability of response for the subjects with the jth covariate pattern $(1 \le j \le I)$. Then, $E_{j1} = n_{j+}\widehat{p}_j$ is the total of the fitted probabilities of response and $E_{j2} = n_{j+}(1 - \widehat{p}_j) = n_{j+} - E_{j1}$ is the total of the fitted probabilities of no response for the n_{j+} subjects with the jth covariate pattern. They are the expected cell counts for the response $(y = 1)$ and nonresponse $(y = 0)$ categories of y in the jth covariate pattern. This results in a table of expected cell counts corresponding to the observed counts in the data:

	Observed (expected)		Total
Covariate pattern	$y = 1$	$y = 0$	
X_1	$n_{11}\ (E_{11})$	$n_{12}\ (E_{12})$	n_{1+}
X_2	$n_{21}\ (E_{21})$	$n_{22}\ (E_{22})$	n_{2+}
$\vdots$	$\vdots$	$\vdots$	$\vdots$
X_I	$n_{I1}\ (E_{I1})$	$n_{I2}\ (E_{I2})$	n_{I+}

The Pearson chi-square goodness-of-fit test compares the observed to the expected counts using the following formula:

$$\chi_P^2 = \sum_{j=1}^{I}\sum_{k=1}^{2} \frac{(n_{jk} - E_{jk})^2}{E_{jk}}.$$

Under the null hypothesis of the correct model, the above statistic χ_P^2 has an asymptotic chi-square distribution with $I - l$ degrees of freedom, where l is the number of (independent) parameters that need to be estimated.

Since $n_{j2} + n_{j1} = E_{j2} + E_{j1} = n_{j+}$, it is straightforward to check that

$$\sum_{k=1}^{2} \frac{(n_{jk} - E_{jk})^2}{E_{jk}} = \frac{(n_{j1} - E_{j1})^2}{n_{j+}\widehat{p}_j} + \frac{(n_{j2} - E_{j2})^2}{n_{j+}(1 - \widehat{p}_j)} = \frac{(n_{j1} - E_{j1})^2}{n_{j+}\widehat{p}_j(1 - \widehat{p}_j)} = \frac{(\Sigma r_i)^2}{n_{j+}},$$

where in the last term the summation of the Pearson residual is over all the subjects in the jth covariate pattern. Thus, the Pearson chi-square statistic is also the sum of squares of the Pearson residuals for the aggregated data.

Example 4.9

In the DOS data, consider modeling the binary response of depression diagnosis as a function of gender x_{i1} ($= 1$ for female and $= 0$ for males) and dichotomized CIRS, $x_{i2} = 1$ if CIRS > 6 and $x_{i2} = 0$ if otherwise. The parameter estimates based on the grouped data are as follows:

Variable	Estimate	Standard Error	P-value
Intercept	-1.5157	0.1800	<0.0001
x_1	0.8272	0.1708	<0.0001
x_2	0.5879	0.1642	0.0003

In this case, $I = 4$ and there are four possible patterns. The observed and expected counts in the pattern can be summarized in the following table:

Covariate pattern	Observed (expected)		Total
	Depressed	No depression	
X_1 $(x_1 = 0, x_2 = 0)$	11(19.81)	99(90.19)	110
X_2 $(x_1 = 0, x_2 = 1)$	55(46.19)	108(116.81)	163
X_3 $(x_1 = 1, x_2 = 0)$	71(62.19)	115(123.81)	186
X_4 $(x_1 = 1, x_2 = 1)$	127(135.81)	159(150.19)	286

where E_{j1} (E_{j2}) are computed based on the fitted probabilities of response. For example,

$$E_{41} = n_4 \widehat{\pi} \left(x_{i1} = 1, x_{i2} = 1 \right),$$

where $n_4 = 286$ is the total number of subjects in the pattern X_4 and

$$\widehat{\pi} \left(x_{i1} = 1, x_{i2} = 1 \right) = \frac{\exp(-1.5157 + 0.8272 + 0.5879)}{1 + \exp(-1.5157 + 0.8272 + 0.5879)} = 0.4749.$$

Thus, the Pearson chi-square goodness-of-fit statistic is

$$\chi^2 = \frac{(11 - 19.81)^2}{19.81} + \frac{(99 - 90.19)^2}{90.19} + \frac{(55 - 46.19)^2}{46.19} + \frac{(108 - 116.81)^2}{116.81}$$
$$+ \frac{(71 - 62.19)^2}{62.19} + \frac{(115 - 123.81)^2}{123.81} + \frac{(127 - 135.81)^2}{135.81} + \frac{(159 - 150.19)^2}{150.19} = 10.087,$$

which follows a χ^2 distribution with $4 - 2 - 1 = 1$ degree of freedom. The corresponding p-value is 0.0015. Since the p-value is very small, we would reject the null. Note that if we include the interaction between gender and CIRS groups in the logistic model, then the number of parameters will equal the number of cells, and there will not be any unexplained variability left in the data, yielding a "saturated" model. Thus, the small p-value indicates that there is a significant interaction between gender and CIRS groups. ☐

Note that this test in principle is similar to several of the other chi-square tests we have discussed before, such as the test for single proportions and the chi-square test for general row by column associations for contingency tables. In fact, there is a general version of the Pearson chi-square test for assessing goodness of fit, which includes all these tests as special cases.

4.4.1.3   General Pearson chi-square tests

In general, given a sample for an outcome X, suppose we need to test whether it follows some parametric distribution model such as Poisson. If X is discrete with a finite number of categories, say K, a general Pearson chi-square statistic is defined as

$$\chi^2_{GP} = \sum_{j=1}^{K} \frac{(n_j - E_j)^2}{E_j}, \tag{4.43}$$

where n_j (E_j) is the observed (expected) number of subjects falling into the jth pattern $(1 \leq j \leq K)$. The expected number E_j is estimated based on the assumed model in most applications. Then, the statistic in (4.43) follows an asymptotic chi-square distribution with $K - 1 - s$ degrees of freedom, where s is the number of parameters that need to be estimated from the sample. As seen below, many chi-square statistics and tests can be derived from this general test.

1. Test the goodness of fit for a multinomial distribution with K levels. Under the null hypothesis, the multinomial distribution is given, i.e., $\mathbf{p} = \mathbf{p}_0$, where $\mathbf{p} = (p_1, \ldots, p_K)^\top$ is the parameter vector for the multinomial distribution and $\mathbf{p}_0$ is a $K \times 1$ vector of known constants. In this case, we can compute E_j without having to estimate $\mathbf{p}$, and the chi-square statistic, $\chi^2_{GP} = \sum_{j=1}^{K} \frac{(n_j - E_j)^2}{E_j}$, follows asymptotically a chi-square distribution with $K - 1$ degrees of freedom. As a special case for a binomial variable, $K = 2$.

2. Test row by column associations for 2×2 tables. Under the null hypothesis of no association, two parameters need to be estimated, and the chi-square distribution has one degree of freedom: $2 \times 2 - 2 - 1 = 1$.

3. Test row by column associations for general $r \times s$ contingency tables. Under the null hypothesis of no association, we need to estimate $r - 1$ parameters for the marginal distribution of the row variable and $s - 1$ parameters for the column variable, and the asymptotic chi-square distribution has $(s - 1)(r - 1)$ degrees of freedom: $r \times s - (r - 1 + s - 1) - 1 = (s - 1)(r - 1)$.

4. Test goodness of fit for the logistic regression model with k parameters. To compute the expected counts in each cell, we need to estimate $I - 1$ parameters for the marginal distribution of the X patterns and k parameters from the model. Thus, based on the general Pearson chi-square test theory, the asymptotic chi-square distribution has $I - k$ degrees of freedom: $2I - (I - 1 + k) - 1 = I - k$.

One problem with using the test is that the result may not be reliable if some expected cell counts are small. This may become inevitable if there are numerous covariates, since the number of possible patterns will increase exponentially with the number of covariates. Exact methods may be used, but they are usually computationally intensive and may not be of practical use.

The Pearson chi-square goodness-of-fit test may also be applied to general situations when some of the components of X are continuous. A technique often used is to group those continuous variables into groups using some cut-points. If we treat the grouped patterns as original data, then the chi-square statistics described above can be readily applied. But, as pointed out by Chernoff and Lehmann (1954), if the original observations are available, one would wish to use more efficient estimates, such as the MLEs based on all the data, rather than their grouped counterparts. If we compute the estimated cell counts using the original observations of X instead of just the grouped data, the resulting chi-square statistic follows an asymptotic distribution between chi-square distributions with $K - 1 - s$ and $K - 1$ degrees of freedom. Actually, Chernoff and Lehmann (1954) proved that the asymptotic distribution is a linear combination of chi-square variables with the coefficients of the linear combination ranging in $(0, 1)$. The main obstacles of using the approach in real study data are how to group the data and that it is in general difficult to compute the coefficients in the asymptotic distribution.

4.4.1.4 The deviance test

The likelihood-based deviance statistic can also be applied to assess the goodness of fit. Similar to the Pearson test statistic, the deviance statistic also compares the observed to the expected counts, but instead uses the following formula:

$$\chi^2_D = 2 \sum_{j=1}^{I} \sum_{k=1}^{2} n_{jk} \log \frac{n_{jk}}{E_{jk}}. \tag{4.44}$$

Since $\log \frac{n_{jk}}{E_{jk}} = \log \frac{n_{jk}}{n_{j+}} - \log \frac{E_{jk}}{n_{j+}} = \log \frac{n_{jk}}{n_{j+}} - \log \widehat{p}_j$, it is straightforward to verify that the the deviance statistic (4.44) is twice the difference between the likelihood of the current model and that of the saturated model where the fitted probability of response for the jth covariate pattern is exactly $\frac{n_{j1}}{n_{j+}}$. The asymptotic distribution of χ_D^2 is the same as the Pearson chi-square test, i.e., χ_D^2 has an asymptotic χ^2 with $I - l$ degrees of freedom (see Problem 4.19).

Example 4.10

The deviance statistic for the model in Example 4.9 is

$$2\left(11\log\frac{11}{19.81} + 99\log\frac{99}{90.19} + 55\log\frac{55}{46.19} + 108\log\frac{108}{116.81} + 71\log\frac{71}{62.19}\right.$$
$$\left. +115\log\frac{115}{123.81} + 127\log\frac{127}{135.81} + 159\log\frac{159}{150.19}\right) = 10.703,$$

and the associated p-value is 0.0011.   ◻

The deviance test also suffers from the sparse data problem. If some expected cell counts are small, it is not reliable.

4.4.1.5   The Hosmer–Lemeshow test

One important limitation of the Pearson and deviance tests is that they are not appropriate when there are continuous independent variables (if we don't group them). Although the observed values for a continuous variable are finite (at most equal to the sample size), none of the tests will work if the cell size is 1 for each of the covariate patterns. The latter is quite likely since a continuous outcome in theory has a zero chance to yield identical values. Even if some of the cells have more than one observation, the test statistic will not follow an asymptotic or approximate chi-square distribution for either the Pearson or the deviance test. This is because for a continuous variable, the number of distinct values will grow at the rate of sample size n.

Continuous independent variables are not the only obstacle to the two tests. For example, if we have four binary independent variables, we will have a total of 32 different combinations or patterns. This number jumps to 64 when an additional binary variable is added. Thus, neither test will work well if the logistic model contains a large number of covariates unless the sample size is extremely large.

Thus, to obtain a reasonably small number of patterns for analysis, we may need to take all variables into consideration when grouping them. Suppose there are s patterns after grouping; then, the Pearson chi-square test may be applied to the grouped $s \times 2$ contingency table based on the grouped data. However, as mentioned in Section 4.4.1.3, Chernoff and Lehmann (1954) also proposed a test to use the original data, which might be more efficient. Although elegant, the theory has two major problems that prevent it from being used in practice. The first is that the division of the X-region is usually arbitrary. The other is that it is generally difficult to compute the coefficients of the linear combination of the chi-square variates to obtain the asymptotic distribution of this statistic. To address these limitations, Hosmer and Lemesbow (1980) developed a procedure to create a set of patterns of covariates by grouping the values of these variables using the fitted probabilities.

The Hosmer and Lemeshow approach first orders the subjects according to the fitted probabilities of the responses and then groups them into ten (or possibly less) groups of comparable sizes. Model-based expected cell counts are then computed in the same way as before, and the test statistic is constructed as

$$\chi^2_{HL} = \sum_{j=1}^{g} \sum_{k=1}^{2} \frac{(n_{jk} - E_{jk})^2}{E_{jk}}. \qquad (4.45)$$

Based on simulation studies, this test statistic has approximately a chi-square distribution with $g - 2$ degrees of freedom (Hosmer and Lemesbow, 1980). The Hosmer–Lemeshow test is widely used in logistic regressions. It circumvents the difficulty in computing the Chernoff and Lehman test statistic and is easy to implement. However, we would like to emphasis that the degrees of freedom $g - 2$ for the Hosmer–Lemeshow test is based on empirical evidence from limited simulation studies, with no theoretical justification. In particular, the chi-square distribution of the statistic in (4.45) is approximate, rather than asymptotic as in the case of the distributions of the general Pearson chi-square and deviance statistics.

Example 4.11

For the models in Example 4.8 DOS, there are several independent variables. Since age, education, and CIRS are continuous covariates, we may use the Hosmer–Lemeshow test to assess goodness of fit of the model. To apply the Hosmer–Lemeshow test, we first divide the subjects into ten groups and then compute the χ^2_{HL} statistic and compare it against a chi-square distribution with eight degrees of freedom to determine the p-value. Shown in the table below are the observed (O), expected (E), and related quantities needed to compute this statistic when the logit link function is applied.

Group	Total	Dep = 1			Dep = 0		
		O	E	$(O-E)^2/E$	O	E	$(O-E)^2/E$
1	73	5	10.01	2.507502	68	62.99	0.398478
2	73	13	14.39	0.134267	60	58.61	0.032965
3	74	21	18.06	0.478605	53	55.94	0.154516
4	73	25	20.68	0.902437	48	52.32	0.356697
5	73	26	23.60	0.244068	47	49.40	0.116599
6	73	28	26.47	0.088436	45	46.53	0.050309
7	73	33	29.43	0.433058	40	43.57	0.292515
8	73	28	32.73	0.683559	45	40.27	0.555572
9	73	34	37.49	0.324889	39	35.51	0.343005
10	73	47	47.15	0.000477	26	25.85	0.000870

The Hosmer–Lemeshow statistic is 8.0974 with a p-value = 0.4240. Thus, we will not reject the good fit of the model. If the probit and complementary log-log link function are used, the Hosmer–Lemeshow statistics are 7.0048 and 8.3057 with p-values 0.5361 and 0.4042, respectively. Thus, no issue of goodness of fit is detected by the Hosmer–Lemeshow test for the three link functions. □

Note that there may be different ways to divide the subjects; thus it is possible to obtain different numbers. But usually the numbers are similar. However, the number of groups may have a large impact on the test. Note also that the Hosmer–Lemeshow test applies when there are continuous or many discrete covariates to form a large number of patterns. However, if there are only a few patterns in covariates as in the case of a few binary covariates, the Hosmer–Lemeshow test is not appropriate. If you request the Hosmer–Lemeshow test in statistic packages such as SAS in this case, it may happen that the statistic computed is actually the Pearson statistic. However, the p-value is computed against the chi-square with degrees of freedom $s - 2$, where s is the number of patterns assumed under the Hosmer–Lemeshow test. Hence, the p-value may not be correct.

4.4.1.6   Tests for sparse data

Instead of grouping sparse data as in the Hosmer–Lemeshow test, asymptotic tests which can be applied to address sparsity were developed. For example, Osius and Rojek (1992) derived the mean and variance of the Pearson chi-square statistic under the null hypothesis and proposed a test by standardizing the Pearson chi-square statistic. Copas (1989) and Spiegelhalter (1986) derived the asymptotic distributions of the unweighted total of the residual squares. These tests rely on the total sample size to be large, but do not require that the sizes of all the different covariate patterns be large.

Another common technique that may also be applied to assess general model goodness of fit is adding terms to the model and testing whether they are significant. Usually terms derived from the fitted values are added to assess the overall goodness of fit. For example, one may add a factor based on the fitted values, i.e., decompose the data according to the estimated linear predictor η_j as in Hosmer–Lemeshow test, and then test whether the factor is significant. Horton (1999) proposed the approach for longitudinal data. Stukel (1988) suggested adding two covariates $\eta_j^2 I (\eta_j \geq 0)$ and $\eta_j^2 I (\eta_j < 0)$ and testing whether the two terms are significant. Pregibon (1980) suggested checking whether η_j^2 is significant in a new model for the same binary response and the same link function, but with only intercept, η_j and η_j^2 in the linear predictor. This is generally referred to as *Pregibon's link test*. Terms involving specific variables may be added to test specific aspects of the models; see Section 4.4.2.1 for an example.

4.4.2   Model diagnosis

There are various data issues that can cause lack of fit. It is possible that the model is good, but there are outliers in the data that cause the issue of lack of fit. Since there are only two levels in the response variable, the outliers, if any, must be for covariates. When there are observations with covariate values far away from others, we may need to double check whether the values are correct. Such values often make the linearity assumption suspect and may have huge impact on inference. We will discuss how to identify such observations in the next section.

The most common cause of lack of fit is model misspecification. It may be that some important covariates are missing in the model or the linear assumption in the linear predictor is not satisfied. When a continuous independent variable is included in the linear predictor, a linearity relation is assumed for the covariate. However, this may not be true for the covariate on its original scale. Many constructs can be measured in different ways, resulting in measurements in different scales. For example, the kinetic energy of an object may be equivalently measured by its speed, and the energy is a quadratic function of the speed. If one scale satisfies the linearity relation, then the other cannot. In such cases, we need to transform the observation to the correct scale. In practice, covariates are often transformed before they are modeled. Commonly used transformation includes the logarithm, square, and square-root transformations. However, limiting transformation to these choices may not be sufficient.

4.4.2.1   Test for linearity

In general, we may test the linearity assumption of a variable by adding functions of the variable as terms in the model and testing whether these additional terms are significant. These additional terms, if significant, will make the relation nonlinear. The commonly used *Box-Tidwell test* simply involves adding the product of a continuous variable and its logarithm. More precisely, to assess the linearity of a variable x_1 in the logistic model

$$\text{logit}(\Pr(y_i = 1 \mid \mathbf{x}_i)) = \beta_0 x_{i0} + \beta_1 x_{i1} + \ldots + \beta_p x_{ip} = \boldsymbol{\beta}^\top \mathbf{x}_i, \qquad (4.46)$$

we add $x_1 \cdot \log x_1$ to the linear predictor to obtain the following model

$$\text{logit}(\Pr(y = 1 \mid \mathbf{x})) = \alpha x_1 \cdot \log x_1 + \boldsymbol{\beta}^\top \mathbf{x}.$$

If the coefficient of $x_1 \cdot \log x_1$, α, is significantly different from zero, we conclude that the relation is not linear and (4.46) must be modified.

Example 4.12
To assess the linearity assumption for CIRS in the model in Example 4.11, we test the coefficient of the added term CIRS * log (CIRS). The point estimate of the coefficient is 0.1213, with standard error 0.1023, which produces a p-value of 0.2357. Thus, no violation of the linearity assumption is detected by the Box-Tidwell test. □

If the linearity assumption is not satisfied, one may group the continuous variables into categorical variables. However, we may have issues such as how to determine the number of and cut-points for groups. Furthermore, this is usually companied by information loss. Another approach is to find some transformations of the variables so that the linearity assumption may be approximately true after the transformation.

4.4.2.2 Scale selection

In practice, logarithm and power functions, such as quadratic and square root, are common transformations. However, it may not be sufficient to limit choices to these transformations. In this section, we discuss some techniques that may be useful in searching for appropriate transformations.

4.4.2.2.1 Power transformations
A simple approach to identify an appropriate transformation is to search among all the functions in a family such as power functions of the form x^p. By treating p as a parameter, we can estimate p based on some optimal criteria. Thus, we have the following "non-linear" models

$$\text{logit}(\Pr(y_i = 1 \mid \mathbf{x}_i)) = \beta_0 x_{i0} + \beta_1 x_{i1}^p + \beta_2 x_{i2} + \ldots + \beta_s x_{is}, \qquad (4.47)$$

where x_1 is the predictor for which an appropriate transformation is to be determined. The model (4.47) with p as a parameter does not belong to the family of GLMs, as the right-side of (4.47) is not a linear function. In theory, we can use maximum likelihood methods to find the optimal p. However, there are several issues with this approach. First, this greatly increases the complexity of the estimation problem. Second, the estimated optimal power p in general will be a decimal number which may complicate interpretation. For example, if we assess the optimal power of CIRS in Example 4.11, the optimal power estimated by SAS PROC NLMIXED is 2.1841. It would be strange to use such a transformation. The 95% confidence interval of the optimal power is $(-0.5712, 4.9394)$, indicating that $p = 1$, or the original linear relation, is also good. Thus, testing whether $p = 1$ based on (4.47) provides us another test for linearity.

Another issue with power transformations with power as a parameter is that commonly used transformations such as the logarithmic transformation are not included. Further, we may need to use multiple terms of the variable to properly model its relation with the outcome. For example, suppose the continuous variable is the only predictor in the model; then no matter which power we adopt, its relation with the outcome must be monotone. Multiple terms are required to model nonmonotone relations.

4.4.2.2.2 Fractional polynomials Royston and Altman (1994) proposed fractional polynomial selection to search through a small useful set of possible values for powers in the fractional polynomials. Consider the logistic regression model

$$\text{logit}\,(\text{Pr}(y = 1 | u, \mathbf{x})) = \alpha u + \mathbf{x}^\top \boldsymbol{\beta},$$

where u is a continuous predictor for which we determine an appropriate transformation. The log-odds of the outcome is linear in the variable u. One way to generalize this logistic model is to replace the linear term in u by a function with J fractional power and/or log terms in u,

$$\text{logit}\,(\text{Pr}(y = 1 | u, \mathbf{x})) = \sum_{j=1}^{J} \alpha_j F_j(u) + \mathbf{x}^\top \boldsymbol{\beta},$$

where each $F_j(u)$ is a particular type of power function and/or log function of u. In theory, the power could be any number, but in most applied settings we try to use simple power functions. Royston and Altman (1994) propose restricting the power to $-2, -1, -0.5, 0, 0.5, 1, 2$, and 3, with power 0 indicating $\log(u)$. Ignoring the coefficients α_j, a J-term fractional polynomial is determined by its J powers, $p_1, ..., p_J$. We may order the power terms by the powers, say from smallest to largest. The powers can be repetitive, in the understanding that each repetitive term signifies a product of $\log(u)$ with the term. Thus, if $p_j = p_{j-1}$, then the jth term $F_j(u) = F_{j-1}(u)\log u$. If $p_j \neq p_{j-1}$, then $F_j(u) = u^{p_j}$. By this definition, a J-term polynomial truly has J different terms. Some of the fractional polynomials are given in the following table.

J	p	Fractional Polynomials
1	0.5	$\alpha_1 \sqrt{u}$
1	0	$\alpha_1 \log u$
2	$-0.5, 0$	$\frac{\alpha_1}{\sqrt{u}} + \alpha_2 \log u$
2	$2, 2$	$\alpha_1 u^2 + \alpha_2 u^2 \log u$
2	$1, 2$	$\alpha_1 u + \alpha_2 u^2$

In most applied settings, we will limit the use of fractional polynomials to $J = 1$ or 2, although one can certainly use higher Js in theory. If the powers are further limited to those eight numbers, then there are eight models when $J = 1$. We may determine the optimal one based on the deviance or log-likelihood. The best model is the one with the largest likelihood or, equivalently, the smallest deviance, -2 times the likelihood. Similarly, there are 36 models when $J = 2$, including 28 pairs where $p_1 \neq p_2$ and eight pairs where $p_1 = p_2$, and the one with the largest log partial likelihood will be treated as the best two-term fraction polynomial.

Models using two-term fractional polynomials are more complicated than one-term fractional polynomials. As we will discuss in Chapter 7 in more detail, we may want a balance between model fit and model simplicity. In other words, we would choose a two-term fractional polynomial only if it is significantly better than the one-term fractional polynomials. Similarly, because of easy interpretation, we would choose a one-term transformation over the original scale (linear) only if it is significantly better than the original linear scale.

Let $L(1)$ be the estimated log likelihood for the linear model ($J = 1$ and $p_1 = 1$), $L(p_1)$ be the log likelihood for the best $J = 1$ model, the one with the largest estimated log likelihood, and $L(p_1, p_2) = $ the log likelihood for the best $J = 2$ model. Based on simulation studies, each term in the fractional polynomial model contributes approximately two degrees of freedom to the model: one for the power and one for the coefficient (Royston and Altman, 1994). Thus, we compare $L(p_1, p_2) - L(p_1)$ with χ_2^2, $L(p_1) - L(1)$ with χ^2, and $L(p_1, p_2) - L(1)$ with χ_3^2 to decide which scale to adopt.

Example 4.13

For the model in Example 4.11, there are three continuous independent variables: age, Edu and CIRS. We use the multivariate fractional polynomial approach to examine their scales. The deviance $(-2L)$ for the different models for education (gender, age, and CIRS are included in the model) is summarized in the following table.

Function	m	p1	p2	Deviance	diffra2	pdiffdev
Omitted	−1			898.066	7.269	0.12235
Linear	0			896.774	5.977	0.11274
First Degree	1	−2		891.364	0.567	0.75307
Second Degree	2	−2	−2	890.797	0.000	1.00000

The table contains information from four models; "Omitted" is the model when education is not included; "Linear" is the model using the original scale of education, number of academic years a person completed; "First Degree" is the best one-term polynomial model, and the corresponding power in the best one-term model is given as p1; "Second Degree" is the best second-degree polynomial model, and the corresponding powers in the best two-term model are given as p1 and p2. Thus, for education, the best one-term polynomial is $(1 + \text{education})^{-2}$, and the best two-term polynomial is $(1 + \text{education})^{-2} + (1 + \text{education})^{-2} \log(1 + \text{education})$. Note that since the the minimum value of the education variable is 0, which is not accommodated by fractional polynomials, we shifted it by 1 and use fraction polynomials of $(1 + \text{education})$ instead.

The column "diffra2" contains the differences in deviances for each model vs. the last model, the best second degree model here. Compared with chi-square with four, three, and two degrees of freedom, we obtain the p-values in the last column "pdiffdev." Thus, the second degree model is not significantly better than any of the other models. Comparing the one-term model with the linear model, the difference in the deviance is $896.774 - 891.364 = 5.41$, resulting in a p-value of 0.02 when referred to a chi-square distribution with one degree of freedom. Thus, the one-term model using education^{-2} is significantly better than the original linear scale.

For age, the best one-term model is age^3, and the best two-term model is $\text{age}^3 + \text{age}^3 \times \log(\text{age})$. However, they are not significantly better than the original scale, age. Thus, we will use the original scale for age. For CIRS, the best one-term model is CIRS^2, and the best two-term model is $\text{CIRS}^2 + \log(\text{CIRS})$. Again, none of them is significantly better than the original scale. Thus, we stay with the original scales for age and CIRS. □

Models may be misspecified even when the scales for continuous covariates are correctly specified. For example, it often occurs that there are interactions between two or more covariates, but such interaction terms are often not included in the model. The link function may also be incorrect. In such cases, it is important to refine the model. Selecting a model that fits the data well is an age-old problem that continues to present challenges for statisticians as research questions and models for addressing them become more complex. We will discuss general *model selection* methods that apply to regression models in Chapter 7.

4.4.2.3   Influence analysis

Not all observations in a data set play an equal role in inference about the model. In some cases, a few particular observations may have an unusually large influence on the results of analysis. This does not necessarily mean that the analysis is inappropriate, but it is important to identify such observations and assess the validity of the analysis accordingly. The general idea of influence analysis is to assess changes in some statistics when one observation is removed from analysis. In general, we would like to have similar conclusions if a subject is excluded from analysis. Thus, we would pay special attention to those subjects whose removal may cause a huge difference on the inference. All statistics including those for goodness of fit, such as Pearson goodness-of-fit statistics, and residual deviance statistics as well as parameter estimation may be examined in influence analysis.

For example, Pregibon's *dbeta* is a standardized measure of the difference $\widehat{\beta}^i - \widehat{\beta}$, where $\widehat{\beta}$ is the estimate using the whole sample and $\widehat{\beta}^i$ is the estimate of the parameter when the ith observation is deleted from the data. Thus, it summaries the overall influence on estimates of all parameters by each individual observation. Note that Pregibon's *dbeta* is also defined to assess the influence of covariate patterns where all subjects that share the same covariate patterns are deleted. Larger values of dbeta indicate higher influence. We can also check for specific parameters. For example, the jth component of $\widehat{\beta}^i - \widehat{\beta}$ measures the impact of the ith observation on the estimation of this specific parameter. How large a change that may constitute a concern is generally subjective. For example, for linear regression, Belsley, Kuh, and Welsch (1980) suggested that observations with the standardized changes, $\left|\widehat{\beta}_j^i - \widehat{\beta}_j\right|/\widehat{\sigma}$, larger than $2/\sqrt{n}$ may deserve special attention, where $\widehat{\sigma}$ is the estimated standard error for $\widehat{\beta}_j$ based on the full sample and n is the sample size, while Bollen and Jackman (1990) suggested a cut-point of 1 instead. It is more important to visualize plots of the influence statistics to assess if there are subjects with much higher influence than others.

When there are highly influential observations, we should double check the data validity and perform model diagnosis such as checking the linearity of related predictors. If none of these checks helps with the situation, then it is generally recommended to report both sets of results based on the whole sample and the one with the highly influential observations is deleted.

Note that strictly speaking, influence analysis needs to be performed for each subsample resulting from the removal of each observation. This can be time-consuming if the sample size is large. Some approximate procedures which do not require model estimation for each subsample are commonly utilized in practice (Storer and Crowley, 1985).

Example 4.14

For the model in Example 4.11, observation number 388 has the largest dbeta, 0.1546. Comparing the estimates with and without the observation included, the results are very similar. Thus, we may conclude that there is no highly influential observation affecting parameter estimation in this example. A plot of the dbeta vs. the subjects (index in the data set) also shows no sign of high-influential subjects. We may also look at the influence on individual parameter estimates. For example, the standardized changes in the estimate of the coefficient of age vs. the subjects are plotted in Figure 4.3. Observation number 153 has the largest influence on this coefficient, with a standardized change score of 0.1794. This is well under the threshold recommended for linear regression by Bollen and Jackman (1990), but still above the one recommended by Belsley, Kuh, and Welsch (1980). We can again compare the estimates with and without this observation. As the results are similar, we may conclude that there is no highly influential observation in this example.   ⬜

4.4.3   Predictability and calibration

The procedures for goodness of fit test and model diagnosis only assess the relation between the outcome and the covariates. Thus, a model with good fit may not necessarily mean a good model, if important predictors are missing from the model. For example, for the logistic regression for the depression outcome for the DOS, if we only include gender,

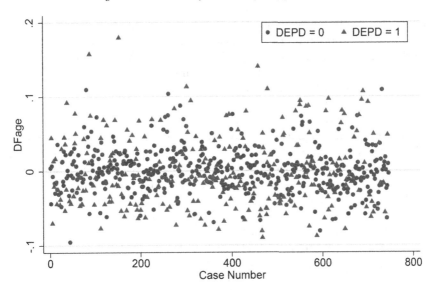

FIGURE 4.3: Influence on the coefficient of age.

then the model will fit the data perfectly as it is a *saturated model*, i.e., a model that has as many parameters as there are values to be fitted. The observed counts will be exactly the same as the expected ones for each of the covariate patterns, and the Pearson and deviance chi-square statistics will both be zero, indicating, again, perfect fit. However, the model is only of very limited use, as important predictors such as age and medical burden are not included in the model. This indicates the importance of assessing the predictability of a model.

The predictability of a model can in general be assessed by evaluation of how well model-based predictions are in line with the observed data. Measures on the agreement between two variables such as the correlation coefficient and concordance-based indices such as Somers' D may be applied to the linear predictor and the binary outcome to indicate the predictability of the model. The higher these indices, the better the model. However, there are no cut-points for these indices to identify good or bad models. Note that for linear regression, the coefficient of determination, or R^2, which can be interpreted as the proportion of the variance in the dependent variable that is predictable from the independent variables, is a common measure of predictability. Such an R^2 does not exist for binary regression models. However, pseudo R^2s have been proposed. For example, the slope of a linear regression of predicted probabilities of event on the binary event status, called the *discrimination slope*, is a popular measure of model performance. This discriminant slope, which Tjur (2009) terms the coefficient of discrimination, is the difference between the mean fitted probabilities when $y = 1$ and $y = 0$ (Yates, 1982). See also several other asymptotically equivalent versions of pseudo R^2s in Tjur (2009).

Because of the discrete nature of the response, another approach to assess model predictability is to check its ability to correctly classify the subjects' responses based on the linear regressor. The ability of the linear predictor in classifying binary responses is usually described by the receiver operating characteristic (ROC) curve, with the area under the ROC curve (AUC) as a common summarizing index. In such discrimination assessment of a prediction model, we focus on the ranking or relative risks of the individuals, without caring about the absolute probability of having an event. We defer the discussion of ROC curves to Chapter 10.

Considering the absolute probability of having an event, *model calibration* compares the observed and predicted probabilities. The goodness of fit tests we discussed above are examples of model calibration. They are usually referred to as *internal calibration* as the sample for the validation, the comparison of the observed and predicted probabilities, is the same as the one in which the model was developed. In practice, *external calibration* is also important in model validation. External calibration uses different samples from the one in which the model was developed. In the analysis of large data sets, it is common to randomly separate the data into training and validation samples, where the training data set is used to fit the model and the validation data set is used to provide an external evaluation of the model fit (sometimes a third, test data set, may be utilized to assess the model after validated by the training and validation data sets). A common reason for the lack of external calibration for models that fit the training data well is the overfitting of the original model (Steyerberg et al., 2010). For example, a saturated model which fits the training data perfectly in general will not have perfect fit in external calibration.

A common procedure for calibration is logistic regression of the observed binary outcomes on the fitted log odds, proposed by Cox (1958). More precisely, let $\widehat{p}_i$ be the predicted probability of $y = 1$; then the logistic model

$$\log \frac{\Pr(y_i = 1)}{\Pr(y_i = 1)} = \beta_0 + \beta_1 \left(\log \frac{\widehat{p}_i}{1 - \widehat{p}_i} \right)$$

will have slope $\beta_1 = 1$ and intercept $\beta_0 = 0$ under perfect calibration, i.e., when the observed and predicted probabilities agree perfectly. The intercept compares the mean predicted probability of the occurrence of $y = 1$ with the mean outcome; a positive (negative) intercept indicates overall under(over)-estimate of the probability. The slope, commonly referred to as the *calibration slope*, indicates the direction of miscalibration. A slope bigger than 1 suggests underestimation of high risk and overestimation of low risk; on the other hand, a slope between 0 and 1 indicates underestimation of low risk and overestimation of high risk. The slope may be negative if the predicted probabilities are in the wrong direction; if $\widehat{p}_i$ are the complements of the true probabilities, the slope will be -1. For a model developed in a relatively small data set, the slope is often between 0 and 1. Note that we need to use both the intercept and slope to assess the calibration (Stevens and Poppe, 2020). Some graphical methods are also available to assess calibration (Austin and Steyerberg, 2013).

Example 4.15
For the model in Example 4.11, we have the following measures for the association of predicted probabilities and observed responses: Somers' D 0.354, Gamma 0.354, and AUC 0.677. On the other hand, for the model with gender as the only predictor, the corresponding values are Somers' D 0.180, Gamma 0.388, tau-a 0.083, and AUC 0.588. Thus, the model in Example 4.11 has much better predictability than the gender-only model.   ▯

4.5   Models for Aggregated Binary Response

For discrete data, many subjects will have identical values in all the variables, and we often aggregate these subjects by counting the total number of the subjects with a specific covariate pattern. Contingency tables we studied in Chapter 2 are such examples. If the data aggregation is performed over independent subjects with the same covariate patterns, then the aggregated counts follow the binomial distribution, and binomial regression may be applied. This binomial regression will be equivalent to the binary regression on the original data before the aggregation. However, we must be mindful that if the aggregation is carried out over correlated observations, it is not appropriate to apply the binary regression to the unaggregated data and treat them as independent. Because of the correlation, the aggregated counts will not follow the binomial distribution, and bias will occur if a binomial regression is applied. While advanced models are available to model correlated binary outcomes (see Chapter 9), we can still apply binomial regressions with some ad hoc corrections. In this section, we will first discuss binomial regression and then discuss potential issues for modeling correlated outcomes under this model and how to deal with such issues.

4.5.1 Binomial regression

Suppose the data are in the format of $(\mathbf{x}_i, y_i, m_i)$, which represents m_i subjects with covariates $\mathbf{x}_i$, where y_i of them experienced outcome 1 and the remaining $m_i - y_i$ subjects had outcome 0. If the original binary outcome follows the same Bernoulli(p_i), with $p_i = g\left(\mathbf{x}_i^\top \boldsymbol{\beta}\right)$, then y_i follows a binomial distribution with size m_i and success rate $p_i = g\left(\mathbf{x}_i^\top \boldsymbol{\beta}\right)$. The GLM for binomial outcomes can be expressed as

$$y_i \mid \mathbf{x}_i \sim BI(m_i, p_i), p_i = g\left(\mathbf{x}_i^\top \boldsymbol{\beta}\right), \tag{4.48}$$

where g is a link function such as the logit function. Note that the binomial size m_i does not need to be the same for all i. The interpretation of the parameters is the same as those in the binary model.

Example 4.16
For the Catheter Study, all the patients were followed up to one year, and the patients were asked bimonthly whether they experienced any urinary tract infections (UTI), catheter blockages, and catheter displacements during the last two months. Thus, each patient was asked up to six times about the three binary outcomes. Due to death or dropout, some patients were asked less than six times, so the total number of times asked varies from 1 to 6. The total number answering "yes" can be analyzed by a binomial regression model. Including variables age, gender, education, and treatment group as predictors, we consider the following binomial regression model, holding the question of correlation for the moment:

$$y_i \mid \mathbf{x}_i \sim BI(m_i, p_i), \text{logit}(p_i) = \beta_0 + \beta_1 Age_i + \beta_2 Gender_i + \beta_3 Education_i + \beta_4 Group_i, \tag{4.49}$$

where m_i is total number of being asked for the questions.
Estimation for the UTI outcome is summarized in the following table.

Parameter	Estimate	Standard Error	Wald Chi-Square	Pr > ChiSq
Intercept	-0.3042	0.3174	0.9186	0.3378
Age	-0.0177	0.00445	15.8372	$< .0001$
Gender	0.3128	0.1536	4.1474	0.0417
Education	-0.0241	0.0471	0.2612	0.6093
Group	0.1947	0.1514	1.6537	0.1985

The interpretation of the parameter is the same as that for binary regression. For example, here gender is a significant predictor at 5% significance level, with estimated odds ratio between females and males being $\exp(0.3128) = 1.3672$. In fact, the results of the binomial regression are equivalent to reformatting the data as $\sum_{i=1}^{n} m_i$ individual binary responses, which would treat all the individual answers as independent.

The Hosmer–Lemeshow test for goodness of fit has a chi-square statistic 16.1324, with p-value 0.0405, indicating that the model does not fit the data well. For binomial regression, another common cause of lack of fit is overdispersion, usually caused by data clustering. □

4.5.2 Overdispersion issues

An implicit assumption underlying the binomial distribution is that it models the sum of independent Bernoulli variables. If these Bernoulli components are correlated, the variance of their sum may be very different from the one based on the binomial model. For example, suppose $y_i = \sum_{j=1}^{n_i} y_{ij}$, where the $y_{ij} \sim Bernoulli\,(p_i)$ and are positively correlated with $corr\,(y_{ij}, y_{ik}) = \alpha > 0$ for $k \neq j$. Then, the variance of y_i is $Var\,(y_i) > n_i p_i(1 - p_i)$ (see Problem 4.18). This phenomenon is known as *overdispersion*. When overdispersion occurs, the expectation of y_i is still the same as that of the binomial model, but the variance exceeds that of the binomial, causing bias when making inferences using the maximum likelihood method. Since the variance is underestimated, false significant findings may result if the overdispersion issue is ignored.

4.5.2.1 Detection of overdispersion

Overdispersion can be detected using goodness-of-fit tests. As we discussed previously, both the Pearson and deviance chi-square statistics can be applied to assess the goodness of fit for these models. If after carefully assessing the scale, the linearity form, and the link function and we are confident of the model for the mean response, then the lack of fit detected by the two statistics may be attributed to overdispersion. If there is no overdispersion, both statistics follow a chi-square with $m - p$ degrees of freedom, where m is the number of covariate patterns and p is the number of parameters in the model. Thus, if the test statistic divided by $m - p$ is significantly larger than 1, then we may conclude that overdispersion is present. One may use the asymptotic chi-square distribution to calculate a p-value for testing the nulll hypothesis of no overdispersion. However, since the asymptotic distribution is derived based on large sample theory, its validity requires a sufficiently large number of subjects in each of the covariate patterns. Thus, it may not apply for sparse data. In general, a large difference between the Pearson and deviance statistics may indicate that the data may be too sparse.

In theory, overdispersion may exist if the variance is a function of the mean as for both binary and binomial outcomes. Thus, it does also apply to Poisson regression, where the variance is also determined by the mean. However, it does not apply to linear regression, as the variance under linear regression is generally unknown and treated as a parameter. Note that we did not discuss overdispersion for binary responses, when they are not grouped. This is because overdispersion is often caused by data clustering. If data clusters (groups) are known, we can compare the empirical variance among the clusters from the observed data and the variance under the binomial model based on the mean responses (probabilities). However, for ungrouped data, we will not be able to observe the variation in each of the clusters and hence unable to detect the overdispersion. In other words, overdispersion should not be an issue for i.i.d. samples of binary responses.

4.5.2.2 Scaled variance for overdispersion

One common approach to deal with overdispersion is to derive inference using score-like equations that include a parameter ϕ to account for overdispersion, i.e.,

$$Var\,(y_i) = Var\left(\sum_{j=1}^{n_i} y_{ij}\right) = \phi\, n_i p_i(1 - p_i).$$

Thus, we assume the same conditional mean as in the binomial regression, but the conditional variance of y_i given $\mathbf{x}_i$ is scaled by ϕ. If $\phi = 1$, the modified variance reduces to the one under the binomial model. In the presence of overdispersion, $\phi > 1$ and $Var(y_i \mid \mathbf{x}_i) > n_i p_i (1 - p_i)$, thus accounting for overdispersion.

Under this scaled-variance approach, we may first estimate $\boldsymbol{\beta}$ using the MLE approach. Then, we estimate the scale parameter ϕ by $\hat{\phi} = \frac{S}{m-p}$, where S is the Pearson or deviance chi-square statistic. Inference will then be performed with the scaled variances. Note that this approach can be interpreted with a quasi-likelihood function. See Section 6.3.2.2 for a discussion on quasi-likelihood for overdispersion.

Williams (1982) suggested another approach similar to the one above, but taking the size n_i of the binomial variable into consideration. Instead of using a common dispersion parameter, an overdispersion parameter that depends on the size of each binomial observation is assumed. More precisely, we assume

$$Var(y_i) = n_i \pi_i (1 - \pi_i) [1 + (n_i - 1) \phi].$$

An iteratively reweighted least squares method was further proposed to fit the resulting model (Williams, 1982).

Example 4.17
Because of the overdispersion of the binomial model in Example 4.16, we refit the data with the same binomial regression, but with the inference corrected for overdispersion, using the scaled variance.

The analytic results with the dispersion scale estimated by the Pearson statistic are summarized in the following table:

Parameter	Estimate	Standard Error	Wald Chi-Square	Pr > ChiSq
Intercept	−0.3042	0.4768	0.4071	0.5234
Age	−0.0177	0.00669	7.0186	0.0081
Gender	0.3128	0.2307	1.8380	0.1752
Education	−0.0241	0.0708	0.1158	0.7337
Group	0.1947	0.2274	0.7329	0.3920

Compared with the results in Example 4.16, it is clear that the point estimates remain unchanged, but their variance estimates are different: The estimated variance of the parameters are scaled by the estimated scale. The scale estimated by the Pearson chi-square statistic is the value of the square root of Pearson $\chi^2 / (n - p) = 2.2565$, where $n = 193$ (number of subjects used to fit the model) and $p = 5$ (number of parameters in the model). Thus, all the estimated standard errors are scaled by a factor of $\sqrt{2.2565}$. For example, the standard error for gender is now $0.1536 * \sqrt{2.2565} = 0.2307$. Along with the larger estimated standard errors, we obtain larger p-values for all the parameters.

We can also apply Williams' approach to deal with the overdispersion. The analysis results by the William's approach are summarized in the following table:

Parameter	Estimate	Standard Error	Wald Chi-Square	Pr > ChiSq
Intercept	−0.2500	0.4850	0.2658	0.6062
Age	−0.0170	0.00672	6.3651	0.0116
Gender	0.3117	0.2328	1.7934	0.1805
Education	−0.0378	0.0717	0.2779	0.5981
Group	0.1353	0.2302	0.3451	0.5569

We obtain different point estimates of the parameters, but the conclusions are similar. The estimate of the scale is 0.30802.

□

4.5.3 Beta binomial regression models

Instead of using a scale parameter to ad hoc adjust the variance, we can also directly model overdispersion. Suppose that Y_i follows the Bernoulli(p_i) distribution, with a random Bernoulli parameter. If the Bernoulli parameter p_i follows a Beta(α, β) distribution with density function $\frac{p^{\alpha-1}(1-p)^{\beta-1}}{B(\alpha,\beta)}$, where $B(\alpha, \beta) = \int_0^1 t^{\alpha-1}(1-t)^{\beta-1} dt$ is the beta function, then the resulting distribution for $Y = Y_1 + \cdots + Y_n$ follows the beta binomial distribution $bb(n, \alpha, \beta)$ given by

$$\Pr(Y = y) = \binom{n}{y} \frac{B(y+\alpha, n-y+\beta)}{B(\alpha, \beta)}.$$

Since the mean and variance of Beta(α, β) are $\frac{\alpha}{\alpha+\beta}$ and $\frac{\alpha\beta}{(\alpha+\beta)^2(\alpha+\beta+1)}$, respectively, it follows that $bb(n, \alpha, \beta)$ has mean $\frac{n\alpha}{\alpha+\beta}$ and variance $\frac{n\alpha\beta(\alpha+\beta+n)}{(\alpha+\beta)^2(\alpha+\beta+1)}$. The variance is larger than the one that would be expected for a binomial variable with size n and probability $\frac{\alpha}{\alpha+\beta}$, which would be $\frac{n\alpha\beta}{(\alpha+\beta)^2}$. Thus, the distribution can be applied to model overdispersed aggregated binary data. In beta binomial regression models, we often reparameterize the beta distribution with parameters $p = \frac{\alpha}{\alpha+\beta}$ and $\sigma = \frac{1}{\alpha+\beta}$. The parameter σ is usually assumed constant, and the parameter p can be modeled the same way as in binomial models. For example, a logistic regression may be applied and the coefficients can be similarly interpreted using odds ratios.

Example 4.18
To address the overdispersion of the binomial model in Example 4.16, we refit the data with the following beta binomial regression

$$y_i \mid \mathbf{x}_i \sim bb(m_i, p_i, \sigma), \operatorname{logit}(p_i) = \beta_0 + \beta_1 Age_i + \beta_2 Gender_i + \beta_3 Education_i + \beta_4 Group_i.$$

The parameter estimates for the Beta-Binomial model are summarized in the following table:

Effect	Estimate	Standard Error	z Value	Pr > |z|
Intercept	−0.4334	0.4493	−0.96	0.3348
Age	−0.01515	0.006132	−2.47	0.0135
Gender	0.3418	0.2151	1.59	0.1121
Education	−0.03478	0.06557	−0.53	0.5958
Group	0.2449	0.2132	1.15	0.2507
Scale Parameter	2.7186	0.5427		

The parameters are interpreted the same way as in regular logistic models. For example, the estimated coefficient of 0.3418 for gender means that the estimated odds ratio for females over males is $\exp(0.3418) = 1.4075$. The results are in fact very similar to one based on scaled variance; only age is significant at the 5% significance level. ▯

Note that it is not uncommon in practice that there may be a lot of observations with a particular value as their outcomes, more than would be expected under a binomial regression model, due to population heterogeneity. For example, if there is a subpopulation of non-drinkers, then their values of "days of any drinking in the past week" will always be zero. If the membership of the subpopulation is known, then we may model the subpopulation separately. Otherwise, a model for mixtures is necessary to fit such zero-inflated binomial outcomes. We defer the discussion of such mixture models to Chapter 6.

Exercises

4.1 Consider a random variable x following the standard logistic distribution with the CDF and PDF given in (4.3) and (4.4).

    (a) Show that the PDF in (4.4) is symmetric about 0.

    (b) Show that CDF in (4.3) is strictly increasing on $(-\infty, \infty)$.

    (c) Plot the CDF in (4.3) and verify that it is S shaped.

4.2 Prove that if

$$\text{logit}\left(\Pr\left(y_i = 1 \mid \mathbf{x}_i\right)\right) = \beta_0 + \mathbf{x}^\top \boldsymbol{\beta}$$

and

$$\text{logit}\left(\Pr\left(y_i = 0 \mid \mathbf{x}_i\right)\right) = \alpha_0 + \mathbf{x}^\top \boldsymbol{\alpha},$$

then $\beta_0 = -\alpha_0$ and $\boldsymbol{\beta} = -\boldsymbol{\alpha}$.

4.3 If Σ is an $n \times n$ invertible matrix and K is a $k \times n$ matrix with rank k ($k \leq n$), show that $K\Sigma K^\top$ is invertible.

4.4 Show that the Wald statistic in (4.15) does not depend on the specific equations used. Specifically, suppose that K and K' are two equivalent systems of equations for a linear hypothesis, i.e., the row spaces generated by the rows of the two matrices are the same, then the corresponding Wald statistics are the same, i.e.,

$$\left(K\widehat{\boldsymbol{\beta}}\right)^\top \left(K\Sigma_{\boldsymbol{\beta}} K^\top\right)^{-1} \left(K\widehat{\boldsymbol{\beta}}\right) = \left(K'\widehat{\boldsymbol{\beta}}\right)^\top \left(K'\Sigma_{\boldsymbol{\beta}} K'^\top\right)^{-1} \left(K'\widehat{\boldsymbol{\beta}}\right).$$

4.5 Prove that the Wald statistic defined in (4.15) follows asymptotically a chi-square distribution with l degrees of freedom.

For Problems 4.6 and 4.7, we use the data from the DOS. We need the following variables:

Depd = 1 if major/minor depression, and Depd = 0 if no depression; R1, R2, and R3: the indicator variables for Race = 1, 2, and 3, respectively; MSD: indicator variable for MS = 1 (married and living with spouse); CIRSD: = 1 if CIRS < 6, = 2 if 6 ≤ CIRS < 10, and = 3 if CIRS ≥ 10.

4.6 Use a logistic model to assess the relationship between CIRSD and Depd, with Depd as the outcome variable.

    (a) Write down the logistic model.

    (b) Write down the null hypothesis that CIRSD has no effect.

    (c) Write down the null hypothesis that there is no difference between CIRSD = 1 and CIRSD = 2.

    (d) Test the null hypotheses in part (b) and (c). Summarize your findings.

4.7 Based on a logistic regression of Depd on some covariates, we obtained the following prediction equation:

$$\text{logit}\left[\widehat{\Pr}(Depd = 1)\right] = 1.13 - 0.02 Age - 1.52 MSD + 0.29 R1 + 0.06 R2$$

$$+ 0.90 MSD * R1 + 1.79 MSD * R2 \tag{4.50}$$

(a) Carefully interpret the effects. Explain the interaction by describing the race effect at each MS level and the MS effect for each race group.

(b) What is the predicted odds ratio of depression of a 50-year-old with MSD $= 0$ and Race $= 2$ versus a 35-year-old with MSD$=1$ and Race $= 5$?

4.8 In suicide studies, alcohol use is found to be an important predictor of suicide ideation. Suppose the following logistic model is used to model the effect:

$$\text{logit} \left[\Pr(\text{has suicide ideation}) \right] = \beta_0 + \beta_1 * Drink \qquad (4.51)$$

where $Drink$ is the daily alcohol usage in drinks.

(a) If we know that the odds ratio of having suicide ideation between a subject who drinks two drinks daily with a subject who drinks one drink daily is 2, compute β_1.

(b) $Drink'$ is a measure of alcohol use under a new scale where two drinks are considered as one unit of drink. Thus, $Drink' = \frac{1}{2} Drink$. If the same logistic model is fitted, $\text{logit}[\Pr(\text{has suicide ideation})] = \beta_0' + \beta_1' * Drink'$. How are β_1 and β_1' related to each other?

(c) If data are applied to test whether alcohol use is a predictor of suicide ideation, does it matter which scale is used to measure the alcohol use?

4.9 Consider the logistic regression in (4.25). Show that for each j $(1 \le j \le p)$, $T_j(x) = \sum_{i=1}^n y_i x_{ij}$ is a sufficient statistic for β_j.

4.10 Use the fact that $x \log x = (x-1) + (x-1)^2/2 + o((x-1)^2)$ to show that the deviance test statistic D^2 in (4.43) has the same asymptotic distribution as the general Pearson chi-square statistic in (4.44).

4.11 For the DOS, we are interested in how MS and gender are related with the depression outcome. Based on the logistic model: Dep $\sim$ MS | Gender (MS, Gender, and their interaction), answer the following questions:

(a) Test the goodness of fit of the model.

(b) Is the interaction significant?

(c) Test the goodness of fit of the model with the interaction removed.

4.12 We add age as a continuous covariate to the model in the last problem and consider the logistic model: Dep $\sim$ MS+ Gender+Age.

(a) Test the goodness of fit of the model.

(b) Use Box-Tidwell test to test the linearity assumption for Age.

(c) Apply Pregibon's link test to the model.

4.13 For the exponential family of distributions defined in (4.41), show

(a) $E(y) = \frac{d}{d\theta} b(\theta)$.

(b) $Var(y) = a(\phi) \frac{d^2}{d\theta^2} b(\theta)$.

(c) Assume that $y \sim N(\mu, \sigma^2)$ and σ^2 is known. Show that the canonical link for the mean μ is the identity function $g(\theta) = \theta$.

(d) Show that the canonical link for Bernoulli distribution is the logistic function.

4.14 Prove that a sufficient statistic for the parameter β_j in the model (4.25) is given by $T_j(x) = \sum_{i=1}^{n} y_i x_{ij}$ $(1 \le j \le p)$.

4.15 Prove the conditional distribution of T_0 given T_1 in Section 4.2.3 is $BI(\sum_{i=1}^{n}(1 - x_i),$ $0.5)$. (Hint: $\sum_{i=1}^{n} y_i = T_0$ is equivalent to $\sum_{i=1}^{n} y_i(1 - x_i) = T_0 - t_1$, conditional on $T_1 = t_1$.)

4.16 This problem illustrates why exact inference may not behave well when conditional on continuous covariates.

(a) Consider the following equation where $a_1, a_2, \ldots, a_n$ are some known numbers and y_i are binary variables,

$$\sum_{i=1}^{n} y_i a_i = 0, \quad y_i \in \{0, 1\}, \quad 1 \le i \le n. \tag{4.52}$$

If the trivial solution, $y_i = 0$ $(1 \le i \le n)$, is the only set of y_i satisfying (4.52), show that for any binary $z_i \in \{0, 1\}$, the following equation

$$\sum_i y_i a_i = \sum_i z_i a_i$$

has a unique solution $y_i = z_i$ $(1 \le i \le n)$. When applied to exact logistic regression, this result implies that if $x_i = a_i$, the observed y_i's are the only possible outcomes that produce the sufficient statistic $\sum_i y_i x_i$, making it impossible to perform exact inference.

(b) Let $n = 5$ and give example of $a_1, \ldots, a_5$, such that (4.52) is true.

4.17 Verify the conditional likelihood (4.29).

4.18 Suppose $y_i = \sum_{j=1}^{n_i} y_{ij}$, where $y_{ij} \sim Bernoulli(p_i)$ and are positively correlated with $cor(y_{ij}, y_{ik}) = \alpha > 0$ for $k \ne j$. Prove $Var(y_i) > n_i p_i(1 - p_i)$.

4.19 Prove that the deviance and Pearson chi-square test statistics are asymptotically equivalent.

4.20 The following 2×2 table is from a hypothetical random sample.

	y		
x	No (0)	Yes (1)	Total
No (0)	12	0	12
Yes (1)	9	11	20
Total	21	11	32

A logistic regression model, $\text{logit}[\Pr(y = 1|x)] = \alpha + \beta x$, is used to assess the relationship between x and y.

(a) Find the MLE of β. Explain your results.

(b) Find the MUE of β. Explain your results.

(c) Test the null hypothesis that x and y are independent.

4.21 Plot and compare the CDFs of logistic, probit, and complementary log-log variables after they are centered at their medians and scaled to unit variances.

4.22 Let $y_i^* = \beta_0 + \boldsymbol{\beta}^\top \mathbf{x}_i + \varepsilon_i$, where $\varepsilon_i \sim N(0,1)$ (a standard normal with mean 0 and variance 1) and y_i is determined by y_i^* as an indicator for whether this latent variable is positive, i.e.,

$$y_i = \begin{cases} 1 & \text{if } y_i^* > 0, \text{ i.e., } -\varepsilon_i < \beta_0 + \mathbf{x}_i^\top \boldsymbol{\beta}, \\ 0 & \text{if otherwise.} \end{cases}$$

Show that $\Pr(y_i = 1 \mid \mathbf{x}_i) = \Phi(\beta_0 + \mathbf{x}_i^\top \boldsymbol{\beta})$, where Φ is the CDF of standard normal.

4.23 Prove that if

$$\Pr(y_i = 1 \mid \mathbf{x}_i) = \Phi(\beta_0 + \mathbf{x}^\top \boldsymbol{\beta}), \quad \Pr(y_i = 0 \mid \mathbf{x}_i) = \Phi(\alpha_0 + \mathbf{x}^\top \boldsymbol{\alpha}),$$

where Φ is the CDF of standard normal, then $\beta_0 = -\alpha_0$ and $\boldsymbol{\beta} = -\boldsymbol{\alpha}$.

4.24 Fit complementary log-log models to DOS data, using dichotomized depression as response and gender as the predictor. Comparing the results between modeling the probability of No depression and modeling the probability of Depression. This confirms that the complementary log-log link function is not symmetric.

4.25 Prove that if

$$\Pr(y_i = 1 \mid \mathbf{x}_i) = \beta_0 + \mathbf{x}^\top \boldsymbol{\beta}, \quad \text{and } \Pr(y_i = 0 \mid \mathbf{x}_i) = \alpha_0 + \mathbf{x}^\top \boldsymbol{\alpha},$$

then $\beta_0 = 1 - \alpha_0$ and $\boldsymbol{\beta} = -\boldsymbol{\alpha}$.

4.26 Use the baseline information from the Catheter Study. We model the binary catheter blockage outcome with age, gender, and education as predictors. Assuming additive effect for all these covariates, we may have five models using the four link functions: logit, probit, c-log-log, and identity. Note that by using the c-loglog link, we have two different models by modeling the probabilities of death $= 1$ or death $= 0$.

(a) Assess the goodness of fit for all the models.

(b) Perform an overall test on the association between age groups and the outcome. What is the p-value?

(c) Calculate the correlation matrix the fitted probabilities under the five models.

4.27 Show that the discrimination slope equals to the differences of the mean fitted probabilities between $y = 1$ and $y = 0$.

4.28 For the Catheter Study, apply the binomial regression model (4.49) with the UTI response replaced by the catheter blockage.

(a) What conclusions you may have based on the model? Is there overdispersion?

(b) Apply the scaling approach to deal with the overdispersion issue. Are there any changes in your conclusions?

4.29 In this problem, we perform a simulation study about clustered binary outcomes with sample size 1000.

(a) Generate random variable X from N(0,1).

(b) For each X, generate five binary responses with response probability $\frac{\exp(x)}{1+\exp(x)}$ in two scenarios and sum up the five binary to obtain the response variable Y. In the first, all the five binary responses are independent. In the second, all the five binary responses are positively correlated with the same correlation.

(c) Fit logistic regression models with Y as the response and X as the predictor:

$$y_i \mid x_i \sim binomial(5, p_i), logit(p_i) = \beta_0 + \beta_1 x_i,$$

Assess the goodness of fit and overdispersion in both scenarios.

(d) Test the null hypothesis $\beta_1 = 1$ without correcting the overdispersion. Repeat 1000 times, what are the empirical type I error (the proportions of times you reject the null hypothesis)?

(e) For the correlated data, test the null hypothesis $\beta_1 = 1$ again, with the overdispersion corrected by the Pearson chi-square statistics. Repeat 1,000 times, what is the empirical type I error?

Chapter 5

Regression Models for Polytomous Responses

In the last chapter, we considered models for binary responses under the rubric of generalized linear models to study the effects of independent variables on the mean response. In this chapter, we examine generalized linear models when applied to polytomous outcomes with more than two levels. In practice, we often encounter multi-level responses. For example, the original depression diagnosis in the Depression of Seniors Study (DOS) in fact consisted of three levels: none, minor, and major depression. We dichotomized the outcome by combining minor and major depression into a single depression level in the last chapter. While such a grouping technique is common in practice, there are two issues associated with the approach.

First, it may not be obvious how to condense the multi-level outcome into a two-level binary variable. For example, consider outcomes consisting the three colors: red, green, and blue. There may not be an apparent choice on which two to combine. Second, there is always information loss associated with any grouping. Dichotomizing polytomous outcomes is usually done for the purpose of simplifying modeling. It would be better to model the original outcome if efficient models are available. Furthermore, modeling approaches often depend on the nature of multi-level responses. For example, different models will be used for responses with a natural order than those without such an order.

In this chapter, we will first discuss different data types and general modeling approaches in Section 5.1. In Section 5.2, we discuss models for nominal responses, where all the levels are unordered and play a symmetric role. In Section 5.3, we discuss models for ordinal responses, where all the levels are ordered. We conclude the chapter with a discussion of model evaluation in Section 5.4.

5.1 Modeling Polytomous Responses

Before developing statistical models for polytomous responses, we first need to distinguish different types of polytomous responses or measurement scales. For example, in automobile marketing research we may want to know the distribution of different car manufacturers. In this example, the response levels are exchangeable with no ordering structure among them. In other examples, different response categories are ordered in some fashion, which may be of interest in their own right. For example, in the DOS, the depression diagnosis has three response levels (major, minor, or no depression). Here, we are not only interested in whether a person is sick, but also the severity of the disease. In this case, major and minor depression may not be exchangeable in their positions in the response levels from a modelling point of view, as the former is a more severe medical condition than the latter. Consideration of the structures of the responses generally leads to qualitatively different classes of models.

5.1.1 Scales of polytomous responses

The most notable feature of polytomous responses is whether the levels are ordered in some fashion. Thus, we can broadly identify polytomous responses as one of the two following types:

1. *Nominal scale.* The categories of such a response are regarded as exchangeable and totally devoid of structure. For example, the type of disciplines for graduate studies usually belongs to such a response type.

2. *Ordinal scale.* The categories of this type of response variable are ordered in terms such as preference or severity. So, for any two levels A and B, we can order them as A < B or B < A, where the order "<" may indicate, for example, less preferred than or less seriously ill than, etc. The order satisfies the transitive property so that for any three levels A, B, and C, A < B and B < C imply A < C. Thus, we may order all the categories from the lowest level to the highest level. The depression diagnosis with three levels, non depression, minor, and major depression, is an example of such an ordinal response.

In applications, the distinction between nominal and ordinal scales may not always be clear. For example, severity of depression is clearly an ordinal outcome, while study disciplines naturally form the categories of a nominal response. However, hair color can be ordered to a large extent on the grey scale from light to dark and therefore may be viewed as ordinal, although the relevance of the order and treatment of such a response may well depend on the application contexts. For binary variables, the distinction between the ordinal and nominal does not exist.

Note that there may exist other structures for the responses. For example, between nominal and ordinal, there may be other structures among the response categories. In comorbidity studies of hypertension and depression, the level "healthy," which refers to healthy subjects with neither disease, is a healthier level when compared to the level "hypertension only" or "depression only." However, between "hypertension only" and "depression only," it is difficult to impose an order, as both are considered equally important in such studies. Although more complicated models are proposed to exploit such *partially ordered* structures, we can simply treat them as nominal.

In other cases, the categories may be clustered. For example, traditional college majors may be roughly grouped into liberal arts, sciences and engineering, etc. Thus, subjects like English and history fall in the same subgroup of liberal arts majors, while mathematics and physics belong to the same subgroup of science majors. This distinction may be important if the polytomous outcomes are the results of sequential selection, which we may model accordingly. For example, if the students first decide between science and liberal arts and then choose a science or liberal arts major in the second step, we may first model the choice between science and liberal arts, followed by the choice of a major within the subject's category.

For some ordinal variables, it is sensible to consider the distance among the different levels. Such interval scales often arise from discretizing (latent) continuous outcomes either because of a failure to observe the original scale directly or for the purpose of modeling and interpretation. To keep the interval scale in modeling such variables as the response, a common approach is to treat them as continuous variables; robust inference is then usually required as the common normality assumption for the error is not satisfied. Another common type of interval scale concerns aggregated binary variables. In such cases, the methods discussed for aggregated binaries in the last chapter apply. In this chapter, we model polytomous responses without such interval scale.

5.1.2 GLM for polytomous responses

Under the framework of GLM, we may model polytomous responses by specifying the multinomial distribution for the random component and appropriate linear predictors and link functions for the deterministic component. Suppose the polytomous response of interest y has J categories, denoted $1, ..., J$. For a set of independent variables $\mathbf{x}$, let

$$\pi_j(\mathbf{x}) = \Pr(y = j \mid \mathbf{x}), \quad j = 1, \ldots, J. \tag{5.1}$$

Given $\mathbf{x}$, y follows a multinomial distribution $MN(1, \boldsymbol{\pi}(\mathbf{x}))$, with the vector of response probabilities given by $\boldsymbol{\pi}(\mathbf{x}) = (\pi_1(\mathbf{x}), \ldots, \pi_J(\mathbf{x}))^\top$. Since $\sum_{j=1}^{J} \pi_j(\mathbf{x}) = 1$, only $J - 1$ of the $\pi_j(\mathbf{x})$ are needed for modeling. Following the framework of GLM, we may specify a GLM for polytomous responses as follows:

$$y_i \mid \mathbf{x}_i \sim MN(1, \boldsymbol{\pi}(\mathbf{x})), (\pi_1(\mathbf{x}), \ldots, \pi_{J-1}(\mathbf{x})) = g(\mu_1, \ldots, \mu_{J-1}), \tag{5.2}$$

where g is a link function and $\mu_1, \ldots,$ and μ_{J-1} are $J - 1$ linear predictors. Note that in special cases such as those arising from binomial and truncated Poisson discussed in Section 2.1, the multinomial distribution may be determined by fewer parameters. However, we consider general polytomous responses, without assuming any additional structure for the multinomial distribution. Thus, $J - 1$ conditional means with $J - 1$ linear predictors are needed to describe the multinomial distribution.

The link function g is a 1:1 function from $\mathbf{R}^{J-1}$ to $\mathbf{R}^{J-1}$. A proper link function should only take values with $\pi_1(\mathbf{x}) > 0, \ldots, \pi_{J-1}(\mathbf{x}) > 0$ and $\pi_1(\mathbf{x}) + \ldots + \pi_{J-1}(\mathbf{x}) < 1$. Further requirements for the link functions are necessary and depend on the nature of the response. For example, because the response levels are exchangeable when the response is a nominal variable, a proper model should be invariant with respect to the choice of the reference levels. Thus, it is more complicated than the binary cases. In general, it would be difficult and unintuitive to specify the link function directly. See Peyhardi et al. (2015) for a systematic discussion on the topic.

In practice, we often transform the modeling task into modeling various outcomes with two levels by first merging some levels and/or subsetting the levels and then applying binary regression models discussed in the last chapter. For example, we may simply consider all the pairs of levels for nominal responses. For ordinal responses, we have more options; for example, we may dichotomize them using various cut-points or simply compare the adjacent levels. Another approach naturally arises when all the response levels are alternative options that one may choose from. In such cases one assumes some utility function associated with each alternative, and the chosen alternative will then naturally be the one with the maximum utility.

We will mainly focus on the first approach of building models based on various binary models. Depending on the nature of the response variables and study aims, different modeling approaches may be applied, and the link function g can then be derived accordingly. Since such link functions can be quite complicated and unintuitive to interpret, in practice it is more natural to use the binary configuration in the model for parameter interpretation. Of course, the fitted values for the original scale of the multinomial outcomes can be easily calculated using the link function. We will discuss the different approaches according to the data types in more detail in the next sections.

5.1.3 Inference for models for polytomous response

Based on (5.2), we assume that $\mu_j(\mathbf{x}_i) = \mathbf{x}_i \boldsymbol{\beta}_j^\top$, $j = 1, ..., J - 1$, and apply ML method for inference about $\boldsymbol{\beta}_j$. Usually we need to estimate all the parameters $\boldsymbol{\theta} = (\boldsymbol{\beta}_1, ..., \boldsymbol{\beta}_{J-1})$

together. Consider a sample of n individuals. To write the likelihood function, we first express the original response variable y using a vector of binary responses. For each individual i, let $\mathbf{y}_i = (y_{i1}, \ldots, y_{iJ})^\top$ be a vector of binary responses y_{ij}, with $y_{ij} = 1$ if the response is in category j and $y_{ij} = 0$ otherwise $(1 \leq j \leq J)$. The likelihood function for a regression model for the polytomous response y has the form

$$L(\boldsymbol{\theta}) = \prod_{i=1}^{n} \left[\prod_{j=1}^{J} [\pi_j(\mathbf{x}_i)]^{y_{ij}} \right] = \prod_{i=1}^{n} \left[\left(1 - \sum_{j=1}^{J-1} \pi_j(\mathbf{x}_i) \right)^{y_{iJ}} \prod_{j=1}^{J-1} [\pi_j(\mathbf{x}_i)]^{y_{ij}} \right], \quad (5.3)$$

and the log-likelihood of the sample is given by

$$l(\boldsymbol{\theta}) = \sum_{i=1}^{n} \left[y_{iJ} \log \left(1 - \sum_{j=1}^{J-1} \pi_j(\mathbf{x}_i) \right) + \sum_{j=1}^{J-1} y_{ij} \log \pi_j(\mathbf{x}_i) \right]. \quad (5.4)$$

Thus, the score equation is given by

$$l'(\boldsymbol{\theta}) = \sum_{i=1}^{n} \sum_{j=1}^{J-1} \left[\frac{y_{ij}}{\pi_j(\mathbf{x}_i)} - \frac{y_{iJ}}{\pi_J(\mathbf{x}_i)} \right] \frac{\partial \pi_j}{\partial \mu_k} \mathbf{x}_i = \mathbf{0}, \quad k = 1, \ldots, J-1. \quad (5.5)$$

In general, the Newton–Raphson method may be used to numerically obtain the maximum likelihood estimate (MLE). Wald, score, and likelihood ratio (LR) tests can all be applied to provide inference about the parameters. In general, we may need a large sample size in each category to obtain reliable inference about $\boldsymbol{\theta}$ under large sample theory. For small samples, exact inference may be considered (Hirji, 1992).

5.2   Models for Nominal Responses

For nominal variables, we use numbers from 1 to J to denote the levels in (5.1). Since numeric values do not carry any specific meaning such as orders, it may not be natural to merge some of the levels. In such cases, comparing all the pairs of levels sounds like a natural approach. For any two categories, i and j, we may specify a binary regression. However, this kind of specification cannot be carried out freely, as there are $\binom{J}{2}$ pairs of categories while only $J - 1$ of the $\pi_j(\mathbf{x})$ in (5.1) are independent.

5.2.1   Generalized logit models

The *generalized logit* model designates one category as a reference level and then pairs each other response category to this reference category. Usually the first or the last category is chosen to serve as such a reference category. Of course, for nominal responses, the "first" or "last" category is not well defined as the categories are exchangeable. Thus, the selection of the reference level is arbitrary and is typically based on convenience. However, we will see that since the generalized logit model uses the logit link function, it does not depend on the choice of the reference level.

To appreciate the specification of the generalized logit model, let us first review the logistic model for binary responses. Let

$$\pi_1(\mathbf{x}) = \Pr(y = 1 \mid \mathbf{x}), \quad \pi_0(\mathbf{x}) = \Pr(y = 0 \mid \mathbf{x}) = 1 - \pi_1(\mathbf{x}).$$

The log odds or logit of response is given by

$$\log\left(\frac{\pi_1(\mathbf{x})}{\pi_0(\mathbf{x})}\right) = \log\frac{\pi_1(\mathbf{x})}{1 - \pi_1(\mathbf{x})} = \text{logit}(\pi_1(\mathbf{x})).$$

For multinomial responses, we have more than two response levels and as such cannot define odds or log odds of responses as in the binary case. However, upon selecting the reference level, say the last level J, we can define the "odds" ("log odds") of response in the jth category as compared to the Jth response category by $\frac{\pi_j(\mathbf{x})}{\pi_J(\mathbf{x})}\left(\log\left(\frac{\pi_j(\mathbf{x})}{\pi_J(\mathbf{x})}\right)\right)$ $(1 \leq j \leq J - 1)$. Note that since $\pi_j(\mathbf{x}) + \pi_J(\mathbf{x}) \neq 1$, $\frac{\pi_j(\mathbf{x})}{\pi_J(\mathbf{x})}\left(\log\left(\frac{\pi_j(\mathbf{x})}{\pi_J(\mathbf{x})}\right)\right)$ is not an odds (log odds) in the usual sense. However, we have

$$\log\frac{\pi_j}{\pi_J} = \log\frac{\pi_j/(\pi_j + \pi_J)}{\pi_J/(\pi_j + \pi_J)} = \log\frac{\pi_j/(\pi_j + \pi_J)}{1 - \pi_j/(\pi_j + \pi_J)} = \text{logit}\left(\frac{\pi_j}{\pi_j + \pi_J}\right).$$

Thus, $\log\left(\frac{\pi_j(\mathbf{x})}{\pi_J(\mathbf{x})}\right)$ has the usual log odds interpretation if we limit our interest to the two levels j and J, giving rise to the name of generalized logit model.

Under the generalized logit model, we model the log odds of responses for each pair of categories as follows:

$$\log\left(\frac{\pi_j(\mathbf{x})}{\pi_J(\mathbf{x})}\right) = \alpha_j + \boldsymbol{\beta}_j^\top \mathbf{x} = \eta_j, \quad j = 1, \ldots, J - 1. \tag{5.6}$$

Since

$$\log\frac{\pi_i(\mathbf{x})}{\pi_j(\mathbf{x})} = \log\frac{\pi_i(\mathbf{x})}{\pi_J(\mathbf{x})} - \log\frac{\pi_j(\mathbf{x})}{\pi_J(\mathbf{x})},$$

these $J - 1$ logits determine the parameters for any other pairs of the response categories. Thus, generalized logistic regression does not depend on the reference level in the model specification. In other words, if a different level is chosen as the reference level, we are considering different odds ratios and different parameters α_j and β_j, but the model is essentially the same.

Each of the linear predictors, η_j, only determines the relative likelihood between the jth and reference level. To obtain the probability of responses in the jth category, we need all the linear predictors in (5.6). The probability of responses in the jth category is

$$\pi_j = \frac{\exp\left(\alpha_j + \boldsymbol{\beta}_j^\top \mathbf{x}\right)}{1 + \sum_{k=1}^{J-1}\exp\left(\alpha_j + \boldsymbol{\beta}_j^\top \mathbf{x}\right)} = \frac{\exp(\eta_j)}{1 + \sum_{k=1}^{J-1}\exp(\eta_j)}, \quad j = 1, \ldots, J - 1.$$

By setting $\alpha_J = 0$ and $\boldsymbol{\beta}_J = \mathbf{0}$ and including the level $j = J$ in the above representation, we can express the probability of response for all J categories symmetrically as

$$\pi_j = \frac{\exp\left(\alpha_j + \boldsymbol{\beta}_j^\top \mathbf{x}\right)}{1 + \sum_{k=1}^{J-1}\exp\left(\alpha_j + \boldsymbol{\beta}_j^\top \mathbf{x}\right)} = \frac{\exp(\eta_j)}{\sum_{k=1}^{J}\exp(\eta_k)}, \quad j = 1, \ldots, J. \tag{5.7}$$

Note that as a special case when $J = 2$, $\pi_2(\mathbf{x}) = 1 - \pi_1(\mathbf{x})$, and $\log\left(\frac{\pi_1(\mathbf{x})}{\pi_2(\mathbf{x})}\right)$ becomes the logit, or log odds, of response and the generalized logit model reduces to logistic regression for a binary response.

5.2.2   Inference for Generalized Logit Models

Based on (5.7) and (5.3), the computation of the likelihood function for model (5.6) is straightforward. In fact, the score equation (5.5) has the following simple format:

$$\sum_{i=1}^{n}(y_{ij} - \pi_{ij}) = 0, \quad \sum_{i=1}^{n}(y_{ij} - \pi_{ij})\,\mathbf{x}_i = \mathbf{0}, \quad j = 1, ..., J-1.$$

The Newton–Raphson method is readily applied to numerically locate the MLE $\widehat{\boldsymbol{\theta}}$. By applying the asymptotic normal distribution of the MLE for i.i.d. data, $\widehat{\boldsymbol{\theta}} \sim_a N\left(\boldsymbol{\theta}, \mathbf{I}_n^{-1}(\boldsymbol{\theta})\right)$, where $\mathbf{I}_n(\boldsymbol{\theta})$ is the observed information matrix with an estimate given by $\mathbf{I}_n\left(\widehat{\boldsymbol{\theta}}\right)$.

The number of parameters for the generalized logit model increases with the number of categories. Thus, we typically need the sample size of each category to be large to obtain reliable inferences. For relatively small samples, data separation may occur as in the case of logistic regression. The exact conditional logistic method has been extended to the generalized logit models (Hirji, 1992). Akin to the logistic model for binary responses, a sufficient statistic for the parameter β_{jk} is given by $T_{jk}(x) = \sum_{i=1}^{n} y_{ij}x_{ik}$ $(1 \leq j \leq J$ and $1 \leq k \leq p)$ (see Problem 5.1). Similar to the binary case, when making inferences about a parameter such as β_{jk}, we treat all others as nuisance parameters (see Hirji (1992) for details).

Example 5.1

Let us apply the generalized logit model to the DOS, using the three-level depression diagnosis (DEP = 0 for nondepression, DEP = 1 for minor depression, and DEP = 2 for major depression) as the response and Gender (indicator variable for females) as the only independent variable.

If DEP = 0 is selected as the reference level, then the generalized logit model has the following form:

$$\log\frac{\pi_j(\text{Gender})}{\pi_0(\text{Gender})} = \alpha_j + \beta_j \cdot \text{Gender}, \quad j = 1, 2.$$

Letting, e.g., "female" below represents being a female, it then follows that for $j = 1, 2$

$$\beta_j = \log\frac{\pi_j(\text{male})}{\pi_0(\text{male})} - \log\frac{\pi_j(\text{female})}{\pi_0(\text{female})} = \log\left[\frac{\pi_j(\text{male})/\pi_0(\text{male})}{\pi_j(\text{female})/\pi_0(\text{female})}\right].$$

Thus, we may interpret β_1 as the log "odds ratio" of minor vs. no depression for comparing the male and female subjects and β_2 as the log "odds ratio" of major vs. no depression diagnosis for comparing the male and female subjects. The log "odds ratio" of minor vs. major would be $\beta_1 - \beta_2$.

The estimates of the parameters are summarized in the following table.

Category	Intercept (α) Estimate	SE	Gender (β) Estimate	SE	p-value
2	−1.7774	0.1828	0.6968	0.2186	0.0014
1	−1.8987	0.1926	0.9396	0.2242	< .0001

From the table, $\log\left[\frac{\pi_2(\text{male})/\pi_0(\text{male})}{\pi_2(\text{female})/\pi_0(\text{female})}\right]$ is estimated to be 0.6968. The positive sign indicates that males are more likely to be majorly depressed than females, and the difference is significant with p-value of 0.0014. Similarly $\log\left[\frac{\pi_1(\text{male})/\pi_0(\text{male})}{\pi_1(\text{female})/\pi_0(\text{female})}\right]$ is estimated to be 0.9396, indicating that males are more likely to be minorly depressed than females, and the difference is significant with p-value $< .0001$. The estimate of $\log\left[\frac{\pi_2(\text{male})/\pi_1(\text{male})}{\pi_2(\text{female})/\pi_1(\text{female})}\right]$ is $0.6968 - 0.9396$. Based on the estimated variance of $\left(\widehat{\beta}_1, \widehat{\beta}_2\right)$, $\begin{pmatrix} 0.04780 & 0.00848 \\ 0.00848 & 0.05026 \end{pmatrix}$, we can obtain an estimate of the variance of $\widehat{\beta}_1 - \widehat{\beta}_2$

$$\begin{pmatrix} 1 \\ -1 \end{pmatrix}^\top \begin{pmatrix} 0.04780 & 0.00848 \\ 0.00848 & 0.05026 \end{pmatrix} \begin{pmatrix} 1 \\ -1 \end{pmatrix} = 0.0811$$

and the Wald chi-square test statistic for testing $\beta_1 - \beta_2 = 0$:

$$\frac{(0.6968 - 0.9396)^2}{0.0811} = 0.7269.$$

Comparing with the chi-square distribution with one degree of freedom, we obtain a p-value of 0.3939.

In this model, the association between gender and the depression outcome is summarized in two parameters, β_1 and β_2. Thus, to assess whether they are associated, we need to perform a composite test of the hypothesis: $\beta_1 = \beta_2 = 0$. The Wald test statistic is given by

$$\begin{pmatrix} 0.6968 \\ 0.9396 \end{pmatrix}^\top \begin{pmatrix} 1 & 0 \\ 0 & 1 \end{pmatrix}^\top \begin{pmatrix} 0.04780 & 0.00848 \\ 0.00848 & 0.05026 \end{pmatrix}^{-1} \begin{pmatrix} 1 & 0 \\ 0 & 1 \end{pmatrix} \begin{pmatrix} 0.6968 \\ 0.9396 \end{pmatrix} = 23.814,$$

and when compared with the chi-square with two degrees of freedom, we obtain a p-value smaller than 0.00001. Thus, there is an overall significant association between gender and the outcome. The estimated maximum log-likelihoods with and without gender as a predictor are -654.52 and -667.20, respectively. Thus, the LR statistic for the composite test is $2 \times (-654.52 - (-667.20)) = 25.353$. Thus, we obtain a slightly larger statistic than the Wald test and again find significant association between gender and the response.

Fitted values or the predicted probabilities for each of the levels for, e.g., males can be calculated by (5.7)

$$\pi_0 = \frac{1}{1 + \exp(-1.8987 + 0.9396) + \exp(-1.7774 + 0.6968)} = 0.5805,$$

$$\pi_1 = \frac{\exp(-1.8987 + 0.9396)}{1 + \exp(-1.8987 + 0.9396) + \exp(-1.7774 + 0.6968)} = 0.2225,$$

$$\pi_2 = \frac{\exp(-1.7774 + 0.6968)}{1 + \exp(-1.8987 + 0.9396) + \exp(-1.7774 + 0.6968)} = 0.1970.$$

$\square$

5.2.3 Multinomial probit models

For generalized logit models, we arbitrarily specify a level as the reference and model every other level vs. the "reference" by a logistic model. It turns out that the choice of the reference does not matter. However, if we change the link function in (5.6) to other link functions such as probit and complementary log-log, then the model does depend on the reference level. Thus, it is not appropriate to define probit models in a similar fashion unless a particular level can be designated as the reference.

To avoid this issue, a multinomial probit model may be defined in terms of latent utility functions. Suppose there is an unobservable continuous variable $y_{ij}^* = \mu_{ij} + \varepsilon_{ij}$, where $\varepsilon_{ij} \sim N(0, 1)$ for each level $j = 1, \ldots, J$ and for subjects $i = 1, \ldots, n$. Further assume that

$\{\varepsilon_{i,1}, \ldots, \varepsilon_{i,J}\}$ are independent. The response for the ith subject is the level with highest utility; i.e., the response is k if $y_{ik}^* > y_{ij}^*$ for $j \neq k$. To address identifiability issues, we may arbitrarily select one level as the reference and set the corresponding linear predictor to 0. It is straightforward to verify that the model does not depend on the choice of the reference level (see Exercise 5.3).

Let the Jth level be the reference, then $\mu_{iJ} = 0$ and

$$z_{ij} = y_{ij}^* - y_{iJ}^* = \mu_{ij} + \varepsilon_{ij} - \varepsilon_{iJ}, \tag{5.8}$$

where $\boldsymbol{\epsilon}_i = (\varepsilon_{i1} - \varepsilon_{iJ}, \ldots, \varepsilon_{iJ-1} - \varepsilon_{iJ}) \sim N(\mathbf{0}, \Sigma)$ and $\Sigma = \begin{pmatrix} 2 & 1 & \ldots & 1 \\ 1 & 2 & \ldots & 1 \\ \vdots & \vdots & \ddots & \vdots \\ 1 & 1 & \ldots & 2 \end{pmatrix}$. The response is

J if $z_{ij} < 0$ for $j = 1, \ldots, J-1$ and is k if $z_{ik} > 0$ and $z_{ik} > z_{ij}$ for $j = 1, \ldots, J-1$. Thus, the probability that the outcome is J is $\Pr(\varepsilon_{i1} - \varepsilon_{iJ} > \mu_1, \ldots, \varepsilon_{iJ-1} - \varepsilon_{iJ} > \mu_{J-1}) = \Phi_\Sigma(-\mu_1, \ldots, -\mu_{J-1})$, where Φ_Σ denotes the CDF of $N(\mathbf{0}, \Sigma)$. When $J = 2$, this model reduces to the probit model for binary responses.

Example 5.2

Let us apply the multinomial probit model to Example 5.1. If DEP $= 0$ is selected as the reference level, then the multinomial probit model has the following form:

$$\mu_j = \alpha_j + \beta_j \cdot \text{Gender}, \quad j = 1, 2.$$

Thus, if the ith subject is male, then he has utility $y_{i0}^* \sim N(0, 1)$ and $y_{ij}^* \sim N(\alpha_j, 1)$, $j = 1, 2$. Similarly, the utility for a female follows $N(0, 1)$ for level 0 and and $N(\alpha_j + \beta_j, 1)$ for level j, $j = 1, 2$. The actual observed response is the level with the maximum utility. The parameter estimates are summarized in the following table:

Category	Intercept (α) Estimate	SE	Female (β) Estimate	SE	p-value
2	-1.4293	0.1333	0.5647	0.1621	$< .0001$
1	-1.5088	0.1373	0.7312	0.1642	$< .0001$

The parameters β_j compare the utility function between the genders for the respective outcome levels. The significant p-values indicate that there are significant gender differences when the responses of major and minor depression are compared with no depression, similar to the generalized logistic model in Example 5.1.

The predicted probabilities can be computed with CDF of the multivariate normal distribution based on the linear predictors. For example, for females, $\mu_1 = -1.5088 + 0.7312 = -0.7776$ and $\mu_2 = -1.4293 + 0.5647 = -0.8646$. Thus, the probability for DEP $= 0$ (non depression) is $\Phi_\Sigma(0.7776, 0.8646) = 0.581$. Note that $\mu_1 - \mu_2 = -0.7776 - (-0.8646) = 0.0870$, the probability for DEP $= 1$ (minor depression) is $\Phi_\Sigma(-0.7776, 0.0870) = 0.222$, and the probability for DEP $= 2$ (major depression) is $\Phi_\Sigma(-0.08703, -0.8646) = 0.197$. The fitted values are the same as those based on the generalized logit model in Example 5.1. This is expected, as both models are saturated models in the sense that we have the same numbers of parameters as there are values to be fitted. $\square$

In theory, it is not necessary to assume the error terms ε_{ij} for different alternatives to be independent and/or homoscedastic. Thus, one may relax the assumption to allow Σ to be any positive definite matrix. However, in such cases choice alternative-specific variables are necessary for identifiability issues. For example, one may choose to commute by walking, cycling, and driving based on the cost and time needed for each of the alternatives. When there are such alternative-specific variables, the data set usually consists of multiple observations for each subject: one for each alternative, selected or not. We will limit ourselves in the book to cases without alternative-specific variables. Interested readers may consult books on discrete choice theory (see, e.g., Train (2009)).

5.2.4 Independence of irrelevant alternatives

Similarly, generalized logistic models can also be introduced using utility functions. In fact, if we assume there is an unobservable continuous variable $y_{ij}^* = \mu_{ij} + \varepsilon_{ij}$, where ε_{ij} are i.i.d. with extreme value distributions for each response level $j = 1, \ldots, J$ and subjects $i = 1, \ldots, n$, then choosing the response with maximum utility results in the generalized logit model (see Problem 5.5). Under this assumption, the ratio of the probabilities for any two alternatives j and k is $\exp(\mu_j - \mu_k)$ and thus the ratio does not depend on which other alternatives are available. This is commonly referred to as the *independence of irrelevant alternatives* (IIA) assumption in the theory of discrete choice models. Tests for IIA can be performed by comparing the estimated odds ratio between two alternatives based on all the data with those based on subsamples when some other alteratives are excluded (Cheng and Long, 2007). The IIA assumption indicates that the elimination of one alternative forces individuals who would take that alternative to choose other options, but the ratios of the probabilities among these other alternatives will remain unchanged. This is generally viewed as a major limitation of the generalized logistic model, since it is often not true. This IIA property follows only from the utility assumptions for choice models; in theory any alternative can be deleted and subjects who would choose that alternative are forced to choose other alternatives. However, in many applications, alternatives cannot be arbitrarily deleted. For example, for mental health studies, patients cannot choose disease status such as major or minor depression. Thus this limitation will not cause any problem, if we simply view the generalized logistic model as a statistical model to fit observed multi-level categorical outcomes.

For multinomial probit models, the utility function assumption does not imply the IIA. Thus, the availability of other alternatives has an impact on the relative likelihood of the two response levels. However, this does not necessary mean that multinomial probit models are more general than generalized logistic models. As in the binary case, the generalized logistic and probit models are very similar and difficult to distinguish from each other based on data.

Example 5.3
Let us include additional variables: Age, CIRS, Education, MS, and Gender, to the generalized logit model in the above example. To represent the two nominal variables, MS and Gender, let $u = 1$ if gender is female and 0 otherwise, and $x_j = 1$ if MS $= j$ and 0 otherwise, $j = 1, 2$ (so, $j = 3$ is the reference level for MS). Then, the generalized logit model has the following form:

$$\log \frac{\pi_j(\mathbf{x}_i)}{\pi_0(\mathbf{x}_i)} = \alpha_j + \beta_{j1}\text{Age}_i + \beta_{j2}\text{CIRS}_i + \beta_{j3}\text{Education}_i + \beta_{j4}x_{i1} + \beta_{j5}x_{i2} + \beta_{j6}u_i + \beta_{j7}\text{Race}_i.$$

For the multinomial probit model, we use similar linear predictors. While coefficients are different as expected, the fitted values are in fact very close. The correlation between the estimated probability for non, minor, and major depression under the two models is 0.9996, 0.9987, and 0.9988, respectively. □

Generalized logistic models are easy to implement and are commonly available in most statistical software packages. On the other hand, the multinomial probit model is generally more computationally complicated as it involves high-dimensional integration of multivariate normal distributions. Since the two models generally yield similar results, we recommend the generalized logistic model for most applications unless there are alternative specific covariates.

5.3 Models for Ordinal Responses

Modeling for ordinal responses is more flexible. For example, there are natural choices for the reference level, i.e., the lowest or highest level. Thus, we can model each pair of levels with the reference. It also makes sense to compare adjacent pairs of levels. Furthermore, we can merge levels based on their order and model merged levels as a whole. For example, by merging all the levels below and including level j into one category and merging all the levels above level j into the other category, we can model the cumulative response probabilities

$$\gamma_j(\mathbf{x}) = \Pr(y \leq j \mid \mathbf{x}), \quad j = 1, \ldots, J-1, \tag{5.9}$$

rather than the probabilities of individual levels:

$$\pi_j(\mathbf{x}) = \Pr(y = j \mid \mathbf{x}), \quad j = 1, \ldots, J-1. \tag{5.10}$$

The two sets of probabilities in (5.9) and (5.10) are equivalent, i.e., one completely determines the other. However, when we apply binary models to the two different approaches, we actually obtain different models. In general, models based on cumulative probabilities are easier to interpret for ordinal responses than models based on the categorical probabilities of individual responses.

5.3.1 Cumulative models

By dichotomizing the original response using different levels as cut-points, we can model the cumulative responses using models for binary responses. All the link functions for binary responses can be applied to model the cumulative response probabilities (5.9). We focus on the most commonly used ones, the logit, probit, and complementary log-log link functions.

5.3.1.1 Proportional odds models

As in the binary case, the most popular link function for ordinal responses is still the logit function. A cumulative logit model is specified as

$$\log\left(\frac{\gamma_j(\mathbf{x})}{1 - \gamma_j(\mathbf{x})}\right) = \alpha_j + \boldsymbol{\beta}^\top \mathbf{x}, \quad j = 1, \ldots, J-1. \tag{5.11}$$

This is usually called the *proportional odds model* for reasons described below. The probability of each response category $\pi_j(\mathbf{x})$ is readily calculated from $\gamma_j(\mathbf{x})$ as follows:

$$\pi_1(\mathbf{x}) = \gamma_1(\mathbf{x}) = \frac{\exp\left(\alpha_1 + \boldsymbol{\beta}^\top \mathbf{x}\right)}{1 + \exp\left(\alpha_1 + \boldsymbol{\beta}^\top \mathbf{x}\right)},$$

$$\pi_j(\mathbf{x}) = \gamma_j(\mathbf{x}) - \gamma_{j-1}(\mathbf{x})$$

$$= \frac{\exp(\boldsymbol{\beta}^\top \mathbf{x})\{\exp(\alpha_j) - \exp(\alpha_{j-1})\}}{\left\{1 + \exp\left(\alpha_j + \boldsymbol{\beta}^\top \mathbf{x}\right)\right\}\left\{1 + \exp\left(\alpha_{j-1} + \boldsymbol{\beta}^\top \mathbf{x}\right)\right\}}, \quad 2 \le j \le J-1,$$

$$\pi_J(\mathbf{x}) = 1 - \gamma_{J-1}(\mathbf{x}) = \frac{1}{1 + \exp\left(\alpha_{J-1} + \boldsymbol{\beta}^\top \mathbf{x}\right)}. \tag{5.12}$$

Since $\gamma_j(\mathbf{x})$ increases as a function of j, the logit transform of $\gamma_j(\mathbf{x})$ also becomes a monotone increasing function of j. Thus, the α_j's satisfy the constraint

$$\alpha_1 \le \alpha_2 \le \cdots \le \alpha_{J-1}.$$

The ratio of the odds of the event $y \le j$ between any two values of the vector of independent variables $\mathbf{x} = \mathbf{x}_1$ and $\mathbf{x} = \mathbf{x}_2$ is independent of the choice of category j. Under the model assumptions in (5.11), it is readily checked that

$$\frac{\gamma_j(\mathbf{x}_1)/(1-\gamma_j(\mathbf{x}_1))}{\gamma_j(\mathbf{x}_2)/(1-\gamma_j(\mathbf{x}_2))} = \exp\left(\boldsymbol{\beta}^\top(\mathbf{x}_1 - \mathbf{x}_2)\right), \quad j = 1,\ldots,J-1. \tag{5.13}$$

Thus, the odds of the cumulative response probabilities are proportional to each other, giving rise to the name of the proportional odds model. The proportionality property follows from the model assumption that the coefficient $\boldsymbol{\beta}$ does not depend on level j. In other words, under the proportional odds model, the two log odds ratios (or odds ratios) are assumed to be the same across all levels j. This is in contrast with the generalized logit model where such generalized log odds ratios are allowed to vary across the response categories. Thus, in applications, we may want to check the proportionality assumption to make sure that it applies to the data at hand. We can assess this assumption by assuming a different $\boldsymbol{\beta}_j$ under the following more general model:

$$\log\left(\frac{\gamma_j(\mathbf{x})}{1-\gamma_j(\mathbf{x})}\right) = \alpha_j + \boldsymbol{\beta}_j^\top \mathbf{x}, \quad j = 1,\ldots,J-1 \tag{5.14}$$

and then testing the null of a common $\boldsymbol{\beta}$:

$$H_0 : \boldsymbol{\beta}_j = \boldsymbol{\beta} \ (1 \le j \le J-1).$$

Wald, score, and LR tests may all be applied to test the null. The Wald and LR tests may not be applicable if the MLE of (5.14) does not exist, in which case the score test may still be applied. Alternatively, Brant (1990) proposed to fit the $J-1$ logistic models included in (5.14) separately and then compare the estimated $\boldsymbol{\beta}_j$s. More precisely, for each $j = 1,\ldots,J-1$, let $\widetilde{\boldsymbol{\beta}}_j$ be the MLE of $\boldsymbol{\beta}_j$ based on the logistic model

$$\log\left(\frac{\gamma_j(\mathbf{x})}{1-\gamma_j(\mathbf{x})}\right) = \alpha_j + \boldsymbol{\beta}_j^\top \mathbf{x}. \tag{5.15}$$

In general, $\widetilde{\boldsymbol{\beta}}_j$ is different from and easier to be computed than $\widehat{\boldsymbol{\beta}}_j$, the MLE of $\boldsymbol{\beta}_j$ when the cumulative model (5.14) is fitted. Brant (1990) derived the asymptotic joint distribution of $\widetilde{\boldsymbol{\beta}}_j$, $j = 1,\ldots,J-1$ and developed a Wald-type test for the proportionality. If the proportional odds assumption is violated, one may relax the assumption by assuming a so-called *partial proportional odds* model, which assumes only some of the predictors have common coefficients (Peterson and Harrell Jr, 1990).

Example 5.4

We applied the generalized logit to the DOS data in Example 5.1. Since the response is ordinal, we may consider fitting the proportional odds model to reduce number of parameters. The proportional odds model has the following form:

$$\log \left(\frac{\gamma_j \text{(Gender)}}{1 - \gamma_j \text{(Gender)}} \right) = \alpha_j + \beta \cdot \text{Gender}, \quad j = 0, 1. \tag{5.16}$$

It follows that

$$\beta = \log \left(\frac{\gamma_j \text{(male)}}{1 - \gamma_j \text{(male)}} \right) - \log \left(\frac{\gamma_j \text{(female)}}{1 - \gamma_j \text{(female)}} \right) = \log \left(\frac{\gamma_j \text{(male)} / \{1 - \gamma_j \text{(male)}\}}{\gamma_j \text{(female)} / \{1 - \gamma_j \text{(female)}\}} \right).$$

For $j = 0$, β is the log odds ratio of No vs. Minor/Major depression for comparing the male and female subjects, while for $j = 1$, β is the log odds ratio of No/Minor vs. Major depression for comparing the male and female subjects.

The parameter estimates are summarized in the following table:

Parameter	Estimate	SE	Wald Chi-square	Pr > ChiSq
α_0	1.1132	0.1393	63.8885	< .0001
α_1	2.1079	0.1556	183.5878	< .0001
β	−0.7663	0.1660	21.3182	< .0001

Based on the parameter estimates, the estimated ORs for No vs. Minor/Major depression and No/Minor vs. Major depression for comparing the female and male subjects are both $\exp(-0.7663) = 0.4647$. As this OR is significantly less than 1, we conclude that males are more likely to be severely depressed.

We can find the predicted values by first calculating the cumulative probabilities. For example, for females the two linear predictors are $1.1132 - 0.7663 = 0.3469$ and $2.1079 - 0.7663 = 1.3416$, respectively. Thus, the corresponding cumulative probabilities are $\Pr(\text{Dep} = 0) = \frac{\exp(0.3469)}{1+\exp(0.3469)} = 0.5859$ and $\Pr(\text{Dep} = 0 \text{ or } 1) = \frac{\exp(1.3416)}{1+\exp(1.3416)} = 0.7928$. It follows that $\Pr(\text{Dep} = 1) = 0.7928 - 0.5859 = 0.2069$ and $\Pr(\text{Dep} = 2) = 1 - 0.7928 = 0.2072$.

Without the proportional odds assumption, we may fit the following cumulative logit model:

$$\log \left(\frac{\gamma_j \text{(Gender)}}{1 - \gamma_j \text{(Gender)}} \right) = \alpha_j + \beta_j \cdot \text{Gender}, \quad j = 0, 1.$$

The point estimates for the parameters for gender are $\widehat{\beta}_0 = -0.8182$ and $\widehat{\beta}_1 = -0.5120$. Thus, the two ORs of No vs. Minor/Major depression and No/Minor vs. Major depression for comparing the female and male subjects are different. We can assess the proportional odds property assumed in (5.16) by testing the null of a common β:

$$H_0 : \beta_0 = \beta_1.$$

The estimated variance of $\widehat{\beta}_0$ and $\widehat{\beta}_1$ is $\begin{pmatrix} 0.02868 & 0.02367 \\ 0.02367 & 0.04616 \end{pmatrix}$. Thus, the variance of $\widehat{\beta}_0 - \widehat{\beta}_1$ is estimated as

$$\begin{pmatrix} 1 & -1 \end{pmatrix} \begin{pmatrix} 0.02868 & 0.02367 \\ 0.02367 & 0.04616 \end{pmatrix} \begin{pmatrix} 1 \\ -1 \end{pmatrix} = 0.0275$$

and the Wald test statistic is given $\frac{(0.8182 - 0.5120)^2}{0.0275} = 3.4094$. Compared with a chi-square with one degree of freedom, we obtain a p-value of 0.0648.

The estimated maximum log-likelihoods with and without the proportional odds assumption are -654.5215 and -656.075, respectively. Thus, the LR statistics for the composite test is $2 \times (-654.5215 - (-656.075)) = 3.107$. This is a slightly smaller statistic than the Wald test. The score and Brant tests yield a chi-square values of 3.474 and 3.409 with p-values 0.062 and 0.065, respectively. Thus, the proportional odds assumption is acceptable at the 5% type I error level.

5.3.1.2 Cumulative probit models

Using the probit link, we have the *cumulative probit model*

$$\gamma_j(\mathbf{x}) = \Phi \left(\alpha_j + \boldsymbol{\beta}^\top \mathbf{x} \right), \quad j = 1, \ldots, J - 1, \tag{5.17}$$

where Φ is the CDF of a standard normal. Similar to the case in the proportional odds models, the α_j's satisfy the same constraint $\alpha_1 \leq \alpha_2 \leq \cdots \leq \alpha_{J-1}$. Similar to the binary case, this model is a natural choice if the ordinal response y is the result of discretizing a normally distributed latent continuous variable by $J-1$ cut-points. More precisely, suppose there is an unobservable continuous variable $y_i^* = \mathbf{x}_i^\top \boldsymbol{\beta} + \varepsilon_i$, where $\varepsilon_i \sim N(0,1)$, and y_i is determined by y_i^* according to latent cut-points γ_j:

$$y_i = \begin{cases} 1 & \text{if } y_i^* < \gamma_1, \text{ i.e., } \varepsilon_i < \gamma_1 - \mathbf{x}_i^\top \boldsymbol{\beta} \\ 2 & \text{if } \gamma_1 \leq y_i^* < \gamma_2, \text{ i.e., } \gamma_1 - \mathbf{x}_i^\top \boldsymbol{\beta} \leq \varepsilon_i < \gamma_2 - \mathbf{x}_i^\top \boldsymbol{\beta} \\ \vdots & \vdots \\ J-1 & \text{if } \gamma_{J-2} \leq y_i^* < \gamma_{J-1}, \text{ i.e., } \gamma_{J-2} - \mathbf{x}_i^\top \boldsymbol{\beta} \leq \varepsilon_i < \gamma_{J-1} - \mathbf{x}_i^\top \boldsymbol{\beta} \\ J & \text{if } \gamma_{J-1} \leq y_i^*, \text{ i.e., } \varepsilon_i \geq \gamma_{J-1} - \mathbf{x}_i^\top \boldsymbol{\beta}. \end{cases}$$

Thus, similar to the binary case, the interpretation of coefficients for the cumulative probit model is usually carried out through the Z-scores $\alpha_j + \mathbf{x}_i^\top \boldsymbol{\beta}$.

In the cumulative probit model (5.17), we assumed that the coefficients are the same because all the response categories are obtained by discretizing the same latent variable. One can similarly model ordinal responses with different $\boldsymbol{\beta}$'s and test whether they are the same.

Example 5.5
We change the link function in Example 5.4 to probit and fit the following cumulative probit model to the DOS data:

$$\gamma_j(\mathbf{x}) = \Phi \left(\alpha_j + \beta \cdot \text{Gender} \right), \quad j = 0, 1.$$

The estimate of the parameter for gender is -0.4344 with standard error 0.0966. Thus, the chi-square statistic for testing whether $\beta = 0$ is 20.2231 with a p-value $<.0001$. The parameter β can be interpreted using z-scores as in the binary cases.

To assess whether the coefficients of Gender should be the same for the two cumulative probabilities, we may assume the following model

$$\gamma_j(\mathbf{x}) = \Phi \left(\alpha_j + \beta_j \cdot \text{Gender} \right), \quad j = 0, 1.$$

and then test the null of a common β:

$$H_0 : \beta_0 = \beta_1.$$

The score test statistic is 5.4539, with a p-value 0.0195; thus the common coefficient assumption should be rejected at the 5% type I error level. We leave the Wald and likelihood ratio tests as an exercise.

5.3.1.3 Cumulative complementary log-log models

Another popular model for ordinal response is based on an extension of the complementary log-log link for binary responses to the current setting:

$$\log\left[-\log\{1 - \gamma_j(\mathbf{x})\}\right] = \alpha_j + \boldsymbol{\beta}^\top \mathbf{x}, \quad j = 1, \dots, J - 1. \tag{5.18}$$

Similar to the case in the proportional odds and cumulative probit models, the α_j's also satisfy the constraint $\alpha_1 \le \alpha_2 \le \cdots \le \alpha_{J-1}$. However, unlike the other models, where it is equivalent to model $\Pr(y \le j)$ or $\Pr(y \ge j)$, the choice of modeling $\Pr(y \le j)$ or $\Pr(y \ge j)$ under (5.18) does yield two different models (see also Section 4.3.1.1).

Example 5.6
We change the link function in Example 5.4 to complementary log-log and fit the following cumulative complementary log-log model to the DOS data:

$$\log\left[-\log\{1 - \gamma_j(\mathbf{x})\}\right] = \alpha_j + \beta \cdot \text{Gender}, \quad j = 0, 1.$$

The estimate of the parameter for gender is -0.3575 with SE 0.0868. Thus, the chi-square statistic for testing whether $\beta = 0$ is 16.9661 with a p-value $<.0001$.
If we assume the following model

$$\log\left[-\log\{1 - \gamma_j(\mathbf{x})\}\right] = \alpha_j + \beta_j \cdot \text{Gender}, \quad j = 0, 1,$$

and then test the null of a common β:

$$H_0 : \beta_0 = \beta_1,$$

the score test statistic is 10.0047, with a p-value 0.0016. Thus, the common coefficient assumption should be rejected at the 5% type I error level. Again, we leave the Wald and LR tests as an exercise. ∎

5.3.2 Continuation ratio models

Instead of considering dichotomized outcomes, we may compare each level with all the levels above it combined. More precisely, we compare outcomes $(j + 1, \dots, J)$ vs. j, for $j = 1, \dots, J - 1$. Thus, we model the conditional probability $\gamma_j' = \Pr(y = j \mid y \ge j)$. The multinomial distributions can be determined by all these conditional probabilities:

$$\pi_1 = \gamma_1', \pi_J = \prod_{j=1}^{J-1} \left(1 - \gamma_j'\right), \text{ and } \pi_k = \gamma_k' \prod_{j=1}^{k-1} \left(1 - \gamma_j'\right) \text{ for } k = 2, \dots, J - 1.$$

This model naturally arises when the outcome is related to some kind of survival process. For example, suppose levels $j = 1, \dots, J$ each represent a discrete survival time. Then, at any time point j there is a binary response: whether an at-risk subject will die at that time point. The binary outcome at the time point would be j and $\ge j + 1$. Subjects with outcomes before j died previously and are not at risk at time point j. In these situations, it is natural to consider the conditional probabilities.
Conditioning on the outcome falling in $j, \dots, J$ (where the subject is at risk at time j), the probability of the outcome being exactly j can be modeled by models for binary responses. For example, using a logit link function, we may have the following continuation ratio model

$$\log\left(\frac{\gamma_j'(\mathbf{x})}{1 - \gamma_j'(\mathbf{x})}\right) = \alpha_j + \boldsymbol{\beta}_j^\top \mathbf{x}, \quad j = 1, \dots, J - 1. \tag{5.19}$$

5.3.2.1 Inference

Let $y_{ij} = 1$ if $y_i = j$ for $j = 1, ..., J$ and all subjects $i = 1, ..., n$. Then, the likelihood of the sample is given by

$$
L(\boldsymbol{\theta}) = \prod_{i=1}^{n} \left[\prod_{j=1}^{J} [\pi_j(\mathbf{x}_i)]^{y_{ij}} \right] = \prod_{i=1}^{n} \left[\prod_{j=1}^{J-1} \left[\gamma'_{ij} \prod_{k=1}^{j-1} (1 - \gamma'_{ik}) \right]^{y_{ij}} \cdot \left[\prod_{k=1}^{J} (1 - \gamma'_{ik}) \right]^{y_{iJ}} \right]
$$

$$
= \prod_{j=1}^{J-1} \left[\prod_{i=1}^{n} \left([\gamma'_{ij}]^{y_{ij}} \cdot [(1 - \gamma'_{ij})]^{\Sigma_{j=k}^{J} y_{ij}} \right) \right]. \tag{5.20}
$$

We add the subscript i in γ'_{ij} to emphasis that it changes with the subjects. The last term in (5.20) is the product of likelihood with binary responses at each at-risk level j from 1 to $J - 1$. Based on this property, we can use regular software packages for binary responses to implement the continuation ratio models if specialized software is not directly available.

For the binary response for subjects at-risk at j, a response of 1 represents those subjects with outcome j (i.e., the subjects who died at the time point j) and a response of 0 represents those subjects who survived time j, i.e., those subjects with outcomes above j. Subjects with outcomes $k < j$ are not at risk for the level of j and thus do not contribute to the likelihood for the binary response at j. In others words, one subject with response j will contribute a "death" to level j and a "survivor" to each level $k < j$, but will not contribute to levels $k > j$ since they are no longer at risk. Thus, if there are n_j subjects with response j, $j = 1, ..., J$ in the data, then at level j there would be a sample of $n - n_1 - n_2 - \cdots - n_{j-1}$ with n_j deaths and $n - n_1 - n_2 - \cdots - n_j$ survivors. In summary, the continuation ratio model is equivalent to $J - 1$ stratified binary models with a total of $n + (n - n_1) + \cdots + (n - n_1 - n_2 - \cdots - n_{J-2})$ observations. Note that this number depends on the direction of the continuation ratio, indicating that applying continuation ratio models in the other direction by comparing each level with all lower levels will result in a different model.

At each level j, we model the binary response, i.e., the outcome is exactly j, among all those at risk for level j using a regression model for binary responses such as the logistic model:

$$
\log \left(\frac{\Pr(\text{response} = 1)}{1 - \Pr(\text{response} = 1)} \right) = \alpha_j + \boldsymbol{\beta}_j^\top \mathbf{x}, \quad j = 1, ..., J - 1. \tag{5.21}
$$

Note that at level j, $\Pr(\text{response} = 1) = \gamma_j(\mathbf{x})$ is the conditional probability of having response j, giving that the response is at least j. We may analyze these binary regression models separately. However, it may be necessary to analyze them together if comparison of the parameters across the models is needed. For example, we may test whether the effects of the covariates are similar across the different levels by comparing the $\boldsymbol{\beta}_j$'s. Let r_j be the indicator for the risk level j, then the models in (5.21) can be integrated into a single model:

$$
\log \left(\frac{\Pr(\text{response} = 1)}{1 - \Pr(\text{response} = 1)} \right) = \alpha + \boldsymbol{\beta}^\top \mathbf{x} + \sum_{j=1}^{J-2} \left(\alpha_j r_j + r_j \boldsymbol{\beta}_j^\top \mathbf{x} \right), \tag{5.22}
$$

if the reference coding is applied to the risk level and the last level is treated as the reference. Based on the model, we can test whether the $\boldsymbol{\beta}_j$'s are the same.

If it is expected that the effects of the covariates are similar across the different levels, we may assume common coefficients

$$
\log \left(\frac{\gamma_j(\mathbf{x})}{1 - \gamma_j(\mathbf{x})} \right) = \alpha_j + \boldsymbol{\beta}^\top \mathbf{x}, \quad j = 1, ..., J - 1.
$$

Usually we would not assume the intercepts α_j are the same. Thus, an indicator variable for the levels is included in the overall analysis.

Example 5.7
Let us apply the continuation ratio models to the DOS. Conceptually, we may think that, during disease development, the minor conditions such as minor depression occur first, based on which more serious conditions may further develop. For illustrative purposes, we use gender as the only predictor. Assuming that the disease is developed from no depression to minor and then major depression, we cumulate the levels starting from no depression. The continuation ratio model is specified as

$$\log\left(\frac{\gamma_j'}{1 - \gamma_j'}\right) = \alpha_j + \beta_j \cdot \text{Gender}, \ j = 0, 1, \tag{5.23}$$

where $\gamma_0' = \Pr(\text{Dep} = 0)$ and $\gamma_1' = \Pr(\text{Dep} = 1 \mid \text{Dep} \geq 1)$.

The two models (5.23) can be analyzed simultaneously by setting up the data as described above. Thus, the data consist of information for two risk levels. All the subjects are at risk at level 0 with response of 1 for non depression and response of 0 for major/minor depression. Only those subjects with major or minor depressions are at risk at level 1 with a response of 1 for minor depression and a response of 0 for major depression. Then, we apply the logistic model

$$\log\left(\frac{\Pr(\text{response} = 1)}{1 - \Pr(\text{response} = 1)}\right) = \beta_0' + \beta_1' \cdot \text{Gender} + \beta_2' \cdot \text{Level} + \beta_3' \cdot \text{Gender} \cdot \text{Level}, \tag{5.24}$$

where Level is the indicator variable for risk level 1. The interaction β_3' in (5.24) is related to the different β_j's in models (5.23). Thus, we may test whether the interaction is significant for the purpose of testing whether the two β_j's in models (5.23) are the same. The Wald test chi-square statistic for the null hypothesis that $\beta_0 = \beta_1$ in models (5.23) is 10.2521 with p-value 0.0014; thus the gender effects are indeed different at different stages.

For an overall test of whether there are gender difference in the depression outcome, we need to test the null hypothesis that $\beta_0 = \beta_1 = 0$ for models (5.23). This is equivalent to testing $\beta_1' = \beta_3' = 0$ in model (5.24). The Wald chi-square statistic for the test is 24.0665. Compared with a chi-square distribution with two degrees of freedom, the p-value is $<.0001$. Thus, overall there are gender differences. Further, the estimate of β_0 in (5.23) is -0.8182 with standard error 0.1694. The Wald test of $\beta_0 = 0$ has a p-value $<.0001$. Thus, females are less likely to be non depressed. Among all those who are depressed, the estimate of β_1 is 0.2427 with standard error 0.2848. While the positive sign may indicate that females are less likely to advance to major depression, the p-value for gender differences is 0.3941, and thus the gender difference is not significant. ⬜

5.3.3 Adjacent categories models

For ordinal outcomes, we can also model the adjacent categories. For each of the $J - 1$ pairs of adjacent categories, we may specify a binary response model and the overall multinomial distribution is determined by all $J - 1$ pairs. More precisely, for each level from 1 to $J - 1$, we may compare it with the next level by a binary regression model; any link function for binary responses may be applied. If a logit link function is applied, we assume

$$\log\left(\frac{\pi_j(\mathbf{x})}{\pi_{j+1}(\mathbf{x})}\right) = \alpha_j + \boldsymbol{\beta}_j^\top \mathbf{x} = \eta_j, \quad j = 1, \ldots, J - 1. \tag{5.25}$$

If all of the different pairs have different parameters, the adjacent categories logit model is actually equivalent to the generalized logit model. However, when specifying the model in this way, it makes sense to have different pairs share some parameters. For example, if all the β_j's in (5.25) are the same across different levels j, then the effect of a predictor on the odds ratio $\frac{\pi_j(\mathbf{x})}{\pi_{j+1}(\mathbf{x})}$ is the same for all j. In such situations, we may use the following model with common slopes:

$$\log\left(\frac{\pi_j(\mathbf{x})}{\pi_{j+1}(\mathbf{x})}\right) = \alpha_j + \boldsymbol{\beta}^\top \mathbf{x} = \eta_j, \quad j = 1, \ldots, J-1. \tag{5.26}$$

One may test whether the model is appropriate by testing whether all the β_j's in (5.25) are the same. It is also possible to assume that only some of the predictors have common slopes. The adjacent categories logit model (5.26) can be naturally applied to study the uniform association of Goodman (1979, 1985) (See Exercise 5.16).

Example 5.8

Let us apply the adjacent category logit model to the DOS, using the three-level depression diagnosis (DEP $= 0$ for nondepression, DEP $= 1$ for minor depression, and DEP $= 2$ for major depression) as the response and Gender as the only independent variable. Then, the adjacent category logit model models the two pairs of adjacent levels: 0 vs. 1 and 1 vs. 2.

$$\log\frac{\pi_j(\text{Gender})}{\pi_{j-1}(\text{Gender})} = \alpha_j + \beta_j \cdot \text{Gender}, \quad j = 1, 2.$$

Thus, this model is exactly the generalized logit model using Minor Depression as the reference. The coefficients here are simple combinations from the estimates in Example 5.1. The parameter estimates below use the next higher ordered value as the reference category:

Category	Intercept (α) Estimate	SE	Gender (β) Estimate	SE	p-value
2	0.1214	0.2466	−0.2427	0.2848	0.3941
1	−1.8987	0.1926	0.9396	0.2242	$< .0001$

Based on the estimates, we can see that there is a significant difference between females and males when comparing no depression and minor depression. The positive sign indicates males are more likely to be minorly depressed than females (p-value $< .0001$). The estimated log odds ratio is 0.9396.

We can test whether $\beta_1 = \beta_2$ to test whether the gender differences, measured by odds ratios, between non depression vs. minor depression and minor depression vs. major depression, are the same. The null hypothesis is rejected at the 5% type I error level based on the likelihood ratio test, and thus a model with common slope should not be applied. $\quad\square$

5.3.4 Comparison of different models

We introduced several different classes of models for ordinal responses. Although we mainly used logit link functions, it is obvious that other link functions such as probit and complementary log-log link functions can also be applied, as all the model settings are based on various binary responses. In the above, we used gender as the only predictor. Thus, if the coefficients for gender are assumed to be different, then all the models are actually saturated models. The fitted probabilities based on all the models are the same. Of course, differences in the fitted probabilities are expected if different unsaturated models are utilized.

Example 5.9

In Example 5.3 we ignored the order of the response. Now, let us apply the model for ordinal responses discussed above. Using the same form for the linear predictors, the cumulative logit model has the following form:

$$\log \frac{\gamma_j(\mathbf{x})}{1 - \gamma_j(\mathbf{x})} = \alpha_j + \beta_{j1}\text{Age} + \beta_{j2}\text{CIRS} + \beta_{j3}\text{Education} + \beta_{4j}x_1 + \beta_{5j}x_2 + \beta_{6j}u, \ j = 1, 2,$$

where $\gamma_1 = \Pr(\text{Dep} = 0 \mid \mathbf{x})$ and $\gamma_2 = \Pr(\text{Dep} = 0 \text{ or } 1 \mid \mathbf{x})$.

The continuation ratio model is given by

$$\log \frac{\gamma_j'(\mathbf{x})}{1 - \gamma_j'(\mathbf{x})} = \alpha_j + \beta_{j1}\text{Age} + \beta_{j2}\text{CIRS} + \beta_{j3}\text{Education} + \beta_{4j}x_1 + \beta_{5j}x_2 + \beta_{6j}u, \ j = 1, 2,$$

where $\gamma_1' = \Pr(\text{Dep} = 2 \mid \mathbf{x})$ and $\gamma_2' = \Pr(\text{Dep} = 1 \mid \text{Dep} \leq 1, \mathbf{x})$.

The adjacent category model is given by

$$\log \frac{\pi_{j-1}(\mathbf{x})}{\pi_j(\mathbf{x})} = \alpha_j + \beta_{j1}\text{Age} + \beta_{j2}\text{CIRS} + \beta_{j3}\text{Education} + \beta_{4j}x_1 + \beta_{5j}x_2 + \beta_{6j}u, \ j = 1, 2,$$

where $\pi_j = \Pr(\text{Dep} = j \mid \mathbf{x})$ for $j = 0, 1, 2$.

The analyses show that Age, CIRS and Gender are all significant predictors under these models. Note that linear composite hypotheses are needed to obtain the test statistics and p-values to test whether each of these variables is associated with the outcomes. For example, to test whether Age affects the depression outcome, we need to test $H_0 : \beta_{j1} = 0, \ j = 1, 2$. MS is a three-level nominal variable, and thus testing the null hypothesis of no MS effect on the depression outcome requires the linear hypothesis consisting of four equations: $H_0 : \beta_{14} = \beta_{15} = \beta_{24} = \beta_{25} = 0$.

We can also check if a covariate has differential effects at different stages by comparing the parameters at $j = 1$ with that at $j = 2$. For example, to test whether the Age effects on depression outcome are similar at the two stages, we need to test $H_0 : \beta_{11} = \beta_{21}$. Different conclusions are obtained based on different models. For example, the effect of CIRS is significantly different at the two stages based on the continuation ratio model with p-value $< .0001$, but not significant under the other two models. This is understandable as the effect at each stage has a different meaning under the different models. $\Box$

5.4   Goodness of fit

Most of the techniques we discussed for binary responses in Chapter 4 can be generalized to models for polytomous responses. When the covariate pattern is discrete, we may apply the Pearson and deviance chi-square statistics to assess the goodness of fit. When there are continuous covariates or there are too many covariate patterns, the Hosmer–Lemeshow test can be applied.

5.4.1   The Pearson chi-Square statistic

Given a sample of n subjects with a categorical response and a number of categorical covariates, we can construct an $I \times J$ table for the observed counts with the rows consisting of all I possible patterns of the categorical predictors, and the columns representing the categories of the different responses. With estimates from the fitted model, we can then construct model-based expected counts for each pattern of the independent variables as we did in the binary case.

Let $\widehat{p}_j = \Pr(y = 1 \mid X_j)$ be the fitted probability of responses for the subjects with the jth covariate pattern $(1 \leq j \leq I)$. Let n_{jk} be the total number of subjects in the jth covariate pattern and response k; then n_{j+} will be the total number of subjects in the jth covariate pattern. Hence, $E_{jk} = n_{j+}\widehat{p}_k$, the total of the fitted probabilities of the response, is the expected cell counts for the response $(y = k)$. This results in a table of expected cell counts corresponding to the observed counts in the data:

	Observed (expected)			Total
Covariate pattern	$y = 1$	$\cdots$	$y = J$	
X_1	$n_{11}\ (E_{11})$	$\cdots$	$n_{1J}\ (E_{1J})$	n_{1+}
X_2	$n_{21}\ (E_{21})$	$\cdots$	$n_{2J}\ (E_{2J})$	n_{2+}
$\vdots$	$\vdots$	$\cdots$	$\vdots$	$\vdots$
X_I	$n_{I1}\ (E_{I1})$	$\cdots$	$n_{IJ}\ (E_{IJ})$	n_{I+}

The Pearson chi-square goodness-of-fit test compares the observed to the expected counts using the following formula:

$$\chi_P^2 = \sum_{j=1}^{I} \sum_{k=1}^{J} \frac{(n_{jk} - E_{jk})^2}{E_{jk}}.$$

Under the null hypothesis of the correct model, the above statistic χ_P^2 has an asymptotic chi-square distribution with $I(J-1) - l$ degrees of freedom, where l is the number of (independent) parameters needed to be estimated. Note this chi-square approximation requires large expected cell sizes and so is vulnerable to the sparse data problem.

Example 5.10
For the proportional odds model in Example 5.4, there is a single binary predictor, and thus the number of covariate patterns is 2. The predicted probabilities are

Gender	No depression	Minor depression	Major depression
Female	0.5858485	0.2068960	0.2072555
Male	0.7527294	0.1389433	0.1083273

There are 472 females and 273 males in the data, and the observed and expected counts for each pattern can be summarized in the following table:

	Observed (expected)			
Covariate	No depression	Minor depression	Major depression	Total
Female	274(276.52049)	105(97.65491)	93(97.82460)	472
Male	207(205.49513)	31(37.93152)	35(29.57335)	273

Thus, the Pearson chi-square goodness-of-fit statistic is

$$\chi^2 = \frac{(274 - 276.52049)^2}{276.52049} + \frac{(105 - 97.65491)^2}{97.65491} + \frac{(93 - 97.82460)^2}{97.82460}$$
$$+ \frac{(207 - 205.49513)^2}{205.49513} + \frac{(31 - 37.93152)^2}{37.93152} + \frac{(35 - 29.57335)^2}{29.57335} = 3.0868.$$

The statistic follows the χ^2 distribution with $2 \times (3-1) - 3 = 1$ degree of freedom. The corresponding p-value is 0.079. Thus, lack of fit is not detected with a type I error of 5%.

Note that the only difference between the model and the saturated model is the proportional odds assumption in this simple example, so the goodness-of-fit test is equivalent to the test for proportional odds assumption in this special example. In Example 5.4, we obtained similar conclusion based on Wald, score, and likelihood ratio tests.   ☐

5.4.2   The deviance test

The deviance statistic can be similarly defined as in the binary case. It also compares the observed to the expected counts, but instead uses the following formula:

$$\chi_D^2 = \sum_{j=1}^{I} \sum_{k=1}^{J} n_{jk} \log \frac{n_{jk}}{E_{jk}}. \tag{5.27}$$

The asymptotic distribution of χ_D^2 is the same as that of Pearson chi-square test statistic, i.e., χ_D^2 has an asymptotic χ^2 with $I(J-1) - l$ degrees of freedom. The deviance test also suffers from the sparse data problem. If some expected cell counts are small, it is not reliable.

Example 5.11
The deviance statistic for the model in Example 5.10 is

$$2 \left(274 \log \frac{274}{276.52049} + 105 \log \frac{105}{97.65491} + 93 \log \frac{93}{97.82460} \right.$$

$$\left. + 207 \log \frac{207}{205.49513} + 31 \log \frac{31}{37.93152} + 35 \log \frac{35}{29.57335} \right) = 3.1067,$$

and the associated p-value is 0.078. Thus, the test based on the deviance statistic is similar to that based on the Pearson chi-square statistic for this example.   ☐

5.4.3   The Hosmer–Lemeshow test

When there are continuous independent variables or there is a large number of covariate patterns, Fagerland, Hosmer, and Bofin (2008) generalized the Hosmer–Lemeshow test to generalized logistic models and Fagerland and Hosmer (2013, 2016) generalized the Hosmer–Lemeshow test to the various models for ordinal responses. The basic idea is similar to that of the Hosmer–Lemeshow test in binary case by dividing the covariate space into several groups and then comparing the expected and observed number of subjects in each of the covariate groups.

In regression models for polytomous responses, there are multiple linear predictors, and we need to decide how to divide the covariate space. For generalized logistic models for nominal responses, Fagerland, Hosmer, and Bofin (2008) proposed to simply use the fitted probabilities for the reference response level. For ordinal responses, Fagerland and Hosmer (2013, 2016) proposed to use scores combining all the predicted values: $\sum_{k=1}^{J} k\widehat{p}_{ik}$, where $\widehat{p}_{ik}$ is the predicted probabilities for response level k. Based on these score values, the subjects are grouped into g groups of comparable sizes. Model-based expected cell counts are then computed in the same way as before and the test statistic is constructed as

$$\chi_{HL}^2 = \sum_{j=1}^{g} \sum_{k=1}^{J} \frac{(n_{jk} - E_{jk})^2}{E_{jk}}. \tag{5.28}$$

According to simulation studies, the test statistic has approximately a chi-square distribution with $(g-2) \times (J-1)$ degrees of freedom for generalized logistic models (Fagerland, Hosmer, and Bofin 2008) and $(g-2) \times (J-1) + (J-2)$ degrees of freedom for ordinal models with equal slopes (Fagerland and Hosmer 2013, 2016).

Example 5.12
For the generalized logistic model in Example 5.3, since there are several continuous covariates, Pearson and deviance statistics cannot be applied to assess the goodness of fit of the model. We can instead apply the Hosmer–Lemeshow test. Based on the fitted probabilities for DEP = 0, we separate the subjects into ten groups of similar sizes and obtain the following table.

		Observed DEP			Expected DEP		
Group	Total	2	1	0	2	1	0
1	73	2	3	68	4.22	5.71	63.1
2	73	7	7	59	6.35	8.04	58.6
3	73	7	13	53	7.88	9.99	55.1
4	74	11	14	49	9.76	11.3	53.0
5	73	9	18	46	10.0	13.5	49.5
6	73	12	13	48	12.6	13.9	46.6
7	74	20	14	40	13.7	16.1	44.3
8	73	17	14	42	15.9	16.8	40.3
9	73	20	16	37	19.3	18.2	35.5
10	72	21	22	29	26.3	20.5	25.2

Partition for the Hosmer and Lemeshow Test

The Hosmer–Lemeshow statistic is 13.5446. Compared with a chi-square distribution with $(g-2) \times (J-1) = 16$ degrees of freedom, we obtain a p-value 0.6326. Thus, we will not reject the fit of the model and conclude that the model is consistent with the data.

If a proportional odds model with the same predictors is applied, the Hosmer–Lemeshow statistic is 15.5425. Comparing it with the chi-square distribution with $(g-2) \times (J-1) + (J-2) = 17$ degrees of freedom, we have a p-value of 0.5565 and we will not reject the model under the Hosmer–Lemeshow test either. Thus, the Hosmer–Lemeshow test cannot distinguish between the generalized logistic model and the proportional odds model. Given the relatively large sample size, this may indicate that power may be an issue for the Hosmer–Lemeshow test as in the binary case. □

Exercises

5.1 Show that for a generalized logit model, $T_{jk}(x) = \sum_{i=1}^{n} y_{ij} x_{ik}$ is a sufficient statistic for parameter β_{jk} ($1 \le j \le J$ and $1 \le k \le p$).

5.2 Compute the fitted probabilities for females based on the generalized logistic model in Example 5.1.

5.3 Prove that the multinomial probit model defined in (5.8) does not depend on the selection of the reference level.

5.4 Compute the fitted probabilities for females based on the multinomial probit model in Example 5.2.

5.5 Prove that the generalized logit model (5.6) can be defined with utility functions $y_{ij}^* = \alpha_j + \boldsymbol{\beta}_j^\top \mathbf{x} + \varepsilon_{ij}$, where ε_{ij} follows the standard Gumbel distribution for each level $j = 1, \ldots, J$ and for subjects $i = 1, \ldots, n$, and that $\{\varepsilon_{i,1}, \ldots, \varepsilon_{i,\,J}\}$ are independent.

5.6 For the DOS data set, treat the three-level depression diagnosis as a nominal outcome, and compare the results.

   (a) Fit a generalized logistic model with the three-level depression diagnosis as the response and Age, Gender, CIRS, and MS as covariates. How does the model fit the data? Explain your results.

   (b) What is the estimated probability ratio of major depression of a widowed male with Age $= 75$ and CIRS $= 15$ to a married female at the same Age but CIRS $= 10$? Explain your results.

   (c) Based on the parameter estimates, compute the odds ratio of being depressed (major and minor depression) vs. nondepression between males and females (other characteristics such as Age, CIRS, and MS are holding constant). Explain your results.

   (d) Fit a multinomial probit model and repeat the analyses (a), (b), and (c) above.

5.7 Verify (5.13) for the proportional odds model defined in (5.11).

5.8 Find the Wald and LR test statistics for testing equality of slopes in Example 5.5.

5.9 Find the Wald and LR test statistics for testing equality of slopes in Example 5.6.

5.10 For the DOS data set, fit a cumulative logit model with the three-level depression diagnosis as the ordinal response and Age, Gender, CIRS, and MS as covariates.

   (a) Repeat the analyses (a), (b), and (c) in Problem 5.6.

   (b) Test whether the equal slopes assumption is true.

   (c) Based on the equal slope assumption, test the linear hypothesis: $H_0 : \beta_{Age} = 1$ vs. $H_a : \beta_{Age} \neq 1$ and explain your results. Note that the above is not a linear contrast, and thus you may use an offset term.

5.11 Repeat the analyses in Problem 5.10, replacing the logit link with the probit and complementary log-log link functions.

5.12 Repeat the analyses in Problem 5.10 with continuation ratio models with logit, probit, and complementary log-log link functions.

5.13 Repeat the analyses in Problem 5.10 with adjacent category logit models.

5.14 Compute the correlation matrix of the fitted probabilities for having major depression based on the models in Problems 5.6, 5.10, 5.11, 5.12, and 5.13.

5.15 For an $I \times J$ contingency table with ordinal column variable y $(= 1, \ldots, J)$ and ordinal row variable x $(= 1, \ldots, I)$, consider the model

$$\text{logit}\left[\Pr\left(y \leq j | x\right)\right] = \alpha_j + \beta x, j = 1, ..., J - 1.$$

   (a) Show that $\text{logit}\left[\Pr\left(y \leq j | x = i+1\right)\right] - \text{logit}\left[\Pr\left(y \leq j | x = i\right)\right] = \beta$.

(b) Show that this difference in logit is log of the odds ratio (cumulative odds ratio) for the 2×2 contingency table consisting of rows i and $i+1$ and the binary response having cut-point following category j.

(c) Show that independence of x and y is the special case when $\beta = 0$.

(d) A generalization of the model replaces βx by unordered parameters $\{\mu_i\}_{i=1}^{I-1}$, i.e., treat x as nominal and consider the model

$$\text{logit} \left[\Pr (y \leq j | x = i) \right] = \alpha_j + \mu_i, \ i = 1, \ldots, I - 1.$$

For ith and Ith rows, show that the log cumulative odds ratio equals μ_i for all $J - 1$ cut-points.

5.16 For an $I \times J$ contingency table with ordinal column variable $y \ (= 1, \ldots, J)$ and ordinal row variable $x \ (= 1, \ldots, I)$, consider the adjacent category model

$$\log \left[\frac{\Pr (y = j + 1 | x)}{\Pr (y = j | x)} \right] = \alpha_j + \beta x, j = 1, \ldots, J - 1.$$

(a) Show that any 2×2 subtable consisting of adjacent rows and adjacent columns has the same odds ratio, $\exp(\beta)$. Thus, this is the uniform association model of Goodman (1979, 1985). See Section 7.2.2.

(b) Show that independence of x and y is the special case when $\beta = 0$.

(c) A generalization of the model replaces βx with unordered parameters $\{\mu_i\}_{i=1}^{I-1}$, i.e., treats x as nominal variable and considers the model

$$\text{logit} \left[\frac{\Pr (y = j + 1 | x)}{\Pr (y = j | x)} \right] = \alpha_j + \mu_i, \ i = 1, \ldots, I - 1.$$

For any two rows, show that all the 2×2 subtables consisting of adjacent columns have the same odds ratio. Note that these common odds ratios may be different for different pairs of rows. This is the row-effect model of Goodman (1979, 1985) (see Section 7.2.2).

5.17 For the adjacent category logit and continuation ratio logit models in Example 5.9, assess the goodness of fit of the models.

Chapter 6

Regression Models for Count Response

This chapter discusses regression models for count responses. Like nominal and ordinal responses, count responses, such as the number of heart attacks, suicide attempts, abortions, or birth defects, arise quite often in studies in the biomedical, behavioral, and social sciences. However, unlike the other discrete responses we have studied thus far, count responses cannot be expressed in the form of several proportions. As the upper limit to the number is infinite, the range of count responses is theoretically unbounded, and thus methods for categorical responses do not apply. Count responses and models for such a response type are the focus of this chapter. In practice, ordinal variables that have too many levels to be effectively modeled by the multinomial distribution may also be treated as a count response and modeled by regression models for count responses.

The Poisson log-linear model is the most popular for modeling the number of events observed. Similar to the normal distribution for continuous variables, the Poisson distribution is fundamental to count responses. We discuss Poisson regression in Section 6.1 and model diagnosis and goodness of fit in Section 6.2. One common reason for lack of fit for Poisson regression models is overdispersion, which usually occurs when the sample is heterogeneous. We discuss how to detect overdispersion and methods for correcting this problem in Section 6.3. In Section 6.4, we discuss parametric models that extend Poisson regression to account for heterogeneity in the data. In particular, we consider negative binomial, generalized Poisson, and Conway–Maxwell–Poisson models. In Section 6.5 we focus on zero modifications, where there are too many, too few, or no zeros at all in count response. We conclude the chapter with a discussion on the comparison of different models.

6.1 Poisson Regression Model for Count Response

We briefly discussed the Poisson distribution in Section 2.1 of Chapter 2 as a model for the number of events, without considering any explanatory variables. Recall that a Poisson distribution is determined by one parameter, μ, which is both the mean and variance of the distribution. When μ varies from subject to subject, this single-parameter model can no longer be used to address the variation in μ.

The Poisson log-linear regression is an extension of the Poisson distribution to account for such heterogeneity. The name *log-linear* model stems from the fact that it is the logarithm of μ rather than μ itself that is being modeled as a linear function of explanatory variables. Thus, it follows from the discussion in Section 4.3 of Chapter 4 that this is a special case of a generalized linear model (GLM) in which the conditional mean of the distribution is linked to the linear predictor consisting of covariates.

More precisely, consider a sample of n subjects, and let y_i be some count response and $\mathbf{x}_i = (x_{i1}, \ldots, x_{ip})^\top$ a vector of independent variables for the ith subject ($1 \leq i \leq n$). The *Poisson log-linear regression model* is specified as follows.

DOI: 10.1201/9781003109815-6

(1) Random component. Given $\mathbf{x}_i$, the response variable y_i follows a Poisson distribution with mean μ_i:

$$y_i \mid \mathbf{x}_i \sim \text{Poisson}(\mu_i), 1 \le i \le n. \tag{6.1}$$

(2) Systematic component. The conditional mean μ_i of y_i given $\mathbf{x}_i$ is linked to the linear predictor by the log function:

$$\log(\mu_i) = \mathbf{x}_i^\top \boldsymbol{\beta} = \beta_1 x_{i1} + \ldots + \beta_p x_{ip}, \tag{6.2}$$

where $\boldsymbol{\beta} = (\beta_1, \ldots, \beta_p)^\top$ is the parameter vector of primary interest.

Thus, by the Poisson log-linear model defined above, we can model the variation in the mean of a count response that is explained by a vector of covariates. The Poisson distribution is a member of the exponential family, and the log function in (6.2) is the canonical link for the Poisson model in (6.1) (see Problem 6.1).

Note that we may let $x_{i1} \equiv 1$ so that β_1 will be the intercept. In the remainder of the book, we may simply use $\mathbf{x}_i^\top \boldsymbol{\beta}$. Whether an intercept is included or not will be clear from the context or will not matter. When we want to emphasize there is an intercept, we may use $\beta_0 + \mathbf{x}_i^\top \boldsymbol{\beta}$, so that all the terms in $\mathbf{x}_i$ are covariates.

6.1.1　Parameter interpretation

The interpretation of parameters is similar to that of a logistic regression. If an intercept is specified in the model, then the intercept β_0 is equal to $\log(\mu)$, where μ is the expected mean response when all the covariates are 0. Thus, the intercept indicates the absolute level of the count response. On the other hand, coefficients of covariates suggest the relative association of the covariates with the count response. Consider first the case in which a covariate, say x_1, in the Poisson regression model in (6.2) is an indicator, with β_1 denoting the coefficient of that covariate. The mean response for $x_1 = 1$ is $\exp\left(\beta_1 + \widetilde{\mathbf{x}}^\top \widetilde{\boldsymbol{\beta}}\right)$, where $\widetilde{\mathbf{x}}$ $(\widetilde{\boldsymbol{\beta}})$ denotes the vector $\mathbf{x}$ $(\boldsymbol{\beta})$ with the component of x_1 (β_1) removed. The mean response for $x_1 = 0$ is $\exp\left(\widetilde{\mathbf{x}}^\top \widetilde{\boldsymbol{\beta}}\right)$. Thus, the ratio of the mean response for $x_1 = 1$ to that for $x_1 = 0$ is $\exp(\beta_1)$. If x_1 is continuous, then the mean response for $x_1 = a$ is $\exp\left(\beta_1 a + \widetilde{\mathbf{x}}^\top \widetilde{\boldsymbol{\beta}}\right)$. For one unit increase in this covariate, i.e., $x_1 = a + 1$, with the remaining components of $\mathbf{x}$, $\widetilde{\mathbf{x}}$, held fixed, the mean response is $\exp\left(\beta_1(a+1) + \widetilde{\mathbf{x}}^\top \widetilde{\boldsymbol{\beta}}\right)$. Thus, the ratio of the mean responses per unit increase in x_1 is $\frac{\exp\left(\beta_1(a+1)+\widetilde{\mathbf{x}}^\top \widetilde{\boldsymbol{\beta}}\right)}{\exp\left(\beta_1 a+\widetilde{\mathbf{x}}^\top \widetilde{\boldsymbol{\beta}}\right)} = \exp(\beta_1)$. If β_1 is positive (negative), higher values of x_1 yield higher (lower) mean responses, provided that all other covariates are held fixed. If $\beta_1 = 0$, then the response y_i is independent of x_1. Thus, testing whether a variable is a predictor is equivalent to testing whether its coefficient is 0. Also, similar to a logistic regression, the coefficient β_1 will generally change under a different scale of x_1. However, inferences such as p-values about whether a coefficient is zero remain the same regardless of the scale used (see Problem 6.2).

6.1.2　Inference about model parameters

As in the general case of GLM, we can readily use the method of maximum likelihood to estimate $\boldsymbol{\beta}$ for the Poisson log-linear model. The log-likelihood function for i.i.d. data is

$$l(\boldsymbol{\beta}) = \sum_{i=1}^{n} \{y_i \mu_i - \exp(\mu_i) - \log y_i!\} = \sum_{i=1}^{n} \{y_i \mathbf{x}_i^\top \boldsymbol{\beta} - \exp\left(\mathbf{x}_i^\top \boldsymbol{\beta}\right) - \log y_i!\}.$$

Thus, the score function is

$$\frac{\partial}{\partial \boldsymbol{\beta}} l\left(\boldsymbol{\beta}\right) = \sum_{i=1}^{n} \left\{ y_i \mathbf{x}_i^\top - \exp\left(\mathbf{x}_i^\top \boldsymbol{\beta}\right) \mathbf{x}_i^\top \right\}. \tag{6.3}$$

Since the second-order derivative is negative definite, i.e.,

$$\frac{\partial^2}{\partial \boldsymbol{\beta} \partial \boldsymbol{\beta}^\top} l\left(\boldsymbol{\beta}\right) = -\sum_{i=1}^{n} \exp\left(\mathbf{x}_i^\top \boldsymbol{\beta}\right) \mathbf{x}_i \mathbf{x}_i^\top < 0, \tag{6.4}$$

there is in general a unique solution to the score equation and thus the MLE of $\boldsymbol{\beta}$ is well defined. The MLE can be obtained by the Newton–Raphson method.

It follows from (6.4) that the MLE $\widehat{\boldsymbol{\beta}}$ is asymptotically normal, $\widehat{\boldsymbol{\beta}} \sim_a N\left(\boldsymbol{\beta}, \mathbf{I}_n^{-1}\left(\boldsymbol{\beta}\right)\right)$, where $\mathbf{I}_n\left(\boldsymbol{\beta}\right)$ is the observed information matrix. For the Poisson log-linear model, $\mathbf{I}_n\left(\boldsymbol{\beta}\right) = \sum_{i=1}^{n} \mu_i \mathbf{x}_i \mathbf{x}_i^\top$. Thus, the asymptotic variance of the MLE $\widehat{\boldsymbol{\beta}}$ is given by $Var_a\left(\widehat{\boldsymbol{\beta}}\right) = \left(\sum_{i=1}^{n} \mu_i \mathbf{x}_i \mathbf{x}_i^\top\right)^{-1}$. We may plug in the MLE $\widehat{\boldsymbol{\beta}}$ for inference.

The ML theory requires the sample sizes to be large. In the Poisson regression, however, this may mean that n is large or that all the μ_is are large. The latter situation occurs in cases where there are a fixed number of subjects, but they are observed over a long period of time.

In practice, we are primarily interested in testing whether a covariate is associated with the response. If the variable is continuous or binary and there is only one parameter involving the variable in the model, we may simply test whether the parameter is zero, which can be carried out based on the MLE of $\boldsymbol{\beta}$ and its asymptotic normal distribution. If the variable is categorical with more than two levels, we can introduce indicator variables to represent the variable in the model. Similar to logistic regression models, we may test linear hypotheses using Wald, score, and likelihood ratio tests.

When the sample size is small, inference based on large sample theory may not be reliable. In such cases, we may use exact methods. Similar to logistic regression, it 19 is easy to confirm that $\sum_{i=1}^{n} y_i x_{ij}$ is a sufficient statistic for β_j (see Problem 6.3). To make inference about β_j, we may use the (exact) distribution of its sufficient statistic, $T_j = \sum_{i=1}^{n} y_i x_{ij}$, conditional on those for the remaining parameters, as we did for logistic regression. Likewise, median unbiased estimates (MUEs) can be computed and used for inference (see Problem 6.5).

Example 6.1

For the Sexual Health pilot study, we are interested in modeling the number of occurrence of protected vaginal sex (VCD) during the three-month period as a function of three predictors, HIV knowledge (assessed by a questionnaire with higher scores indicating more informed about HIV and associated risks), depression (CESD score), and baseline value of this outcome in the past three months (VAGWCT1).

Let y denote VCD, and x_1, x_2, and x_3 represent the three covariates HIVKQTOT, CESD, and VAGWCT1, respectively. By fitting the following Poisson model

$$\log\left(E\left(y \mid x_1, x_2, x_3\right)\right) = \beta_0 + \beta_1 x_1 + \beta_2 x_2 + \beta_3 x_3, \tag{6.5}$$

we obtain the estimates

Parameter	Estimate	Standard Error	P-value
Intercept(β_0)	1.1984	0.1780	<0.0001
HIVKQTOT(β_1)	0.0577	0.0103	<0.0001
CESD(β_2)	0.0089	0.0043	0.0401
VAGWCT1(β_3)	0.0259	0.0017	<0.0001

With a type I error of 0.05, all three variables are predictors of the outcome of number of encounters of protected vaginal sex in the next three months. ▯

Note that as in the case of a binomial distribution, the variance of a Poisson variable is also determined by its mean (actually, it is the same as the mean). Thus, overdispersion also arises for Poisson regression models. For this reason, it is common to add an additional dispersion parameter λ to accommodate the extra variation, i.e., the variance is assumed to be $\lambda^2 \exp\left(\mathbf{x}_i^\top \boldsymbol{\beta}\right)$, rather than $\exp\left(\mathbf{x}_i^\top \boldsymbol{\beta}\right)$ based on the Poisson model. Under such an assumption, the above ML approach is problematic since the data no longer strictly follow a Poisson distribution. We may still use quasi-likelihood or estimating equations to derive inference. We discuss these approaches in detail in Section 6.3.

6.1.3 Incidence rates and offsets in log-linear models

When the count outcome is the number of events observed during a period of time as in Example 6.1, then the model parameter can also be interpreted with incidence rates. Consider a sample of n subjects. Let t_i denote the length of observation time for the ith subject. Suppose that the rate of events for the count response of interest (number of events per unit of time) follows a Poisson process. Then, we can model such an event rate by a log-linear model with $r_i = \exp\left(\beta_0 + \mathbf{x}_i^\top \boldsymbol{\beta}\right)$. When the observation times t_i vary across subjects, the number of events y_i for each individual i over time t_i still has a Poisson distribution with mean $\mu_i = t_i r_i = t_i \exp\left(\mathbf{x}_i^\top \boldsymbol{\beta}\right)$. Thus, in this case we can still model the mean response μ_i using the log-linear model:

$$\log \mu_i = \log t_i + \log r_i = \log t_i + \log\left(\exp\left(\mathbf{x}_i^\top \boldsymbol{\beta}\right)\right)$$
$$= \log t_i + \beta_0 + \beta_1 x_{i1} + \ldots + \beta_p x_{ip}. \tag{6.6}$$

As a special case, if $t_i = t$, i.e., the observation time is the same for all individuals, one can absorb $\log t_i$ into β_0 so the above reverts back to the log-linear model defined in (6.2). The parameters in (6.6) can be interpreted in terms of incidence rate ratio (IRR). For example, since $\beta_1 = \log\left(\frac{\exp\left(\beta_1(a+1)+\tilde{\mathbf{x}}^\top \tilde{\boldsymbol{\beta}}\right)}{\exp\left(\beta_1 a+\tilde{\mathbf{x}}^\top \tilde{\boldsymbol{\beta}}\right)}\right)$, it is the log of the IRR per unit increase in x_1. Note that this IRR does not depend on the time scale utilized.

In many studies, the observation time often varies across subjects, even for controlled clinical trials. For example, most studies recruit patients over a period of time. Such staggered entries cause varying lengths of observation time as patients enter the study at different time points. Consequently, we must address the issue of heterogeneous observation times across patients when modeling the occurrence of an event as a count response, as those with longer observation times are likely to have more events. In fact, we can include $\log t_i$ as in (6.6) to take care of the heterogeneity in observation times. When t_i are different across subjects, $\log t_i$ cannot be absorbed into β_0 and rather should be regarded as a covariate in the log-linear model. However, since its corresponding coefficient is always one, it should not be treated the same way as other covariates. In the nomenclature of GLM, $\log t_i$ is called *offset*. We have discussed offsets earlier within the context of testing the general linear hypothesis.

If a different time scale $t_i' = c \cdot t_i$ is utilized, then the offset term would be $\log\left(t_i'\right) = \log c + \log t_i$. Since $\log c$ may be absorbed into the intercept, the coefficients of the covariates won't change. This is expected as they are related to the relative incidence rate. However, the intercept does change; the intercept β_0 is the log of the incidence rate for the subjects with all covariates being zero in a unit time; the time unit depends on the time scale.

Example 6.2

In the SIAC, information was collected on parents' psychiatric symptoms (BSI), stressful life events (LEC), and family conflict (PARFI) as well as child illnesses (number of times the child was sick) over the study. As the observation times varied across children, we need to take this into account when modeling child illnesses as a function of parent's stress and family conflict levels.

This is a longitudinal study with at most seven visits for each child, with approximately six months between visits. The first visit happened about one year following the entry of the study. Here we consider only the period between the entry and the first visit. In the data set, DAY is the number of days from entry to the study until the first visit, which ranges from 5 to 275 days with a mean of 169 days, indicating a substantial amount of variability across the subjects.

We are interested in the number of fevers that occurred for each child during this period. This outcome has a range from 0 to 10 with a mean of 2.7. We assume that the rate of occurrence of fever follows a Poisson process. Then, the number of illnesses that occur within a period of time still follows a Poisson distribution with the mean proportional to the length of the period. Since the observation period varies substantially from child to child, it is inappropriate to ignore the effect of observation time and assume the form of the linear predictor in (6.2).

In this example, we study the effects of Gender, Age, and BSI on the response of number of fevers from entry into the study to the end of the first visit. To examine the effects of the offset, we fit the data using two different approaches: (1) Poisson model without considering the effect of length of the follow-up period and (2) Poisson model with the logarithm of length added as an offset term.

The parameter estimates based on the first model are summarized in the following table:

Parameter	Estimate	SE	95% CI		Wald χ^2	Pr $> \chi^2$
Intercept	1.4197	0.2578	0.9137	1.9245	30.32	<0.0001
Gender	0.0878	0.0942	−0.0970	0.2726	0.87	0.3516
Age	−0.0832	0.0320	−0.1465	−0.0211	6.77	0.0093
BSI	0.2229	0.1038	0.0164	0.4233	4.62	0.0317

The estimated parameters from the second model are given in the following table:

Parameter	Estimate	SE	95% CI		Wald χ^2	Pr $> \chi^2$
Intercept	−4.2180	0.2544	−4.7178	−3.7203	274.84	<0.0001
Gender	0.0394	0.0942	−0.1452	0.2241	0.18	0.6753
Age	−0.0713	0.0315	−0.1335	−0.0101	5.14	0.0234
BSI	0.1446	0.1028	−0.0601	0.3432	1.98	0.1597

With the offset term added, BSI is no longer significant at the 5% type I error. ▯

6.2 Goodness of Fit

Departures from the Poisson assumption are not unusual even in well-designed and controlled laboratory experiments. Thus, it is important to assess how good the model is in terms of fitting the data. In this section, we discuss two goodness-of-fit tests for Poisson regression models. These two tests are not new and have been used in different settings in the earlier chapters of the book. One is the Pearson χ^2 statistic, defined as the sum of the normalized squared differences of the expected and observed counts, while the other is the scaled deviance statistic based on the log-likelihood.

6.2.1   Pearson chi-Square statistic

We have seen variations of the Pearson statistic for different applications in the previous chapters. This statistic is generally defined as the sum of normalized squared residues between the observed and model-fitted values of the response variable. In general, the asymptotic theory of this statistic requires large cell sizes. For this reason, it can be applied to assess goodness of fit for logistic regression models only when the covariates patterns are limited and the sample size is large. Similarly, the Pearson statistic for Poisson regression models also requires large cell sizes. For a Poisson regression, large cell sizes may mean a large number of the observations in the cell or a large expected count response. Further, this statistic is useful for assessing the closely related concept of overdispersion, which will be our focus in Section 6.3 of this chapter.

Let y_i be the count response and $\widehat{\mu}_i = \exp\left(\mathbf{x}_i^\top \widehat{\boldsymbol{\beta}}\right)$ the fitted value under the log-linear model obtained by substituting $\widehat{\boldsymbol{\beta}}$ in place of $\boldsymbol{\beta}$ in the mean response in (6.2) ($1 \leq i \leq n$). Since the mean and variance are the same for the Poisson distribution, we may also estimate the variance by $\widehat{\mu}_i$. Thus, the Pearson residue for the ith subject is $\frac{y_i - \widehat{\mu}_i}{\sqrt{\widehat{\mu}_i}}$, and the Pearson statistic is simply $P = \sum_{i=1}^{n} \frac{(y_i - \widehat{\mu}_i)^2}{\widehat{\mu}_i}$. It can be shown that the Poisson distribution converges to a normal distribution when the mean μ grows unbounded (see Problem 6.6), i.e.,

$$\frac{y_i - \mu_i}{\sqrt{\mu_i}} \sim_a N(0,1), \quad \text{as } \mu_i \to \infty, \quad 1 \leq i \leq n.$$

It follows that for fixed n,

$$P = \sum_{i=1}^{n} \frac{(y_i - \widehat{\mu}_i)^2}{\widehat{\mu}_i} \sim_a \chi^2_{n-p}, \quad \text{as } \mu_i \to \infty \text{ for all } 1 \leq i \leq n, \tag{6.7}$$

where p is the number of parameters or the dimension of $\boldsymbol{\beta}$.

Thus, if $y_i \sim \text{Poisson}(\mu_i)$ and μ_i are all large, Pearson statistic approximately follows a chi-square distribution with degrees of freedom $n - p$. The degrees of freedom follow the same pattern as in other similar applications, such as contingency tables to account for the loss of information when estimating the parameter vector $\boldsymbol{\beta}$.

Note that the asymptotic distribution of the Pearson statistic is obtained based on the assumption that $\mu_i \to \infty$ while n is fixed. This is a bit unusual, as most asymptotic results are typically associated with a large sample size n. In practice, if some μ_i are not large, inferences based on the asymptotic chi-square distribution may be invalid. Asymptotic normal distributions of this statistic when $n \to \infty$ are available for inference (McCullagh, 1986). However, given the complex form of the asymptotic variance, this asymptotic normal distribution is not frequently used.

A particularly common violation of the Poisson distribution is overdispersion due to data clustering. Recall that under the Poisson law, the mean and variance are the same. This is actually a very stringent restriction, and in many applications, the variance $\sigma^2 = Var(y)$ often exceeds the mean $\mu = E(y)$, causing *overdispersion* and making it inappropriate to model such data using Poisson models. When overdispersion occurs, the standard errors of parameter estimates of the Poisson log-linear model are artificially deflated, leading to exaggerated effect size estimates and false significant findings. One approach to this problem is to use a GLM that assumes the same log-linear structure for the mean response, but a different variance model $Var(y_i) = \lambda^2 \mu_i$ $(\lambda^2 > 1)$ to account for overdispersion. The common factor λ is called the *scale* or *dispersion* parameter. For a fixed dispersion parameter λ, the scaled Pearson chi-square statistic is given by $P = \sum_{i=1}^{n} \frac{(y_i - \hat{\mu}_i)^2}{\lambda^2 \hat{\mu}_i}$. Again, this statistic is approximately chi-square with degrees of freedom $n - p$ under the assumed model, provided that the μ_i are large for all $1 \le i \le n$.

6.2.2 Scaled deviance statistic

The deviance statistic, as in the case of a logistic regression for binary responses, is defined as twice the magnitude of the difference between the maximum achievable log-likelihood and the log-likelihood evaluated at the MLE of the model parameter vector.

Let $\mathbf{y} = (y_1, \ldots, y_n)^\top$ denote the vector of responses from the n subjects. The deviance statistic of a model is defined by

$$D(\mathbf{y}, \boldsymbol{\theta}) = 2 \left[l(\mathbf{y}, \mathbf{y}) - l(\mathbf{y}, \boldsymbol{\theta}) \right],$$

where $l(\mathbf{y}, \mathbf{y})$ is the log-likelihood that would be achieved if the model gave a perfect fit to the data and $l(y, \boldsymbol{\theta})$ the log-likelihood of the model under consideration. For Poisson log-linear regression, the deviance statistic has the form

$$D(\mathbf{y}, \boldsymbol{\theta}) = 2 \sum_{i=1}^{n} \left[y_i \log\left(\frac{y_i}{\hat{\mu}_i}\right) - (y_i - \hat{\mu}_i) \right], \quad \hat{\mu}_i = \exp\left(\mathbf{x}_i^\top \hat{\boldsymbol{\beta}}\right).$$

If the Poisson model under consideration is correct, $D(\mathbf{y}, \boldsymbol{\theta})$ is approximately a chi-squared random variable with degrees of freedom $n - p$. When the deviance divided by the degrees of freedom is significantly larger than 1, there is evidence of lack of fit.

As mentioned above for the Pearson χ^2 statistic, overdispersion is a common cause of violation of the assumptions of Poisson regression. In such cases, a GLM with a dispersion parameter λ may be used to address this issue. We can use a deviance statistic to test this revised model. In particular, for a fixed value of the dispersion parameter λ, the *scaled deviance* is defined by

$$D(y, \boldsymbol{\theta}) = \frac{2}{\lambda^2} \sum_{i=1}^{n} \left[y_i \log\left(\frac{y_i}{\hat{\mu}_i}\right) - (y_i - \hat{\mu}_i) \right].$$

If the latter GLM model is correct, D follows approximately a chi-squared distribution with degrees of freedom $n - p$. As in the case of the Pearson statistic, valid inference again requires large μ_i for all $1 \le i \le n$.

Note that as these statistics are defined based on the likelihood function, they are not valid for estimating equations based distribution-free inference, which will be discussed in the next section to obtain robust estimates.

Example 6.3

In Example 6.1 we modeled the number of encounters of protected vaginal sex during the three-month period as a function of three predictors: HIV knowledge, depression, and baseline value of this outcome in the past three months, using a Poisson regression model.

The nonscaled deviance and Pearson statistics are given by

$$\text{Deviance} \quad D = 1419.94, \quad \text{and} \quad \frac{D}{n-p} = \frac{1419.94}{94} = 15.106,$$

$$\text{Pearson} \quad P = 1639.94, \quad \text{and} \quad \frac{P}{n-p} = \frac{1639.94}{94} = 17.446.$$

Since $\frac{D}{n-p}$ and $\frac{P}{n-p}$ are both much larger than 1, we may conclude that the model does not fit the data well. Thus, the Poisson model may not be appropriate for modeling the count response in this example. ☐

6.3   Overdispersion

As discussed in Section 6.2, a particularly common cause of violation of Poisson model assumption is overdispersion. Overdispersion may occur for many reasons. For example, overdispersion is very likely to occur if the observations are based on time intervals of varying lengths. Another common factor responsible for overdispersion is data clustering. The existence of these data clusters invalidates the usual independent sampling assumption, and as a result, statistical methods developed based on independence of observations are no longer applicable to such data. Clustered data often arise in epidemiological and psychosocial research where subjects sampled from within a common habitat (cluster) such as families, schools, and communities are more similar than those sampled across different habitats, leading to correlated responses within a cluster.

If the nature of data clustering is well understood, then refined models may be developed to address overdispersion. We discuss some of these models in Section 6.4. On the other hand, if the precise mechanism that produces overdispersion is unknown, a common approach is to use a modified robust variance estimate in the asymptotic distribution of parameter estimates and make inference based on that corrected asymptotic distribution. In this section, we first introduce some statistics for detecting overdispersion and then discuss how to correct this problem using the robust variance estimate.

Note that in some applications, the variance of the count variable may be less than the mean, producing *underdispersion*. In such cases, the standard errors of parameter estimates are artificially inflated when modeled using Poisson regression, giving rise to underestimated effect size estimates and missed opportunities for significant findings. Since overdispersion is much more common, we focus on this common type of dispersion throughout the chapter, though similar considerations may also apply to the case of underdispersion.

6.3.1   Detection of overdispersion

Detection of overdispersion is closely related to the goodness-of-fit test. In fact, the two goodness-of-fit tests, the deviance and Pearson chi-square tests, discussed in the last section can also be used to provide indications of overdispersion. As noted earlier, both the deviance and Pearson statistics have approximately a chi-square distribution with degrees of freedom $n-p$ under the Poisson log-linear model. Thus, overdispersion is indicated if the normalized version of each statistic, i.e., the respective statistic divided by the degrees of freedom $n - p$, is significantly larger than 1. For example, overdispersion may have occurred in the Sexual Health pilot study data, as indicated by the large values of the Pearson and deviance statistics in Example 6.3. Note that the deviance and Pearson statistics can be quite different if the mean is not large. Simulation studies seem to indicate that Pearson statistics are better in detecting overdispersion (Hilbe, 2011).

Dean and Lawless (1989) discussed score statistics to check overdispersion based on $T = \frac{1}{2} \sum_{i=1}^{n} \left((y_i - \widehat{\mu}_i)^2 - y_i \right)$ under the assumption of a correct specification of the mean response μ_i. This statistic is motivated by assuming some form of extra Poisson variation, $Var(y_i) = \mu_i + \tau \mu_i^2$, and then testing the null hypothesis that the model is Poisson: $H_0 : \tau = 0$. When the sample size $n \to \infty$, the following normalized version of the statistic,

$$T_1 = \frac{\sum_{i=1}^{n} \left[(y_i - \widehat{\mu}_i)^2 - y_i \right]}{\sqrt{2 \sum_{i=1}^{n} \widehat{\mu}_i^2}},$$

follows approximately the standard normal distribution under H_0. If the sample size n is fixed, but $\mu_i \to \infty$ for all $1 \le i \le n$, then T is asymptotically equivalent to

$$T_2 = \frac{\sum_{i=1}^{n} (y_i - \widehat{\mu}_i)^2}{\frac{1}{n} \sum_{i=1}^{n} \widehat{\mu}_i}.$$

Dean and Lawless (1989) showed that the limiting distribution of T_2 is a linear combination of chi-squares as $\mu_i \to \infty$ $(1 \le i \le n)$.

6.3.2 Correction for overdispersion

When overdispersion is indicated and the appropriateness of the Poisson model is in serious doubt, the model-based asymptotic variance no longer indicates the variability of the MLE and inference based on the likelihood approach may be invalid. One can use a different and more appropriate model for fitting the data, if the overdispersion mechanism is known. We will discuss such approaches in Section 6.4. Alternatively, we can use a different variance estimate to account for overdispersion, such as the sandwich estimate discussed in Chapter 1.

As noted in Chapter 1, the most popular alternative to the MLE is the sandwich variance estimate derived based on the estimating equations (EE). Along with EE, this variance estimate provides robust inference regardless of the distribution of the response. We describe this approach for the class of generalized linear models (GLMs), and in particular we derive the sandwich estimate for the Poisson log-linear model.

6.3.2.1 Sandwich estimate for asymptotic variance

Consider the class of GLM defined in Section 4.3.1 of Chapter 4. The mean response $\mu_i = E(y_i \mid \mathbf{x}_i)$ is related with the linear predictor through a link function; more precisely we assume that

$$g(\mu_i) = \mathbf{x}_i^\top \boldsymbol{\beta}. \tag{6.8}$$

Let $D_i = \frac{\partial \mu_i}{\partial \boldsymbol{\beta}}$; then it is straightforward to check that the following EE:

$$\sum_{i=1}^{n} D_i V_i^{-1} (y_i - \mu_i) = \mathbf{0}, \tag{6.9}$$

is unbiased, for any function of μ_i, V_i. As reviewed in Chapter 1, the solution $\widetilde{\boldsymbol{\beta}}$ is consistent and asymptotically normal regardless of the distribution of y_i and choice of V_i, as long as the mean relation (6.8) is correct. The asymptotic variance of $\widetilde{\boldsymbol{\beta}}$ is given by

$$\boldsymbol{\Phi}_{\widetilde{\boldsymbol{\beta}}} = \frac{1}{n} B^{-1} E \left(\frac{1}{n} \sum_{i=1}^{n} D_i V_i^{-2} Var(y_i \mid \mathbf{x}_i) D_i^\top \right) B^{-1}, \tag{6.10}$$

where $B = E\left(D_i V_i^{-1} D_i^\top\right)$. Further, if y_i given $\mathbf{x}_i$ is modeled parametrically using a member of the exponential family, the estimating equations in (6.9) can be made to equal the score equations of the log-likelihood with an appropriate selection of V_i, thus yielding the MLE of $\boldsymbol{\beta}$. The EE inference based on (6.10) only requires the mean response assumption (6.8), but does not require any specific parametric distribution, thus providing robust inference when the model assumptions are violated.

For example, EE can be used to deal with overdispersion for Poisson regression. For a Poisson log-linear model, we assume that $\mu_i = E\left(y_i \mid \mathbf{x}_i\right) = \exp\left(\mathbf{x}_i^\top \boldsymbol{\beta}\right)$; thus $D_i = \frac{\partial \mu_i}{\partial \boldsymbol{\beta}} = \mu_i \mathbf{x}_i$. If we choose $V_i = \mu_i$, the variance of y_i given $\mathbf{x}_i$ when it follows a Poisson with mean μ_i, then it is readily checked that the EE in (6.9) is identical to the score equations of the log-likelihood of the Poisson log-linear regression in (6.3) and the asymptotic variance of the EE estimate in (6.10) simplifies to

$$\Phi_{\widehat{\boldsymbol{\beta}}} = \frac{1}{n} B^{-1} \left(E\left(\frac{1}{n} \sum_{i=1}^{n} Var\left(y_i \mid \mathbf{x}_i\right) \mathbf{x}_i \mathbf{x}_i^\top\right)\right) B^{-1}, \tag{6.11}$$

where $B = E\left(\mu_i \mathbf{x}_i \mathbf{x}_i^\top\right)$. Thus, by using $V_i = Var\left(y_i \mid \mathbf{x}_i\right) = \mu_i$, the EE yields the same estimate as the MLE $\widehat{\boldsymbol{\beta}}$ of $\boldsymbol{\beta}$. If the conditional distribution of y_i given $\mathbf{x}_i$ follows a Poisson with mean $\mu_i = \exp\left(\mathbf{x}_i^\top \boldsymbol{\beta}\right)$, then $Var\left(y_i \mid \mathbf{x}_i\right) = \mu_i$ and the asymptotic variance of the EE estimate can further be simplified to

$$\Phi_{\widehat{\boldsymbol{\beta}}} = \frac{1}{n} B^{-1} E\left(\frac{1}{n} \sum_{i=1}^{n} \mu_i \mathbf{x}_i \mathbf{x}_i^\top\right) B^{-1} = \frac{1}{n} I^{-1}\left(\boldsymbol{\beta}\right), \tag{6.12}$$

the same as the asymptotic variance of the MLE $\widehat{\boldsymbol{\beta}}$. However, as the EE estimate $\widetilde{\boldsymbol{\beta}}$ and its associated asymptotic variance $\Phi_{\widetilde{\boldsymbol{\beta}}}$ in (6.10) are derived independent of such distributional models, it still provides valid inference even when the conditional distribution of y_i given $\mathbf{x}_i$ is non-Poisson distributed. For example, in the presence of overdispersion, $Var\left(y_i \mid \mathbf{x}_i\right)$ is larger than μ_i, biasing the asymptotic variance in (6.12) based on the MLE of $\widehat{\boldsymbol{\beta}}$. By correcting this bias, the EE-based asymptotic variance $\Phi_{\widetilde{\boldsymbol{\beta}}}$ in (6.10) still provides valid inference about $\boldsymbol{\beta}$.

By estimating $\widehat{Var}\left(y_i \mid \mathbf{x}_i\right)$ and $\widehat{B}$ by their respective moments

$$\widehat{Var}\left(y_i \mid \mathbf{x}_i\right) = \left(y_i - \widehat{\mu}_i\right)^2, \quad \widehat{B} = \frac{1}{n} \sum_{i=1}^{n} \widehat{\mu}_i \mathbf{x}_i \mathbf{x}_i^\top,$$

we obtain the *sandwich variance estimate*

$$\widehat{\Phi}_{\boldsymbol{\beta}} = \left(\sum_{i=1}^{n} \widehat{\mu}_i \mathbf{x}_i \mathbf{x}_i^\top\right)^{-1} \left(\sum_{i=1}^{n} \widehat{Var}\left(y_i \mid \mathbf{x}_i\right) \mathbf{x}_i \mathbf{x}_i^\top\right) \left(\sum_{i=1}^{n} \widehat{\mu}_i \mathbf{x}_i \mathbf{x}_i^\top\right)^{-1}. \tag{6.13}$$

Thus, if overdispersion is detected, we can still use the ML approach to estimate $\boldsymbol{\beta}$, but we need to use the sandwich variance estimate in (6.13) to ensure valid inference.

6.3.2.2 Scaled variance and quasi-likelihood

For the Poisson log-linear model, another popular alternative for correcting overdispersion is to use scaled variance as we did for overdispersion for binomial models in Chapter 4. Under this approach, we assume an additional scale parameter to inflate the Poisson-based variance of y_i. Specifically, we assume the same conditional mean as in the Poisson log-linear regression, but a scaled conditional variance of y_i given $\mathbf{x}_i$ as follows:

$$\mu_i = \exp\left(\mathbf{x}_i^\top \boldsymbol{\beta}\right), \quad Var\left(y_i \mid \mathbf{x}_i\right) = \lambda^2 \mu_i.$$

If $\lambda^2 = 1$, $Var(y_i \mid \mathbf{x}_i) = \mu_i$ and the modified variance reduces to that under the Poisson model. In the presence of overdispersion, $Var(y_i \mid \mathbf{x}_i) > \mu_i$, then $\lambda^2 > 1$ and the scale parameter with $\lambda^2 > 1$ can account for the overdispersion.

Under this scaled-variance approach, we first estimate β using either the MLE or EE approach. Then, based on the Pearson chi-square statistic we may estimate the scale parameter λ^2 by

$$\widehat{\lambda}^2 = \frac{1}{n-p} \sum_{i=1}^n \frac{(y_i - \widehat{\mu}_i)^2}{\widehat{\mu}_i} = \frac{P}{n-p}.$$

The estimate of the asymptotic variance of the estimate of β based on the Pearson scale is $\widehat{\lambda}^2 \left(\sum_{i=1}^n \widehat{\mu}_i \mathbf{x}_i \mathbf{x}_i^\top \right)^{-1}$. Alternatively, we can substitute the deviance statistic to estimate λ^2 in (6.14) to obtain a slightly different consistent estimate of the asymptotic variance of the estimate of β. Note that if we substitute $\widehat{\lambda}^2 \widehat{\mu}_i$ in place of $\widehat{Var}(y_i \mid \mathbf{x}_i)$ in the sandwich variance estimate $\widehat{\Phi}_\beta$ in (6.13), we obtain the scale variance estimate of β:

$$\widetilde{\Phi}_\beta = \left(\sum_{i=1}^n \widehat{\mu}_i \mathbf{x}_i \mathbf{x}_i^\top \right)^{-1} \left(\sum_{i=1}^n \widehat{\lambda}^2 \widehat{\mu}_i \mathbf{x}_i \mathbf{x}_i^\top \right) \left(\sum_{i=1}^n \widehat{\mu}_i \mathbf{x}_i \mathbf{x}_i^\top \right)^{-1} = \widehat{\lambda}^2 \left(\sum_{i=1}^n \widehat{\mu}_i \mathbf{x}_i \mathbf{x}_i^\top \right)^{-1}. \quad (6.14)$$

Unlike the sandwich estimate $\widehat{\Phi}_\beta$ in (6.13), the estimate $\widetilde{\Phi}_\beta$ is derived based on a particular variance model for overdispersion, $Var(y_i \mid \mathbf{x}_i) = \lambda^2 \mu_i$. If this variance model is incorrect for the data, inference based on this asymptotic variance estimate $\widetilde{\Phi}_\beta$ is likely to be wrong.

This scaled variance approach is also called *quasi-likelihood* approach because a quasi-likelihood function can be defined to play the roll of the likelihood function for inference. For both binomial and Poisson outcomes, the score function has the form $U = \frac{y-u}{v(\mu)}$, when the mean μ is treated as the parameter and $v(\mu)$ is the variance. It has the following properties: $E(U) = 0$, $Var(U) = \frac{1}{v(\mu)}$, and $-E\left(\frac{dU}{d\mu}\right) = \frac{1}{v(\mu)}$, and the asymptotic theory of the MLE mostly relies on these properties (McCullagh and Nelder, 1989). Thus, a log quasi-likelihood which gives rise to a "score" $U = \frac{y-u}{\phi v(\mu)}$ is of interest. We may *reverse-engineer* such a likelihood by integrating $\frac{y-u}{\phi v(\mu)}$ to obtain

$$q(\mu, y) = \int_y^\mu \frac{y-t}{\phi v(t)} dt.$$

This $q(\mu, y)$ is commonly referred to as a log quasi-likelihood, or simply a quasi-likelihood (McCullagh and Nelder, 1989). For the "score function" based on the quasi-likelihood, the only difference compared with the score function based on the likelihood is the constant scale parameter ϕ. So, under scaled-variance, or quasi-likelihood, we may first estimate β using the MLE or EE approach. However, we cannot estimate the scale based on the "score equation" of the quasi-likelihood. So, we still estimate the scale parameter ϕ by $\widehat{\phi} = \frac{S}{n-p}$, where S is the Pearson or deviance chi-square statistic, and make inference based on the scaled variance.

Example 6.4

For the Sexual Health pilot study, there is overdispersion, as illustrated in Example 6.3. We may use the sandwich estimates and overdispersion methods to correct the overdispersion.

The analysis results with the overdispersion scale estimated by the deviance statistic are summarized in the following table:

Parameter	Estimate	SE	95% CI		Wald χ^2	Pr $> \chi^2$
Intercept	1.1984	0.6917	−0.1683	2.5472	3.00	0.0832
VAGWCT1	0.0259	0.0065	0.0125	0.0381	15.78	<0.0001
HIVKQTOT	0.0577	0.0401	−0.0196	0.1378	2.07	0.1498
CESD	0.0089	0.0168	−0.0257	0.0402	0.28	0.5974
Scale	3.8866	0.0000	3.8866	3.8866		

Shown in the table below are the results when the overdispersion scale is estimated by the Pearson statistic.

Parameter	Estimate	SE	95% CI		Wald χ^2	Pr $> \chi^2$
Intercept	1.1984	0.7434	−0.2715	2.6474	2.60	0.1070
VAGWCT1	0.0259	0.0070	0.0114	0.0390	13.66	0.0002
HIVKQTOT	0.0577	0.0431	−0.0253	0.1440	1.80	0.1802
CESD	0.0089	0.0180	−0.0284	0.0425	0.24	0.6231
Scale	4.1769	0.0000	4.1769	4.1769		

The results based on EE are summarized in the following table:

Parameter	Estimate	SE	95% CI		Wald χ^2	Pr $> \chi^2$
Intercept	1.1984	0.5737	0.0739	2.3228	2.09	0.0367
VAGWCT1	0.0259	0.0048	0.0164	0.0353	5.35	<0.0001
HIVKQTOT	0.0577	0.0393	−0.0194	0.1348	1.47	0.1424
CESD	0.0089	0.0202	−0.0308	0.0485	0.44	0.6613

Based on these results, it is clear that all the covariates' coefficients, including the intercepts, are the same, regardless of the approaches used, although the variance estimates and p-values are slightly different. Thus, all three approaches yield quite close results in this example. $\square$

6.4 Models for Overdispersed Count Responses

In addition to using the sandwich and scaled variance estimates, we can also correct for overdispersion by using different statistical models for count response, if the overdispersion mechanism is well understood. The most popular models for overdispersed count data are the negative binomial, generalized Poisson, and Conway-Maxwell-Poisson models. We discuss these models in detail in this section.

6.4.1 Distributions for overdispersed count variables

Although the Poisson distribution is commonly used for modeling count variables, it is also very restrictive as its mean and variance are the same. As we mentioned in the last section, overdispersion may occur if the observations are based on time intervals of different lengths. Of course, if the interval lengths are known, we can remove the overdispersion by using offsets in the Poisson regression models to account for their effect. Otherwise, we can model the effect of the varying length of observation period using latent variables, a common statistical approach to deal with unobserved heterogeneity in study data. Within the context of count data, a popular implementation of this approach is to treat the mean of the Poisson distribution for each subject as a random variable (λ), drawn from some distributions. Since

$$Var\,(y) = E\,(Var\,(y|\lambda)) + Var\,(E\,(y|\lambda)) = E\,(\lambda) + Var\,(\lambda) = E\,(y) + Var\,(\lambda)\,,$$

overdispersion always occurs for a mixture of Poisson distributions when $Var\,(\lambda) \neq 0$. Based on the distributions of λ, we can derive distribution for overdispersed count outcomes.

6.4.1.1 Negative binomial distribution

The most commonly used model for overdispersed count data is the *negative binomial* (NB) distribution. $NB(\alpha, \beta)$ is defined by

$$Pr\,(y = k \mid \alpha, \beta) = \frac{\Gamma(k + \alpha)}{k!\Gamma(\alpha)} \left(\frac{\beta}{1 + \beta}\right)^k \left(\frac{1}{1 + \beta}\right)^\alpha, \tag{6.15}$$

where $\alpha > 0$, $\beta > 0$, and $\Gamma(\cdot)$ is the Gamma function defined by $\Gamma(x) = \int_0^\infty t^{x-1} e^{-t} dt$ for $x > 0$. Note that $\Gamma(n + 1) = n!$ for a whole number n. When α is an integer, the NB distribution models the number of fails, k, before the rth success, where each trial is independent and has a probability of success $\frac{1}{1+\beta}$. In such cases, it is common to use another parameterization, $p = \frac{1}{1+\beta}$ and $r = \alpha$. The NB distribution is then given by

$$f_{NB}\,(y = k \mid p, r) = \binom{r + k - 1}{r - 1} p^r\,(1 - p)^k\,. \tag{6.16}$$

The NB distribution can be used to address the overdispersion issue because it is a mixture of Poisson distributions, and hence its variance is always larger than its expectation. In fact, if $y \sim \text{Poisson}(\lambda)$, and the parameter λ itself is a random variable following the Gamma distribution, then y, as a mixture of Poissons, follows a NB distribution.

Gamma distributions are defined with the density function

$$f(x \mid \alpha, \beta) = \frac{1}{\Gamma(\alpha)\beta^\alpha} x^{\alpha-1} e^{-x/\beta}, \quad 0 < x < \infty, \quad \alpha > 0, \quad \beta > 0,$$

where the parameter α is the shape parameter, and β is the scale parameter. Gamma distributions are a flexible family of distributions on $[0, \infty)$. The exponential distribution is the special case when $\alpha = 1$ and chi-square distributions χ_n^2 is the special case when $\alpha = \frac{n}{2}$ and $\beta = 2$. The mean and variance of $\text{Gamma}(\alpha, \beta)$ are $\alpha\beta$ and $\alpha\beta^2$, respectively. It is straightforward to show that if $\lambda \sim \text{Gamma}(\alpha, \beta)$, then the Poisson-Gamma mixture follows the $NB(\alpha, \beta)$ (6.15) (see Problem 2.9). $NB(\alpha, \beta)$ has mean $\alpha\beta$ and variance $\alpha\beta + \alpha\beta^2$. Thus, its variance is indeed always larger than its mean, and it may be applied to model overdispersed data.

6.4.1.2 Generalized Poisson distribution

Another distribution that generalizes the Poisson distribution and is gaining some popularity in recent years is the generalized Poisson (GP) distribution. The distribution function of the GP has the following form

$$f_{GP}\,(y = k \mid \alpha, \mu) = \left(\frac{\mu}{1 + \alpha\mu}\right) \frac{(1 + \alpha k)^{k-1}}{k!} \exp\left(-\frac{\mu(1 + \alpha k)}{1 + \alpha\mu}\right), \tag{6.17}$$

where μ is the mean and $\alpha \geq 0$ is a dispersion parameter. In the literature, a different parameterization with $\alpha = \frac{\omega}{\lambda}$ and $\mu = \frac{\lambda}{1-\omega}$ is also used.

The GP distribution has mean μ and variance $\mu(1 + \alpha\mu)^2$. Thus, the variance is also larger than the mean when $\alpha > 0$, and hence may be applied to model overdispersion in Poisson models. The larger the positive value of α, the more variability there is in the data over and beyond that explained by the Poisson model. Joe and Zhu (2005) showed that it is a Poisson mixture without providing an explicit distribution for the clustering. When $\alpha = 0$, $f_{GP}(y_i \mid \mu, 0) = f_P(y \mid \mu)$, the GP distributions reduce to the Poisson distributions. As in the NB cases, one may assume a GP model and assessing overdispersion by testing whether $\alpha > 0$.

Note that GP distributions can be defined when $\alpha < 0$. In such cases, it is necessary to require that $1 + \alpha y > 0$, or $y < -\frac{1}{\alpha}$. Thus, the outcome is truncated, and it does not define a distribution as the total probabilities of all the possible y's do not add up to 1. However, it has been shown that the total error of truncation is negligible if the absolute value of α is small (Consul and Shoukri, 1985). In such cases, the variance is actually smaller than the mean, so GP distributions may be applied to model under dispersed data. However, it appears that most of the applications of GP models focus on the overdispersion cases where $\alpha > 0$.

6.4.1.3 Conway–Maxwell–Poisson distribution

For the Poisson, NB, and GP distributions introduced above, all of them have a linear decay rate in their probability mass function: the ratio of successive probabilities, $\Pr(Y = k - 1) / \Pr(Y = k)$, is asymptotically linear in k. The Conway–Maxwell–Poisson (CMP) distributions can have nonlinear decay rates, thus they may be applied to model under and overdispersed count data with nonlinear decay rates. The distribution was originally proposed by Conway and Maxwell (1962), but it appears that Shmueli et al. (2005) is the first to use it to model under and overdispersed count data. The Conway–Maxwell–Poisson distribution is defined as

$$P(Y = k; \lambda, v) = \frac{1}{Z(\lambda, v)} \frac{\lambda^k}{(k!)^v}, \quad k = 0, 1, 2, \ldots, \tag{6.18}$$

where $Z(\lambda, v) = \sum_{n=0}^{\infty} \frac{\lambda^n}{(n!)^v}$ is the normalization factor to make (6.18) a probability distribution. Note that

$$\frac{P(Y = k - 1)}{P(Y = k)} = \frac{k^v}{\lambda},$$

thus, the parameter v governs the rate of decay of successive ratios of probabilities. When $v = 1$, $Z(\lambda, v) = \sum_{n=0}^{\infty} \frac{\lambda^n}{(n!)^v} = \exp(\lambda)$, the distribution reduces to Poisson(λ), and the ratio is linear in the outcome Y. When $v > 1$, the probability $\Pr(Y = k)$ will decay faster than the Poisson situation, and the variance is smaller than the mean (underdispersion). As v goes to infinity, the probability $\Pr(Y = k)$ converges to 0 for $k > 1$, and the limiting distribution is a Bernoulli distribution (Problem 6.10). On the other hand, when $v < 1$, the probability $\Pr(Y = k)$ will decay slower than the Poisson distributions and we will have overdispersion. In the special case when $v = 0$, the distribution is $P(Y = k) = \frac{\lambda^k}{1-\lambda}$, the geometric distribution.

There is no closed formula for the expectation and variance of CMP(λ, ν). However, Shmueli et al. (2005) provided the following approximation for the mean:

$$E(X) = \lambda \frac{d[\log\{Z(\lambda, \nu)\}]}{d\lambda} \approx \lambda^{1/\nu} - \frac{\nu - 1}{2\nu}. \tag{6.19}$$

They showed that the approximation is especially good for $\nu \leqslant 1$ or $\lambda > 10^\nu$ (Shmueli et al., 2005).

6.4.2 Regression models

Under the framework of GLM, we may replace the Poisson distribution with one of the distributions discussed in the last section to model counts with over or under-dispersion. For all these distributions, there is a dispersion parameter. Thus there is no need to use quasi-likelihood, and it is straightforward to use MLE for the inference.

6.4.2.1 Negative binomial regression model

Since there are two parameters for the Gamma distribution, the additional parameter allows NB to address the lack of fit of Poisson in the presence of overdispersion. However, when the NB distribution is applied to model count responses, we reparameterize the parameters with the mean and a dispersion parameter instead of directly model the parameters α and β. The mean is commonly modeled with a log-linear model, assuming that the log of the mean response is linear function of the covariates. The dispersion parameter is often assumed a constant that needs to be estimated. The most popular approach is using $(1/\alpha, \mu\alpha)$ in place of (α, β), i.e., assuming $\lambda \sim \text{Gamma}(1/\alpha, \mu\alpha)$. Thus, μ is the mean and α is the dispersion parameter. The NB distribution is given by

$$\Pr\left(y = k \mid \mu, \alpha\right) = \frac{\Gamma(k + 1/\alpha)}{k!\Gamma(1/\alpha)} \left(\frac{\alpha\mu}{1 + \alpha\mu}\right)^k \left(\frac{1}{1 + \alpha\mu}\right)^{1/\alpha}. \tag{6.20}$$

This dispersion parameter α, called the *size* of the NB, may address the overdispersion issue. In fact, the mean and variance of the NB distribution are given by

$$E\left(y \mid \mu, \alpha\right) = \mu, \quad Var\left(y \mid \mu, \alpha\right) = \mu(1 + \alpha\mu).$$

Thus, the NB model adds a quadratic term $\alpha\mu^2$ to the Poisson variance to account for extra-Poisson variation or overdispersion. For this reason, α is known as the dispersion or shape parameter.

More generally, Carmeron and Trivedi (1986) considered a general class of negative binomial regression models, assuming $\lambda \sim \text{Gamma}(\frac{1}{\alpha}\mu^{2-p}, \alpha\mu^{p-1})$, where p is a constant. Under this reparameterization, the NB_p distribution is given by:

$$\Pr\left(y = k \mid \mu, \alpha\right) = \frac{\Gamma(k + \frac{1}{\alpha}\mu^{2-p})}{k!\Gamma(\frac{1}{\alpha}\mu^{2-p})} \left(\frac{\alpha\mu^{p-1}}{1 + \alpha\mu^{p-1}}\right)^k \left(\frac{1}{1 + \alpha\mu^{p-1}}\right)^{\frac{1}{\alpha}\mu^{2-p}} \tag{6.21}$$

with mean μ and variance $\mu + \alpha\mu^p$. Thus, the overdispersion amount is $\alpha\mu^p$. Since α is positive, the variance is always larger than the mean, which is expected for Poisson mixtures. For any fixed p, when $\alpha \to 0$, the NB distribution converges to Poisson(μ). The NB distributions get closer to the Poisson if α becomes smaller, i.e., $\Pr_{NB}(y_i) \to \Pr_P(y_i)$ as $\alpha \to 0$. Thus, the larger the value of α, the more variability there is in the data over and beyond that explained by the Poisson.

By testing the null $H_0 : \alpha = 0$, we can detect overdispersion in the data, if our other model assumptions are correct. However, since $\alpha \geq 0$, $\alpha = 0$ under H_0 is a boundary point. In this case, the asymptotic distribution of the MLE, $\widehat{\alpha}$ of α, cannot be used directly for inference about α, as 0 is not an interior point of the parameter space. This problem is known in the statistics literature as inference under *nonstandard conditions* (Self and Liang, 1987). Under some conditions about the parameter space, which are satisfied by most applications including the current context, inference about boundary points can be based on a modified asymptotic distribution (e.g., Self and Liang (1987)). For testing the null, $H_0 : \alpha = 0$, the revised asymptotic distribution is an equal mixture consisting of a point mass at 0 and the positive half of the asymptotic normal distribution of $\widehat{\alpha}$ under the null H_0. Intuitively, the negative half of the asymptotic distribution of $\widehat{\alpha}$ is "folded" into a point mass concentrated at 0, since negative values of α are not allowed under the null.

Note that p determines the order of the overdispersion in terms of the mean. When $p = 2$, the NB_2 distribution is the classical NB distribution in (6.20). Under the NB_2, the overdispersion $(\alpha\mu^2) = (\alpha\mu)\mu$. We call $\alpha\mu$ the dispersion factor. Since the dispersion factor is a multiple of the mean, we call the corresponding overdispersion as *mean dispersion*. When $p = 1$, the NB_1 regression has an overdispersion of $\alpha\mu$, a constant times the mean, so it is also called *constant dispersion*. NB_2 and NB_1 are the most commonly used members in the family.

The negative binomial regression model is commonly specified by assuming a log-linear model of the mean response and a constant dispersion:

$$y_i \sim NB_p\left(\mu_i, \alpha\right), \log\left(\mu_i\right) = \log\left(E\left(y_i \mid x_i\right)\right) = x_i^\top \beta, \quad 1 \le i \le n. \tag{6.22}$$

As in loglinear Poisson regression models, we model the logarithm of the mean response as linear function of the covariates. Thus, the coefficients β can be similarly interpreted as the log of the mean response ratio per unit increase in the corresponding covariates while all others are held fixed. The intercept is the log of the expected mean response for subjects with all covariates being zero.

Note that the distribution functions of NB_p distributions are the same after reparameterization. However, as we model the parameters differently, NB_p regression models with different ps represent different regression models.

The family of NB regressions is derived by assuming the parameter of the Poisson follows a Gamma distribution. Of course, the parameters may be clustered in other ways such as lognormally, see Hilbe (2011) for additional examples of such distributions. However, no matter how the parameters are clustered, Poisson mixture models can only be applied in overdispersion cases.

Example 6.5

In Example 6.3 we applied the Poisson regression model to the Sexual Health pilot study data, using several approaches to correct the possible bias in inference due to overdispersion. We may also use NB regression models to address this issue.

Shown in the following table are the results based on the classical NB (NB_2) model:

Parameter	Estimate	SE	95% CI		Wald χ^2	Pr > χ^2
Intercept	0.1045	1.1035	−2.0591	2.3147	0.01	0.9245
VAGWCT1	0.0473	0.0161	0.0181	0.0819	8.65	0.0033
HIVKQTOT	0.0862	0.0556	−0.0245	0.1962	2.41	0.1208
CESD	0.0323	0.0268	−0.0188	0.0878	1.45	0.2287
Dispersion	1.9521	0.3152	1.4308	2.6962		

The regression coefficients for the covariates from the NB model are comparable to those corrected for overdispersion discussed in Example 6.3, with only the frequency of protected vaginal sex in the three months prior to the survey remaining significant in the model. Note that the dispersion parameter α is part of the NB model, and it must be estimated or set to a fixed value in order to estimate other parameters. In this sense, it is different from the dispersion parameter in a GLM, since in the latter case this parameter is not part of the GLM. As a result, estimates of the dispersion parameter in GLM are not required for estimating the regression parameters β, but they do play an important role in estimating the asymptotic variance of the MLE of β under likelihood inference.

If the corresponding NB_1 regression model is applied, we obtain the following parameter estimates.

Parameter	DF	Estimate	SE	t-value	Appox Pr > $|t|$
Intercept	1	1.028460	0.620076	1.66	0.0972
VAGWCT1	1	0.033897	0.005254	6.45	<.0001
HIVKQTOT	1	0.063404	0.036370	1.74	0.0813
CESD	1	0.006592	0.014515	0.45	0.6497
_Alpha	1	20.728723	4.362721	4.75	<.0001

The regression coefficients for the covariates are also comparable to those corrected for overdispersion discussed before, with only the frequency of protected vaginal sex in the three months prior to the survey remaining significant in the model. The dispersion parameter α is also significantly different from 0: this confirms the overdispersion. ⬚

6.4.2.2   Generalized Poisson regression

Switching the distribution in (6.22) to GP distributions, we have the following log-linear GP regression model.

$$y_i \sim GP\left(\mu_i, \alpha\right), \ \log\left(\mu_i\right) = \log\left(E\left(y_i \mid \mathbf{x}_i\right)\right) = \mathbf{x}_i^\top \boldsymbol{\beta}, \quad 1 \le i \le n, \qquad (6.23)$$

Similar to NB regression, we usually assume that α is a constant. We can also test the null $H_0 : \alpha = 0$ to assess overdispersion. The interpretation of the coefficients is similar to interpretation of Poisson models.

Example 6.6
If we change the NB distributions to GP for the model in Example 6.5, the parameter estimates of the generalized Poisson (GP) model are summarized in the following table.

Parameter	Estimate	SE	z value	Pr > $|z|$
Intercept	0.8984	0.6127	1.47	0.1426
VAGWCT1	0.03727	0.004791	7.78	<.0001
HIVKQTOT	0.06882	0.03588	1.92	0.0551
CESD	0.006041	0.01382	0.44	0.6621
Scale Parameter	1.7248	0.1696		

The regression coefficients for the covariates are similar to those from NB models, and we obtain similar conclusions. For example, the higher the VAGWCT1, the larger the outcome. The log of mean responses per one unit increase is 0.03727. The dispersion parameter α is significantly different from 0, and this confirms the overdispersion. ⬚

6.4.2.3   Conway–Maxwell–Poisson regression

When Shmueli et al. (2005) first applied the CMP distribution for regression analysis, they applied the log-linear model to the parameter λ, assuming $\lambda_i = \exp\left(\mathbf{x}_i^\top \boldsymbol{\beta}\right)$. They further proposed to model v_i via $v_i = -\exp\left(\mathbf{z}_i^\top \boldsymbol{\delta}\right)$, where $\mathbf{z}_i$ is a covariate vector that may be different from $\mathbf{x}$. Since λ in the CMP distribution is not the mean, thus the interpretation is different from common log-linear models. However, based on (6.19), λ and the mean response μ change in the same direction. Thus, the sign of the parameter coefficients may indicate the direction of the association of the corresponding covariates and the mean responses.

Guikema and Goffelt (2008) proposed to directly model the log of the mean response μ_i as a linear function of the covariate, so the parameter interpretation is similar to that of the log-linear models discussed above.

Example 6.7

Now, we change the NB distributions in Example 6.5 to the CMP distribution. Assuming a log-linear mean response and a constant ν, the estimates of the CMP regression model are summarized in the following table.

Parameter	Estimate	Std.Err	Z value	Pr($>|z|$)
Intercept	0.232969	0.650071	0.358	0.7201
VAGWCT1	0.044690	0.007985	5.597	<0.0001
HIVKQTOT	0.082325	0.036000	2.287	0.0222
CESD	0.029775	0.016841	1.768	0.0771

The dispersion parameter ν is estimated to be 1.944×10^{-7}, very close to 0. Thus, the analysis also confirms the overdispersion. Note that for CMP distributions, $\nu < 1$ indicates overdispersion. □

6.4.2.4   EE for overdispersed regression models

Similar to the Poisson case, we may use EEs to relax the parametric assumptions for NB, GP, and CMP regressions. In the regression models for overdispersed count responses, we specify the same mean as that under the Poisson model:

$$\mu_i = E\left(y_i \mid \mathbf{x}_i\right) = \exp(\mathbf{x}_i^\top \boldsymbol{\beta}).$$

Thus, the EEs for the corresponding Poisson using $V_i = \mu_i$ can also be applied, as the choice of V_i does not affect the validity of the inference. However, V_i that are close to the true variances would be preferred for efficiency considerations. For example, we have variance $V_i = \mu_i(1 + \alpha\mu_i)$ for a NB_2 distribution. Thus, we may use the following NB_2-based EE:

$$\sum_{i=1}^{n} D_i V_i^{-1} S_i = \mathbf{0}. \tag{6.24}$$

where $D_i = \frac{\partial \mu_i}{\partial \boldsymbol{\beta}}$, $S_i = y_i - \mu_i$, and $V_i = \mu_i(1 + \alpha\mu_i)$ based on the NB_2 assumption, instead of $V_i = \mu_i$ under the Poisson. Note that V_i involves a dispersion parameter α. Unless it is known, which is seldom the case, we need to estimate the parameter. A simple approach is to estimate it using the MLE and plug it in as if it is known. Alternatively, one may estimate it based on the second moment. For example, under NB_2,

$$Ey_i^2 = \mu_i + (1 + \alpha)\mu_i^2,$$

one can generate an equation for α in a similar way. Stacking this equation with (6.24), we can simultaneously make inferences about $\boldsymbol{\beta}$ and α. This approach makes it straightforward to assess the overdispersion by testing the null $H_0 : \alpha = 0$. Note that here $\alpha = 0$ is not a boundary point under EE inference since we can allow $\alpha < 0$ as long as $1 + \alpha\mu_i > 0$.

6.5   Zero-Modification Models

Another common reason for overdispersion in Poisson regression is an excessive amount of zeros in the responses. In biomedical and psychosocial research, the distribution of zeros often exceeds the expected frequency of the Poisson model. For example, the distribution of the frequency of protected vaginal sex in the Sexual Health pilot study is given in Figure 6.1.

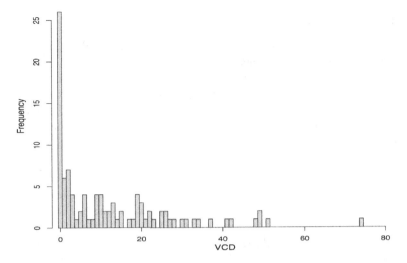

FIGURE 6.1: Distribution of VCD.

The distribution is dominated by zeros. Within the context of this example, the presence of excess zeros reflects a proportion of subjects who were either abstinent from sex or only had other types of sex such as unprotected vaginal sex, inflating the sampling zeros under the Poisson distribution. As will be seen shortly, the excessive zeros also cause overdispersion, precluding the application of the Poisson model. However, overdispersion in this case is caused by a totally different phenomenon, which has other serious ramifications. Thus, the methods discussed above do not address the underlying cause of overdispersion. We must consider new models to explicitly account for the excessive zeros.

On the other hand, it is not uncommon for a count variable to have only positive responses in practice. For example, in studies comparing the length of hospitalization among different hospitals for some disease of interest, the number of days hospitalized is typically recorded only for those patients who end up being hospitalized. Thus, the distribution of this count variable does not include the value 0 and as a result, the Poisson and other log-linear models considered in the last several sections do not directly apply to such data. By truncating the distribution at 0, we can readily modify these models to accommodate such zero-truncated count data.

In this section, we first discuss zero-inflated models to address the issue of excessive zeros and then discuss models for zero-truncated data.

6.5.1 Structural zeros and zero-inflated models

In this section, we discuss zero-inflated models to address the issue of excessive zeros. Although not always, excessive zeros are often caused by the presence of the so-called *structural zeros*, subjects who are not "at-risk" of the event for the count response. If the count responses for "at-risk" subjects follow a Poisson distribution, then structural zeros will cause excessive zeros overall, under the Poisson distribution. In such cases, the whole population is a mixture of two heterogenous subpopulations. Zero-inflated models are based on the notion of *mixture distributions*, so we will start with a brief introduction to such distributions.

6.5.1.1 Mixture distributions

Most of the commonly used parametric distributions are unimodal, i.e., there is one unique mode in the distribution. For example, consider a Poisson distribution with mean μ. This distribution has a mode around the mean μ. Now, let y_i be an i.i.d. sample from a mixture of two Poissons with a mixing probability p. More specifically, y_i is from Poisson(μ_1) with a probability p and from Poisson(μ_2) with a probability $1 - p$. To derive the distribution of y_i, let z_i be a binary indicator with $z_i = 1$ if y_i is sampled from Poisson(μ_1) and $z_i = 0$ if otherwise. Then, the distribution function of the mixture is given by

$$y_i \sim p f_P\left(y \mid \mu_1\right) + (1 - p) f_P\left(y \mid \mu_2\right), \tag{6.25}$$

where $f_P\left(y \mid \mu_k\right)$ denotes the probability distribution function of Poisson(μ_k) ($k = 1, 2$).

Distributions with multiple modes arise in practice all the time. For example, consider a study comparing the efficacy of two treatments for some disease of interest using some count variable of interest y, with larger values indicating worse outcomes. At the end of the study, if one treatment is superior to the other, y will show a bimodal distribution with the subjects from the better treatment condition clustering around the left mode and those from the inferior treatment clustering around the right mode of the mixture. Given the treatment assignment codes, we can identify each component of the bimodal mixture for each treatment condition. If y follows a Poisson(μ_k) for each treatment condition k, the parameter of interest for comparing treatment effects is the vector $\theta = (\mu_1, \mu_2)^\top$.

Now, suppose that the treatment assignment codes are lost. Then, we do not have information to model the distribution of y for each treatment group. For a given subject's response, y_i, we do not know whether it is from Poisson(μ_1) or Poisson(μ_2), then we have to model y_i using a mixture of the two Poissons. If we know the number of subjects who have been assigned to the treatment conditions, then the mixing proportion p is known and the parameters of interest are again $\theta = (\mu_1, \mu_2)^\top$. Otherwise, we also need to estimate p by including p as a component of the parameter vector, i.e., $\theta = (p, \mu_1, \mu_2)^\top$.

Based on (6.25), the log-likelihood is

$$l(\theta, \mathbf{y}) = \sum_{i=1}^{n} \left[p f_P\left(y_i \mid \mu_1\right) + (1 - p) f_P\left(y_i \mid \mu_2\right) \right].$$

We can then make inferences about θ using the MLE of θ and the associated asymptotic distribution.

6.5.1.2 Zero-inflated Poisson distribution

The concept of a mixture distribution is the basis for the zero-inflated Poisson (ZIP) regression model. Specifically, the ZIP model is based on a two-component mixture consisting of a Poisson(μ) and a degenerate distribution of the constant 0:

$$f_{ZIP}\left(y \mid \rho, \mu\right) = \rho f_0\left(y\right) + (1 - \rho) f_P\left(y \mid \mu\right), \quad y = 0, 1, \ldots, \tag{6.26}$$

where $f_0\left(y\right)$ denotes the probability distribution function of the constant 0 with a point mass at 0, i.e., $f_0\left(0\right) = 1$ and $f_0\left(y\right) = 0$ for all $y \neq 0$. The distribution in (6.26) may be expressed as

$$f_{ZIP}\left(y \mid \rho, \mu\right) = \begin{cases} \rho + (1 - \rho) f_P\left(0 \mid \mu\right) & \text{if } y = 0 \\ (1 - \rho) f_P\left(y \mid \mu\right) & \text{if } y > 0. \end{cases} \tag{6.27}$$

So, at $y = 0$, the Poisson probability $f_P\left(0 \mid \mu\right)$ is inflated by ρ to account for structural zeros:

$$f_{ZIP}\left(0 \mid \rho, \mu\right) = \rho + (1 - \rho) f_P\left(0 \mid \mu\right).$$

For example, if $\rho = 0$, then $f_{ZIP}(0 \mid \rho, \mu) = f_P(0 \mid \mu)$, and the ZIP distribution reduces to Poisson, i.e., $f_{ZIP}(y \mid \rho, \mu) = f_P(y \mid \mu)$.

Since the probability of 0, $f_{ZIP}(0 \mid \rho, \mu)$, must be constrained between 0 and 1, it follows from (6.27) that

$$\frac{-f_P(0 \mid \mu)}{1 - f_P(0 \mid \mu)} \le \rho \le 1.$$

Thus, ρ is bounded between $\frac{-f_P(0\mid\mu)}{1-f_P(0\mid\mu)}$ and 1. For the Poisson-based mixture ZIP, this implies that $\frac{1}{1-\exp(\mu)} \le \rho \le 1$. When $\frac{-f_P(0\mid\mu)}{1-f_P(0\mid\mu)} < \rho < 0$, then the expected number of zeros is actually fewer than what would be expected under a Poisson distribution and the resulting distribution becomes a *zero-deflated* Poisson. In the extreme case when $\rho = \frac{-f_P(0\mid\mu)}{1-f_P(0\mid\mu)}$, we have

$$f_{ZIP}(y \mid \rho, \mu) = \begin{cases} 0 & \text{if } y = 0 \\ \frac{\mu^y}{[1-\exp(-\mu)]y!}\exp(-\mu) & \text{if } y > 0. \end{cases}$$

In this case, there are no zeros, either structural or random, and $f_{ZIP}(y \mid \rho, \mu)$ in (6.27) becomes a Poisson truncated at 0. This happens in some studies where zero is not part of the scale of the response variable, and the zero-truncated Poisson regression may be used for such data. We will discuss zero-truncated Poisson and zero-truncated negative binomial regression models in more detail later in this section. Note that the zero-truncated Poisson is not a mixture model. Also, since $\rho < 0$ in the case of zero-deflated Poisson, the zero-deflated Poisson is not a mixture model either.

When $0 < \rho < 1$, ρ represents the number of structural zeros above and beyond the sampling zeros expected by the Poisson distribution $f_P(y \mid \mu)$. In the presence of such structural zeros, the distribution of ZIP, $f_{ZIP}(y \mid \rho, \mu)$, is defined by the parameters ρ and μ. The mean and variance for the ZIP model are given by

$$E(y) = (1 - \rho)\mu, \quad Var(y) = \mu(1-\rho)(1+\mu\rho). \tag{6.28}$$

If $0 < \rho < 1$, $E(y) < Var(y)$ and vice versa. Further, if $0 < \rho < 1$, $E(y) < \mu$. Thus, unlike data clustering that typically impacts only the variance of the model estimate, structural zeros also affect the mean estimate itself because of the downward bias in modeling the mean response as well, giving rise to a far more serious consequence than overdispersion.

When applying the ZIP model within a regression setting, we must model both ρ and μ as a function of explanatory variables $\mathbf{x}_i$. We can still use the log link to relate μ to such variables. For ρ, we can use the logit link since $0 < \rho < 1$. Also, as each parameter may have its own set of predictors and covariates, we use $\mathbf{u}_i$ and $\mathbf{v}_i$ (which may overlap) to represent the two subsets of the vector $\mathbf{x}_i$ that will be linked to ρ and μ, respectively, in the ZIP model. The ZIP regression is defined by the conditional distribution, $f_{ZIP}(y_i \mid \rho_i, \mu_i)$, along with the following specifications for ρ_i and μ_i:

$$\text{logit}(\rho_i) = \mathbf{u}_i^\top \boldsymbol{\beta}_u, \quad \log(\mu_i) = \mathbf{v}_i^\top \boldsymbol{\beta}_v, \quad 1 < i < n.$$

Thus, the presence of structural zeros gives rise not only to a more complex distribution but also creates an additional link function for modeling the effect of explanatory variables on the occurrence of such zeros.

6.5.1.3 Inference

Likelihood-based inference for the ZIP regression model is again straightforward. Let $\boldsymbol{\theta} = \left(\boldsymbol{\beta}_u^\top, \boldsymbol{\beta}_v^\top\right)^\top$. For ML inference, the probability distribution function is

$$f_{ZIP}(y_i \mid \mathbf{x}_i, \boldsymbol{\theta}) = \rho_i f_0(y_i \mid \rho_i) + (1 - \rho_i)f_P(y_i \mid \mu_i), \quad \text{logit}(\rho_i) = \mathbf{u}_i^\top \boldsymbol{\beta}_u, \quad \log(\mu_i) = \mathbf{v}_i^\top \boldsymbol{\beta}_v.$$

Thus, the log-likelihood is given by

$$l\left(\mathbf{\theta}\right) = \sum_{i=1}^{n} \log\left[\rho_i f_0\left(y_i \mid \rho_i\right) + \left(1 - \rho_i\right) f_P\left(y_i \mid \mu_i\right)\right].$$

The MLE of $\mathbf{\theta}$ along with its asymptotic distribution provides the basis for inference about $\mathbf{\theta}$.

Distribution-free inference for mixture models is generally more complicated. In particular, the standard EE cannot be used for inference about $\mathbf{\theta}$. For example, in the absence of explanatory variables, the mean of the ZIP-distributed count variable y is given by $E\left(y\right) = \left(1 - \rho\right)\mu$. Since different ρ and μ can yield the same mean of y, the mean response itself alone does not provide sufficient information to identify ρ and μ. Alternative approaches must be used to provide distribution-free inference about ρ and μ, or $\mathbf{\beta}_u$ and $\mathbf{\beta}_v$ in the regression setting. For example, Tang et al. (2015) utilized both the count mean response and the probability of zero outcome to set up the estimating equations.

Example 6.8

As noted earlier, there is evidence of structural zeros in the outcome of the number of protected vaginal sex encounters (VCD) during the three-month study period. Thus, we fit a ZIP model with three predictors: HIV knowledge (HIVKQ, a higher score means more informed about HIV knowledge), depression (CESD, a higher score means more depressed), and baseline number of protected vaginal sex encounters in past 3 months (VAGWCT1).

The ZIP model consists of two components; the logistic model component for structural zeros in VCD:

$$\text{logit}\left(\Pr\left(\text{structural zero of } VCD\right)\right) = \alpha_0 + \alpha_1 VAGWCT1,$$

and the Poisson model component for the "at-risk" subpopulation:

$$\log\left(E\left(VCD \mid \text{at risk}\right)\right) = \beta_0 + \beta_1 VAGWCT1 + \beta_2 HIVKQ + \beta_3 CESD.$$

Shown in the table below are the estimates from the ZIP model.

Parameter	Estimate	SE	t Value	$\Pr > \lvert t \rvert$
Logistic model for predicting structural zeros				
Intercept	0.2585	0.3244	−0.80	0.4274
VAGWCT1	−0.3952	0.1205	3.28	0.0014
Poisson model for count response for at-risk subjects				
Intercept	2.1249	0.1831	11.60	< 0.0001
VAGWCT1	0.01324	0.001861	7.11	< 0.0001
HIVKQ	0.03386	0.01032	3.28	0.0014
CESD	0.000895	0.004428	0.20	0.8402

Based on the logistic model for structural zeros, VAGWCT1 is a significant predictor for structural zeros. The larger the count of baseline protected sex encounters VAGWCT1, the less likely the subject is to have a structural zero in protected sex at the follow-up. The log of the odds ratio per unit increase in VAGWCT1 is −0.3952. The predicted probability of having a structural zero can also be computed from the estimated coefficients. For example, for a subject with VAGWCT1= 10, the probability of having a structural zero is $\frac{\exp(0.2585 - 0.3952 \times 10)}{1 + \exp(0.2585 - 0.3952 \times 10)} = 0.02428$.

In the count component, both VAGWCT1 and HIVKQ are significant predictors for the count response; the higher VAGWCT1 and HIVKQ, the higher the protected sex VCD. However, CESD is not a significant predictor.   □

6.5.1.4   Testing of zero-inflation for Poisson regression

ZIP models are a natural generalization of Poisson regression models when there are structural zeros, or zero-inflation. However, whether or not there are inflated zeros cannot simply be determined by the amount of observed zeros alone, as it is relative to the models adopted for the count data. Thus, formal tests of whether there are inflated zeros may be desired before we apply the zero-inflated models. Note that the existence of structural zeros is not equivalent to inflated zeros, as the name "structural zeros" bears the meaning of being not at risk. However, we use the existence of structural zeros and inflated zeros interchangeably from statistical modeling point of view.

To test for inflated zeros for Poisson regression models, we may simply compare the amount of zeros observed with that expected under the Poisson regression model. Suppose we want to test whether there is zero-inflation for the following Poisson regression model:

$$y_i \mid \mathbf{x}_i \sim \text{i.d. Poisson}(\mu_i), \quad \log(\mu_i) = \mathbf{x}_i^\top \boldsymbol{\beta}. \tag{6.29}$$

Under (6.29), the probability of $y_i = 0$ with covariate $\mathbf{x}_i$ is $p_i = \exp\left(-\exp\left(\mathbf{x}_i^\top \boldsymbol{\beta}\right)\right)$. He et al. (2019) proposed a test based on difference of the amount of zeros observed with that which would be expected under the Poisson regression model:

$$s = \frac{1}{n} \sum_{i=1}^{n} (r_i - E(r_i)) = \frac{1}{n} \sum_{i=1}^{n} (r_i - p_i), \tag{6.30}$$

where r_i is the indicator of $y_i = 0$, i.e., $r_i = 1(= 0)$ if $y_i = 0 \ (> 0)$. Let $\widehat{\boldsymbol{\beta}}$ be the MLE of the Poisson regression model, then $\widehat{\mu}_i = \exp(\mathbf{x}_i^\top \widehat{\boldsymbol{\beta}})$ is the fitted mean outcome, and $\widehat{p}_i = \exp(-\exp(\mathbf{x}_i^\top \widehat{\boldsymbol{\beta}}))$ is the fitted probability of y_i being a zero under the Poisson regression. Then, s can be estimated by $\widehat{s} = \frac{1}{n} \sum_{i=1}^{n} (r_i - \widehat{p}_i)$. He et al. (2019) originally applied the theory of estimating equations to estimate the variation of $\widehat{s}$ to test whether $s = 0$. The test statistic can be expressed as

$$\sqrt{n} \frac{\frac{1}{n} \sum_{i=1}^{n} (r_i - \widehat{p}_i)}{\left(\frac{1}{n} \sum_{i=1}^{n} \left[\widehat{p}_i(1 - \widehat{p}_i) - (\widehat{\mathbf{p}}\widehat{\boldsymbol{\mu}})^\top X \left(X^\top diag(\widehat{\boldsymbol{\mu}}) X\right)^{-1} X^\top (\widehat{\mathbf{p}}\widehat{\boldsymbol{\mu}})\right]\right)^{1/2}},$$

where $X = (\mathbf{x}_1, ..., \mathbf{x}_n)^\top$ is the design matrix, $\widehat{\boldsymbol{\mu}} = (\widehat{\mu}_1, ..., \widehat{\mu}_n)^\top$, $\widehat{\mathbf{p}} = (\widehat{p}_1, ..., \widehat{p}_n)^\top$, and $\widehat{\mathbf{p}}\widehat{\boldsymbol{\mu}} = (\widehat{p}_1 \widehat{\mu}_1, ..., \widehat{p}_n \widehat{\mu}_n)^\top$ (Tang and Tang, 2019). Under the null hypothesis that the Poisson regression is true, the statistic follows the standard normal distribution asymptotically. Both one-sided and two-sided tests for $\rho = 0$ can be performed based on the statistic.

Another approach to test for inflated zeros for Poisson regression models is to start from a zero-inflated Poisson (ZIP) model. One may test for zero-inflation by checking if the ZIP model fits the data significantly better than the corresponding Poisson model. Since generalized linear models with proper link functions are often utilized for the zero-inflation component of the ZIP models, the probability of being a structural zero is guaranteed to be positive, meaning the ZIP and Poisson models are not nested. However, as illustrated in He et al. (2019), the Poisson regression model is still a limit of the ZIP models, so utilization of the Vuong test is not appropriate here. If we are only interested in testing whether there are inflated zeros, it is not necessary to model the structural zeros. We can simply assume a constant probability of y_i being a structural zero for all the subjects, and then the Poisson models are the special case when the probability is zero.

Consider a ZIP model with the probability of being a structural zero, ρ_i, being a constant:

$$y_i \mid \mathbf{x}_i \sim \text{i.d. ZIP}(\rho_i, \mu_i), \quad \rho_i = \rho, \quad \log(\mu_i) = \mathbf{x}_i^\top \boldsymbol{\beta}. \tag{6.31}$$

Thus, the Poisson model (6.29) is nested in the ZIP model, and hence the Wald, score, and likelihood ratio tests may be applied to test whether $\rho = 0$. Note that under (6.31), ρ can even be negative, corresponding to the so-called *zero-deflated* Poisson cases where the amount of observed zeros are less than that would be expected under a Poisson model. Thus, $\rho = 0$ is actually an interior point and we don't have the non standard condition issue when ρ can only take non negative values.

It appears that the score test of Van den Broek (1995) based on (6.31) was the first test formally proposed for testing zero-inflation. The score statistic given by Van den Broek (1995) is

$$\sqrt{n}\frac{\frac{1}{n}\sum_{i=1}^{n}\frac{1}{\widehat{p}_i}\left(r_i - \widehat{p}_i\right)}{\left(\frac{1}{n}\sum_{i=1}^{n}\left[\frac{(1-\widehat{p}_i)}{\widehat{p}_i} - \widehat{\boldsymbol{\mu}}^\top X\left(X^\top diag(\widehat{\boldsymbol{\mu}})X\right)^{-1}X^\top\widehat{\boldsymbol{\mu}}\right]\right)^{1/2}} \sim N\left(0,1\right).$$

Based on (6.31), one may also apply Wald and LR tests. Note that since the MLE for ρ under (6.31) is estimated over both sides of 0, the likelihood ratio test can only be applied for a two-sided test, while the Wald test may be applied for both one- and two-sided tests. Also note that since maximum likelihood estimates are required under both models, this approach also suffers from the issue that the MLEs under the ZIP models may not exist.

Both the He and score tests only require fitting of the Poisson model, and thus are computationally simpler compared with the Wald and LR tests, as they require the fitting of a relatively more complicated ZIP model. Given the fact that ZIP models with the identity link for zero-inflation are not commonly implemented in statistical software packages, this can be a useful advantage. More importantly, the He and score tests may still be applicable when the MLE of the ZIP model (6.31) does not exist. Thus, although starting from different perspectives, the He statistic bears many semblances with the score statistic. In fact, the score function with respect to ρ in model (6.31) under the null hypothesis $\rho = 0$ is $\frac{1}{p_i}(r_i - p_i)$, with $r_i - p_i$ weighted by p_i^{-1}, the inverse of the probability being zero for each observation y_i. However, the weighting approach may suffer from the issue that some subjects may dominate the test. Tang and Tang (2019) compared all these tests and found that in general the He test performs the best in controlling for type I error. On the other hand, regarding their performance in terms of power, there is no one test that appears to perform uniformly better than others. Given the fact that the comparison of the power is only meaningful after type I error is successfully controlled, the He test is in general recommended (Tang and Tang, 2019).

Example 6.9

In Example 6.1 we applied the following Poisson regression model for the Sexual Health study:

$$\log\left(E\left(VCD\right)\right) = \beta_0 + \beta_1 VAGWCT1 + \beta_2 HIVKQ + \beta_3 CESD.$$

The He statistic for testing zero-inflation for this model is 206.8 with a p-value of < 0.0001, thus we may conclude that there is zero-inflation. Based on the corresponding ZIP model assuming a constant probability for the zero-inflation, the score, Wald, and LR tests can also be applied. The score statistic is 0.002 with a p-value of 0.998 for zero-modification. The Wald and LR statistics are 7.2 and 452.5, respectively, both with p-value < 0.0001. Thus, the score test failed to detect zero-modification, while all of the other three tests detected significant zero-inflation in this example. $\Box$

Jansakul and Hinde (2002) considered testing zero-inflation based on ZIP models with covariates included in the zero-inflation components using the identity link. As expected, their test would be more powerful if the model of the zero-component is correctly specified. However, it would be difficult to develop such a model in practice if we are even not sure about the existence of zero-inflation. Thus, we would recommend utilizing the tests discussed above before developing more complicated ZIP models.

6.5.1.5 Zero-inflated overdispersed count responses

It is not necessary to always use the Poisson model in the mixture, although ZIP is by far the most popular in such applications. For example, if the count response from the at-risk subpopulation shows overdispersion and can be modeled with distributions such NB, generalized Poisson (GP) or CMP, then the corresponding mixture of structural zeros and the overdispersed count response should be modeled by the corresponding zero-inflated models. Such zero-inflated overdispersed models have been applied in many fields (Yau et al., 2003, Moghimbeigi et al., 2008, Famoye and Singh, 2006, Sim et al., 2018). For all these models, the differences compared with the corresponding ZIP models are the additional dispersion parameters utilized to handle the overdispersion.

Before applying these models, one may need to assess the dispersion parameter in the at-risk component to check if it can be reduced to ZIP models (Ridout et al., 2001) and/or test whether there is zero-inflation (Ye et al., in press). For example, both NB (without zero-modifications) and ZIP models are nested in ZINB models. Thus, we may assess the ZINB models by testing whether the additional parameters are significant. Jansakul and Hinde (2002) developed a score test for the overdispersion by comparing the ZINB and ZIP models. The Wald and LR tests can also be applied. On the other hand, assuming a constant probability for zero-inflation, score, Wald and LR tests for zero-inflation can be applied for the NB, GP, and CMP models. By directly comparing the observed and expected zero counts, the He test can also be developed without any assumptions on the zero-inflation (Ye et al., in press).

Example 6.10
If ZINB is applied instead in Example 6.8, then the parameter estimates are given in the following table.

Parameter	Estimate	SE	Wald Chi-Square	Pr > ChiSq
Logistic model for predicting structural zeros				
Intercept	0.1787	0.3634	0.24	0.6230
VAGWCT1	−0.5329	0.3293	2.62	0.1055
NB model for count response for at-risk subjects				
Intercept	1.7042	0.8714	3.83	0.0505
VAGWCT1	0.0197	0.0102	3.74	0.0531
HIVKQTOT	0.0428	0.0444	0.93	0.3356
CESD	0.0097	0.0203	0.23	0.6314
Dispersion	0.8971	0.2058		

Based on this model, none of the covariates is significantly associated with the response, although the p-value for VAGWCT1 is very close to the 5% level (p-value 0.0531). Note that the dispersion parameter is significantly different from zero; thus there is overdispersion for the count component for the at-risk subjects. □

Note that the mixture phenomena of zero-inflation are not limited to zeros. For example, there are also studies of one-inflation and zero-and-one inflation for count response (Zhang et al., 2016). For other data types, similar phenomena and modeling also apply. As we mention at the end of Section 4.5.3, we may have the issue of inflated outcomes at either or both of the end points for binomial regression models. In such cases, endpoint-inflated binomial regression models can be similarly developed (Deng and Paul, 2000, Tian et al., 2015, Ye et al., 2021). For ordinal outcomes, there are studies of middle category inflation (Harris and Zhao, 2007).

6.5.1.6   Zero-deflated count outcomes

Although count variables with excessive zeros or positive probability of structural zeros are most common, data with fewer zeros than those expected by Poisson or "negative" probability of structural zeros also occasionally arise in practice. For example, in an HIV prevention study, condom use over a period of time may have positive structural 0's before the intervention, since many subjects may never consider using condoms. After the intervention, many or all nonusers may start using condoms and such a trend will reduce the number of zeros so that at posttreatment assessment, the outcome of number of condom-protected sex may exhibit either none or even "negative" structural zeros. In the latter zero-deflated case, there are fewer zeros than expected by the Poisson model. Since this model is not frequently used in practice, we will not discuss it in detail.

6.5.2   Data without zeros

In practice, it is not uncommon for a count variable to have only positive responses. For example, in studies comparing the length of hospitalization among different hospitals for some disease of interest, the number of days hospitalized is typically recorded only for those patients who end up being hospitalized. Thus, the distribution of this count variable does not include value 0 and as a result, the Poisson and NB log-linear models considered earlier do not directly apply to such data. By truncating the distribution at 0, we can readily modify both models to accommodate such zero-truncated count data.

6.5.2.1   Zero-truncated models

If a count variable y follows a Poisson model, the subset of the response of y with 0 excluded will follow a truncated Poisson, with the distribution function given by

$$f_{ZTP}\left(y \mid \lambda\right) = \frac{\lambda^y \exp\left(-\lambda\right)}{y!\left[1 - \exp\left(-\lambda\right)\right]}, \quad \lambda > 0, \quad y = 1, 2, \dots.$$

Similarly, for regression analysis, the zero-truncated Poisson (ZTP) log-linear model is specified as

$$y_i \mid \mathbf{x}_i \sim \text{ZTP}\left(\mu_i\right), \quad \log(\mu_i) = \mathbf{x}_i^\top \boldsymbol{\beta}, \quad 1 \le i \le n. \tag{6.32}$$

Note that the mean μ_i is the expected value of the Poisson outcome, the outcome before the zero-truncation. Thus, we must be mindful when we interpret the parameters as their effects are on the mean response before zero-truncation.

As in the case of Poisson log-linear regression, inference for $\boldsymbol{\beta}$ under the zero truncated Poisson model can be based on maximum likelihood or EE. By replacing the zero-truncated Poisson with the following truncated NB,

$$f_{ZTNB}\left(y \mid \mu, \alpha\right) = \frac{\Gamma\left(y + \frac{1}{\alpha}\right)}{y!\Gamma\left(\frac{1}{\alpha}\right)\left[1 - \left(\frac{1}{1+\alpha\mu}\right)^{1/\alpha}\right]}\left(\frac{\alpha\mu}{1+\alpha\mu}\right)^y\left(\frac{1}{1+\alpha\mu}\right)^{1/\alpha}, \tag{6.33}$$

for $y = 1, 2, \ldots$, we obtain a zero-truncated NB (ZTNB) model. It is straightforward to make ML inference for both ZTP and ZTNB models.

Example 6.11

To illustrate these methods, consider the Sexual Health pilot study again. By removing the zeros from the original data, we obtained a data set with 72 observations. We assume the same loglinear model (6.5) before the zero-truncation and apply the corresponding ZTP model to the model. To take of the potential overdispersion issue, we use the sandwich estimate of the variance for robust inference. The analysis results are summarized in the following table:

Parameter	Estimate	SE	95%CI		z	Pr > |z|
VAGWCT1	0.01324	0.004175	0.005057	0.021422	3.17	0.002
HIVKQTOT	0.03386	0.0358	−0.03631	0.104025	0.95	0.344
CESD	0.000895	0.019213	−0.03676	0.038551	0.05	0.963
Intercept	2.124854	0.519087	1.107462	3.142245	4.09	< 0.0001

Based on this model, we again find only VAGWCT1 to be a significant predictor for the number of protected sex encounters in a three-month period, consistent with the findings derived using different models in the previous examples. ⬜

6.5.3   Hurdle models

Models for truncated count responses discussed in the last section provide another approach for the analysis of count data with modified zeros. Unlike the zero-inflated models, the zero-truncated count response does not include zero as a valid observation, either sampling or structural zeros. These different models and associated applications demonstrate a distinct construct underlying the number zero in the count response.

Indeed, the structural and sampling zeros arise from heterogeneous samples comprising two conceptually and clinically different groups. In the applications discussed in Section 6.5.1, the two types of zeros unfortunately cannot be distinguished from each other, because of the lack of sufficient information to identify the two subpopulations underlying the different types of zeros. In some studies, structural zeros may be the only ones present, eliminating the need to use zero-inflated models.

For example, attendance to individual counseling and/or group sessions for psychotherapy studies often exhibits an excessive number of zeros, akin to the zero-inflated distributions such as ZIP. However, most psychotherapy studies have a large number of sessions planned, and patients receive reminders before each scheduled session. If a patient attends none of the sessions, it is difficult to argue that he/she has any intention to receive counseling. In such cases, all zeros may be viewed as structural zeros, representing a group of subjects who completely lack interest in attending any of the counseling sessions, rather than a mixture of structural and sampling zeros. Although some of the zeros might be of sampling type, constituting a group of patients interested in counseling sessions, the fraction of such subjects must be so small that it can be ignored for all practical purposes.

Although the disappearance of sampling zeros in this case obviates the need to use zero-inflated models to identify the types of zeros, the mixture nature of the study population still remains. Thus, we may still be interested in modeling both the between-group (at- vs. non risk to be positive) differences and within-group (at-risk) variability. For example, in the session attendance example, it is of great interest to characterize the subjects with absolutely no interest in counseling, as well as those interested in psychotherapy, but with limited attendance.

To this end, let z_i be the indicator for the non risk subgroup ($z_i = 1$). We then jointly model z_i (between group differences) and the positive count response y_i given $z_i = 0$ (variability within at-risk group) as follows:

$$z_i \mid \mathbf{x}_i \sim \text{Bernoulli}(p_i), \quad f(p_i) = \mathbf{u}_i^\top \boldsymbol{\alpha},$$

$$y_i \mid z_i = 0, \mathbf{x}_i \sim \text{ZTP}(\mu_i), \quad g(\mu_i) = \mathbf{v}_i^\top \boldsymbol{\beta}, \quad 1 \le i \le n, \tag{6.34}$$

where f (g) is a link function such as logit (log), and $\mathbf{u}_i$ and $\mathbf{v}_i$ (which may overlap) are two subsets of the vector $\mathbf{x}_i$. We can obtain a variety of models from the above by selecting different f (e.g., logit, probit link) and changing the Poisson to distributions for count responses such as the negative binomial. In the literature, (6.34) is known as the *hurdle* model. Under the assumptions of (6.34), it is readily checked that the likelihood of the hurdle model is the product of the likelihood for the binary component involving $\boldsymbol{\alpha}$ only, and the one for the truncated part involving $\boldsymbol{\beta}$ alone. Thus, although jointly modeled, inference can actually be performed separately for each of the model components. In principle, $\boldsymbol{\alpha}$ and $\boldsymbol{\beta}$ may contain overlapping parameters. However, in practice, the two sets of parameters are typically assumed nonoverlapping, as we have adopted in the formulation above.

The hurdle model addresses a special case of the two-component mixture involving no sampling zero. In some cases, both types of zeros are present and observed. For example, in the counseling example above, if a sizable number of subjects within the non risk group did miss the scheduled sessions due solely to reasons other than unwillingness to attend sessions, their zeros may be considered random. Standard models such as the Poisson regression may be applied when restricted to the at-risk subgroup of the study population. If our interest is in both the at- and non risk groups, we can apply the hurdle model in (6.34) by using a regular Poisson rather than the zero-truncated version.

6.6   Model Comparisons

As the Poisson distribution is a special case of the NB, GP, and CMP distributions when the dispersion parameter is a specific value (zero for NB and GP and one for CMP), we may be interested in assessing whether the more complicated models are necessary; in other words, are those models significantly better than the Poisson model? In general, we may compare the NB, GP, and CMP regression models with Poisson models by assessing the dispersion parameters. For example, one may easily apply score tests to the dispersion parameter to test whether there is overdispersion. In such cases, one need only to fit the Poisson model, and then calculate the score statistics according to the different alternative overdispersion models. If the NB, GP, and CMP regression models can be efficiently estimated, then likelihood ratio and Wald tests may also be applied for their comparison with the corresponding Poisson regression (Deng and Paul, 2000, Jansakul and Hinde, 2002).

The NB, GP, and CMP regression models may all be applied to address overdispersion. Thus, there is the question of which of these models should be adopted. Since these models are nonnested, the commonly applied score, Wald, and LR tests are not applicable. We can apply Vuong's test to compare these nonnested models (Vuong, 1989).

Let $f_1(y_i \mid \boldsymbol{\theta}_1)$ and $f_2(y_i \mid \boldsymbol{\theta}_2)$ denote the distribution functions of two models. Under the classic testing paradigm, the form of the correct distribution is given and only the true vector of parameters is unknown. Under Vuong's setup, the form of the distribution is also not specified. The null hypothesis under Vuong's test is that the two models fit the data equally well. So, it is possible that neither of f_1 and f_2 is a correct model for the data. The idea of Vuong's test is to compare the likelihood functions under the two competing models. If the two models fit the data equally well, then their likelihood functions would be identical. Let θ_1^* be a pseudo-true value of θ_1 at which $E\left[\log\left(f_1\left(y_i \mid \theta_1\right)\right)\right]$ achieves the maximum. It is worth noting that the expectation is computed with respect to the true distribution, which may not be a member of the models considered. Thus, the pseudo-true values can be viewed as the best choice of θ_1 for $f_1\left(y_i \mid \theta_1\right)$ to model the population. Similarly, let θ_2^* denote the pseudo-true value for θ_2. Vuong's test is to compare the best likelihood functions that may be achieved between the two models, i.e., $E\left[\log\left(f_1\left(y_i \mid \theta_1^*\right)\right)\right] - E\left[\log\left(f_2\left(y_i \mid \theta_2^*\right)\right)\right] = E\left[\log\left(\frac{f_1(y_i \mid \theta_1^*)}{f_2(y_i \mid \theta_2^*)}\right)\right]$. Since the true distribution is unknown, the sampling analogue $\log\left(\frac{f_1(y_i \mid \theta_1^*)}{f_2(y_i \mid \theta_2^*)}\right)$ computed based on the empirical distribution is used to derive the test statistic.

In general, the MLE $\hat{\theta}_k$ based on the sample will converge to θ_k^* as the sample size $n \to \infty$ ($k = 1, 2$) (White, 1982). Thus, by substituting the MLEs in place of the pseudo-true parameters, let

$$g_i = \log\left[\frac{f_1\left(y_i \mid \hat{\theta}_1\right)}{f_2\left(y_i \mid \hat{\theta}_2\right)}\right], \quad \bar{g} = \frac{1}{n}\sum_{i=1}^{n} g_i, \quad s_g^2 = \frac{1}{n-1}\sum_{i=1}^{n}(g_i - \bar{g})^2. \tag{6.35}$$

Vuong's test for comparing f_1 and f_2 is defined by the statistic $V = \frac{\sqrt{n}\bar{g}}{s_g}$, which has an asymptotic standard normal distribution. Because of the symmetry of the two competing models, the test is directional. If the absolute value $|V|$ is small, e.g., the corresponding p-value is bigger than a prespecified significance level such as 0.05, then we will say that the two models fit the data equally well, with no preference given to either model. If $|V|$ yields a p-value smaller than the threshold such as 0.05, then one of the models is better; f_1 (f_2) is better if V is positive (negative). Thus, testing the null $H_0 : E\left(g_i\right) = 0$ by the Vuong statistic is the same as testing for a zero mean of the variable g_i.

Example 6.12
We applied ZIP and NB_2 models to sexual health study in Examples 6.8 and 6.5, respectively. We may apply Vuong's test for nonnested models to compare these two models. The Vuong test statistic is -3.8308 with p-value < 0.0001. The negative sign of the test statistic indicates that the second model (NB_2) is better than the first model (ZIP). We can also compare the Akaike information criteria (AIC) of the two models; see Chapter 7 for a discussion of the information criteria. The AIC for the NB_2 model is 663.3049 and that for the ZIP model is 1234.2045, confirming that the NB_2 fits the data much better than the ZIP model. If ZINB if further applied, the corresponding AIC is 635.6121; thus, the ZINB is even better than the NB model. So, it appears that there are both zero-inflation and over-dispersion in the count component. ⧠

In this chapter, we have discussed different models for count responses characterized based on the presence of dispersion as well as types of zeros, and the contexts within which they are applicable. To help appreciate the differences and similarities between these models and data settings, we would like to highlight the main features of the different data types and associated modeling strategies in the following table.

Data	Model
No zero (zero deleted)	ZTP, ZTNB, ZTGP, ZTCMP
No structural zero	Poisson, NB, GP, CMP
No random zero	Hurdle models with zero-truncated models forthe at-risk component
Structural and random zeros	
structural zero unobserved	ZIP, ZINB, ZIGP, and ZICMP
structural zero observed	Hurdle models with Poisson, NB, GP, and CMP for the at-risk component

Exercises

6.1 Show that the log function in (6.2) is the canonical link for the Poisson model in (6.1).

6.2 Consider a Poisson regression model for a count response y with a single continuous covariate x, $E(y \mid x) = \exp(\alpha_0 + \alpha_1 x)$. If x is measured on another scale x' such that $x' = kx$, and the model expressed in terms of x' is $E(y \mid x') = \exp(\alpha'_0 + \alpha'_1 x')$, show that $\alpha'_0 = \alpha_0$ and $\alpha'_1 = \alpha_1/k$.

6.3 Show that for the Poisson regression model in (6.2), $\sum_{i=1}^{n} y_i x_{ij}$ is a sufficient statistic for β_j $(1 \le j \le p)$.

6.4 In Example 6.2, we used an offset term to account for the heterogeneity in the duration of the follow-up periods. In this problem, we compare it with an alternative approach by treating the duration as a covariate.

   (a) Fit a main effect Poisson regression with NUM_ILL as the response variable, DAY, BSI, Age, and Gender as covariates.

   (b) Compare the model in part (a) with the one in Example 6.2.

   (c) Use the deviance and the Pearson chi-square statistics to check whether there is overdispersion in the data.

   (d) Refit the log-linear model in (a), but use the deviance and the Pearson chi-square statistics in (b) to estimate the scaled-variance to account for overdispersion, and compare the results from this model with those in (a).

   (e) Repeat (c) for the model in Example 6.2 and compare the results with those in (d).

6.5 Similar to logistic regression, give a definition of median unbiased estimate (MUE) of a parameter based on the exact conditional distribution.

6.6 Let $y \sim \text{Poisson}(\mu)$.

   (a) If $\mu = n$ is an integer, show that the normalized variable $\frac{y-\mu}{\sqrt{\mu}}$ has an asymptotic normal distribution $N(0,1)$, i.e., $\frac{y-\mu}{\sqrt{\mu}} \sim_a N(0,1)$ as $\mu \to \infty$.

   (b) Generalize the conclusion in (a) to an arbitrary constant μ.

6.7 Show that the asymptotic result in (6.7) still holds if β is replaced by the MLE $\widehat{\beta}$.

6.8 For the Sexual Health pilot study, consider modeling the number of unprotected vaginal sex behaviors during the three month period of the study as a function of three predictors, HIV knowledge, depression, and baseline number of unprotected vaginal sex (VAGWOCT1).

   (a) Fit the Poisson log-linear model.

   (b) Use the deviance and Pearson statistics to examine whether there is overdispersion in the data.

   (c) Does the model have an inflated zero issue?

6.9 Use the intake data for the Catheter Study to study the association between urinary tract infection (UTI) and demographic characteristics including age, gender, and marital status (ms). There are seven levels for the ms in the original data, and we combine levels 2 to 6 to make it a categorical variable with three levels. Fit a log-linear Poisson regression model for the UTI counts with age, gender, and the three-level marital status as predictors (main effects only, and no interactions)

   (a) Write down the null hypothesis that there no differences in the UTI outcome among the three marital statuses in terms of the model parameters.

   (b) Fit the model using the MLE and test the null hypothesis in part (a).

   (c) Fit the model using EE and test the null hypothesis in part (a).

   (d) What is the predicted count of UTIs of a 50-year-old unmarried (original ms = 1) female?

6.10 Prove that the CMP distribution $CMP(\lambda, v)$ converges to

   (a) a Bernoulli distribution as v goes to infinite. Find the parameter for the limiting Bernoulli distribution;

   (b) a geometric distribution as $v \to 0$. Find the parameter for the limiting geometric distribution.

6.11 Use the intake data for the Catheter Study to study the association between catheter blockage and demographic characteristics including age, gender, and marital status (ms). Fit a log-linear Poisson regression model for the blockage counts with age, gender, and the three-level marital status (see Problem 6.9) as predictors (main effects only, and no interactions).

   (a) Assess if there is an overdispersion issue.

   (b) Test whether there is zero-inflation based on the Poisson model.

   (c) Compare the ZIP model with the NB model.

6.12 For Problem 6.8,

   (a) Fit the corresponding negative binomial model with the same linear predictor.

   (b) Compare the analysis between part (a) and that from Problem 6.8.

6.13 Consider the Poisson log-linear model

$$y_i \mid \mathbf{x}_i \sim \text{Poisson}(\mu_i), \quad \log(\mu_i) = \mathbf{x}_i^\top \boldsymbol{\beta}, \quad 1 \leq i \leq n.$$

Show that the score equations have the form (6.9), with $D_i = \frac{\partial \mu_i}{\partial \boldsymbol{\beta}}$ and $V_i = \mu_i$.

6.14 Show that inference about $\boldsymbol{\beta}$ based on EE is valid even when the NB model does not describe the distribution of the count variable y_i, provided that the systematic component $\log(\mu_i) = \mathbf{x}_i^\top \boldsymbol{\beta}$ specified in (6.22) is correct.

6.15 Let y follow the negative binomial distribution (6.20). Show that $E(y) = \mu$ and $Var(y) = \mu(1 + \alpha\mu)$, where α is the dispersion parameter for the negative binomial distribution.

6.16 Let y follow a mixture of structural zeros of probability p and a Poisson distribution with mean μ of probability $q = 1 - p$. Show that $E(y) = q\mu$, and $Var(y) = q\mu + pq\mu^2$. Thus, the variance of a ZIP outcome variable is always larger than its mean, and the phenomenon of overdispersion occurs if a Poisson regression is applied to such data.

Chapter 7

Log-Linear Models for Contingency Tables

In this chapter, we discuss how to apply the log-linear models introduced in the last chapter to model cell counts in contingency tables. We considered methods for two-way contingency tables in Chapter 2 and stratified two-way tables by a categorical variable in Chapter 3. Although easy to understand, these methods are too specific to the questions considered; different statistics need to be constructed depending on the nature of the problems. Furthermore, their generalizations to higher-dimensional tables are quite complex and difficult. Although regression analysis, introduced in Chapters 4 and 5, may be applied in that regard, we must select one variable to serve as the dependent, or response, variable, which may not always be reasonable or even possible.

In this chapter, we introduce a general alternative to facilitate the analysis of higher-dimensional contingency tables by utilizing log-linear Poisson regression models. Under this alternative paradigm, the response is the cell count in the contingency tables. The relationships among the categorical variables are studied by investigating how they work together to predict the cell count. For example, the hypothesis that all the categorical variables are independent is equivalent to the hypothesis that there are no interaction terms in the log-linear model in predicting the cell counts. Thus, testing the independence assumption is the same as testing whether the log-linear model contains any interaction term. Further, by framing such problems in the context of regression, we are ready to explore much more complex relationships among the variables using the familiar inference theory for regression models.

In Section 7.1, we introduce the ideas and groundwork that allow us to connect log-linear models to analysis of cell counts and discuss the inference theories for log-linear models in such a setting. Illustrative examples are given for two-way and three-way contingency tables in Sections 7.2 and 7.3. In Section 7.4, we study the phenomenon of structural zeros within the context of contingency tables. As the number of potential model candidates increases quickly with the number of variables, we discuss how to find an appropriate model and compare it with other competing alternatives in Section 7.5.

7.1 Analysis of Log-Linear Models

Since categorical variables are described by the multinomial distribution, it is natural to study contingency tables based on this distribution. Methods discussed in Chapters 2 and 3 prove to be very effective for the analysis of two-way contingency tables. However, since the number of possible combinations increases exponentially with the number of variables, analyses will become quite complex if the same or similar approaches are applied to three- or higher-dimensional contingency tables. Fortunately, the close relationship between the Poisson and multinomial distributions makes it possible to apply the Poisson log-linear regression model to facilitate analysis of such higher-dimensional tables.

7.1.1 Motivation

At first glance, it may seem strange that log-linear models can be used for the analysis of cell counts, since Poisson distribution is not bounded above, while its multinomial counterpart is. However, Poisson and multinomial distributions have the following interesting relationship which enables us to treat the sample size as random and apply log-linear models.

Suppose that $y_1, y_2, \ldots, y_k$ are independent Poisson random variables with parameters λ_1, $\lambda_2, \ldots, \lambda_k$. Then conditioning on $y_1 + y_2 + \ldots + y_k = n$, the y_i's jointly have a multinomial distribution

$$\mathbf{y} = (y_1, y_2, \ldots, y_k)^\top \sim MN(n, \boldsymbol{\pi}), \quad \boldsymbol{\pi} = (\pi_1, \pi_2, \ldots, \pi_k)^\top, \tag{7.1}$$

$$\pi_i = \frac{\lambda_i}{\sum_{j=1}^k \lambda_j}, \quad i = 1, \ldots, k.$$

where $MN(n, \boldsymbol{\pi})$ denotes a multinomial distribution with sample size n and probability vector $\boldsymbol{\pi}$.

This relation between Poisson and multinomial distributions lays the foundation for modeling cell counts using the Poisson distribution. To illustrate with a simple example, consider a one-way table with I cells, designated as cell $1, 2, \ldots, I$. If we treat the sampling as an ongoing process, then the final data set, including the total sample size, depends on when the sampling process is stopped, and thus is itself random. For the ith cell, denote the observed count as n_i and the expected as μ_i. If we assume the cell counts follow Poisson distributions, we can model the logarithm of the mean count for the ith cell as

$$\log \mu_i = \lambda + \beta_i, \quad i = 1, 2, \ldots, I. \tag{7.2}$$

By choosing one cell, say the last one, as the reference level, we have $\beta_I = 0$. The log-likelihood (constant terms omitted) based on the above log-linear model is

$$l(\lambda, \boldsymbol{\beta}) = \log(L(\lambda, \boldsymbol{\beta})) = \sum_{i=1}^I n_i(\lambda + \beta_i) - \sum_{i=1}^I \exp(\lambda + \beta_i)$$

$$= \left\{ \sum_{i=1}^I n_i \beta_i - n \log \sum_{i=1}^I \exp(\beta_i) \right\} + (n \log \tau - \tau). \tag{7.3}$$

where $\boldsymbol{\beta} = (\beta_1, \ldots, \beta_I)^\top$ and $\tau = \sum_{i=1}^I \mu_i = \sum_{i=1}^I \exp(\lambda + \beta_i)$. The log-likelihood function can then be maximized separately by maximizing the first term as a function of $\boldsymbol{\beta}$ and the second term as a function of τ; note that τ can be treated as an independent variable as λ occurs only in the second term.

Conditional on $\sum_{i=1}^I n_i = n$, the multinomial distribution is given by

$$\pi_i = \frac{\mu_i}{\sum_{k=1}^I \mu_k} = \frac{\exp(\lambda + \beta_i)}{\sum_{k=1}^I \exp(\lambda + \beta_k)} = \frac{\exp(\beta_i)}{\sum_{k=1}^I \exp(\beta_k)}, \quad i = 1, \ldots, I.$$

Since the likelihood for the multinomial distribution is $\prod_{i=1}^I \pi_i^{n_i}$, it follows that the corresponding log-likelihood is

$$\sum_{i=1}^I n_i \beta_i - n \log \sum_{i=1}^I \exp(\beta_i). \tag{7.4}$$

This is exactly the first term of L in (7.3). Since $\boldsymbol{\beta}$ only enters the first term in (7.3), the maximum likelihood based on the Poisson distribution produces the same inference as that based on the corresponding multinomial model. Thus, the log-linear and multinomial approaches are equivalent, so long as our interest focuses on the parameter vector $\boldsymbol{\beta}$.

Note that the Poisson model has one more parameter τ, which designates the random total sample size, $\sum_i^I n_i = n$, of the distribution of the sum of I independent Poissons. The second term of the likelihood in (7.3), $n \log \tau - \tau$, is maximized at $\tau = n$ (see Problem 7.2).

In the case of independent sampling within each different group such as those in case-control studies, each group follows a multinomial, and hence, the entire sample follows a product of multinomial, distributions. For example, the data from a binary variable in a case–control study can be displayed using a 2×2 table as discussed in Chapter 2. The table can be described by the product of two binomials, with one for the case and the other for the control group. If each of the group sizes is treated as random as in the above, the same loglinear methodology can be applied (Birch, 1963).

The equivalence between the multinomial and Poisson approaches enables us to study contingency tables using standard regression analysis tools. This switch of paradigm allows us to take advantage of the power of regression methodology to facilitate analysis of high-dimensional contingency tables, for which the traditional multinomial approach is very cumbersome at best.

7.1.2 Log-linear models for contingency tables

Consider an m-way contingency table, where the m factors are denoted as x_i, $i = 1, \ldots, m$. Suppose that each x_i takes on levels $j = 1, 2, \ldots, m_i$ $(i = 1, \ldots, m)$. Then there are $M = \prod_{i=1}^m m_i$ distinct possible combinations. By indexing the cell defined by the m factors $x_i = k_i$ $(i = 1, \ldots, m)$ using a vector $(k_1, \ldots, k_m)$, we can denote the observed and expected cell counts as $n_{k_1 \ldots k_m}$ and $\mu_{k_1 \ldots k_m}$. As in standard regression analysis for categorical outcomes, we create indicator variables to represent the different levels of each factor x_i. Let $x_j^i = 1$ if $x_i = j$ and $= 0$ otherwise $(j = 1, \ldots, m_i; i = 1, \ldots, m)$. For each i, only $m_i - 1$ of these x_j^i are independent. If we use the last level m_i as a reference, then x_j^i $(j = 1, \ldots, m_i - 1)$ are the predictors in the regression model.

Log-linear models for contingency tables are specified as follows.

(1) Random component. The observed cell count $n_{k_1 \ldots k_m}$ follows a Poisson distribution with mean $\mu_{k_1 \ldots k_m}$.

(2) Systematic component. The expected count $\mu_{k_1 \ldots k_m}$ for cell $(k_1, \ldots, k_m)$ is linked to the linear predictor by the log function. The linear predictor containing all variables and their interactions is a linear function of the indicator variables x_j^i, $j = 1, \ldots, m_i - 1$; $i = 1, \ldots, m$, and their products:

$$\log(\mu_{k_1 \ldots k_m}) = \lambda + \sum_{i=1}^m \sum_{j=1}^{m_i-1} \beta_j^i x_j^i + \sum_{i<i'} \sum_{j=1}^{m_i-1} \sum_{j'=1}^{m_{i'}-1} \beta_{jj'}^{ii'} x_j^i x_{j'}^{i'} + \cdots.$$

The above may simply be expressed as

$$\log(\mu_{k_1 \ldots k_m}) = \lambda + \sum_{i=1}^m \beta_{k_i}^i + \sum_{i<i'} \beta_{k_i k_{i'}}^{ii'} + \cdots, \tag{7.5}$$

with the understanding that $\beta = 0$ for any term involving a reference level.

In log-linear models for contingency tables, each cell $(k_1, \ldots, k_m)$ (a possible combination of the different levels of the factors involved) provides a single observation. Thus, the sample size for the contingency tables is fixed at M, the total number of cells, no matter how many subjects may be sampled. Thus, increasing the number of subjects will only grow the expected cell count $\mu_{k_1 \ldots k_m}$, not the sample size M. This is exactly the case of small sample size with large expected counts, for which the corresponding asymptotic inference applies (see Section 6.2).

7.1.3 Parameter interpretation

The same interpretation as that discussed in Section 6.1.1 for log-linear models for count responses applies within the current context. For example, $\exp(\lambda)$ is the expected count of the cell corresponding to the reference levels of all factors, i.e., $x_i = m_i$ $(i = 1, \ldots, m)$, in model (7.5). The parameters β indicate differences among different levels in both the logarithms of mean counts and the cell probabilities. Based on (7.1), each cell probability will involve all the parameters, all β's and λ, as the denominator is the sum of all mean counts. The ratio of any two such probabilities, however, will only depend on the difference in logarithms of the corresponding cell mean counts, or only β's, since the same denominator in each cancels out when forming the ratio.

For example, consider the simple log-linear model (7.2) for a one-way contingency table with I levels. In this case, β_i represents the logarithm of the ratio of the expected count in the ith level to the reference level, i.e., $\exp(\beta_i) = \mu_i/\mu_I$. Based on (7.1), this is also the ratio of the corresponding probabilities, i.e., $\exp(\beta_i) = \pi_i/\pi_I$. Hence, the coefficient of each main effect term has the odds interpretation.

The coefficient for a two-way interaction term measures the logarithm of the ratio of two odds and thus has the odds ratio interpretation. We discuss this in detail in Section 7.2.1. For higher-dimensional contingency tables, the parameter interpretation can be more complex if there are higher-order interactions. For example, coefficients of three-way interaction terms measure the difference of odds ratios, and their absence may indicate homogeneity of association (see Section 7.3.2). However, one can always interpret them based on the (product of) multinomial distribution of the contingency tables according to (7.1).

For ordinal variables, we may ignore the internal ordering and simply treat them as categorical. But, if we want to keep the ordering, we may assign some scores based on the ordered levels of the variable as we did in Chapter 2. For example, for ordinal variables with interval scales, their inherent scales may be important and as such may be used for the ordered levels. Otherwise, a common approach is to assign consecutive numbers to the ordinal levels. In log-linear models, such ordinal variables are treated as continuous. For example, a log-linear model for a two-way table with one categorical (x) and one ordinal (y) variable may be expressed as

$$\log(\mu_{ij}) = \lambda + \beta_i^x + \beta^y c_j + \beta_i^{xy} c_j, \quad i = 1, \ldots, I; \quad j = 1, \ldots, J. \tag{7.6}$$

where c_j is the score assigned to the jth level of the ordinal variable, and I (J) denotes the number of levels in x (y).

The main advantage of such a treatment is being able to model some trend effect across the ordinal levels. In addition, the number of parameters used in the model may also be significantly reduced, especially when some variables have a large number of levels. For example, the model in (7.6) has only $1 + (I - 1) + 1 + (I - 1) = 2I$ parameters, rather than IJ if y is treated as nominal. The coefficients of such terms can be interpreted as the difference in the logarithms of the expected cell means per unit change of y.

Since log-linear models involving ordinal variables such as the one in (7.6) can be subsumed into the general form in (7.5) with some restrictions on the parameters, we will focus the discussion on categorical variables. Note that this distinction between the nominal and ordinal variables does not apply to the binary variable, as the same log-linear model results whether the variable is treated as categorical or ordinal.

7.1.4   Inference

As illustrated in Section 7.1.1, the likelihood based on the log-linear Poisson is equivalent to the one based on the (product) multinomial distribution. For stratified sampling such as in case–control studies where the size is fixed for some subgroups, it is important to account for different group sizes by including additional parameters, just as λ does for the total sample size for the log-linear model. For example, in the hypothetical example of comparing success rates of some surgery between two hospitals in Chapter 3, the data are stratified by patient disease severities. To account for this stratification factor, we need to set aside a parameter for the size of each stratum. As the Poisson and multinomial approaches are equivalent within each stratum, the resulting stratified log-linear model, with the likelihood being in the form of the product of two Poisson distributions, provides the same inference as the product-multinomial model based on the stratified data. If we do not include the stratification information in the model, we will then model the pooled data and obtain biased estimates because of the Simpson paradox.

Inferences based on likelihood for contingency tables is then exactly the same as described in the last chapter for the count response. However, goodness-of-fit tests play a particularly important role in the current context. In most regression analyses, we are typically interested in the regression coefficients, since their magnitudes and signs answer the primary questions of causality or association, and the degrees and directions of the relationship, for which regression models are being employed in the first place. Goodness-of-fit tests are also used to either confirm an a priori relationship or to help find an appropriate model for such a relationship. In other words, such tests are not our primary objectives. However, when using log-linear models for contingency table analysis, our priority is to see whether some association among the variables is lacking, and as such goodness-of-fit tests are employed to help facilitate the investigation of this primary question.

Methods such as the Pearson chi-square and likelihood-ratio-based deviance statistics discussed in the last chapter are readily applied for testing goodness of fit within the current context. We can also check model fit via approaches that start with a broader and more plausible (fit the data well) model and then make their way toward a parsimonious one by trimming off the redundant terms (often those not statistically significant). For categorical variables, there is always this saturated model that fits the data perfectly. One may start with this omnibus model, which contains all the interaction terms as shown in (7.5), and then successively remove terms to derive a final parsimonious model based on some criteria such as some level of type I error. However, one potential problem for such a top-down approach is the large number of parameters, creating difficulty for reliable parameter estimation and inference, especially with small sample sizes.

Note that the sample size for the log-linear approach is the number of cells. Application of asymptotic theory to log-linear models requires a large expected cell count for each cell. Since the number of cells is fixed, this is equivalent to the requirement that the sample size (the total number of individuals) n be sufficiently large. When the sample size is small, or more precisely, if the expected counts of some cells are small (typically less than 5), the asymptotic results may not be reliable. In such cases, exact methods may be applied. We may also compare the likelihood ratio statistic and the corresponding Pearson chi-square statistic to see whether they give rise to the same inferential conclusion. As the two tests are asymptotically equivalent, we may feel more comfortable with the asymptotic results, if the two statistics yield similar p-values. Otherwise, the sample size may not be large enough to arrive at a reliable conclusion.

Example 7.1

For a one-way table with m levels, there are n_i subjects in the ith level, with a total of $\sum_{i=1}^{m} n_i = n$ subjects. Suppose we want to check if the distribution is homogeneous across the levels, i.e., if $p_i = \frac{1}{m}$ for all $1 \leq i \leq m$. The corresponding log-linear model is

$$\log \mu_i = \lambda, \quad i = 1, \ldots, m. \tag{7.7}$$

In the above, the expected counts are the same across all the levels.

It is straightforward to write down the likelihood and find the MLE (see Problem 7.4). The expected cell count based on the MLE is $\frac{n}{m}$, and the Pearson chi-square statistic is $\sum_{i=1}^{m} \frac{(n_i - n/m)^2}{n/m}$, which follows a chi-square distribution with $m - 1$ (number of cells minus number of free parameters in the model) degrees of freedom asymptotically. This is exactly the same statistic as that given in (2.7) for testing homogeneity in one-way frequency tables.

We may also consider more general hypotheses for one-way tables. For example, we can apply log-linear models to Example 2.1. In the Metabolic Syndrome study, the prevalence of MS is 0.4, implying that the ratio of probabilities of MS to non-MS is $0.4/0.6 = 2/3$. Thus, $\mu_1/\mu_0 = 2/3$, where μ_0 (μ_1) are the expected counts for non-MS (MS). The hypothesis that the prevalence is 0.4 can be tested by assessing the log-linear models, $\log \mu_1 = \lambda$ and $\log \mu_0 = \lambda + \log(2/3)$, to see whether it fits the data. The Pearson chi-square statistic is 5.2258. Comparing it with the chi-square distribution (with one degree of freedom), we obtain the p-value $= 0.0223$, similar to that obtained in Example 2.1.

We may also test the hypothesis by working on the saturated model (7.2), where all μ_i can be different. Since $\beta_i = \log \frac{p_i}{p_m}$, $i = 1, \ldots, m - 1$, to test whether the one-way table follows a specific multinomial distribution with $\boldsymbol{\pi} = (p_1, ..., p_m)$, we can simply test whether $\beta_i = \log \frac{p_i}{p_m}$, $i = 1, \ldots, m-1$. For example, to test $H_0 : \Pr(MS) = 0.4$, we may test whether $\beta_1 = \log(2/3)$ under the saturated model, if non-MS is the reference level. This is a linear contrast with an offset term (see Section 6.1.3). The p-value for the Wald test is 0.0235, also similar to that obtained in Example 2.1. We will provide more examples of higher dimensional tables in later sections. □

7.2 Two-Way Contingency Tables

In this section, we apply the idea of log-linear models to two-way contingency tables. Although they have been thoroughly studied in Chapter 2, the simplicity of this alternative approach will help us appreciate the elegance of log-linear models for contingency tables.

Consider a two-way $I \times J$ contingency table that cross-classifies each of n subjects based on a row and a column categorical response. Let n_{ij} denote the cell count in the ijth cell. If we regard n_{ij} as independently distributed Poisson variables with mean μ_{ij}, then it follows from (7.1) that conditional on $\sum_{i=1}^{I} \sum_{j=1}^{J} n_{ij} = n$, the n_{ij} jointly have a multinomial distribution, $MN(n, \boldsymbol{\pi})$, where

$$\boldsymbol{\pi} = (\pi_{11}, \pi_{12}, \ldots, \pi_{1I}, \ldots, \pi_{I1}, \pi_{I2}, \ldots, \pi_{IJ})^{\top}, \quad \mu = \sum_{k=1}^{I} \sum_{l=1}^{J} \mu_{kl}.$$

$$\pi_{ij} = \frac{\mu_{ij}}{\sum_{k=1}^{I} \sum_{l=1}^{J} \mu_{kl}} = \frac{\mu_{ij}}{\mu}, \quad 1 \leq i \leq I, \quad 1 \leq j \leq J.$$

In Chapter 2, we discussed methods that are based on modeling n_{ij} conditional on the total sample size n, which jointly have a multinomial distribution. As the result of the shift of the modeling approach, the log-linear model focuses on the expected frequencies μ_{ij}, rather than the cell probabilities π_{ij}, as in the traditional approach.

7.2.1 Independence

A primary question for a two-way contingency table is whether the two categorical variables are associated. Nonassociation, or independence, between the two variables is equivalent to

$$\pi_{ij} = \pi_{i+}\pi_{+j}, \quad 1 \le i \le I, \ 1 \le j \le J, \tag{7.8}$$

where π_{i+} and π_{+j} are the marginal probabilities as defined in Chapter 2. By multiplying both sides of (7.8) by μ, we obtain

$$\mu_{ij} = \mu\pi_{i+}\pi_{+j}. \tag{7.9}$$

In other words, the cell mean has the above form under independence. Also, it is straightforward to check that the converse is also true, i.e., (7.9) implies (7.8) (see Problem 7.3). Thus, it follows that the log-linear model has the following form if and only if the row and column are independent:

$$\log(\mu_{ij}) = \lambda + \lambda_i^x + \lambda_j^y, \quad 1 \le i \le I, 1 \le j \le J, \tag{7.10}$$

where $\lambda = \log\mu$, $\lambda_i^x = \log\pi_{i+}$ and $\lambda_j^y = \log\pi_{+j}$. This Poisson log-linear model has additive main effects of the row and column variables but no row by column interaction, with λ_i^x (λ_j^y) indicating the row (column) effect. Since the π_{i+} (π_{+j}) add up to 1, the parameters λ_i^x (λ_j^y) are not free to vary. We may set one for the row (column) to 0, say the last level $\lambda_I^x = 0$ $(\lambda_J^y = 0)$, to identify the remaining λ_i^x (λ_j^y). The model in (7.10) is a GLM that treats cell counts as independent observations from a Poisson, with the mean (expected cell counts) linked to the linear predictor using the log function. This is in stark contrast to the traditional multinomial-based approach that identifies the data as classifications of n individual subjects.

To confirm that (7.10) is indeed a model for testing independence between the rows and columns, and see how the parameters are interpreted, consider the special case where both row and column variables are binary, i.e., $I = J = 2$. The values of the indicator variables and expected cell counts under the model are summarized in the following table:

Cell	x	y	$\mathrm{Log}(\mu)$
$(1,1)$	1	1	$\lambda + \lambda_1^x + \lambda_1^y$
$(1,2)$	1	0	$\lambda + \lambda_1^x$
$(2,1)$	0	1	$\lambda + \lambda_1^y$
$(2,2)$	1	1	λ

Here, λ represents the logarithm of the expected count of cell $(2,2)$, corresponding to the reference levels in both variables. It is clear that λ_1^x represents the difference between cells $(1,1)$ and $(2,1)$, as well as between cells $(1,2)$ and $(2,2)$, i.e., $\lambda_1^x = \log\left(\frac{\mu_{11}}{\mu_{21}}\right) = \log\left(\frac{\mu_{12}}{\mu_{22}}\right)$, and hence $\frac{p_{11}}{p_{21}} = \frac{p_{12}}{p_{22}}$. Likewise, $\lambda_1^y = \log\left(\frac{\mu_{11}}{\mu_{12}}\right) = \log\left(\frac{\mu_{21}}{\mu_{22}}\right)$, implying $\frac{p_{11}}{p_{12}} = \frac{p_{21}}{p_{22}}$. Thus, the two variables are independent under the additive log-linear model, with the coefficients of the variables determining the multinomial distribution. The intercept λ plays no role in the interpretation of the model for the independence hypothesis, reflecting the fact that it is a parameter added to account for the sample size.

The log-linear independence model (7.10) is analogous to the case of two-way ANOVA without interaction. Parameters λ_i^x and λ_j^y represent the main effects of the row and column variables. To capture the dependence between the row and column variables, we include in (7.10) the term for the row by column interaction, yielding

$$\log \mu_{ij} = \lambda + \lambda_i^x + \lambda_j^y + \lambda_{ij}^{xy}, \quad 1 \le i \le I, 1 \le j \le J. \tag{7.11}$$

The added term λ_{ij}^{xy} accounts for the deviation from independence. Like the additive model under independence, we impose constraints $\lambda_{Ij}^{xy} = \lambda_{iJ}^{xy} = 0$ in (7.11) to make the model identifiable. Thus, as in the two-way ANOVA setting, we can view λ_i^x as the coefficients of $I - 1$ binary indicators for the first $I - 1$ categories of the row factor, λ_j^y as the coefficients of $J - 1$ binary indicators for the first $J - 1$ categories of the column factor, and λ_{ij}^{xy} as the coefficients of the $(I - 1)(J - 1)$ product terms of the two sets of indicators. Altogether, there are $1 + (I - 1) + (J - 1) + (I - 1)(J - 1) = IJ$ parameters for the log-linear model in (7.11), which is the total number of cells in the table. As the number of parameters is the same as the number of cells (the sample size under the log-linear model), this model fits the data perfectly, thus it is a saturated model. In practice, unsaturated models are preferable because we would like to summarize the information contained in the data with fewer parameters for easy interpretation.

As in linear regression, the interpretation is more complicated if there are interactions among the variables. Take the 2×2 table as an example again. The values of the indicator variables and expected cell counts under the model are summarized in the following table:

Cell	x	y	$\mathbf{Log}(\mu)$
$(1,1)$	1	1	$\lambda + \lambda_1^x + \lambda_1^y + \lambda_{11}^{xy}$
$(1,2)$	1	0	$\lambda + \lambda_1^x$
$(2,1)$	0	1	$\lambda + \lambda_1^y$
$(2,2)$	1	1	λ

Here,

$$\log\left(\frac{p_{11}}{p_{21}}\right) = \log\left(\frac{\mu_{11}}{\mu_{21}}\right) = \lambda_1^x + \lambda_{11}^{xy}, \quad \log\left(\frac{p_{12}}{p_{22}}\right) = \log\left(\frac{\mu_{12}}{\mu_{22}}\right) = \lambda_1^x.$$

The interaction term λ_{11}^{xy} represents the log of the odds ratio, $\log\left(\frac{p_{11}}{p_{21}} / \frac{p_{12}}{p_{22}}\right)$. Thus, to test for the row and column independence for the contingency table, we can examine whether the interactions in the saturated log-linear model in (7.11) are significant, or whether the additive model fits the data well using goodness-of-fit tests.

In the case of independent sampling according to row levels, we may have different log-linear models for different rows:

$$\log \mu_{ij} = \lambda_i + \beta_j^i, \quad j = 1..., J, \, i = 1, 2, \ldots, I. \tag{7.12}$$

If the distributions of the column variables across the row levels are homogenous, then $\beta_j^1 = \beta_j^2 = \ldots = \beta_j^i$, $j = 1, \ldots, J - 1$, $i = 1, \ldots, I$. It is straightforward to verify that in such case, the model is equivalent to model (7.10); in fact, it is the model for (7.10) when cell-means coding is applied to the row variable. Similarly, (7.12) is equivalent to model (7.11). Thus, we may simply apply models (7.10) and (7.11) to two-way contingency tables without concerning the different data sampling procedures, similar as we discussed in Chapter 2.

Example 7.2

In Table 2.9 of Example 2.13, we investigated whether proband and informant "diagnoses" of proband depression were marginally homogeneous. One might ask first whether they are associated at all. To test the null hypothesis of row (proband) and column (informant) independence using the log-linear model, we can apply goodness-of-fit tests to the additive model. The Pearson chi-square statistic is 47.31. Taken as a goodness-of-fit test of the independence model relative to the saturated model, the deviance statistic defined in Chapter 6 is 46.17. Both test statistics have $3 \times 3 - (3-1) - (3-1) - 1 = 4$ degrees of freedom. The corresponding p-values are both $<.0001$. We can also fit the saturated model, and test whether there is any significant interaction. Since both the row and column variables have three levels, there are four interaction terms. The Wald and likelihood ratio test statistics, for testing the composite hypothesis of all four interaction terms are 0, are 38.94 and 46.17, respectively. The corresponding p-values, obtained by comparing with chi-square distributions with four degrees of freedom, are both $<.0001$. All the tests indicate that there is significant association between the diagnosis of probands and informants, as expected. The use of loglinear models allows us to perform further modeling within the same methodological context. $\Box$

Please note that we used the same deviance statistic in the deviance goodness-of-fit test and the (deviance) test comparing the models with vs. without interactions. In fact, we can compare any two nested models, where one is contained within the other, using the difference in magnitude of their deviances, with degrees of freedom the difference in number of parameters between the two models. This is equivalent to the chi-squared approximation to the likelihood ratio test.

7.2.2   Uniform associations

When two categorical variables are associated, we may further be interested in how they are associated. When both categorical variables are ordered, the association is determined by all the 2×2 subtables formed by adjacent levels. We may be interested in learning if all the odds ratios for these 2×2 subtables are the same, implying *uniform association* (Goodman 1979). More precisely, let $OR_{i,j} = \frac{\pi_{i+1,j+1}/\pi_{i,j+1}}{\pi_{i+1,j}/\pi_{i,j}}$ be the odds ratio of the subtable consists of the ith and $(i+1)$th rows and jth and $(j+1)$th columns. Then, the uniform association means

$$OR_{ij} = \frac{\pi_{i+1,j+1}/\pi_{i,j+1}}{\pi_{i+1,j}/\pi_{i,j}} = c, \quad 1 \le i \le I-1,\ 1 \le j \le J-1, \qquad (7.13)$$

where c is a constant positive number. When $c = 1$, this reduces to the case that the row and column variables are independent. Since the interaction between the row and column variables when each is increased by one level remains the same, we need to treat both the row and column variables as a scalar for the interaction. We may assign consecutive scores for both variables, say $1, ..., I$ for the I levels of the row variable and $1, ..., J$ for the J levels of the column variable. The log-linear model for uniform association is

$$\log(\mu_{ij}) = \lambda_i^x + \lambda_j^y + \lambda \cdot x \cdot y, \quad 1 \le i \le I, 1 \le j \le J, \qquad (7.14)$$

where λ_i^x and λ_j^y are the main contributions and λ represents the common interaction. Note that in this Poisson log-linear model, the rows and columns are treated as categorical for the main effects, and treated as scalars for their interactions. Thus, when the 2×2 subtable increases by one level for the rows or the columns, the corresponding ratio of the odds ratios is the same, $\exp(\lambda)$, confirming the uniform association. To test for uniform association, it is more convenient to assess the goodness of fit of the uniform association model (7.14), although it is also possible to fit a saturated model and test the significance of all the terms not in the uniform association model. For the goodness of fit tests of (7.14), both the Pearson and deviance chi-square statistics have degrees of freedom $IJ - (I-1) - (J-1) - 1 - 1 = IJ - I - J$.

Instead of a uniform odds ratio, we can assume that $OR_{i1} = \ldots = OR_{iJ-1}$, for $i = 1, \ldots, I - 1$. In this case the odds ratios for all the 2×2 subtables in the same two rows are the same, but 2×2 subtables on different rows may have different odds ratios. The model is called the *row-effect association* model (Goodman 1979). The log-linear model for row-effect association is

$$\log(\mu_{ij}) = \lambda_i^x + \lambda_j^y + \lambda_i^{xy} \cdot x_i \cdot y, \quad 1 \leq i \leq I, 1 \leq j \leq J. \tag{7.15}$$

Compared with the log-linear model (7.14), this Poisson log-linear model has the interaction of the row as a categorical variable with the column as a scalar. We leave it as an exercise to verify that (7.15) is indeed the model for the row-effect association (Problem 7.9). We may similarly check the goodness of fit of (7.15) to test the row-effect association. Both the Pearson and deviance chi-square statistics have degrees of freedom $(I-1)(J-2)$. Note that in the definition, the row variable is not required to be ordinal. One may similarly define column-effect association models where the odds ratios for all the 2×2 subtables in the same two columns are the same when the row variable is ordinal.

Example 7.3

Both the row and column variables in Table 2.9 are ordinal, and thus we may use model (7.14) to assess uniform association of proband vs informant "diagnoses". If we apply goodness-of-fit tests to the log-linear model for uniform association, the Pearson chi-square statistic is 1.18, and the deviance statistic is 1.21, both with three degrees of freedom. The corresponding p-values are 0.7587 and 0.7512, respectively. Both tests indicate that there is no sufficient ground to reject the null hypothesis that the association between the diagnoses of probands and informants is uniform. Further, λ is significantly positive, indicating the two diagnoses are positively associated.

If the row variable is not ordinal, we may still be able to assess the row-effect association. If model (7.15) is applied to Table 2.9, the Pearson chi-square statistic is 0.5487, and the deviance statistic is 0.5446, both with two degrees of freedom. The corresponding p-values are 0.760 and 0.762, respectively. Both tests support the row-effect association. □

7.2.3 Symmetry and marginal homogeneity

When the row and column variables have the same number of levels, the two-way contingency table becomes a square. For such tables, we may be interested in testing for symmetry and marginal homogeneity, in which case we may apply the respective Bowker's and Maxwell-Stuart test as discussed in Chapter 2. In this section, we discuss how to apply log-linear models to facilitate such hypothesis testing.

Based on the saturated model in (7.11), the expected counts for the (i, j) and (j, i) cells are $\exp(\lambda + \lambda_i^x + \lambda_j^y + \lambda_{ij}^{xy})$ and $\exp(\lambda + \lambda_j^x + \lambda_i^y + \lambda_{ji}^{xy})$, respectively. If the last cell is set as the reference level for both the row and column variables, the identifiability constraints become $\lambda_I^x = \lambda_J^y = \lambda_{Ij}^{xy} = \lambda_{iJ}^{xy} = 0$ in (7.11). The null hypothesis of equal cell means between cells (i, j) and (j, i), $\mu_{ji} = \mu_{ij}$, for all i and j can be equivalently stated as

$$\lambda_i^x = \lambda_i^y, \quad \lambda_{ji}^{xy} = \lambda_{ij}^{xy}, \quad \text{all} \quad i, j, \tag{7.16}$$

under the saturated model in (7.11).

Thus, the log-linear model for symmetry can be written as

$$\log \mu_{ij} = \lambda + \lambda_i + \lambda_j + \lambda_{ij}, \quad \text{all} \quad i, j, \tag{7.17}$$

with the restriction $\lambda_{ij} = \lambda_{ji}$. Note that the superscripts in the model above have been suppressed because of symmetry. There are a total of $1 + (I - 1) + \frac{I(I-1)}{2} = \frac{1}{2}I(I + 1)$

parameters for the model in (7.17). The Pearson and deviance chi-square statistics for goodness of fit both have $\frac{1}{2}I(I-1)$ degrees of freedom. In the particular case where both variables are binary, the condition in (7.16) simplifies to $\lambda_1^x = \lambda_1^y$.

Note that the interaction λ_{ij}^{xy} is an adjustment of the expected cell counts after the main effect of the row and column variables, and sometimes it is of interest to test whether the adjustment is symmetric, i.e., whether $\lambda_{ji}^{xy} = \lambda_{ij}^{xy}$. More precisely, under the independence between x and y, the expected cell count are determined by the marginal counts: $\mu_{ij} = \frac{1}{\mu}\mu_{i+}\mu_{+j}$. When x and y are dependent, the interaction terms adjust for the discrepancies between the observed and expected cell counts under independence to improve model fit. It is not difficult to check that $\lambda_{ji}^{xy} = \lambda_{ij}^{xy}$ is equivalent to $\frac{\mu_{ij}}{\mu_{i+}\mu_{+j}} = \frac{\mu_{ji}}{\mu_{j+}\mu_{+i}}$. This kind of symmetry is called *quasi-symmetry*, and the log-linear model for quasi-symmetry satisfies the condition $\lambda_{ji}^{xy} = \lambda_{ij}^{xy}$ (though λ_i^x and λ_i^y can be different).

We may also use log-linear models to test for marginal homogeneity. Based on (7.1), this is equivalent to $\mu_{i+} = \mu_{+i}$, $i = 1, \ldots, I$. When framed under the saturated model (7.11),

$$\sum_{j=1}^{I} \exp\left(\lambda_i^x + \lambda_{ij}^{xy}\right) = \sum_{j=1}^{I} \exp\left(\lambda_i^y + \lambda_{ji}^{xy}\right), \quad \text{for } i = 1, 2, \ldots, I. \tag{7.18}$$

As (7.18) involves nonlinear functions of λ_i^x and λ_{ij}^{xy}, the above is not a linear contrast. However, we can apply the delta method to carry out the test (see Chapter 1, Section 1.4.2 for details on the delta method). Some software such as Stata offer support for testing the nonlinear hypothesis in (7.18).

Note that by changing the log function to the identity link to model the cell mean directly, we can express the nonlinear constraint in (7.18) for marginal homogeneity as a linear contrast. However, the problem with such an approach, akin to modeling binary responses using linear regression, is that the range of fitted values may be negative, violating the conceptual requirement of nonnegative response for the cell mean.

Example 7.4

In Example 2.13, if using major depression (MajD) as the reference level for both the row and column variables, symmetry of the two-way table is equivalent to the following restrictions under the saturated model in (7.11):

$$\lambda_{\text{NO}}^{\text{Proband}} = \lambda_{\text{NO}}^{\text{Informant}}, \ \lambda_{\text{MinD}}^{\text{Proband}} = \lambda_{\text{MinD}}^{\text{Informant}}, \ \lambda_{\text{NO,MinD}}^{\text{Proband,Informant}} = \lambda_{\text{MinD,No}}^{\text{Proband,Informant}}.$$

Testing the above linear contrast based on the study data using the Wald statistic yields a p-value 0.0044. Although the p-value is different from the one obtained in Example 2.13, it yields the same conclusion. The small difference here is expected, since the Bowker statistic in Example 2.13 is a linear combination of the discrepancies in counts, while the statistic under the log-linear model is a linear combination of logarithms of the estimated cell counts.

We also tested the marginal homogeneity under the linear model and obtained a p-value 0.0009 by the Wald test. The results are similar to that based on the Stuart–Maxwell statistic in Example 2.13. □

7.3 Three-Way Contingency Tables

In this section, we discuss log-linear models for three-way contingency tables which contain cell counts based on the cross-classification of three factors. Consider three factors x,

y, and z, with their respective levels indexed by $1, \ldots, I$, $1, \ldots, J$, and $1, \ldots, K$. The cell probabilities π_{ijk} are given by

$$\pi_{ijk} = \Pr(x = i, y = j, z = k), \quad i = 1, \ldots, I, \ j = 1, \ldots, J, \ k = 1, \ldots, K.$$

Let μ_{ijk} be the expected cell count under the log-linear model, and $\mu = \sum_{i=1}^{I} \sum_{j=1}^{J} \sum_{k=1}^{K} \mu_{ijk}$ the expected total count. As before, we denote marginal probabilities and marginal expected cell counts by putting the "+" sign in the respective places. We first discuss how to assess association under a log-linear model. Unlike two-way tables, there are several different types of independence for three variables. In addition, if two variables are associated, it is of interest to see whether the association is the same across the different levels of the third variable, a phenomenon known *association homogeneity*.

7.3.1   Independence

Relationships among three variables can be very complicated. Based on the lack of some kind of association, Birch (1963) discussed several different types of independence, all of which are commonly applied in practice.

7.3.1.1   Marginal independence

For any two variables, we may consider their association by ignoring the third variable. In such cases, it is simply a two-way contingency table, and methods described in the last section and Chapter 2 may be applied. Independence within such a context is also called *marginal independence*.

7.3.1.2   Mutual independence

The three variables x, y, and z are *mutually independent,* if each cell probability equals the product of three corresponding marginal probabilities, i.e.,

$$\pi_{ijk} = \pi_{i++}\pi_{+j+}\pi_{++k}, \quad i = 1, \ldots, I, j = 1, \ldots, J, \ k = 1, \ldots, K.$$

This is also called *complete independence* (Wickens, 1989). Since

$$\mu_{i++} = \mu\pi_{i++}, \ \mu_{+j+} = \mu\pi_{+j+}, \ \mu_{++k} = \mu\pi_{++k},$$

it follows that under mutual independence

$$\mu_{ijk} = \gamma\mu_{i++}\mu_{+j+}\mu_{++k},$$

where $\gamma = 1/\mu^2$. Thus, the log-linear model for mutual independence is

$$\log\mu_{ijk} = \log\gamma + \log\mu_{i++} + \log\mu_{+j+} + \log\mu_{++k} = \lambda + \lambda_i^x + \lambda_j^y + \lambda_k^z. \tag{7.19}$$

Hence, similar to two-way contingency tables, mutual independence implies additivity in the corresponding log-linear model.

Based on (7.1), it is easy to verify that under (7.19)

$$\frac{\pi_{ijk}}{\pi_{i'jk}} = \exp\left(\lambda_i^x - \lambda_{i'}^x\right). \tag{7.20}$$

From (7.20), it is clear that the conditional distribution of x does not depend on y and z. Likewise, the same conclusion is obtained for the relationship between the conditional distribution of y (or z) and the two remaining variables. Thus, the additive form of the log-linear model in (7.19) also indicates mutual independence among the three variables.

7.3.1.3 Conditional independence

In practice, people are often interested in the association between two factors while controlling for the third. The distributions of cell counts formed by two factors of interest at different levels of the third factor can be displayed in a two-way table based on each cross section of the three-way table. These cross sections, called *partial tables*, show the relationship between the two variables by holding the third at a given level. The two-way contingency table obtained by combining the partial tables is called the *marginal table* of the two variables. Each cell count in the marginal table is a sum of counts from the same location in the partial tables. The marginal table describes the marginal distribution after integrating out the third factor, and hence it does not contain any information about the controlled variable. If the partial tables exhibit different associations from the marginal table, one may be interested in the association of the two variables, controlling for the third.

Consider the distribution of x and y at the kth level of z. If x and y are independent in the partial table for the kth level of z, then x and y are said to be *conditionally independent* at level k of z $(1 \leq k \leq K)$. Let

$$\pi_{ij|k} = \frac{\pi_{ijk}}{\pi_{++k}}, \quad i = 1, \ldots, I, j = 1, \ldots, J,$$

denote the joint distribution of x and y conditional on level k of z. The conditional independence of x and y at level k of z means $\pi_{ij|k} = \pi_{i+|k}\pi_{+j|k}$ for all i and j, where $\pi_{i+|k} = \Pr(x = i \mid z = k)$ and $\pi_{+j|k} = \Pr(y = j \mid z = k)$ are the marginal distribution of x and y conditional on $z = k$. This is equivalent to

$$\mu_{ijk}\mu_{++k} = \mu_{i+k}\mu_{+jk}, \quad i = 1, \ldots, I, \quad j = 1, \ldots, J.$$

Thus, the log-linear model for conditional independence between x and y, given $z = k$, is

$$\log \mu_{ijk} = \log \mu_{i+k} + \log \mu_{+jk} - \log \mu_{++k}. \tag{7.21}$$

We call x and y conditionally independent if the above holds for all levels of z.

Under the log-linear model, x and y again have additive effects, though the main effects may vary for different levels of z:

$$\log (\mu_{ijk}) = \lambda + \lambda_i^x + \lambda_j^y + \lambda_k^z + \lambda_{ik}^{xz} + \lambda_{jk}^{yz}, \quad i = 1, \ldots, I, \ j = 1, \ldots, J, \ k = 1, \ldots, K. \tag{7.22}$$

Note that when combined with the main effects term x (y), the interaction between x (y) and z serves to accommodate the different main effects of x (y) across the different levels of z. For example, the main effect of x at $z = k$ level is

$$\lambda_i^x + \lambda_{ik}^{xz}, \quad i = 1, \ldots, I,$$

which varies across the levels of z as λ_{ik}^{xz} is a function of k. Thus, the loglinear model (7.22) can also be applied to the case of independent sampling according the levels of z, or two-way $(x \times y)$ tables stratified by z.

7.3.1.4 Joint independence

The variable y is *jointly independent* of x and z, if

$$\pi_{ijk} = \pi_{i+k}\pi_{+j+}, \quad i = 1, \ldots, I, j = 1, \ldots, J, \ k = 1, \ldots, K.$$

This is the ordinary two-way independence between y and a new variable composed of the IK combinations of the levels of x and z. We can likewise define the joint independence of

x from y and z, and of z from x and y. Note that as "joint independence" and "mutual independence" mean the same thing in the nomenclature of probability theory, we should keep in mind their distinctively different connotations within the current context.

If y is jointly independent of x and z, then

$$\mu_{i+k} = \mu \pi_{i+k}, \quad \mu_{+j+} = \mu \pi_{+j+},$$

The above corresponds to the following log-linear models:

$$\log \mu_{i+k} = \lambda + \lambda_i^x + \lambda_k^z + \lambda_{ik}^{xz}, \quad \log \mu_{+j+} = \lambda + \lambda_j^y.$$

Note that the intercept, λ, in the two models are in general different. Under joint independence, $\pi_{ijk} = \pi_{i+k}\pi_{+j+}$, implying

$$\mu_{ijk} = \mu \pi_{ijk} = \frac{1}{\mu}\mu_{i+k}\mu_{+j+}.$$

Thus, if y is jointly independent of x and z, the log-linear model is given by

$$\log \mu_{ijk} = \log\left(\frac{1}{\mu}\right) + \log \mu_{i+k} + \log \mu_{+j+} = \lambda + \lambda_i^x + \lambda_j^y + \lambda_k^z + \lambda_{ik}^{xz}. \tag{7.23}$$

In other words, there is no interaction involving y in the log-linear model, if y is jointly independent with x and z. However, x and z can be associated, which is accounted for by the interaction term λ_{ik}^{xz} in (7.23).

The different types of independence are interrelated. For example, it is clear that mutual independence is the strongest requirement, which implies all the other types of independence. See Problem 7.15 for more results in this direction.

Example 7.5

Consider the relationship among three-level depression diagnosis (DEP), gender (two levels), and marital status (MS: three levels as defined in Section 4.2.2) in the DOS data. Mutual independence means that all the three factors are independent, and thus we may use the additive log-linear model to describe the data

$$\log(\mu_{ijk}) = \lambda + \lambda_i^{dep} + \lambda_j^{gender} + \lambda_k^{MS}.$$

We use the last level of each variable as a reference. For example, for the marital status variable, $\widehat{\lambda}_1^{MS} = 1.1637$, $\widehat{\lambda}_2^{MS} = 0.6762$, and $\widehat{\lambda}_3^{MS} = 0$. In testing the adequacy of the three-way independence model, we can use the difference in deviance statistics, which compares the saturated model (with all two- and three-way interactions) with the reduced independence model above. This statistic is 106.25 with df = 12. The Pearson goodness-of-fit statistic is 109.88 with df = 12. Based on both statistics, the p-values are < 0.0001, suggesting that the independence model is inappropriate.

We can apply model (7.23) to assess whether DEP is jointly independent with MS and gender. The deviance statistic for goodness of fit is 38.7077 with degrees of freedom 10, and p-value < 0.0001, indicating no evidence to support joint independence either.

To check if DEP and gender are independent conditional on MS, we may check if model (7.21) fits the data well. The deviance statistic is 27.3744 with df = 6 and p-value = 0.0001, indicating that gender is associated with depression even after controlling for MS. Note that we can also use the Cochran–Mantel–Haenszel test for stratified tables discussed in Chapter 3 to examine this null hypothesis. This statistic is 16.3818 with degrees of freedom 2 and p-value = 0.0003, yielding the same conclusion as the log-linear model. □

7.3.2   Association homogeneity

When two factors are associated while controlling for the third, we may want to further ascertain if the association is homogeneous across the different levels of the third variable. In Chapter 3, we discussed homogeneous odds ratio when the two variables are binary. Similar concepts can be applied for factors with more than two levels.

Given $z = k$, we may define an odds ratio for two levels i and i' of x and two levels j and j' of y as

$$OR_{ii',jj'}^k = \frac{\pi_{i',j'|k}/\pi_{i,j'|k}}{\pi_{i',j|k}/\pi_{ij|k}} = \frac{\pi_{i',j',k}/\pi_{i,j',k}}{\pi_{i',j,k}/\pi_{ijk}}. \tag{7.24}$$

This odds ratio has the standard odds ratio interpretation, akin to regression analysis of polytomous responses as we discussed in Chapter 5, by restricting the subjects to the corresponding levels, i.e., conditioning on $x = i$ or i', $y = j$ or j', and $z = k$. If the odds ratio in (7.24) is independent of k ($1 \leq k \leq K$), the association is *homogeneous*. Note that under homogeneous association, the odds ratios $OR_{ii',jj'}^k$ may still vary across the different levels of x and y.

To develop the log-linear model for homogeneous association, first consider the saturated model for the three-way table:

$$\log \mu_{ijk} = \lambda + \lambda_i^x + \lambda_j^y + \lambda_k^z + \lambda_{ij}^{xy} + \lambda_{ik}^{xz} + \lambda_{jk}^{yz} + \lambda_{ijk}^{xyz}.$$

As in the case of two-way table, we impose the constraints that any effect involving the last level of any of the three variables is 0, i.e.,

$$\lambda_I^x = \lambda_J^y = \lambda_K^z = \lambda_{Ij}^{xy} = \lambda_{Ik}^{xz} = \lambda_{Jk}^{yz} = \lambda_{jK}^{yz} = \lambda_{iJ}^{xy} = \lambda_{iK}^{xz} = \lambda_{Ijk}^{xyz} = \lambda_{iJk}^{xyz} = \lambda_{ijK}^{xyz} = 0.$$

This saturated model has the same number of parameters as the number of cells (a total of IJK). According to (7.1), the parameters can be interpreted using the cell probabilities

$$\pi_{ijk} = \frac{\exp(\lambda_{ijk})}{\sum_{l=1}^I \sum_{m=1}^J \sum_{t=1}^K \exp(\lambda_{lmt})},$$

where $\lambda_{ijk} = \lambda_i^x + \lambda_j^y + \lambda_k^z + \lambda_{ij}^{xy} + \lambda_{ik}^{xz} + \lambda_{jk}^{yz} + \lambda_{ijk}^{xyz}$.

Thus, the odds of $x = i'$ over $x = i$ for $y = j$ conditional on $z = k$ equals

$$\frac{\pi_{i',j|k}}{\pi_{i,j|k}} = \exp(\lambda_{i'jk} - \lambda_{ijk}) = \exp\left(\lambda_{i'}^x + \lambda_{i'j}^{xy} + \lambda_{i'k}^{xz} + \lambda_{i'jk}^{xyz} - (\lambda_i^x + \lambda_{ij}^{xy} + \lambda_{ik}^{xz} + \lambda_{ijk}^{xyz})\right).$$

Similarly, the odds of $x = i'$ over $x = i$ for $y = j'$ conditional on $z = k$ equals

$$\frac{\pi_{i',j'|k}}{\pi_{i,j'|k}} = \exp(\lambda_{i'j'k} - \lambda_{ij'k}) = \exp\left(\lambda_{i'}^x + \lambda_{i'j'}^{xy} + \lambda_{i'k}^{xz} + \lambda_{i'j'k}^{xyz} - (\lambda_i^x + \lambda_{ij'}^{xy} + \lambda_{ik}^{xz} + \lambda_{ij'k}^{xyz})\right).$$

It follows that the odds ratio in (7.24) is

$$OR_{ii',jj'}^k = \exp\left(\lambda_{i'j'}^{xy} + \lambda_{ij}^{xy} - \lambda_{ij'}^{xy} - \lambda_{i'j}^{xy}\right) \exp\left(\lambda_{i'j'k}^{xyz} + \lambda_{ijk}^{xyz} - \lambda_{ij'k}^{xyz} - \lambda_{i'jk}^{xyz}\right). \tag{7.25}$$

Thus $OR_{ii',jj'}^k$ is independent of k for all $1 \leq k \leq K$ if and only if $\lambda_{i'jk}^{xyz} + \lambda_{ij'k}^{xyz} = \lambda_{ijk}^{xyz} + \lambda_{i'j'k}^{xyz}$ for all i, i', j, j', and k. Hence, the corresponding log-linear model is

$$\log \mu_{ijk} = \lambda + \lambda_i^x + \lambda_j^y + \lambda_k^z + \lambda_{ij}^{xy} + \lambda_{ik}^{xz} + \lambda_{jk}^{yz}. \tag{7.26}$$

Based on the log-linear model (7.26), it is clear that if x and y are homogeneously associated across the different levels of z, then so are y and z across the different levels of

x, and x and z across the different levels of y. The corresponding log-linear models are identical, all excluding three-factor interactions (7.26). This may not be obvious under the alternative multinomial-based approach (see Problem 7.23). The three-way interaction λ_{ijk}^{xyz} describes how the odds ratio between two variables changes across the different categories of the third. For example, in the special $2 \times 2 \times 2$ three-way table case, there is only one parameter in the three-factor interaction, namely, λ_{111}^{xyz}, which equals the logarithm of the ratio of the odds ratios. Note that the Breslow–Day or Breslow–Day–Tarone test introduced in Chapter 3 can only be applied to test association homogeneity for stratified 2×2 tables, while the loglinear model (7.26) can be applied to general stratified two-way tables.

Example 7.6
The analysis in Example 7.5 indicates that the association of MS with DEP is different between males and females. Thus, the association between any two of the three variables may not be the same across the different levels of the third variable. We may formally test this null by applying the following log-linear model:

$$\log(\mu_{ijk}) = \lambda + \lambda_i^{dep} + \lambda_j^{gender} + \lambda_k^{MS} + \lambda_{ij}^{dep,gender} + \lambda_{ik}^{dep,MS} + \lambda_{jk}^{gender,MS},$$

and see how well it fits the data. The Pearson and deviance goodness-fit-statistics are 11.0521 and 10.5717, respectively. Compared with a chi-square distribution with four degrees of freedom, the corresponding p-values are 0.0260 and 0.0318. Since the only difference between this model and the saturated model are the three-way interaction terms, the deviance statistic is also the LR test statistic for the three-way interaction. Thus, we reject the homogenous association assumption. □

7.4   Irregular Tables

We have focused on rectangular tables in the previous sections. Such tables are formed by creating all combinations of the factor levels of the variables. If the cells represent a well-defined possible outcome for a given study, the cell counts will grow as the sample size increases. For a specific sample in a study, it is quite possible that some cells have zero cell counts, especially when there are a large number of cells under a small or moderate sample size. The zero counts in this case occur by chance, and hence are called *sampling zeros* or *random zeros*. They will become positive if the sampling process is repeated, especially with increased sample sizes. Random zeros may trigger a warning message in most software packages, if asymptotic theory is applied. However, they require neither modification in the model nor in the interpretation.

In some studies, zeros are not just the result of a small sample size or a large number of cells, as they represent a category not meaningful for some or all study subjects. A similar issue emerged when discussing the count response in the last chapter, in which the term "structural zero" was used to describe the nonrandom nature of zero for a special group of subjects within the study population. Within that context, structural zeros are really a symbol or designator for a subgroup of subjects for whom the values of the count response are not meaningful. Because of that connotation, we use the same term here to refer to the nonrandom zero in the current text to distinguish it from its sampling counterpart. Thus, if a cell represents a structural zero within the current context, the cell probability is zero, and the cell count stays at zero, regardless of repeated sampling and increased sample size.

It is important to note that unlike the count response, structural zeros here arise for a different reason. The structural-zero cell represents undefined or unmeaningful combinations of the factor levels of the variables of interest, which apply to a subgroup of or even the entire study population. In contrast, this concept within the context of count response only pertains to a proper subpopulation. Further, if a subgroup has structural zeros in the current context, this subgroup is known, whereas the subgroup represented by the structural zero in the count response case is in general unobservable. Thus, unlike the analysis of count response, structural zeros can be safely removed without creating any bias in the estimates.

When cells with structural zeros are deleted, the resulting table often becomes *irregular*, i.e., nonrectangular. Concepts such as independence can become subtle for irregular tables. Thus, structural zeros deserve special attention for analysis of contingency tables.

7.4.1   Structural zeros in contingency tables

Structural zeros can occur when certain combinations of the variables do not represent meaningful characteristics for a subgroup of subjects. For example, consider a table involving gender and history of diabetes during pregnancy. If the variable for diabetes history during pregnancy has three levels, Yes, No, and N/A (not applicable), then no subject will be observed for the cells defined by level Male of gender, and Yes and No of the diabetes history variable. We call this type of structural zeros *inherent zeros*.

Another common situation involving inherent zeros is preference/comparison of different objects. For example, game results among some chess players can be summarized in a contingency table, with the row representing the winner and the column designating the loser. A cell count is the number of the games a player listed in the row wins over his/her opponent in the column. Since a player cannot play the game against himself/herself, the diagonals are structural zeros. For example, we may observe the following (hypothetical) table:

		Loser		
Winner	A	B	C	D
A	-	7	8	8
B	3	-	9	6
C	10	4	-	6
D	0	3	2	-

where structural zeros denoted by "-" represent N/A, and sampling zeros are left intact (e.g., 0 for the cell defined by row D and column A). In the table, A and C played $10 + 8 = 18$ games, with A winning 8 of the total games.

Structural zeros also often occur because of how the data are collected. All modern clinical trials have clear guidelines and strict criteria on the eligibility of subjects for the study trials. These inclusion/exclusion criteria stipulate the type of subjects who should be included/excluded, which are typically based on disease and demographic characteristics. For example, those who meet the exclusion criteria will not be eligible for participation in the study, although such subjects may well represent a sizable group in the population. As this subpopulation is completely excluded from the sampling frame because of study purposes, any inference and/or conclusion drawn from the study does not apply to this subgroup of the population. We call such structural zeros *excluding zeros*.

For example, if a study requires that a patient may participate only if he/she or his/her informant understands written English, then the cells corresponding to illiterate in English of both the proband and informant will have structural zeros, since by the study design such pairs should be excluded. In the Sexual Health study, as only adolescent girls are included, cells for other age groups all have structural zeros. As another example, if the proband/informant pair is excluded from the DDPC when both report that the proband has major depression, then we may have the following table for depression diagnoses:

| | Informant | | | |
Proband	No	MinD	MajD	Total
No	66	13	6	85
MinD	36	16	10	62
MajD	14	12	–	26
Total	116	41	16	173

In the above situations, the presence of structural zeros does not affect the distribution of other cells. However, structural zeros also arise if we force subjects who would fall in some cells to be redistributed to other cells. For example, suppose that a weight-loss program for overweight subjects would let participants finish the program only if they showed sufficient weight loss. More precisely, suppose that overweight subjects are grouped into two categories, OW-I (overweight) and OW-II (obese), and the participants can graduate only if they move over at least one category toward Normal weight (NW) at the end of the program. Then, if we use a two-way contingency table with the row (column) representing the weight level of a participant at the pre-(post-)program, the cells representing post-program weight gains such as (OW-I, OW-II) will have structural zeros. These *redistributing zeros* are the result of shifting the subjects that would have been in the structural-zero cells to the other cells that represent weight loss as required by the program policy. In this example, the cells corresponding to Normal pre-program weight will all have structural zeros (excluding type) because this is a program for overweight people.

The various types of structural zeros do carry different implications for analysis. Cells with inherent zeros do not represent a meaningful or logical outcome. Although cells associated with excluding zeros represent meaningful responses, the subjects that would have fallen into these cells are excluded for study purposes. Thus, we may still make valid inferences based on the observed data, so long as we do not extrapolate our findings to the subpopulation of such subjects. However, analysis of the resulting irregular tables may be a bit complicated, as all the methods discussed up to this point do not apply to nonrectangular tables.

We need to be more cautious with redistributing zeros. The cells for such structural zeros are created by redistributing the subjects that would have fallen into them to other cells as defined by the study purpose. For example, for the weight-loss example, we may observe Table 7.1(a) after one session of training. However, the subjects in and above the diagonals are required to continue training until eventually they get better. Thus, the 23 subjects who stayed at level OW-II during the first session must continue and/or work even harder to leave the program and will be redistributed to cells (OW-II, NW) or (OW-II, OW-I) if they succeed in the program. Likewise, the three subjects in (OW-I,OW-I) and six subjects in (OW-I, OW-II) will stay and eventually be redistributed to the other cells under the diagonal (if we assume every one will reach the goal at last). As a result, we may observe Table7.1 (b) instead. In practice, we must be mindful about redistribution zeros and make inference accordingly.

7.4.2   Models for irregular tables

Modeling tables with structural zeros can be tricky, because even fundamental concepts like independence may need special attention. For example, the row and column variables in a two-way contingency table will never be independent in the usual sense if there are

TABLE 7.1: Distribution of pre- and post-weight categories

	Post				Post		
Pre	**NW**	**OW-I**	**OW-II**	**Pre**	**NW**	**OW-I**	**OW-II**
NW	-	-	-	NW	-	-	-
OW-I	10	3	6	OW-I	19	-	-
OW-II	0	34	23	OW-II	3	54	-
(a). No graduation requirement.				(b). With graduation requirement			

structural zeros. Under independence, cell probabilities are the products of the marginal probabilities and hence are positive. It may still be reasonable to assess independence if the structural zeros are of the excluding type, since such cells would have positive counts if the study inclusion/exclusion criteria were not enforced. However, such zeros need to be excluded in the analysis since there is no subject in the sample falling into the corresponding cells. Hence, we may still apply the additive model:

$$\log \mu_{ij} = \lambda + \lambda_i^x + \lambda_j^y, \tag{7.27}$$

with the understanding that it does not apply to the cells defined by structural zeros.

In general, if (7.27) holds, the row and column variables are said to be *quasi-independent*. It is easy to check that for a quasi-independent irregular table, the row and column variables are independent in the usual sense if restricted to any rectangular subtable. Quasi-independence can be generally viewed as regular independence in cases where we are able to observe the excluded subjects in the structural-zero cells. This is valid for inherent and excluding zeros, since the distributions of other cells are not changed. However, we must be careful when dealing with redistributing zeros; for example, if the policy of the weight-loss program is changed, the distributions of the non structural-zero cells may also change.

The principle of inference about log-linear models stays the same for irregular tables. The only difference is that unlike random zeros, we do not include structural zeros in data analysis. Thus, it is important to distinguish structural zeros from their sampling counterparts, even at the data preparation stage. For example, for the hypothetical weight-loss program example discussed earlier, we may have a data set in which the row contains the record for each individual subject with two variables denoting the pre- and postprogram weight, respectively. We may aggregate the data manually to obtain a new data file by excluding structural zeros but including the random zeros. For example, in Table 7.1(a), the data set will consist of counts for cells with nonzero counts as well as counts for cells with random zeros. A record with count 0 will present for cell (OW-II, NW), but there will be no record for the cells with structural zeros (the three cells with structural zeros in the first row of Table 7.1(a)).

Once an appropriate analysis data file is created, inference for irregular tables using log-linear models follows the same procedure. Parameter estimates may be obtained by MLE and hypotheses of interest may be tested using appropriate linear contrasts or goodness-of-fit tests. However, some formulas often used for regular tables may not apply. For example, under the (quasi-) independence between the row and column variables for a two-way table, the cell probabilities are no longer the product of the marginal probabilities. Further, in terms of cell counts, the cell means are no longer estimated by the formula $\mu_{ij} = \frac{n_{i+}n_{+j}}{n}$.

Example 7.7

Let us check the quasi-independence between the proband and informant based on Table 2.9, after removing the proband/informant pairs when both reported that the proband had major depression. By applying the quasi-independence model in (7.27), we obtain statistics

12.5851 and 12.9852 with three degrees of freedom for the deviance and Pearson chi-square statistics. The p-values under the two goodness-of-fit tests are 0.0056 and 0.0047, respectively. Thus, the proband's and informant's "diagnoses" are associated.

Note that if the random zero was incorrectly removed, we may obtain incorrect conclusions (see Problem 7.29). ▯

7.4.3 Bradley–Terry model

Pairwise comparisons are commonly used to facilitate studies on preference over different objects or assessing strengths of different subjects in a contest. For example, chess players play games in pairs to determine the best player. Here, it is impossible to have more than two players playing a single game. In other situations where multiple subjects or objects can be compared simultaneously, such as ranking preferences over different fruits, it may still be more convenient to make pairwise comparisons. In all such instances, a common task is to rank the subjects (objects) by strengths (preferences) based on the pairwise comparisons. For example, we may rank chess players according to their performances in all pairwise competitions during the tournament. Bradley and Terry studied this class of problems in the 1950s with a series of publications (Bradley and Terry, 1952, Bradley, 1954, 1955).

Suppose there are m players participating in a series of two-player games, with no ties in their competitions. In this scenario, the game results may be summarized in an $m \times m$ table, with the count n_{ij} of the (i, j) cell representing the number of winning games of the ith player over the jth competitor. As no one will play the game against himself/herself, the cells on the diagonal of the table all have structural zeros. When restricted to a given pair (i, j) of players, the data are binomial; out of a total of $n_{ij} + n_{ji}$ games, the ith player won n_{ij} games over and lost n_{ji} games to the jth player. We are interested in the probability π_{ij} that the ith player wins over the jth opponent. Based on the observed binomial data, it is straightforward to obtain an estimate, $\widehat{\pi}_{ij} = \frac{n_{ij}}{n_{ij}+n_{ji}}$. However, we may not be able to rank all the players according to these estimated probabilities; it is possible that player A is more likely to win over player B, player B is more likely to win over player C, and player C is more likely to win over player A in a pairwise competition. It is also possible that two players have never played with each other, but we would still like to rank them. The key idea of the Bradley–Terry model is to assume that there is a latent strength scale for each player, which may be applied to rank the players.

Let λ_i denote the score of the latent strength scale for the ith player. The Bradley–Terry model assumes that

$$\log \frac{\pi_{ij}}{1 - \pi_{ij}} = \lambda_i - \lambda_j, \quad 1 \le i, j \le m, \tag{7.28}$$

where π_{ij} is the likelihood of the ith wins over the jth player. Thus, the difference between the two scores $\lambda_i - \lambda_j$ is the log odds of winning by player i over player j. If $\lambda_i > \lambda_j$, player i is more likely to win when playing with player j. Hence, we may order the players according to their strength scores.

The logistic model (7.28) is a special case of the conditional logistic models (4.29) for matched pair outcomes; each game represents one matched pair. To implement the model, we may define one variable for each of the players, say x_k for the kth player, $k = 1, ..., m$. For a game played between players i and j, we generate a pair of observations, one for each of the two player. For the ith player, the covariates are $x_i = 1$, and $x_k = 0$ for all $k \ne i$. Similarly, for the jth player, the covariates are $x_j = 1$, and $x_k = 0$ for all $k \ne j$. Then $\mathbf{d}_i$ in (4.29) is the difference of the covariates between the winner and loser. For example, if the ith player won over the jth player in this game, then $x_i = 1$, $x_j = -1$, $x_k = 0$ all $k \ne i$. Since $\sum_{k=1}^m x_k = 0$, we need to set the coefficient for one of the players to be zero (reference). The estimated coefficients for other players are then the relative strengths for those players relative to the reference player.

We can also fit the data with the log-linear model, which again requires some rearrangements of the data. The expected count of the ith player's wins over the jth player does not only depend on their respective strength scores λ_i and λ_j, but also the total number of games played between them. In other words, the loglinear model for the count outcome of a player winning over another needs to control the total number of games played between them. We may generate a new table with one column for each pair of players, if they played with each other. More precisely, consider a two-way $m \times n$ table, with its rows identifying each individual player, the columns representing all distinct pairs of players who played with each other (thus $n \leq \frac{1}{2}m(m-1)$), and the entry being the number of the player wins against the other player who shares the column. To distinguish this new table from the original $m \times m$ table above, we use l and k to index its rows and columns. For each column, the cell count in the jth row either represents the number of winning games by the jth player over the other opponent sharing the column or a structural zero, depending on whether the jth player is part of the pair. With this setup, the log-linear model for the pairwise comparison can be expressed as

$$\log \mu_{lk} = \lambda + \lambda_l^{player} + \lambda_k^{games}, \quad 1 \leq l \leq m, \quad 1 \leq k \leq n. \tag{7.29}$$

We may then order the players based on the estimated parameters λ_l^{player}.

Example 7.8

Now let's apply the Bradley–Terry model to the games among the four players presented in Section 7.4.1. By redisplaying the original 4×4 table using the above format, we obtain

	Games					
Winner	AB	AC	AD	BC	BD	CD
A	7	8	8	-	-	-
B	3	-	-	9	6	-
C	-	10	-	4	-	6
D	-	-	0	-	3	2

The games between the two players in a pair are grouped into a column, with the rows designating the winners. The count for the (l, k) cell represents the games won by the lth player over the opponent defined in the kth column. For example, the third column records the information about the games between players A and D. The cell count 0 in the last row indicates that player D did not win any game against A (random zero), and the cell count 8 in the first row shows that A won 8 games over D. Since other players couldn't participate in the games between A and D, all the other entries in the column are structural zeros denoted by "-". The log-linear model (7.29) can then be applied to the table.

Based on both the logistic (7.28) and the log-linear (7.29) models, we obtain that, if D is set as the reference, the strength scores for players A, B, and C are 1.6157, 1.3122, and 1.2624, respectively. Thus, the order based on the estimated parameters suggest that we may rank the players as A > B > C > D, where " > " indicates "stronger than". Thus, the Bradley–Terry model provide a convenient approach to rank subjects. Based on the scores, we can make predictions for games between two players. For example, the estimated probability of A wins over B is $\frac{\exp(1.6157-1.3122)}{1+\exp(1.6157-1.3122)} = 0.5753$.

The overall test of if all the scores are zero has p-value 0.0147, indicating that the players have different skill levels. If the model does not fit the data well, then the assumption of a one-dimensional latent strength score under the Bradley–Terry model may not be correct.

The goodness-of-fit deviance statistic of the Bradley–Terry model (7.29) for this example is 7.5287 with df = 3, and p-value = 0.0568. Thus, the model assumption may be problematic. However, the model may still be applied if we must rank the subjects. ⬚

Note that we only discussed the simplest case which assumes that the probability of the outcome for two players are the same. More complicated models considering factors such as hosting advantage have been developed. Interested readers may consult, e.g., Bradley and Terry (1952), Bradley (1954, 1955).

7.5   Model Selection

In the previous chapters, we have largely focused on building models to test some specific hypotheses. However, there are also situations where we have limited knowledge about the data, and want to find a model that summarizes the data well. In such cases, the unidirectional significance test driven by some specific hypothesis is not of much help, and a bidirectional dynamic process between model evaluation and refinement must be employed to select the best model among competing alternatives. In this section, we first describe some common criteria for model evaluation, followed by a discussion on procedures using such model selection criteria. We conclude this section by discussing an important class of models for contingency tables, graphical models, to help with model selection and interpretation. Restricting to such graphical models may significantly reduce the number of potential candidates for consideration.

7.5.1   Model evaluation

Goodness of fit is an obvious criterion for model evaluation, and has been used for model assessment. After all, we may not feel comfortable to use a model if it does not fit the data well. However, goodness of fit alone may not be sufficient as a criterion for model selection. For example, when comparing two nested models, goodness of fit will always favor the broader and more complex one, since it fits the data better according to this criterion. Thus, model selection based on such a criterion may result in models that *overfit* the data by paying too much attention to the noise. Such models are usually unnecessarily complicated, hard to interpret, and not very useful for prediction purposes. For example, the saturated model is always the best under this criterion. However, since no data reduction is achieved, it is hardly a useful model for describing the data. Further, as power decreases as the number of parameters grows, we may fail to find any significant association when applying overfitted models.

In general, the goal of model selection is to find a comparatively simple model that adequately represents the data. This is the *principle of parsimony* (Box et al., 2008). In this section, we introduce some information-based model selection criteria by taking both goodness-of-fit and model complexities into consideration. For space considerations, our discussion focuses on log-linear models within the current context. More comprehensive treatments of this topic can be found in Burnham and Anderson (2002) and Konishi and Kitagawa (2008). We emphasis that although the techniques described in this section are very helpful in seeking the appropriate models, knowledge of the subject matter may play an even more important role in model selection.

The problem of using estimated likelihood, the value of the likelihood function at the MLE, is that a broader model will always have at least as high a likelihood value as its less complex counterparts, since the MLE is determined over a wider range. Furthermore, as shown by Akaike, the likelihood at the MLE is an upwardly biased estimate of the true likelihood, and the asymptotic bias equals the number of parameters in the model. Thus, as broader models contain more parameters, the larger bias in the likelihood value for such models further exacerbates the problem.

Since this bias is characterized by the number of parameters, we can develop a bias-corrected, likelihood-based criterion for comparing nested models by subtracting off the bias, i.e.,

$$AIC = 2k - 2l\left(\widehat{\theta}\right).$$

The above is called the *Akaike's information criterion* (AIC). Note that as the (log) likelihood above has a negative sign, smaller values of AIC correspond to better models. The AIC takes both the goodness-of-fit and model complexity into consideration and enables us to compare two models, nested or not.

The AIC is *minimax-rate optimal* in the sense that if the true distribution is not among any of the candidate families, the average squared error of the selected model based on AIC is the smallest possible offered by the candidate models asymptotically. A major downside of the AIC is that it is not a *consistent* selection criterion in that its probability of selecting the true model among the candidates does not approach one as the sample size goes to infinity. A popular consistent selection criterion is the Bayesian information criterion (BIC):

$$BIC = 2k - 2\ln(n)l\left(\widehat{\theta}\right).$$

The above is based on the same principle and is also called the *Schwarz information criteria*, in tribute to Schwarz who first developed this model selection index. Although consistent, the BIC is not minimax-rate optimal. In fact, Yang (2005) has proven that a selection criterion cannot be both consistent and minimax-rate optimal. There are many other criteria proposed for model selection, and interested readers may check Burnham and Anderson (2002) and Konishi and Kitagawa (2008) for details.

It is important to point out that it is not the exact value of the AIC or BIC that is of interest, but rather the change of the index across the different models that is informative for ranking models to select the best among the competing alternatives. Further, the comparison of the AIC and BIC is only sensible when the models are applied to the same data set.

7.5.2   Stepwise selection

In practice, the stepwise procedure is perhaps the most popular. It is generally not practical to write down all the possible models. For example, for high-dimensional contingency tables, the number of possible models increases exponentially with the number of categorical variables. When there are continuous variables, the number can be even infinite since higher powers of such variables may also be used in building the model. In practice, a popular procedure, called *stepwise model selection*, is typically employed to make the task manageable. Under this procedure, models are dynamically selected using the *forward selection* and *backward elimination* techniques or a combination. In forward selection, one initiates with a simple model and then adds additional terms to improve the model fit. In contrast, with backward elimination, one starts with a complex model and then tries to simplify it by eliminating redundant terms. In stepwise selection procedures, only a small percentage of possible models are examined, and as such it is quite possible that the best model may not be among the ones considered.

7.5.2.1   Forward selection

In forward selection, we start with a simple model and then add additional variables, albeit one at a time. The beginning model may include only the key variables and some of their interactions that are to be kept in the final model based upon some prior knowledge and/or initial criteria. Note that the initial model may simply contain the intercept term if no such information or criterion exists. To help inform about the variables to be added in the selection process, we may perform at the outset a series of univariate analysis to determine which variables may significantly improve model fit. Because of the exploratory nature of model selection, it is more important to cover than to miss all potentially informative variables by using lenient inclusion criteria, especially at the beginning of the model-building process.

At each step of forward selection, the candidate variables not yet in the model will be compared for selection. For each term (a candidate variable or an interaction) under consideration, we add it to the model, refit the model, and decide whether to accept the revised model. For example, suppose we are interested in studying the association among three factors x, y, and z. Also, suppose we start with a model with only all the main effects. To include additional terms, we may first consider the three two-way interactions, $x \times y$, $x \times z$, and $y \times z$, as candidates for inclusion. Upon fitting the model containing $x \times y$, or $x \times z$, or $y \times z$, in addition to the main effects, we compare the p-values corresponding to the term $x \times y$ in the first, $x \times z$ in the second, and $y \times z$ in the third model. If all the p-values are above the preset level, we stop the selection process and keep the original main effect model.

If the minimum p-value falls below some preset significance level, we select the term with the minimum p-value. We then revise the model by adding the interaction selected, say $x \times y$, and repeat the process with the unselected interactions, $x \times z$ and $y \times z$. After finishing the examination of the two-way interactions, we may consider the three-way interaction $x \times y \times z$. If it is significant, it will be added. Note that once $x \times y \times z$ is added, we arrive at the saturated model. This is the most complex model that can be built with the three factors in this particular example, as there is no parameter left for further improvement of model fit.

In addition to testing the significance of the term added at each step, we may also use goodness-of-fit tests to see whether the revised model significantly improves the model fit. The starting model, such as the one containing only the main factor effects, may not fit the data well, but as more terms are added, the model fit will improve. The selection process continues until either no significant improvement is obtained or the saturated model is reached.

Note that as multiple tests are carried out at each step, the p-value for the significance of a term in the model does not hold the usual meaning of type I error, i.e., the probability for the null hypothesis that the term is truly redundant. For this reason, it is common to set the preset significance level higher than the conventional 0.01 or 0.05. For example, we may use 0.1 or 0.2 as the threshold for selecting variables. Note also that in typical regression analysis, a primary objective is to study the relationship between a set of predictors/covariates and some response. In this case, we may be mainly concerned about the main effects. However, within the present context, our interest in model selection is to investigate association among the variables. To this end, not only the main effects as in standard regression analysis, but two-way and higher-order interactions are of interest as well, as the latter correspond to meaningful hypotheses concerning the relationship across the different factors.

For high-dimensional contingency tables, the total number of potential models can be very large, and the step-wise model selection procedure described above may not be practicable for such studies. One way to deal with this is to restrict efforts to a subset of models. In this regard, we may follow the *hierarchy principle*, which stipulates that if an interaction term of some variables is present, then all lower-order interactions among these variables should also be included. We can easily represent such *hierarchical* models by *terminal* interaction terms, i.e., interactions that are not embedded in any higher-order interaction in the models. For example, we may denote the log-linear model by $[xyz][zw]$:

$$\log \mu_{klmn} = \lambda + \lambda_k^x + \lambda_l^y + \lambda_m^z + \lambda_{kl}^{xy} + \lambda_{lm}^{yz} + \lambda_{km}^{xz} + \lambda_{klm}^{xyz} + \lambda_n^w + \lambda_{mn}^{zw}.$$

This model includes two terminal interactions, $[xyz]$ and $[zw]$, as well as all associated lower-order interactions.

The step-wise model selection can be carried out on the hierarchical models. For example, beginning with the additive model, we may next check whether we need to add some interaction terms. If the variables have more than two levels, the interactions may involve several parameters.

Example 7.9

Consider the relationship among gender (g), three-level marital status (m), two-level education (e), three-level depression diagnosis (d), and two-level medical burden (c) for the DOS. Since we know that depression is associated with gender, and the association is different across the different levels of marital status, we include the three-way interaction of depression, gender, and marital status in the initial log-linear model. We restrict our attention to hierarchical models and thus include all the two-way interactions among the three variables.

In the first step of our forward selection process, candidates for inclusion are interactions between the remaining variables, education and medical burden, as well as their interactions with gender, marital status, and depression. The p-values are summarized in the following table.

Term added	$e \times c$	$c \times g$	$c \times m$	$c \times d$	$e \times d$	$e \times m$	$e \times g$
p-value	0.0133	0.9505	0.0021	0.0019	0.0148	< 0.0001	0.0003

Having the smallest p-value, the interaction between education and gender will first be added. We repeat the procedure based on the revised model with this interaction added. We leave the completion of the procedure as an exercise. $\Box$

7.5.2.2 Backward elimination

In backward selection, we start with a complex model containing all variables of interest. This initial model generally overfits the data, and the objective of backward selection is to trim the model by eliminating the redundant terms one at a time. In addition to the experience of the investigator, exploratory tools such as plotting, univariate analyses, and smoothing methods may be applied to help decide which variables and interactions are to be included in the initial model. For example, we may perform a series of univariate analyses relating the response and each candidate variable to determine the set of variables to include in the starting model. Again, as the level of significance in each univariate model no longer has the usual interpretation because of the multiple analyses performed, more lenient threshold levels such as 0.1 and 0.2 may be used as the inclusion criteria for the initial model. After all, redundant terms will be removed during the backward selection process, if they do not significantly contribute to the final model.

When modeling contingency tables, we can always start the process with the saturated model, as it fits the data perfectly. However, inference may not be reliable if there are too many parameters present in the model. So, depending on the sample size, we may want to start with a more parsimonious alternative.

Regardless of which model to use at the start-up, we begin eliminating terms from the initial model one at a time. In each step along the way, we delete the least significant term, i.e., the term with the highest p-value among the terms being considered for removal, as long as it is above the preset critical level. The elimination process continues until no term in the model meets the elimination criteria. We may also stop the procedure if further trimming significantly degrades the model fit. Further, a term will not be considered for removal at a step if any of its associated higher-order interactions is still in the model. For example, for a model containing both $x \times y$ and $x \times y \times z$, the lower-order interaction $x \times y$ will not be considered for removal until after the associated higher-order term $x \times y \times z$ is purged, because of the difficulty in interpreting the latter in the absence of the former. This is de facto the hierarchy principle applied in the current backward elimination context.

Example 7.10

As an illustrative example, we apply the backward elimination procedure to Example 7.9. Starting from the saturated model, we trimmed the model stepwise, beginning with the least significant term. At each step, we check the terminal terms in the current model, i.e., terms not embedded in any higher-order interaction in the model. For example, when beginning the process, we check the interaction of all the five variables, since all others are lower-order terms in comparison. The following table is a summary of the elimination process.

Step	p-value	Pearson χ^2 / DF	AIC	BIC	Term to be Removed
1	0.9673	0	395.2	559.2	$g \times m \times e \times c \times d$
2	0.9673	0.0921	387.8	542.6	$g \times m \times e \times c$
3	0.5059	0.0664	383.9	534.1	$c \times g \times m \times d$
4	0.6325	0.3143	379.2	520.3	$c \times g \times d \times e$
5	0.5272	0.3166	376.1	512.7	$c \times m \times d \times e$
6	0.4902	0.4151	371.3	498.8	$c \times m \times e$
7	0.3517	0.4487	368.7	491.7	$c \times m \times d$
8	0.3224	0.5727	365.1	479.0	$c \times d \times e$
9	0.2897	0.6223	363.4	472.7	$c \times e \times g$
10	0.1651	0.6425	362.5	469.5	$c \times g \times m$
11	0.0898	0.7206	362.1	464.6	$c \times e$

After eliminating the terms according to the steps in the table from the initial saturated model, the final model based on AIC is $[gmde][cdg][cm]$ when expressed using terminal terms. If 0.15 is used as the threshold for the p-values in the selection procedure, the final model would be $[gmde][cdg][cm][ce]$, since the term $c \times e$ in Step 11 has a p-value less than 0.15. □

It may not be easy to find a nonsaturated model to start the backward elimination procedure. Likewise, for forward selection, adding new variables may cause some variables already in the model to become redundant (nonsignificant). In practice, we may combine the two to allow for refining the model in both directions to take full advantage of the benefits of the two procedures.

7.5.3   Graphical models

When there are many factors under consideration, we may significantly reduce the number of potential models by restricting ourselves to graphical models, which can be easily represent by graphs. By representing factors as points in the diagram, we can denote an interaction between two factors by a edge between the two corresponding points. For example, the following graph shows the two-way interactions in the model $[gmde][cdg][cm]$. However, a model in general is not uniquely determined by its two-way interactions. For example, the model represented by the terminal terms, $[edgm][cdgm]$, have exactly the same two-way interactions as the model $[gmde][cdg][cm]$. However, the two models are not the same, as $[edgm][cdgm]$ contains additional interactions not present in $[gmde][cdg][cm]$ such as $[cdm]$, $[cgm]$, and $[cdgm]$.

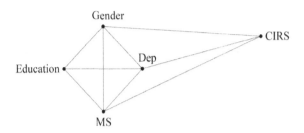

By definition, graphical models refer to the most complicated models involving two-way interactions. In other words, if a graphical model contains all two-way interactions among some variables, then it will also contain all possible (higher-order) interactions among these variables. Thus, the model in Example 7.10 is not a graphical model, since it does not include any three-way interactions among depression, marital status, and CIRS, despite the fact that it contains all possible two-way interactions. A graphical model containing all two-way interactions among the points, e, d, g, and m, and those among c, d, g, and m, will also contain the four-way interactions $edgm$ and $cdgm$ (and all the lower order interactions contained in them). It is easy to check that $[edgm][cdgm]$ is a graphical model represented by the above graph (see Problem 7.31).

Graphical models are important for contingency table data analysis, mainly because they are usually easy to interpret in terms of conditional independence. The absence of an edge between two nodes in the graph indicates the independence of the two variables conditional on the other factors. For example, the graph of the log-linear model in (7.21) is

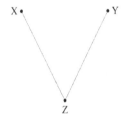

The absence of the edge between x and y indicates that they are independent, conditioning on the other variable z. Also, in the final model of Example 7.10, there is no edge between education and medical burden. Thus, education and medical burden are independent, conditional on the other variables, gender, depression, and marital status.

More generally, consider three sets of variables, S_1, S_2, and S_3. The two sets S_1 and S_2 are said to be *separated* by S_3, if there is no edge connecting S_1 and S_2 once all edges with one of the nodes in S_3 are removed. If S_1 and S_2 are separated by S_3, then variables in S_1 are all independent with those in S_2, conditional on the variables in S_3.

By limiting our attention to graphical models, we may significantly reduce the number of models to be compared in model selection. For example, when we model n-way contingency tables, there is only one graphical model that includes all the two-way interactions; however, there are at least $2^{\binom{n}{3}}$ hierarchical models (see Problem 7.32). See Edwards and Kreiner (1983) for an in-depth discussion of model selection using graphical models, and Edwards (2000) and Lauritzen (1996) for a more comprehensive treatment of graphical models.

Exercises

7.1 Prove (7.1).

7.2 Show that the log-likelihood function in (7.3) can be maximized by maximizing the first term as a function of β and the second term as a function of τ separately, and find the MLE of τ.

7.3 Suppose that $\{\mu_{ij}\}$ satisfy a multiplicative model

$$\mu_{ij} = \mu \alpha_i \beta_j, \quad 1 \le i \le I, 1 \le j \le J, \tag{7.30}$$

where $\{\alpha_i, i = 1,\dots,I\}$ and $\{\beta_j, j = 1,\dots,J\}$ are positive numbers satisfying the constraint

$$\sum_{i=1}^{I} \alpha_i = \sum_{j=1}^{J} \beta_j = 1.$$

(a) Compute the multinomial distribution conditional on $\sum_{i=1}^{I}\sum_{j=1}^{J} n_{ij} = n$, and verify that α_i and β_j are actually the marginal probabilities of the row and column variables of the two-way contingency table, respectively.

(b) Prove that the row and column variables are independent.

7.4 Verify that the log-likelihood of model (7.7) is $\sum_{i=1}^{k} [n_i \lambda - \exp(\lambda)]$.

(a) Compute MLE of λ.

(b) Compute the Pearson chi-square statistic and compare it with (2.7).

7.5 Redo Example 2.1 using log-linear models.

7.6 Use loglinear models to test the hypothesis concerning the distribution of depression diagnosis in Example 2.2.

7.7 For the DOS, use the three-level depression diagnosis and variable MS for marital status as defined in Section 4.2.2 to test for uniform association, assuming that both the depression diagnosis and MS are ordered according their defined values.

7.8 In Example 7.3, we assessed the uniform association between the diagnosis of probands and informants. Perform the following alternative approaches for assessing uniform association and compare the results with that of Example 7.3.

(a) Fit a saturated log-linear model and test whether the association is uniform.

(b) Treating one of row and column variables as the response and the other as the predictor, apply an appropriate generalized (or adjacent) logistic model to assess the uniform association between the two variables.

7.9 Prove that model (7.15) is indeed the model for the row-effect association.

7.10 For the DOS, use the three-level depression diagnosis and variable MS for marital status as defined in Section 4.2.2. If MS is not ordered, we may not be able to talk about the uniform association. However, we may assess MS-effect association, the row-effect association when MS is treated as the row variable, as the depression diagnosis is ordered.

(a) Use goodness-of-fit tests to assess if the row-effect model is acceptable.

(b) Use the saturate model to assess if the row-effect model is acceptable.

(c) Use a generalized (or adjacent) logistic model to assess if the row-effect model is acceptable.

7.11 Redo Example 2.12 using log-linear models.

7.12 A saturated model for contingency tables may be easily specified using cell-means coding. For example, the saturated model for a two-way $I \times J$ contingency table may be simply specified as

$$\log (\mu_{ij} | X = i, Y = j) = \sum_{i=1}^{I} \sum_{j=1}^{J} \beta_{ij}, \quad 1 \le i \le I, \quad 1 \le j \le J. \qquad (7.31)$$

It follows that $\log (\mu_{ij}) = \beta_{ij},$, $1 \le i \le I$, $1 \le j \le J$. Using the cell-means coding, test the distribution symmetry and homogeneity for the contingency table in Example 7.4 (a linear link function may be used in place of the log link in (7.31) for testing the homogeneity of marginal distributions.

7.13 Each of the three random variables x, y, and z has two levels: 0 and 1. The joint distribution of these three variables can be determined from the facts $\Pr(x = 0, y = 0, z = 0) = \frac{1}{4}$, $\Pr(x = 0, y = 1, z = 1) = \frac{1}{4}$, $\Pr(x = 1, y = 0, z = 1) = \frac{1}{4}$, and $\Pr(x = 1, y = 1, z = 0) = \frac{1}{4}$.

(a) Are x, y, and z mutually independent?

(b) Are x and y marginally independent?

(c) Are x and y independent given z?

(d) Is x jointly independent of y and z?

7.14 Prove that under the mutual independence log-linear model (7.19), the three variables are indeed mutually independent.

7.15 Prove that three variables being mutually independent implies that any two of them are marginally independent, conditionally independent, and any one of them is jointly independent with the others.

7.16 Prove that if x is jointly independent with y and z, then x and y are marginally independent.

7.17 Prove that if x is jointly independent with y and z, then x and y are conditionally independent.

7.18 To obtain the log-linear models for association homogeneity, we need the following two key facts:

    (a) Prove (7.25).

    (b) Prove that $\lambda_{i'jk}^{xyz} + \lambda_{ij'k}^{xyz} = \lambda_{ijk}^{xyz} + \lambda_{i'j'k}^{xyz}$ for all i, i', j, j', and k implies $\lambda_{ijk}^{xyz} = 0$ for all i, j, k, i.e., no three-way interaction.

7.19 Verify that under the mutual independent log-linear model (7.23), the variable y is independent with the other two.

7.20 Apply the log-linear models to Example 3.4.

7.21 Verify the numbers of free parameters in the model (7.17).

7.22 Write down the log-linear model for quasi-symmetry, and count the number of free parameters in the model.

7.23 Prove under the paradigm of a multinomial distribution that if x and y are homogeneously associated, then y and z as well as x and z are also homogeneously associated.

7.24 Prove that $\exp(\lambda_{111}^{xyz}) = \frac{\pi_{2,2,2}/\pi_{1,2,2}}{\pi_{2,1,2}/\pi_{1,1,2}} / \frac{\pi_{2,2,1}/\pi_{1,2,1}}{\pi_{2,1,1}/\pi_{1,1,1}}$ for a $2 \times 2 \times 2$ three-way table.

7.25 For the DOS, use the three-level depression diagnosis.

    (a) Use the Poisson log-linear model to test whether depression and gender are independent.

    (b) Use methods for contingency tables studied in Chapter 2 to test the independency between depression and gender.

    (c) Compare parts (a) and (b), and describe your findings.

7.26 For the DOS, use the three-level depression diagnosis and variable MS for marital status as defined in Section 4.2.2 to test

    (a) whether depression, gender, and MS are mutually independent;

    (b) whether depression is independent of MS given gender;

    (c) whether depression is jointly independent of gender and MS.

7.27 Suppose that x and y are conditionally independent given z, and x and z are marginally independent.

    (a) Show that x is jointly independent of y and z.

    (b) Show x and y are marginally independent.

    (c) Show that if x and z are conditionally (rather than marginally) independent, then x and y are still marginally independent.

    (d) Explain that although x and y are conditionally independent given z, they are not necessarily marginally independent. How is this fact related to Simpson's paradox? For a more interesting discussion of this fact, see the paper by Samuels (1993).

7.28 Use the log-linear model to test whether SCID (two levels: no depression and depressed including major and minor depression) and dichotomized EPDS (EPDS $\leq$ 9 and EPDS $>$ 9) are homogeneously associated across the three age groups in the PPD.

7.29 Check that for Example 7.7, you may obtain different (incorrect) results if the random zero is removed from the data set for data analysis.

7.30 Complete the forward model selection in Example 7.9 and compare it with the models selected in Examples 7.10.

7.31 Check that [edgm][cdgm] is a graphical model.

7.32 Check that there are at least $2^{\binom{n}{3}}$ different hierarchical models which contain all two-way interaction terms for an n-way contingency table.

Chapter 8

Analyses of Discrete Survival Time

Survival data analysis is widely used in research studies and investigations when the outcome is time to occurrence of some event of interest. In many applications, especially in studies involving seriously ill patients such as those with cancers and cardiovascular and infectious diseases, we are interested in the patients' survival times (from onset of disease to death). For this reason, such data are often called *survival data*, and the field of study of this type of data is known as *survival analysis*. However, the events of interest are not necessarily negative in nature like death and may instead be comprised of a diverse range of outcomes both good and bad. For example, in a weightloss program, it may be of interest to study the length of time for an overweight person to reduce weight to a desired level. The survival methodology may even be applicable to outcomes not involving time, but which share the properties of time, such as space. Although survival analysis in most applications involves continuous survival times, discrete survival times also arise frequently in practice. We focus on discrete survival times in this chapter.

Survival data present some unique features and issues which we have not considered in the preceding chapters. Understanding and addressing these distinctive features of survival data are crucial for modeling such data. In Section 8.1, we describe the unique features of survival data. In Section 8.2, we discuss models for survival data assuming a homogeneous sample. In Section 8.3, we discuss regression analysis to accommodate covariates.

8.1 Special Features of Survival Data

A survival outcome, or survival time, is the duration from a starting point to the occurrence of an event. For a well-defined survival outcome, we must have both a clear starting point as well as event of interest. Depending on the study purpose, various survival outcomes may be defined. For example, in cancer studies one may be interested in studying the time from different starting points, such as onset, diagnosis, and treatment of a type of cancer, to death due to the cancer. In this example, we have three different starting points, and thus we may accordingly have three different survival times. The event is death due to the cancer, so death due to other reasons such as a car accident will not be counted as the event. For those subjects, we will not be able to observe the event of interest. This is an example of *censoring*, a common phenomenon in survival data analysis. We will discuss it in more detail shortly.

Survival analysis can also be applied to recurrent events such as heart attacks and depression. For example, we may be interested in the time from discharge from a hospital to the first heart attack afterwards. In such applications, it is not sufficient to observe a heart attack; it has to be the first heart attack since the discharge from the hospital. For recurrent events, various survival times can be defined. For example, in the heart attack case, we may be interested in the time from the first heart attack to the second (recurrence), the time

DOI: 10.1201/9781003109815-8

from the first to the third heart attack (second recurrence), or the time from the second
heart attack to the third occurrence (first inter-recurrence time). Thus, it is important to
identify the event of interest, such as the first or second recurrence in such applications,
before applying survival analysis. This often requires following up with the subjects from
the starting point through the event; if there is any lapse in observation during the follow-up
period, then the survival time based only on the observed period may be invalid, as events
may happen during the lapsed period(s) when the subjects are not observed.

Survival data call for special statistical analysis tools because of their special features.
The most notable phenomenon in survival analysis is censoring. It is common that the time
to the occurrence of some event of interest is not observed for some of the subjects, in which
case standard methods such as those discussed in the preceding chapters are not applicable.
Another related issue is *truncation*, which is particularly common in observational studies,
and also threatens the validity of the standard methods when applied to survival data.
Because of these unique features, survival data are more effectively described and modeled
by a new set of concepts and parameters, which are introduced later in this section.

8.1.1   Censoring

For a survival outcome, it is necessary to know both the starting time and the time of the
event. For a prospective study, in order to observe the occurrence of the event of interest,
the time frame of the observation must be sufficiently long to contain the time when the
event occurs. However, it is generally not possible to have such an extended time frame to
observe the events for all the subjects, primarily due to logistics and cost considerations.
For example, many clinical trials last two to five years because of considerations of logistics,
cost constraints, and advances in the knowledge and availability of new medication and
treatments. If the event does not occur when the study is terminated, then the survival time
is not observed and it is *censored*. Censoring may also occur, if there are events that prevent
the occurrence of the event of interest from happening. For example, death of a patient from
a car accident would make the observation of death due to the cancer impossible, no matter
how long the study may last. If a subject withdraws from the study before the event occurs,
we may also not observe the event. This is common in modern clinical trials (see Chapter
9).

The occurrence of an event of interest is often called a *failure* in the nomenclature of
survival analysis. This is because the issue of censoring was tackled initially in the analysis
of life-testing data to determine the life expectancy of certain objects such as lightbulbs.
Because the life of such objects may be longer than the observation time, even under unusual
and extreme conditions to accelerate the failure of the object, not all objects will fail at
the end of life testing, yielding censored failure times. Since the failure is not observed, the
exact survival time is unknown. However, partial information about the survival time is
available; it is longer than the duration from the starting point until the last observed time.
This is commonly referred to as *right censoring*, because the failure can only happen after
the last observed time.

If the duration of a study is fixed in advance, such as in most clinical trials, this kind of
censoring is called *Type I censoring*. Otherwise, we may have *Type II censoring*, which is
sometimes employed to generate a sufficient number of events, which can be an important
consideration in the study of rare diseases and long survival times. Under Type II censoring,
the study termination time is not fixed a priori, but rather depends on whether the number
of failures reaches a threshold level predetermined to ensure sufficient power in data analysis.
For example, a study may be designed to stop after, say, 10% of the subjects develop the
events. Under such a study design, the censoring of a subject depends on the survival times
of other subjects, invalidating the usual independence assumption and making the analysis

much more difficult. Such a dependence structure does not arise under Type I censoring.

Censoring may also arise from other situations which are not caused by a limited follow-up time as discussed above. For example, in AIDS/HIV research, it is often difficult to know the exact time when the HIV infection occurs for an AIDS patient. If a person tests positive, we only know that the time of infection must have occurred before testing. Since the event of HIV infection occurs before the censoring (testing) time, the survival time is shorter than the duration from the starting point to the testing time. We call this *left censoring*, as opposed to right censoring as discussed above when the event of interest occurs after it is censored. *Interval censoring* also occurs in some applications, in which the occurrence of the event of interest is only known to be sandwiched between two observed time points. For example, in AIDS/HIV research, the infection time of HIV for hemophiliac patients is often determined by testing the blood samples from the patient over a period of time, with the infection time censored in an interval defined by the last negative and first positive test. Of all these, right censoring is by far the most common, and we focus on this popular censoring mechanism in this chapter.

8.1.2 Truncation

Another issue arising in the analysis of time to event data is *truncation*. Under truncation, only a portion of the study population is samplable. For example, in the early years of the AIDS epidemic, interest was centered on estimating the latency distribution between HIV infection and AIDS onset. Data from the Centers for Disease Control and Prevention (CDC) and other local (state health departments) surveillance systems are used for this purpose. Since the time of HIV infection is usually unknown due to the lack of screening for HIV during this period, only those of the infected individuals who come down with AIDS symptoms are captured by the surveillance system. Because of the long duration of the latency period (the mean is about 10 years), and the relatively short time span covered by the surveillance database, the AIDS subjects in the surveillance systems during the early years represent a sample from the subgroup of the patients' population whose latency times fall within the surveillance time frame.

Shown in Figure 1.1 is a diagram illustrating the truncation arising from the above example. The surveillance system pictured has a time frame between 0 and M, where 0 denotes the time of the earliest HIV infection case and M designates the length of the observation period determined by the time when the analysis is performed. All HIV-infected individuals with a latency less than M, such as the case depicted, are captured, but those with a latency longer than M, such as the one case shown in the diagram, will be missed, or *right truncated*. If $f(t)$ and $F(t)$ are the PDF and CDF of the latency distribution, we can estimate each only over the interval $[0, M]$, i.e.,

$$f_T(t) = \frac{f(t)}{F(M)}, \quad F_T(t) = \frac{F(t)}{F(M)}, \quad 0 \le t \le M.$$

If $F(M) < 1$, i.e., then M is less than the longest latency, and only $1 - F(M)$ proportion of the HIV-infected population will be captured by the surveillance system, implying that the reported AIDS cases in the database underestimate the true scale of the AIDS epidemic.

Under right truncation, what is missing is the subject, not just the value of the outcome (failure time) as in the case of censoring, thereby restricting the inference to the observable proportion of the study population. Within the context of AIDS/HIV surveillance, a major ramification is the underestimation of the scale of the epidemic.

To further clarify the difference between censoring and truncation, consider a race with 10 athletes. If we stand at the starting line, we see all 10 athletes start the race. Suppose that only the times for the first three crossing the finish line are announced. Then, the finishing times for the other seven are right-censored. Now, suppose that we stand at the finish line and do not know how many start the race. If we leave before all 10 athletes cross the finishing line, we will only observe those who finish the race by the time we leave, with the remaining ones right-truncated by our observation window.

Like left censoring, *left truncation* also occurs in practice. Under left truncation, only subjects with failure times beyond some point are observable. For example, in the 1970s, a study was conducted by the Stanford heart transplant program to see whether a heart transplant would prolong the life of a patient with heart disease (Cox and Oakes, 1984, Lawless, 2002). The patients were admitted to the program if other forms of therapy were unlikely to work, as determined by their doctors. Because each patient had to wait until a suitable donor heart was available, only those who were able to survive the waiting period would receive the operation. In this study, the patients who could not survive the waiting period are left-truncated, resulting in a sample of relatively healthier patients who received the heart transplant in the treatment groups. Because of this selection bias, it is not appropriate to assess the effect of heart transplant by simply comparing the survival times between the two groups with and without a heart transplant.

Truncation usually is not as apparent as censoring. When there are censored subjects, these subjects are still in the data set, although their survival outcomes are only partially observed. However, when there are truncated subjects, they are not in the data set, making it easy to overlook the issue. There are methods developed to deal with truncation, based on modeling the truncation process; see, e.g., Lagakos et al. (1988). However, they are often based on strong assumptions that are difficult or even impossible to verify. Since truncation may be avoided if a study is carefully designed, it is not as popular as censoring, so we will not discuss this issue further in this chapter.

8.1.3 Discrete survival time

As time is inherently continuous, much of the survival analysis literature focuses on continuous survival time. However, discrete survival times are also common. For example, when analyzing data from large survey studies and surveillance systems, it is common to group continuous survival times because the typically huge sample sizes in such databases make it computationally difficult to apply methods for continuous survival times. Another common situation of grouping is interval censoring. If the event status is only assessed at a set of prespecified time points, the occurrence of the event can only be ascertained to be somewhere between two consecutive assessment times. For example, in animal cancer experiments, cages are only checked periodically, such as daily. If an animal is found dead at the next assessment point, we only know that the death had occurred between this and the prior assessment time. The intervals defined by the successive assessment points serve as discrete failure times for the death of the animal. We can improve the uncertainty about the timing of the death and thus the accuracy of model estimates by scheduling more frequent assessments.

Discrete times can also arise if the occurrence of an event of interest is itself not instantaneous or cannot be observed in such a manner. For example, depression is not an instantaneous event, and as such it is usually measured by a coarse scale such as weeks or months, yielding discrete outcomes.

Survival times can also be genuinely discrete. In modern clinical trials, subject retention is an important issue. If many patients drop out of the study prematurely, the study may not be able to draw reliable conclusions. Moreover, such study drop-outs also result in missing values, which not only complicates the analysis but also threatens the integrity of inference. Thus, assessing and improving retention for clinical trials with large attrition is important, and survival analysis can be used to facilitate the investigation of this issue. For many clinical trials such as the DOS, patients are regularly assessed at a set of prescheduled visits such as weekly, monthly, or annually, thereby creating discrete drop-out times.

Although the general modeling principles are the same for the different types of discrete survival times, applications of such models to data in real studies do require careful considerations of the differences between them. For example, when considering genuinely discrete time $T = t_j$ $(j = 1, 2, \cdots)$, a subject censored at a time point t_j implies that the subject has survived up to and including time $T = t_j$, but is not under observation, or at risk for failure, beyond that point. If T is a discretized continuous survival time, a subject censored at time point t_j could arise from many different scenarios. For example, if $t_j = [\tau_{j-1}, \tau_j)$ represents a grouped time interval $(j = 1, 2, \cdots)$ with $\tau_0 = 0$, a subject censored at t_j is at risk at the beginning of the interval, and then becomes censored at some time inside the interval $[\tau_{j-1}, \tau_j)$, which could be anywhere in the interval $[\tau_{j-1}, \tau_j)$. Moreover, the subject could still have failed in the time interval $[\tau_{j-1}, \tau_j)$, had the follow-up been continued after the censoring time. Thus, it may not be appropriate to treat such a censored subject as a survivor for the whole time interval. We discuss some common approaches to address this complex problem in Section 8.2.

8.1.4 Survival and hazard functions

Much of the survival analysis literature focuses on continuous survival time. Although models for continuous survival times are generally not applicable to discrete outcomes, much of the terminology can be carried over directly to describe survival analysis within the context of discrete times. Thus, we give a brief introduction to the concepts and terms commonly used to describe distributions and models for continuous survival times to facilitate the discussion of models for discrete survival time data.

8.1.4.1 Continuous survival time

For a continuous survival time of interest T (≥ 0), the function, $S(t) = \Pr(T > t)$, is called the *survival function*. It is readily seen that $F(t) = 1 - S(t)$, where $F(t) = \Pr(T \leq t)$ is the CDF of the survival time variable T. The notation $S(t)$ is commonly used, as it has a more meaningful interpretation as the probability of having survived longer than time t within the context of survival analysis. The *hazard* function, defined as $h(t) = -\frac{S'(t)}{S(t)}$, measures the instantaneous failure rate or the rate of occurrence of failure within an infinitesimal time interval, given that the subject has survived up to time t, or is still at risk for failing at time t (see Problem 8.3).

Popular parametric models for survival times include the exponential and Weibull distributions. An *exponential* survival function posits an exponential distribution for modeling the survival time, i.e., $S(t) = \exp(-\lambda t)$, and hence the hazard function is $h(t) = \frac{f(t)}{S(t)} = \lambda$, for some parameter $\lambda > 0$. The constant hazard indicates that the risk of failing at any instant is the same, no matter how long the subject has survived. This *memoryless* property of the exponential distribution can also be checked directly (see Problem 8.5). A constant hazard is unrealistic for modeling most survival times, since the risk of failing typically changes as time elapses, but such an assumption may be reasonable over a short period of time.

A *Weibull* survival function has the form $S(t) = \exp\left(-(\lambda t)^k\right)$ $(\lambda > 0, k > 0)$, yielding a hazard of the form $h(t; k, \lambda) = \lambda k (\lambda t)^{k-1}$. The Weibull distribution overcomes the limitations of the exponential by introducing another shape parameter k. If $k < 1$, it yields a very high hazard at the beginning, which then decreases over time. Under this setting, the hazard resembles the risk profile of childbirth or an acute disease, where the subject is initially at high risk for mortality, but with a return to a state of complete-premorbid health if surviving this early critical period. If $k > 1$, the hazard increases with time, depicting a reasonable trajectory of disease progression for most chronic diseases with increased morbidity and mortality over time. In the special case with $k = 1$, Weibull reduces to the exponential $\exp(-\lambda t)$.

8.1.4.2   Discrete survival time

Unlike its continuous counterpart, a discrete survival time T ranges over a set of time points t_j, with $0 < t_1 < t_2 < \cdots$. Since the subjects cannot be followed indefinitely in practice, those who are not observed to fail within some time points will be censored. If t_{k-1} is the last time point of the observation, all subjects who are at risk afterwards are right-censored and the distribution for the discrete survival time can be characterized by a multinomial distribution.

Let $\pi_j = \Pr(T = t_j)$ for $1 \leq j \leq k-1$ and $\pi_k = \Pr(T > t_{k-1})$, then $T \sim MN(1, \boldsymbol{\pi})$, where

$$\boldsymbol{\pi} = (\pi_1, \dots, \pi_k), \quad \sum_{j=1}^{k} \pi_j = 1. \tag{8.1}$$

Denote t_k for $T > t_{k-1}$, then $\pi_k = \Pr(T = t_k)$. Under the multinomial model, the *discrete survival function* is given by

$$S_j = \Pr(T > t_j) = \sum_{l=j+1}^{k} \pi_l, \quad 0 \leq j \leq k-1, \tag{8.2}$$

which is the probability that the failure has not occurred by time t_j. Since all the subjects are at risk at time 0, $S_0 = 1$. The *discrete hazard* is defined, in analogy to continuous times, as

$$p_j = \Pr(T = t_j \mid T \geq t_j) = \frac{\pi_j}{S_{j-1}} = 1 - \frac{S_j}{S_{j-1}}, \quad 1 \leq j \leq k-1. \tag{8.3}$$

The hazard p_j above measures the probability of failing at time t_j, given that the subject is at risk at time t_j. Note that p_k is not defined.

Note that each of the $\{\pi_j\}_{j=1}^{k-1}$, $\{S_j\}_{j=1}^{k-1}$, and $\{p_j\}_{j=1}^{k-1}$ determines the other two (see Problem 8.7). Also, the setup above applies to both discretized continuous and genuinely discrete outcomes. For grouped-continuous data, $T = t_j$ means that the failure occurs within an interval $[\tau_{j-1}, \tau_j)$ $(1 \leq j \leq k-1)$, with $\tau_0 = 0$ and $\tau_k = \infty$. For the genuinely discrete survival time, t_j $(1 \leq j \leq k)$ simply denotes the range of the discrete survival time.

8.2   Life Table Methods

Life tables are commonly used to present information about discrete survival times, especially in actuarial science. For example, the CDC releases information about life expectancies of people in the United States on a regular basis, in which the survival times, grouped in years, are tabulated, along with the hazard, survival, and other related quantities (check the website http://www.cdc.gov/nchs/products/life_tables.htm).

Life tables tabulate the number of failures, survivors, and at-risk subjects in an easy-to-interpret format so that information about the risk of failure such as hazard and survival functions can be readily derived. Life tables can be used to provide such information for both discretized and intrinsically discrete survival times.

8.2.1 Life tables

Consider a random sample of size n from the study population of interest, with events of failures, withdrawals, and number of survivors recorded over the study period. Let t_j $(1 \leq j \leq k)$ denote the observed values of the discrete outcome T, with $t_1 > 0$. A life table typically has the following form:

Time	Failure	Survivor	Withdraw	At Risk	Hazard	Survival
t_1	d_1	s_1	w_1	n_1	p_1	S_1
t_2	d_2	s_2	w_2	n_2	p_2	S_2
$\vdots$	$\vdots$	$\vdots$	$\vdots$	$\vdots$	$\vdots$	$\vdots$

where n_j, d_j, s_j, and w_j denote the number of subjects at risk, failures, survivors, and withdrawals at time t_j, respectively. The table shows that after the start of the study, there are $n_1 = n$ subjects at risk at the first time point t_1, where n is the total sample size. Among the n_1 at risk subjects, we observe d_1 failures, s_1 survivors, and w_1 withdrawals (censored cases) at the first time point t_1. At time t_2, there are $n_2 = n_1 - d_1 - w_1$ subjects at risk. Out of these n_2 subjects, there are d_2 failures, s_2 survivors, and w_2 withdrawals. In general, at time t_j, there are $n_j = n_{j-1} - w_{j-1} - d_{j-1}$ at risk for $j = 1, 2, \ldots, k$.

For genuinely discrete survival times, a subject censored at time t_j means that the subject has survived up to and including time t_j, but is not at risk beyond this point. Thus, the subject is a survivor at time t_j. However, for a grouped continuous time, each point t_j actually represents an interval $[\tau_{j-1}, \tau_j)$. Depending on how the survival outcomes are obtained, subjects at risk at time τ_{j-1} can be censored at any point within the interval $[\tau_{j-1}, \tau_j)$. Thus, it may not be appropriate to completely ignore this variability and simply interpret a censored case in $[\tau_{j-1}, \tau_j)$ in the original scale as being censored at time τ_j (or having survived the whole interval). If the information is collected at the beginning of the interval, then censored subjects are not actually at risk for the whole censoring time interval. On the other hand, if the data are collected at the end of the time interval, then censored subjects have actually been at risk and survived the whole interval; they will be treated as complete survivors, survivors for the whole censoring interval. When the underlying continuous survival time is real time, then generally a censored subject was at risk and a survivor for only part of the censoring time interval. Without further information, a common approach is to assume that the censoring time is uniformly distributed over the time interval, and so on average each censored subject can be treated as half a survivor. Thus, for grouped continuous times, each withdrawal entry w_j in the life table is replaced with $\frac{1}{2}w_j$, and the number of subjects at risk at t_j is adjusted accordingly by $n'_j = n_j - \frac{1}{2}w_j$.

With the information in the life table, we can readily compute statistics of interest such as hazard and survival functions. Before proceeding with such calculations, we need to know how censoring arises so that its effect on such statistics can be accounted for, an issue akin to the different types of missing data mechanisms.

8.2.1.1 Random, or independent, censoring

A unique characteristic of survival data analysis is the possibility that the event of failure may not be observed due to censoring caused by a variety of reasons such as withdrawals, limited follow-up times, and competing events. If such censoring events occur first, the event of interest will not be observed. Thus, we may assume that each subject in the sample has a potential censoring time competing with the survival time of interest to cause censored observations.

Let T_i and V_i denote the failure and censoring time for the ith subject. If the event of failure occurs first, we observe T_i; otherwise we observe V_i. In other words, in the presence of censoring, we only observe the smaller of the two times, $U_i = \min(T_i, V_i)$. Thus, the likelihood consists of the observed time U_i, which is neither the failure time T_i nor the censoring time V_i, but the lower of the two.

Except for the trivial case when $T_i \leq V_i$ for all i, the observed-data likelihood cannot be used directly for inference about the distribution of the survival time T_i. In general, it is necessary to model the censoring time V_i to use the likelihood for inference. However, in most applications, it is quite difficult to model the censoring event because of limited information and the complexity of such a process. Rather than attempting to model V_i, a popular alternative in practice is to assume independence between T_i and V_i, or *random censoring*.

Let $S(t, \beta_T)$ and $f(t, \beta_T)$ ($S_V(t, \beta_V)$ and $f_V(t, \beta_V)$) denote the survival and probability distribution function of the failure time T_i (censoring time V_i), parameterized by β_T (β_V). Under the random censoring assumption, the likelihood for the observed time $u_i = \min(t_i, v_i)$ is given by

$$
L_i = \begin{cases} f(u_i, \beta_T) S_U(u_i, \beta_V) & \text{if } c_i = 1, \\ \\ S(u_i, \beta_T) f_V(u_i, \beta_V) & \text{if } c_i = 0, \end{cases}
$$

where c_i is the event indicator with the value 1 (0) for failure (censoring). It follows that the likelihood for the sample is

$$
L = \prod_{i=1}^{n} [f(u_i, \beta_T)]^{c_i} [S(u_i, \beta_T)]^{1-c_i} \prod_{i=1}^{n} [S_V(u_i, \beta_V)]^{c_i} [f_V(u_i, \beta_V)]^{1-c_i}. \tag{8.4}
$$

Thus, the log-likelihood is the sum of two terms, with the first involving β_T only and the second containing just β_V. Since we are only interested in β_T, we can apply the method of maximum likelihood to the first term for inference about β_T. In this sense, the censoring mechanism is completely ignored, or noninformative.

Thus, we may make inferences based on the likelihood of the survival time distribution. By expressing the probability distribution function in terms of the hazard and survival functions, we obtain the likelihood function for the discrete survival outcome (see Problem 8.8)

$$
L = \prod_{i=1}^{n} [p(t_j, \beta_T)]^{c_i} [1 - p(t_j, \beta_T)]^{1-c_i} S(t_{j-1}, \beta_T). \tag{8.5}
$$

For the discretized continuous time, the likelihood is modified as

$$
L = \prod_{i=1}^{n} [p(t_j, \beta_T)]^{c_i} [1 - p(t_j, \beta_T)]^{\alpha(1-c_i)} S(t_{j-1}, \beta_T), \tag{8.6}
$$

where the power of α for the censored subject reflects how we will treat the censored subjects in their censoring time interval. The convention is to treat such a subject as half a survivor for the censoring time interval in grouped continuous survival times.

It is seen from the likelihood in (8.4) that survival and censoring times are symmetric in the sense that if one is considered as the failure time of interest, the other becomes the censoring time. For example, consider the heart transplant study discussed earlier. If we are interested in the time to death for those patients without a heart transplant, surgery defines the censoring time. On the other hand, if interest lies in the waiting time for the operation from admission to surgery, then those who die before surgery are censored by death. Such a symmetry between failure and censoring is often called *competing risks*.

8.2.1.2 Inference

Under random censoring, it follows from (8.6) that we can estimate the hazard $p(t_j, \beta_T)$ by maximizing the likelihood of the survival time only. It is straightforward to check that at each time point t_j, d_j follows a binomial with mean p_j and sample size n_j, and thus following the discussion in Chapter 2 for binomial distributions, the maximum likelihood estimate of (8.6) is (see Problem 8.9)

$$\widehat{p}_j = \frac{d_j}{n'_j}, \text{ where } n'_j = \begin{cases} n_j - \alpha w_j & \text{for grouped continuous } T, \\ n_j & \text{for discrete } T, \end{cases} \quad 1 \le j \le k-1. \quad (8.7)$$

As discussed earlier, for discretized continuous T, we may need to adjust the number of at-risk subjects to account for the presence of the censored subjects during part of the observation interval. Thus, instead of being treated as a complete survivor as in the case of genuinely discrete times, each censored case is treated as α subject. In practice, we often use $\alpha = \frac{1}{2}$, and thus each censored case is treated as half a subject at risk and half a survivor during the time interval. Under this convention, there are a total of $n'_j = n_j - \frac{1}{2}w_j$ at risk, which is also known as the *effective sample size* for the time interval.

By applying the variance formula for a binomial outcome, we obtain the variance of $\widehat{p}_j$: $Var(\widehat{p}_j) = \frac{\widehat{p}_j(1-\widehat{p}_j)}{n'_j}$, $1 \le j \le k-1$.By substituting $\widehat{p}_i$ in place of p_i in

$$S_j = \prod_{l=1}^{j} (1 - p_l), \, 1 \le j \le k-1, \quad (8.8)$$

we immediately obtain estimates of $\widehat{S}_j$ ($1 \le j \le k-1$). The asymptotic variance (or standard error) of $\widehat{S}_j$ is more difficult to estimate.

Consider $\log \widehat{S}_j = \sum_{l=1}^{j} \log \widehat{p}_l$. It can be shown that $\widehat{p}_l$ are asymptotically independent (See Section 8.2.2 for a discussion on the asymptotic independence among the different $\widehat{p}_l$'s). Thus, it follows that

$$Var\left(\log \widehat{S}_j\right) \approx Var\left(\sum_{l=1}^{j} \log \widehat{p}_l\right) = \sum_{l=1}^{j} Var\left(\log \widehat{p}_l\right).$$

By applying the delta method, we obtain the asymptotic standard error of $\widehat{S}_j$ (see Problem 8.13):

$$\widehat{S}(t_j) \left[\sum_{l=1}^{j} \frac{1 - \widehat{p}_l}{n_l \widehat{p}_l} \right]^{1/2}. \quad (8.9)$$

Example 8.1
In the DOS, we may look at the time from the enrolment in the study to the first onset of major depression during the study period. For this survival analysis, patients with depression at study intake are excluded. Note that the subjects are assessed for depression each year during the study period, and thus we first need to make sure the survival time is well defined. Since the instrument used for assessing depression in this study, SCID, asks for any episode of depression that has occurred during the past year, rather than the depression

status at the time of assessment, a non depression diagnosis by SCID implies no depression experience since the last assessment. Thus, this outcome is well defined because of the use of SCID in this study, despite the recurrent nature of depression. Had the diagnosis only captured depression information at the time of assessment, it could not rule out the possibility that the subject may have experienced depression between the last and current assessment. Thus, it would have been impossible to define such a time to first depression outcome, because of the recurrent nature of the disorder.

For a discrete survival outcome, different coding of the censoring time may be applied and interpreted accordingly. For example, the patients are assessed annually in the DOS. We use the last visit time for the censoring time. Thus, a patient with a censoring time of 1 (year) means that the patient had the first annual visit with a non depression diagnosis, but with no follow-up visit. Thus, the patient will be treated as a complete survivor for the first year; note that as we discussed above, the diagnosis at the annual assessment applies to the whole past year. However, since the patient has absolutely no information after year 1, we treat it as censored right after this visit, or the beginning of the time interval between the first and second year. Thus, the patient is not at risk at time 2. On the other hand, since the patient had the first annual visit, and was censored by the second annual visit, the time variable for censoring may have a value of 2. With a censoring time 2, the subject is treated as a complete survivor for time 1. However, time 2 represents the time interval between year 1 and year 2, the subject was censored at the beginning of the interval and thus should be considered as not at risk for the interval. Thus, consistent inference is obtained regardless of how the time is recorded, as long as censoring is interpreted consistently.

Based on the information given in Table 1.3, it is easy to obtain the following table with the sizes of the failure, censored and at-risk groups, along with estimates of hazard and survival and their standard errors.

Time	Number Failed	Number Censored	Effective Sample Size	Hazard (se)	Survival (se)
1	41	45	370	0.111 (0.014)	0.889 (0.016)
2	16	78	284	0.056 (0.013)	0.839 (0.020)
3	16	138	190	0.084 (0.017)	0.768 (0.025)
4	2	34	36	0.056 (0.013)	0.726 (0.038)
5	0	0	0	-	-

As discussed above, censored subjects are considered survivors for the whole censoring time interval, and thus the sample size is not adjusted. Had the censoring time been considered to be distributed over the whole interval, we would have treated each censored subject as a half survivor for the censoring time interval and obtained the following life table (if a censoring time 2 indicates that the censoring happens between years 1 and 2).

Time	Number Failed	Number Censored	Effective Sample Size	Conditional Probability of Failure (se)		Survival (se)	
[0, 1)	0	0	370	0	0	1.0000	0
[1, 2)	41	45	347.5	0.1180	0.0173	0.8820	0.0173
[2, 3)	16	78	245	0.0653	0.0158	0.8244	0.0213
[3, 4)	16	138	121	0.1322	0.0308	0.7154	0.0314
[4, 5)	2	34	19	0.1053	0.0704	0.6401	0.0577
≥ 5	0	0	0	-	-	-	-

Comparing the two tables, the estimated survival function based on grouped survival times in the second table is smaller because of the reduced effective sample size (the number of failures remains unchanged).   □

8.2.2 The Mantel–Cox test

In survival analysis, the most often asked question is whether a treatment improves the survival time of some event of interest such as recurrence of cancer and death. Sometimes this is carried out by comparing the survival rates across two or more groups of subjects over a period of time such as a 5-year survival rate for cancer patients. In the case of two groups, such as an intervention and a control group in a clinical trial study, the null hypothesis is the equivalence between two survival distributions, or curves, i.e., $S_1(t) = S_2(t)$ for all t. For discrete survival data, this translates into the following null:

$$H_0 : S_{j,1} = S_{j,2}, \quad j = 1, \ldots, k - 1,$$

where $S_{j,g} = S_g(t_j)$ $(g = 1, 2)$ are the survival probabilities of the two groups at time t_j $(j = 1, 2, \ldots, k - 1)$. Since by (8.8) $S_{j,g}$ is determined by the hazard function of the gth group, $p_{j,g} = p_g(t_j)$, it follows that the above is equivalent to

$$H_0 : p_{j,1} = p_{j,2}, \quad j = 1, \ldots, k - 1. \tag{8.10}$$

To find an appropriate test statistic for (8.10), note that the equality $p_{j,1} = p_{j,2}$ implies that there is no association between the treatment condition and failure at time t_j. If there is no censoring, then techniques for contingency table analysis discussed in Chapter 2 are readily applied to test the null. In the presence of censoring, as is the case for most survival analyses, we can readily derive a test to account for censoring by using an argument similar to the life-table construction.

Suppose at each time t_j in the range of the survival time variable, there are $n'_{j,g}$ subjects at risk (effect sample size), $d_{j,g}$ failures, and $w_{j,g}$ censored cases for the gth group. By treating the groups as the row and the status of failure as the column of a 2×2 contingency table, we can display the survival data at time t_j as follows:

	Failure	Non failure
group 1	$d_{j,1}$	$n'_{j,1} - d_{j,1}$
group 2	$d_{j,2}$	$n'_{j,2} - d_{j,2}$

Thus, to test the between-group difference at each time t_j, we can apply the chi-square test by assessing the row by column independence of the 2×2 table above. To test such independence across all time points in the study period, we apply the Cocharan–Mantel–Haenszel test, which generalizes the chi-square statistic for a single table to a set of 2×2 tables defined by the different time points within the current context. This is called the *Mantel–Cox test*. Note that the validity of the test requires that the samples used in each of the 2×2 tables above are representative for both groups. Thus, we assume independent censoring for each group for this purpose.

Based on the null hypothesis of row and column independence, the expected number of failures for group 1 is $m_j = \frac{(d_{j,1} + d_{j,2}) n'_{j,1}}{n'_{j,1} + n'_{j,2}}$. Thus, the Mantel–Cox statistic in our setting has the form $\sum_{j=0}^{k} (d_{j,1} - m_j)$, which can be used to provide inferences about the null hypothesis (see Chapter 3 for details about the inference procedures). More generally, we can use the following class of statistics

$$Z = \sum_{j=0}^{k} W_j (d_{j,1} - m_j), \tag{8.11}$$

where W_j is a known constant weight. The Mantel–Cox test is a special case of the above with $W_j = 1$, which is also called the *log-rank test*. Another popular choice is $W_j = n'_{j,1} + n'_{j,2}$, the total (effective) sample size, which is a generalization of the Wilcoxon statistic for right-censored data (see Gehan (1965)).

Strictly speaking, we need independence across the tables to obtain valid inferences when using the Mantel–Cox test. Within our context, however, this assumption may appear questionable, since the multiple tables are generated by the different time intervals of the discrete survival data, and as such a subject may appear in more than one such table. For example, the sample size for the second time interval depends on the number of survivors in the first. Thus, this assumption of independence cannot be taken for granted.

For a subject last observed at time t_j (either failed or censored), the likelihood for the subject for each time interval conditioning on it is at risk for the time interval is

$$L_{ij} = p_j^{c_i} \left(1 - p_j\right)^{\alpha(1-c_i)} \text{ and } L_{ik} = (1 - p_{k,}) \ \ k = 1, \ldots, j-1, \tag{8.12}$$

where c_i is the event indicator with the value 1 (0) for failure (censored event) and the power α in $(1 - p_j)^\alpha$ for the censored case has a value between 0 and 1, depending on how such a case is treated. Again, for grouped continuous survival times, a censored subject will usually be treated as half a survivor for the censoring time interval ($\alpha = 0.5$). It follows that the likelihood for the ith subject is the product of L_{ij} over all the time points j. This shows that the total likelihood is actually the likelihood of the stratified tables, and thus the inference above based on the assumption of independence among the stratified tables is valid. Intuitively, this is because the above-mentioned dependence of the tables stratified by time intervals affects only the sample size of the table, not the outcome of failure.

Example 8.2
For the DOS, consider testing whether there is any difference in time to the first depression between males and females. The test based on the Mantel–Cox statistic gives a p-value of 0.0060, while that based on Gehan's generalized Wilcoxon test yields a p-value of 0.0086. Both show a significant difference, with the females succumbing to depression sooner than their male counterparts, as indicated by the positive sign of both statistics. ⬜

8.3   Regression Models

As nonparametric methods, the life table estimates and the Mantel–Cox test have the advantage of not relying on any parametric assumptions. However, they are based on strong independent censoring assumptions. For the Mantel–Cox test, the independent censoring assumption is made for each group, thus the censoring can depend on the group. However, in practice, censoring may depend on other covariates, which may make the life table estimates and the Mantel–Cox test invalid. Furthermore, nonparametric methods are generally less powerful than their parametric counterparts, as the latter can describe the data with fewer parameters. In addition, parametric methods are also capable of modeling the survival time as a function of predictors and/or covariates. For regression analysis, the generalized linear models discussed in Chapter 5 for ordinal responses, especially the continuation ratio models, may be applied to model discrete survival times. However, in the presence of censoring, the likelihood becomes more complex, and estimates of model parameters cannot be obtained by applying the functions or procedures developed for fitting ordinal responses.

Regression models for survival times are commonly based on the hazard function. As shown by (8.12) in Section 8.2.2, the switch to hazard not only yields a simple expression, but also a natural interpretation of the likelihood. Moreover, we can also utilize the same functions and procedures for fitting binary responses discussed in Chapter 4 to provide inferences within the current context.

8.3.1 Complementary log-log regression

Consider a discretized continuous event time T with each point t_j representing an interval $[\tau_{j-1}, \tau_j)$ $(1 \leq j \leq k)$. Suppose the underlying continuous survival time is piece-wise exponential, i.e., it has a constant hazard over each interval, but the constant can vary across the different intervals. While the exponential assumption is hardly met in most real study applications, especially over a long period of time as discussed in Section 8.1 , the piece-wise exponential model overcomes this major limitation by assuming a constant hazard over a small interval, a much more reasonable imposition than the exponential model.

Let T_i be the discrete event time and $\mathbf{x}_i$ be a vector of covariates from an ith subject. Let λ_{ij} be the constant hazard in each time interval $t_j = [\tau_j, \tau_{j+1})$. If the underlying continuous survival time follows an exponential with mean λ_{ij}, then by linking this mean to $\mathbf{x}_i^\top \boldsymbol{\beta}_j$ using a log link, we obtain a generalized linear model

$$\log \lambda_{ij} = \mathbf{x}_{ij}^\top \boldsymbol{\beta}_j, \quad 1 \leq j \leq k - 1, \tag{8.13}$$

where i runs over all subjects who are at risk at time j. Note that different $\boldsymbol{\beta}_j$ are specified in (8.13) because λ_{ij} can be different across the different time intervals. Under the discrete survival time model (8.13), the hazard $p_{ij} = 1 - \exp\left(-\exp\left(\mathbf{x}_{ij}^\top \boldsymbol{\beta}_j\right)\right)$ (see Problem 8.14). Thus, when expressed in terms of p_{ij}, the model (8.13) becomes a generalized linear model with a complementary log-log link:

$$\log\left(-\log\left(1 - p_{ij}\right)\right) = \mathbf{x}_{ij}^\top \boldsymbol{\beta}_j, \quad 1 \leq j \leq k - 1, \tag{8.14}$$

where i runs over all subjects who at risk at time j. This is exactly the continuation ratio model with complementary log-log link function we studied in Chapter 5. However, the inference we discussed there does not directly apply because of the censoring.

We can also derive a similar complementary log-log model with a different set of assumptions. Instead of a piece-wise exponential model, assume that the covariate $\mathbf{x}_i$ induces a multiplicative effect on the *baseline* hazard, $h(t, \mathbf{x}_i) = \phi(\mathbf{x}_i; \boldsymbol{\beta}) h_0(t)$, where $h_0(t)$ is the hazard in the absence of $\mathbf{x}_i$ ($\mathbf{x}_i = \mathbf{0}$). This is commonly referred to as the *proportional hazards* model, because the hazard ratio between two subjects with covariates $\mathbf{x}_i$ and $\mathbf{x}_j$, $\frac{h(t, \mathbf{x}_i)}{h(t, \mathbf{x}_j)} = \frac{\phi(\mathbf{x}_i; \boldsymbol{\beta})}{\phi(\mathbf{x}_j; \boldsymbol{\beta})}$, does not change with time (Cox, 1972). In the proportional hazards model, the baseline hazard function is unspecified, thus it may be applied to model hazard functions with any shape as long as the proportional hazard assumption is satisfied. Let $S_0(t) = \exp\left(\int_0^t h_0(u)\, du\right)$ be the *baseline* survival function. Since $S(t, \mathbf{x}_i) = S_0(t)^{\phi(\mathbf{x}_i; \boldsymbol{\beta})}$ (see Problem 8.12), it follows that for each time interval $[t_{j-1}, t_j)$

$$p_j(\mathbf{x}_i) = 1 - \frac{S(t_j, \mathbf{x}_i)}{S(t_{j-1}, \mathbf{x}_i)} = 1 - \frac{S_0(t_j)^{\phi(\mathbf{x}_i; \boldsymbol{\beta})}}{S_0(t_{j-1})^{\phi(\mathbf{x}_i; \boldsymbol{\beta})}} = 1 - \left[1 - p_j(0)\right]^{\phi(\mathbf{x}_i; \boldsymbol{\beta})},$$

where $p_j(0) = 1 - \frac{S_0(t_j)}{S_0(t_{j-1})}$ is the discrete hazard at time t_j for a subject with baseline survival ($\mathbf{x}_i = \mathbf{0}$). If we assume $\phi(\mathbf{x}_i; \boldsymbol{\beta}) = \exp\left(\mathbf{x}_i^\top \boldsymbol{\beta}\right)$, then we can rewrite the above as a complementary log-log model:

$$\log\left(-\log\left(1 - p_j(\mathbf{x}_i)\right)\right) = \alpha_j + \mathbf{x}_i^\top \boldsymbol{\beta}, \quad 1 \leq i \leq n, \quad 1 \leq j \leq k, \tag{8.15}$$

similar to model (8.14). Note that $\boldsymbol{\beta}$ in (8.15) is the same for all j, because of the proportional hazards assumption of the underlying continuous survival time over the time range.

Under the assumption of (8.15), the ratio of the hazards between two subjects with covariates $\mathbf{x}_i$ and $\mathbf{x}_j$, $\frac{\phi(\mathbf{x}_i;\boldsymbol{\beta})}{\phi(\mathbf{x}_j;\boldsymbol{\beta})}$, is also independent of t; thus the model (8.14) is also known as the *discrete proportional hazards* model. However, unlike its continuous time counterpart, the discrete version is parameterized by a finite number of parameters, rather than one with infinite dimensions as in the continuous case (see, e.g., Cox and Oakes (1984), Kalbfleisch and Prentice (2002) and Lawless (2002)), making it possible to compute the MLE of $\boldsymbol{\beta}$ in the present context.

Maximum likelihood inference can be performed based on a weaker version of independent censoring assuming that the censoring and survival times are independent conditioning on observed covariates $\mathbf{x}_i$. The likelihood can again be written as a product of two parts, one for survival time and one for censoring, as in (8.4). The part for survival time is given by

$$\prod_{i=1}^{n} \left(p_{t_i}(\mathbf{x}_i)^{c_i} (1 - p_{t_i}(\mathbf{x}_i))^{\alpha(1-c_i)} \prod_{k=1}^{t_i-1} (1 - p_k(\mathbf{x}_i)) \right) \qquad (8.16)$$

where c_i is, as before, the event indicator with 1 for failure and 0 for a censored event, and the value of α $(0 < \alpha \leq 1)$ depends on how censored subjects are treated.

The discrete proportional hazards model (8.15) is nested within (8.14). The constraint imposed on the parameters $\boldsymbol{\beta}$ for the former model may not be satisfied in some studies. In practice, we may start with the more general model (8.14) and then test the null that all the $\boldsymbol{\beta}_j$ are the same to see whether it can be simplified to the discrete proportional hazards model.

8.3.1.1   Risk adjusted survivor function

Under the weaker version of the independent censoring assumptions that the censoring and survival times are independent conditioning on observed covariates $\mathbf{x}_i$, the survival function estimate introduced in Section 8.2.1 may be invalid as its assumption of independent censoring without conditioning on $\mathbf{x}_i$ may be violated. In practice, people often still use it as a descriptive statistic to summarize the data before adjusting the covariates. However, to obtain valid estimate of survival function for the population, we may use the risk adjusted survival function based on the regression model. The risk adjusted survivor function is simply the average of all these individual estimated survival functions.

Note that intercepts in the regression models (8.14) and (8.15) are related to the conditional survival probability at time t_j for subjects with $\mathbf{x}_i = \mathbf{0}$, and thus their estimates will give us an estimate of the baseline hazard function. Based on the estimated $\boldsymbol{\beta}_j$, we can further estimate the survival function for each individual.

Example 8.3
We may also use the discrete proportional hazards model to assess differential risks for depression between the male and female subjects in the DOS.

We start with the model in (8.14):

$$\log[-\log(1 - p_j(\mathbf{x}_i))] = \alpha_j + \beta_j x_i, \quad 1 \leq j \leq 4,$$

where x_i is a binary indicator for gender with 1 (0) for female (male). Procedures discussed in Chapter 4 for fitting generalized linear models with binary responses may be applied for inference about the parameters. However, some rearrangement of the data is necessary before these procedures can be applied to the present context. The original survival data typically have a single outcome containing the survival or censoring time, whichever is observed, an indicator for the type of the observed time (survival or censoring), and a set of covariates. To fit the models in (8.14) and (8.15) using procedures for ordinal responses, however, we need to transform the information to create a data set with multiple observations per subject, one for each time point at which the subject is at risk. Thus, the data set contains a new time variable to identify each time point and an indicator for the status of the subject at the corresponding time point. For example, we need j records to recode a subject with a failure or censoring event at time t_j, one for each of the time point t_l, $1 \leq l \leq j$. The subject survived the first $j-1$ time points, and thus there is no event for these time points. For the jth point, the subject has an event if the subject failed, or no event if the subject was censored. However, if a censored case is treated as half a survivor, the jth time point is given a weight of 0.5 (all others have weight 1). Other covariates for each subject are replicated across all the observations within the subject. Models (8.14) and (8.15) can then be applied to the data set. Of course, the individual records in the new data file are not really different subjects as in the original survival data but can be treated so for inference purposes because of the property of conditional independence discussed in Section 8.2.2.

The test of the null hypothesis of proportional hazards $H_0 : \beta_0 = \cdots = \beta_k$ in this example yields a p-value 0.6837. Since it is not significant, the proportional hazard assumption is reasonable. Estimates of the corresponding discrete proportional hazards model are summarized in the following table.

Parameter	Estimate	SE	Wald 95% CI		Wald χ^2	Pr > χ^2
α_1	−2.5785	0.2395	−3.0479	−2.1091	115.89	< .0001
α_2	−3.2674	0.3057	−3.8666	−2.6682	114.23	< .0001
α_3	−2.8473	0.3051	−3.4452	−2.2493	87.09	< .0001
α_4	−3.3589	0.7352	−4.7999	−1.9179	20.87	< .0001
Gender	0.6830	0.2539	0.1854	1.1807	7.24	0.0071

The estimated coefficient for gender is 0.6830 with a p-value 0.0071, leading to the same conclusion that females become depressed sooner than males in this study. The estimates of α_j provide the estimate of the hazard for the reference group, males. For example, the estimated hazard for males at time 3 is

$$1 - \exp\left(-\exp\left(-2.8473\right)\right) = 0.05635.$$

The estimates of the hazard and survival function for males and females at times 1 to 4 are summarized in the following table.

		Time			
Function	Gender	1	2	3	4
Hazard	Female	0.1395	0.0727	0.1085	0.0665
	Male	0.0731	0.0374	0.0564	0.0342
Survival	Female	0.8605	0.7980	0.7114	0.6641
	Male	0.9269	0.8923	0.8420	0.8132

Since there are 160 males and 210 females in the survival analysis, the risk adjusted survival function at time 1 to 4 is estimated by the weighted average

$$\frac{160}{370}(0.9269, 0.8923, 0.8420, 0.8132) + \frac{210}{370}(0.8605, 0.7980, 0.7114, 0.6641)$$
$$= (0.8892,\ 0.8388,\ 0.7679,\ 0.7286).$$

The estimates are very close to those in Example 8.1, especially for the first three time points. This may indicate that the censoring mechanism is similar for both genders.

□

8.3.2   Discrete proportional odds model

As in modeling general discrete responses, we may use different link functions to create different models. For example, we may assume that the hazard-based odds ratio is independent of time, i.e.,

$$\frac{p_j(\mathbf{x}_i)}{1 - p_j(\mathbf{x}_i)} = \phi(\mathbf{x}_i; \boldsymbol{\beta}) \frac{p_j(\mathbf{0})}{1 - p_j(\mathbf{0})}, \quad 1 \le j \le k. \tag{8.17}$$

Under (8.17), the odds ratio of failure at time j, $\frac{\phi(\mathbf{x}_i; \boldsymbol{\beta})}{\phi(\mathbf{x}_j; \boldsymbol{\beta})}$, is independent of time. Thus, under the above assumptions, we can immediately model the binary failure outcome at each point j using the familiar logistic regression. As in the case of the proportional hazards model, we may test such a proportionality assumption using a model with different parameters for different time intervals, i.e., replacing $\boldsymbol{\beta}$ in (8.17) with a time-varying $\boldsymbol{\beta}_j$, and testing whether $\boldsymbol{\beta}_j$ are the same under the null.

If $p_j(\mathbf{x}_i)$ is small, then $\frac{p_j(\mathbf{x}_i)}{1 - p_j(\mathbf{x}_i)} \approx p_j(\mathbf{x}_i)$ and hence $p_j(\mathbf{x}_i) \approx \phi(\mathbf{x}_i; \boldsymbol{\beta}) p_j(\mathbf{0})$ (see Problem 8.16). Thus, the discrete proportional odds model in (8.17) yields similar estimates as the discrete proportional hazards model discussed above. Also, when fitting this logistic model, we may need to recode the survival information to create a new data file amenable to the software package used.

Example 8.4
Using the logit link function instead of the complementary log-log in Example 8.3, i.e., consider the proportional odds

$$\text{logit}(p_{lj}) = \alpha_j + \beta x_i, \quad 1 \le j \le k.$$

To test the null of the proportional odds assumption, we can test whether there is an interaction between gender and time interval j, i.e., whether $\beta_0 = \cdots = \beta_k$ under the model

$$\text{logit}(p_{lj}) = \alpha_j + \beta_j x_i,$$

which in this example yields a p-value 0.6912. Since it is not significant, the proportional odds assumption is reasonable. By applying the corresponding discrete proportional odds model, we obtain an estimate -0.7157 for the coefficient of gender indicator and a p-value 0.0051, leading to the same conclusion that females become depressed sooner than males in this study.

The estimate of the hazard and survival function for males and females at times 1 to 4 based on the discrete proportional odds model are summarized in the following table.

Function	Gender	Time			
		1	2	3	4
Hazard	Female	0.1394	0.0729	0.1088	0.0665
	Male	0.0733	0.0370	0.0563	0.0337
Survival	Female	0.8606	0.7979	0.7111	0.6638
	Male	0.9267	0.8924	0.8421	0.8138

Estimates of the risk adjusted survival function based on the proportional odds model at times 1 to 4 are 0.8892, 0.8388, 0.7678, and 0.7287, respectively. All the results are similar to those based on the proportional hazards model in Example 8.3.   ⬚

Exercises

8.1 In a study to determine the distribution of time to the occurrence of cancer after exposure to certain type of carcinogen, a group of mice is injected with the carcinogen, and then sacrificed and autopsied after a period of time to see whether cancer cells have been developed. Define the event of interest, and determine if censoring is present. If the event is censored, is it left, right, or interval censoring?

8.2 Under the race set up in Section 8.1.2, think a scenario when you may have left truncation issue.

8.3 Given that a subject survives up to and including time t, how likely is the failure to occur within the next infinitesimal time interval $(t, t + \Delta t)$?

8.4 For a continuously differentiable survival function $S(t)$, the hazard function is defined as $h(t) = -\frac{S'(t)}{S(t)}$. Prove that $S(t) = \exp\left(-\int_0^t h(s)ds\right)$.

8.5 For $T \sim$ exponential(λ), the conditional distribution of $T - t_0$, given $T \geq t_0$, again follows an exponential(λ) distribution.

8.6 Plot the survival and hazard function for the exponential and Weibull survival times using different parameters and check their shapes.

8.7 Let $\{\pi_j\}_{j=1}^{k-1}$, $\{S_j\}_{j=1}^{k-1}$, and $\{p_j\}_{j=1}^{k-1}$ be defined in (8.1), (8.2), and (8.3). Show that any one of them determines the other two.

8.8 Derive the likelihood (8.5) based on (8.4).

8.9 Prove that (8.7) provides the ML estimates based on the likelihood (8.5) and (8.6).

8.10 For the DOS, we are interested in the time to drop out of the study.

    (a) Create a life table including the number of subjects at risk, the number of failures (new depression), the number of survivors, and the number of the censored subjects for each gender, at each year;

    (b) Create a life table including the estimated discrete hazards and survival functions and their associated standard deviations stratified for each gender at each year. Indicate how the censoring is handled.

8.11 Verify the likelihood (8.16).

8.12 Let $h_0(t)$ denote the hazard function when $x_i = 0$, and $S_0(t) = \exp\left(\int_0^t h_0(u)\,du\right)$ be the corresponding survival function. If $h(t, x_i) = \phi(x_i; \beta) h_0(t)$, show $S(t, x_i) = S_0(t)^{\phi(x_i; \beta)}$.

8.13 Use the delta method to prove (8.9).

8.14 Assume that a survival time T has a constant hazard λ over time interval $[\tau_j, \tau_{j+1})$ with $\log \lambda = x^\top \beta$. Prove $1 - p = \exp\left(-\exp\left(x^\top \beta\right)\right)$, where $p = \Pr(T < \tau_{j+1} \mid T \geq \tau_j)$.

8.15 Fit the following models for genuinely discrete time to drop-out with age and gender as covariates for the DOS:

(a) the proportional hazards models;

(b) the proportional odds models.

8.16 Check that if $p_j(\mathbf{x}_i)$ is small, then $\frac{p_j(\mathbf{x}_i)}{1-p_j(\mathbf{x}_i)} \approx p_j(\mathbf{x}_i)$ and hence (8.17) implies $p_j(\mathbf{x}_i) \approx \phi(\mathbf{x}_i; \boldsymbol{\beta}) p_j(\mathbf{0})$.

8.17 For the Catheter Study, the patients were assessed bimonthly and the measurements about UTIs, catheter blockages, and replacements cover the previous two months. Thus, the patients were under observation the whole time while they were in the study. So, we can study the survival times from the participation of the study until the first UTI, blockage, and replacement, which are defined by their first incident after the intake. Incidents of any of the three at the intake will be included as a covariate.

(a) Estimate the survival functions without adjusting for covariates.

(b) Compare whether there are differences in the survival functions between the two treatment groups.

(c) Fit the proportional hazards and proportional odds models to assess treatment effects while controlling for age, gender, and incident of the same type of event at the intake (main effect only).

(d) Estimate the risk adjusted survival function based on the regression models in (c) and compare them with the estimate in (a).

Chapter 9

Longitudinal and Clustered Data Analysis

In this chapter, we focus on the analysis of longitudinal and clustered data. Unlike cross-sectional studies taking a single snapshot of study subjects at a particular time point, individuals in longitudinal or cohort studies are followed up for a period of time, with repeated assessments during the follow-up time. By taking advantages of multiple snapshots over time, longitudinal studies have the ability to capture both between-individual differences and within-subject dynamics, permitting the study of more complicated biological, psychological, and behavioral processes than their cross-sectional counterparts.

For example, plotted in Figure 9.1 are HIV knowledge scores of a random sample of adolescent girls at baseline (0 month) and 3 months post-baseline in the Sexual Health study. The HIV knowledge scores are from a dimensional scale, with higher scores indicating greater HIV knowledge regarding the transmission and prevention of HIV. We may see that HIV knowledge was elevated among the group as a whole, but the right plot, with the two scores of the same subject at each of the two assessment points connected, clearly indicates differential change patterns within the group; those with lower scores at baseline showed larger improvement. Such dynamic individual differences in response to treatment are unique features of longitudinal studies.

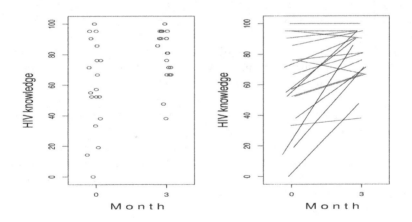

FIGURE 9.1: HIV knowledge scores of a random sample.

Longitudinal data present special methodological challenges for study designs and data analyses because the responses from the same individual are inevitably correlated. Standard statistical methods discussed in the previous chapters for cross-sectional data analysis such as logistic regression do not apply to such data. The DOS data that has been studied intensively so far is in fact a longitudinal study, but mostly only the baseline data have been used up to this point. In the DOS, patients were assessed for depression and other related health conditions such as medical burden annually for up to five years. These repeated assessments of depression and other comorbid health issues on the same individual over time are, of course, correlated. Consequently, we cannot treat these annual observations as data from different individuals and must take into account their correlated nature when analyzing such longitudinal outcomes.

Note that analysis of longitudinal data is different from survival analysis discussed in Chapter 8. Although subjects in survival analysis are also followed up for a period of time, the primary interest is a single outcome of the time to some significant event of interest such as occurrence of a certain cancer or death. In contrast, longitudinal data contain multiple outcomes for each subject over the study period, and time is only used to index the temporal assessments of the subjects, rather than being the primary focus as in survival analysis. Thus, longitudinal models in general bear little resemblance to those in survival analysis methodology. They are also different from analysis of classic time series data, where only the repeated values of a single variable over time, such as daily stock price, monthly unemployment rate, and quarterly earnings by a firm in the Fortune 500 company listing, are examined. In longitudinal data, each individual contributes a time series. Thus, in addition to characterizing temporal changes as in traditional time series analysis, we can also study between–subject variability while accounting for such individual dynamics to understand causes, sources, and factors giving rise to differential treatment effects and disease progression.

Longitudinal studies may be considered as a special type of clustered data. Usually clustered data occur in studies where subjects themselves are grouped, which naturally arise when subjects are randomized in clusters. For example, patients of the same physician are often assigned to the same treatment group for reasons such as feasibility and avoiding potential treatment contaminations. Like repeated measures in the longitudinal study, outcomes from patients with the same doctor are usually correlated, as they are receiving treatment from the same doctor. The clustering may also occur at different levels. For example, patients may be clustered by doctors (level 1), who are in turn clustered by clinics (level 2). Subjects in the same cluster usually do not have any special order. Thus, one may treat the repeated measures from the same subject as a special cluster, where members in the cluster are ordered by measurement times. We concentrate our discussions on longitudinal data in this chapter, but most of the methods are readily applied to more general clustered data arising from multi-center and other related clinical trials and cohort studies.

In Section 9.1, we describe basic aspects of longitudinal data and techniques for exploratory analysis, a stepping stone to building appropriate statistical models for the data. Following this section, we discuss the two most popular approaches for statistical modeling of longitudinal data. We first discuss mixed-effect models for continuous outcomes in Section 9.2 and then generalize it to categorical and count outcomes following the framework of generalized linear models in Section 9.3. In Section 9.4, we discuss marginal models and their inference via generalized estimating equations. In Section 9.5, we focus on model evaluation and diagnosis.

9.1 Data Preparation and Exploration

While repeated measures in longitudinal studies enable us to study within-individual dynamics, they also make the recording and use of such information a bit more challenging. In general, longitudinal data are stored in a data file using one of two common approaches.

One is to record the repeated measures across different columns, with each row containing a complete set of repeated measures for each subject. The other is to designate each variable using a column, and thus unlike the first approach, repeated measures are recorded using different rows. One format may be preferred over the other, depending on the purposes of analysis and software packages used.

Compared with cross-sectional studies, modeling longitudinal data is inevitably more complicated because of the correlation among the serial measurements of the same subject. Thus, it is important to get a sense of as well as to understand the features of the data by performing some exploratory data analysis. Such preliminary work will help guide one to appropriate models to get the most out of the data at hand. In this section, we first give a brief account of the data formats and then discuss some popular tools for exploratory data analysis.

9.1.1 Longitudinal data formats

As repeated assessments in longitudinal studies generate correlated data for each subject, it is important to link such repeated measures to the right subject from whom the data are obtained. One approach is to include all information from each subject in a single row. In this case, we need different variables to represent the repeated measures of the same variable across different assessment times. In most longitudinal studies, assessment times are fixed a priori, and thus all subjects follow the same assessment schedule during the study period. Thus, it is convenient to name the variables by adding the visit number as a suffix. For example, we may use Dep1, Dep2, ..., Depm as variable names for depression status assessed at visit 1, 2, ..., m. Since it is rare that patients come for assessment at exactly the planned times, some variation is expected between the scheduled and actual visit times. In most studies, the actual visits are close enough to their scheduled counterparts that the difference can often be ignored for all practical purposes. In cases the difference is important, we may also create additional variables to record the actual visit times for each assessment by each individual in the study.

For example, in the DOS, each individual is assessed up to 5 times, with one for every year up to 5 years. Thus, we may use five different names for the repeated measures on the same characteristics, as shown in the following table.

TABLE 9.1: Horizontal format for longitudinal data

Subject	Age	⋯	Dep1	Med1	⋯	Dep2	Med2	⋯
⋮	⋮	⋮	⋮	⋮	⋮	⋮	⋮	⋯
n	75	⋯	maj	x	⋯	min	y	⋯
n+1	72	⋯	no	x	⋯	maj	y	⋯
⋮	⋮	⋮	⋮	⋮	⋮	⋮	⋮	⋯

In Table 9.1, demographic information such as gender, race, and age at baseline do not change with time, and are thus recorded using a single variable for each subject as in cross-sectional studies. However, multiple variables are used for each repeatedly measured variable, such as Dep1, Dep2, etc., for depression status at year 1, 2, etc., respectively. The number of variables needed for each measure is the maximum number of assessment such as 5 for the DOS example. A subject with Dep2 = "Major Depression" means that the subject had major depression when assessed at year 2. If the actual assessment time is important, an additional variable such as "VisitTime2" may be added to include such information.

Under this approach, all information about the repeated measures from each subject is recorded in a single row, and thus is often called the *horizontal*, or *wide* format. Also commonly used in practice is the *vertical*, or *long*, format in which each construct for a subject from different visits is recorded in different rows. The advantage of this alternative approach is that we need only one variable for the same characteristic across the different visits. To link data from different rows within the same subject as well as between different individuals, we use a variable that takes the same value in the rows containing the repeated assessments from the same individual, but different values for such rows for different subjects. Note that such a subject index, or ID, variable is not necessary for the horizontal format, since the row serves as a natural demarcation line for each individual's data.

In longitudinal studies, it is common that a subject may miss some visits. In the horizontal format, this may be easily flagged by a symbol or designated value for missing values in the corresponding variables such as a ".", 99, or "NA" depending on the software packages used. When using the vertical format, we need to exercise more caution. As the multiple visits are identified by the different rows, a missing visit may simply be indicated by removing the corresponding row. However, this approach may not work for all software packages since some, such as SAS, expect sequentially ordered rows based on assessment visits. For example, if the second row represents outcomes at the third visit, because of the missing second visit, that row will be interpreted as data from the second visit by the SAS GENMOD procedure, unless there is a variable indexing the visit. To avoid confusion, it is customary to keep the rows for all planned visits, but code the missing visits with a special symbol or value such as "." or 99.

TABLE 9.2:   Vertical format for longitudinal data

Subject	Visit	Age	$\cdots$	Dep	Med	$\cdots$
$\vdots$	$\vdots$	$\vdots$	$\vdots$	$\vdots$	$\vdots$	$\vdots$
n	1	75	$\cdots$	maj	x	$\cdots$
n	2	75	$\cdots$	min	y	$\cdots$
$\vdots$	$\vdots$	$\vdots$	$\vdots$	$\vdots$	$\vdots$	$\vdots$
n+1	1	72	$\cdots$	no	x	$\cdots$
n+1	2	72	$\cdots$	maj	y	$\cdots$
$\vdots$	$\vdots$	$\vdots$	$\vdots$	$\vdots$	$\vdots$	$\vdots$

As an example, shown in Table 9.2 is a vertical version of the DOS data. A single variable is used for each unique construct of the subject such as age (Age) and depression (Dep). Repeated measures appear in different rows, with the order of assessments indexed by the variable "Visit" and linked to each individual subject by the "Subject" id variable. For example, the rows with "Subject = n" indicate that they contain the repeated measures for the nth subject, with the first measure in the row indexed by "Visit = 1," the second by "Visit = 2," etc. In this data set, it is crucial to include both a subject and a time index to distinguish data between individuals and multiple assessments within the same subject. In contrast, none of these variables is necessary for the horizontal format.

Based on the needs of the analysis and the choice of software, we may frequently need to transform data between these two popular formats. Fortunately, most of the statistical software packages have the ability to do the transformation (see Problem 9.1a).

9.1.2 Exploratory analysis

Longitudinal data is generally quite complex to model because of the serial correlation among the repeated measures within the same subject and varying temporal change patterns across different individuals. Thus, it is important to perform some exploratory data analysis before starting the formal model-building process. Common methods for cross-sectional data usually provide useful information about the data to be modeled. For examples, descriptive statistics such as mean and standard deviation for continuous variables and proportions and sample sizes for categorical variables at each of the assessment points can be informative for depicting how the longitudinal measures of a variable change with time. To assess the relationship between two time-varying variables such as depression and medical burden in the DOS, it may be helpful to compute their correlations (for continuous variables) and odds ratios (for binary variables) at each time point. In this section, we describe some of the commonly used exploratory analysis tools. Readers should keep in mind that in addition to the ones considered here any method helpful in understanding the data may be applied.

9.1.2.1 Summary index analysis

As in the case of cross-sectional studies, summary statistics are quite useful for depicting features of a longitudinal outcome. For example, to explore temporal changes, we may use summary indices such as averaged outcomes over the sample (or subsample) at each assessment point over the study period. Summarized in Table 9.3 are the proportions of subjects with major depression at each yearly visit for the DOS.

TABLE 9.3: Proportions of major depression at each visit

Year	0	1	2	3	4
Percent (%)	17.2275	18.5897	18.0428	21.3974	22.6415
Standard Error	1.3854	1.7983	2.1265	2.7161	5.7487

Based on the sample proportion of subjects with major depression at each visit, there seems not much change in the rate of occurrence of this mental disorder over the study period. The trend is more visually depicted by plotting the proportion along with its 95% confidence interval over the study visits, as shown in Figure 9.2. Both the table and plot suggest that time by itself may not be a predictor for depression over the study period. This may be plausible as the study does not involve any intervention to treat depression. However, since the subjects in this cohort are followed up over time, we may still want to include time as a predictor in the initial model to account for this important feature of the outcome. We can, of course, remove this predictor from the model later, if such a time effect is not supported by the data. Note that the standard error increases with time, reflecting the fact that the sample size decreases because of missing values.

9.1.2.2 Pooled data analysis

If we ignore the correlation among repeated measures and treat them as independent, we can pool data from different time points and apply the methods described in the previous chapters for cross-sectional studies to the pooled data. This naive approach usually inflates the sample size and underestimates the standard error. As a result, it may produce falsely significant test results. Despite these potential flaws, it may still be useful to apply this

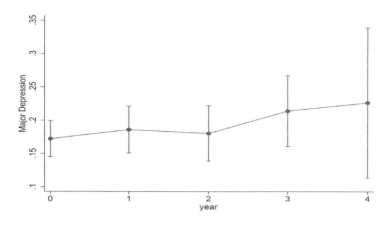

FIGURE 9.2: Proportions of major depression and confidence intervals.

approach as an exploratory analysis tool to garner information about the data such as the direction of association among different variables.

The analysis may be stratified according to some groups of interest based on study purposes. For example, in randomized clinical trials, it is of primary interest to compare different treatments. In such cases, we can compute summary statistics such as the sample mean and standard deviation for each treatment over the different time points. Likewise, we may even want to break down each treatment group based on baseline and demographic variables such as gender and race to see whether such covariates play any moderating role on treatment. We can show and compare the multi-group summary indices the same way as before using either a table with rows representing the groups and columns indexing visits, or a figure consisting of overlaid boxplots for the different groups. With the latter, one may further connect the means within the same group to more visually depict temporal trends in the outcome. With the help of such plots, it is quite easy to spot treatment differences or moderation effects. For example, if the groups within each treatment condition do not show parallel temporal patterns, it suggests an interaction between time and the covariate defining the groups. Such information also helps to determine the functional forms of the predictors for the initial models when starting the formal model-building process.

9.1.2.3 Individual response profiles

To assess time effect at the individual level, we may again plot the repeated responses against time for each subject, rather than summary descriptive statistics as with the pooled data discussed above. Such individual response profiles, or *spaghetti plots*, are helpful not only for discerning temporal trends for each group as a whole but also for assessing individual variabilities around the group mean as well. Such a plot may become too crowded for showing individual response profiles for the entire study sample, especially when the sample size is large. Thus, one may choose to plot data within each subgroup defined by demographic variables such as gender and race. If the overall temporal pattern is desired for the entire study group, we may plot data from a representative subgroup with a reasonable size such as a randomly selected subsample of the study. For example, shown in Figure 9.1 is a random sample of size 20 from the Sexual Health study. For discrete outcomes with a bounded range, such plots may show limited variations because of range restrictions and tied values. To indicate the sample size associated with each plotted value, one may use dots with different sizes to represent the varying number of subjects within each cluster of tied observations. In Figure 9.1, we used another commonly used graphing technique called *jittering* to make each observation visible in the plot by adding a small random error term to each value of the discrete outcome.

9.2 Linear Mixed Effect Models

For analyses of longitudinal data, a key issue is how to address the within-subject correlation arising from repeated measures. Although different approaches have been developed, the two most popular are the marginal models based generalized estimating equations (GEEs) and the mixed-effects models. Mixed-effects models extend the GLM to longitudinal data by explicitly modeling the covariance structure of the repeated assessments. Because inference for mixed-effects models typically relies on the parametric assumptions, biased estimates may arise if the data do not follow the assumed parametric distributions. On the other hand, the marginal models focus on modeling at each assessment time, or the marginal distributions. The GEE models focus on the mean response at each assessment time, with inferences based on a set of estimating equations, similar to the approach used for distribution-free inferences for the generalized linear models for cross-sectional data discussed in Chapters 4-6. Thus, like its cross-sectional counterpart, the marginal models provide valid inferences regardless of data distributions.

We focus on linear mixed-effects models (LMM) for continuous outcomes in this section and generalize it to other data types in the next section. We will take up GEEs in Section 9.4. Note that both LMMs and GEEs model the mean response as a function of a set of predictors/covariates. In some applications, one may also be interested in whether and/or how responses at previous times predict responses at a later time point. Neither model applies to such dependent, or *autoregressive,* response relationships. Readers interested in modeling such relationships may consult Diggle et al. (2002) and Molenberghs and Verbeke (2005) for details.

9.2.1 Multivariate linear regression models

Consider a longitudinal study with n subjects and m assessment times. For notational brevity, we assume a set of fixed assessment times $1 \leq t \leq m$, though this setup, is readily extended to accommodate varying assessment times across subjects. Under this setup, all subjects have exactly m measures from the same assessment times, or balanced panels. When considering longitudinal data with unbalanced panels, i.e., when subjects have varying numbers of assessments (panel sizes), we must be mindful about the possibility that the imbalance may be the result of missed visits or study dropout, in which case the number of assessments as well as the timing of the missed visits and/or dropout may be related to the values of the repeatedly assessed variables of interest, a phenomenon known as *informative missing* data. The methods discussed in this chapter do not apply when unbalanced panels arise from such informative missing data. Analysis of longitudinal data with informative missing values is a complex issue, requiring careful considerations of the causes and their relationships with the variables being modeled. We discuss approaches to address this complex issue in a systematic fashion in Chapter 11.

Let y_{it} denote a scalar response and $\mathbf{x}_{it} = \left(x_{it1} \; x_{it2} \; \cdots \; x_{itp} \right)^{\top}$ a set of explanatory variables of interest from the ith subject at time t. One can apply a linear model

$$y_{it} = \mathbf{x}_{it}^{\top}\boldsymbol{\beta} + \varepsilon_{it}, \quad \varepsilon_{it} \sim \text{i.i.d. } N\left(0, \sigma^2\right), \quad 1 \leq i \leq n,$$

for each time point. In such cases, the independence of ε_{it} may be assumed between the subjects as only one time point is considered. However, the error terms ε_{it} are generally dependent across t because of between-assessment (serial) correlations for repeated measures for the same subject, so the commonly applied i.i.d assumption for ε_{it} will no longer be realistic if we combine the models. In fact, for longitudinal data analysis a central issue is how to deal with the correlations among the repeated measures. One approach would be to model the correlation directly. For continuous outcomes, this is usually modeled using multivariate normal distributions.

A natural generalization of linear models is multivariate linear models for longitudinal data, which assume that the vector $\boldsymbol{\epsilon}_i = (\varepsilon_{i1}, \ldots, \varepsilon_{im})^{\top}$ follows a multivariate normal distribution. The correlation among the repeated measures is addressed by specifying an appropriate covariance structure for $\boldsymbol{\epsilon}_i$. A general multivariate linear regression for longitudinal data may be expressed in a vector and matrix form as:

$$\mathbf{y}_i = \mathbf{x}_i\boldsymbol{\beta} + \boldsymbol{\epsilon}_i, \quad \boldsymbol{\epsilon}_i \sim \text{i.i.d. } N\left(\mathbf{0}, \Sigma\right), \quad 1 \leq i \leq n, \tag{9.1}$$

where

$$\mathbf{y}_i = \begin{pmatrix} y_{i1} \\ y_{i2} \\ \vdots \\ y_{im} \end{pmatrix}, \quad \mathbf{x}_i = \left(\mathbf{x}_{i1} \; \mathbf{x}_{i2} \; \cdots \; \mathbf{x}_{im} \right)^{\top} = \begin{pmatrix} x_{i11} & x_{i12} & \cdots & x_{i1p} \\ x_{i21} & x_{i22} & \cdots & x_{i2p} \\ \vdots & \vdots & \vdots & \vdots \\ x_{im1} & x_{im2} & \cdots & x_{imp} \end{pmatrix}, \quad \boldsymbol{\epsilon}_i = \begin{pmatrix} \varepsilon_{i1} \\ \varepsilon_{i2} \\ \vdots \\ \varepsilon_{im} \end{pmatrix},$$

and $N\left(\mathbf{0}, \Sigma\right)$ denotes the m-dimensional multivariate normal distribution with mean $\mathbf{0}$ and variance-covariance matrix, or simply covariance matrix, Σ.

Without any constraints, the covariance matrix Σ has $\frac{1}{2}m(m+1)$ parameters as it is symmetric. When the dimension m is large, we often need to impose some constraints on the covariance matrix for efficiency considerations. Next we discuss some of the most commonly used covariance structures.

9.2.1.1 Covariance structures

A covariance structure is determined by the marginal variances and correlation structure, because

$$\Sigma = \begin{pmatrix} \sigma_1 & & & \\ & \sigma_2 & & \\ & & \ddots & \\ & & & \sigma_m \end{pmatrix} \cdot \text{Correlation}(\boldsymbol{\epsilon}_i) \cdot \begin{pmatrix} \sigma_1 & & & \\ & \sigma_2 & & \\ & & \ddots & \\ & & & \sigma_m \end{pmatrix},$$

where $\sigma_j^2 = Var\left(\varepsilon_{ij}\right), j = 1, \ldots, m$, is the marginal variance of ε_{ij} at time j and a blank entry in the matrices means zero. For the marginal variance, we are usually interested in learning if it is homogeneous. Thus, for each correlation structure, we consider two corresponding covariance structures; one with homogeneous marginal variances $\left(\sigma_1^2 = \sigma_2^2 = \cdots = \sigma_m^2 = \sigma^2\right)$ and the other with heterogeneous marginal variances where the σ_i^2 can be different. Listed below are some commonly used correlation structures.

- Independence structure: correlation matrix $= \mathbf{I}_m$ (all repeated measures are independent).

- Exchangeable, or compound symmetric correlation structure: correlation matrix = $C_m(\rho)$, where any two different components of $\boldsymbol{\epsilon}_i$ have a common correlation ρ, i.e., $\text{corr}(\varepsilon_{is}, \varepsilon_{it}) = \rho$ for all $s \neq t$. This structure may be a natural choice when members in a cluster have no special order.

- First-order autoregressive correlation structure: correlation matrix = $AR_{m,1}(\rho)$, with $\text{corr}(\varepsilon_{is}, \varepsilon_{it}) = \rho^{|s-t|}$. The correlation decays exponentially in $|s-t|$. This is commonly applied for modeling exponential-decay correlations in elapsed time or distance.

If no constraint is imposed on the covariance or correlation matrix, the covariance or correlation is called *unstructured*.

9.2.1.2 Likelihood functions and score equations

The likelihood function for (9.1) is given by

$$L_n(\boldsymbol{\beta}, \Sigma) = \prod_{i=1}^{n} f(\mathbf{y}_i \mid \boldsymbol{\beta}, \Sigma) = (2\pi)^{-\frac{nm}{2}} |\Sigma|^{-\frac{n}{2}} \exp\left(-\frac{1}{2} \sum_{i=1i}^{n} \boldsymbol{\epsilon}_i^\top \Sigma^{-1} \boldsymbol{\epsilon}_{ii}\right),$$

where $|\Sigma|$ is the determinant of matrix Σ. Thus, the log-likelihood function is given by

$$l_n(\boldsymbol{\beta}, \Sigma) = \log(L_n(\boldsymbol{\beta}, \Sigma)) = -\frac{nm}{2} \log(2\pi) - \frac{n}{2} \log|\Sigma| - \frac{1}{2} \sum_{i=1}^{n} \boldsymbol{\epsilon}_i^\top \Sigma^{-1} \boldsymbol{\epsilon}_i.$$

Taking derivatives of the log-likelihood l_n with respect to $\boldsymbol{\beta}$ and $\Sigma = [\sigma_{st}]_{1 \leq s \leq m, 1 \leq t \leq m}$, we obtain the score equations:

$$\frac{\partial}{\partial \boldsymbol{\beta}} l_n = -2 \sum_{i=1}^{n} \left(\frac{\partial \boldsymbol{\epsilon}_i}{\partial \boldsymbol{\beta}}\right)^\top (\Sigma^{-1} \boldsymbol{\epsilon}_i) = 2 \sum_{i=1}^{n} \mathbf{x}_i^\top \Sigma^{-1} (\mathbf{y}_i - \mathbf{x}_i \boldsymbol{\beta}) = \mathbf{0}$$

$$\frac{\partial}{\partial \Sigma} l_n = -\frac{1}{2} n \Sigma^{-1} + \frac{1}{2} \sum_{i=1}^{n} \Sigma^{-1} \boldsymbol{\epsilon}_i \boldsymbol{\epsilon}_i^\top \Sigma^{-1} = \mathbf{0},$$

where $\frac{\partial}{\partial \Sigma} l_n = \left[\frac{\partial}{\partial \sigma_{st}} l_n\right]_{1 \leq s \leq m, 1 \leq t \leq m}$ is an $m \times m$ matrix. These equations can be transformed to

$$\boldsymbol{\beta} = \left(\sum_{i=1}^{n} \mathbf{x}_i^\top \Sigma^{-1} \mathbf{x}_i\right)^{-1} \left(\sum_{i=1}^{n} \mathbf{x}_i^\top \Sigma^{-1} \mathbf{y}_i\right) \tag{9.2}$$

$$\Sigma = \frac{1}{n} \sum_{i-1}^{n} (\mathbf{y}_i - \mathbf{x}_i \boldsymbol{\beta})(\mathbf{y}_i - \mathbf{x}_i \boldsymbol{\beta})^\top.$$

Note that, unlike linear regression models for cross-sectional data, the MLE $\widehat{\boldsymbol{\beta}}$ for $\boldsymbol{\beta}$ involves the estimation of the covariance matrix Σ. So, we must solve the equations simultaneously to obtains MLE for $\boldsymbol{\beta}$ and Σ. Some iterative techniques may be applied to solve these non linear equations.

Based on (9.2), $\widehat{\boldsymbol{\beta}}$ follows an asymptotic distribution:

$$\widehat{\boldsymbol{\beta}} = \left(\sum_{i=1}^{n} \mathbf{x}_i^\top \widehat{\Sigma}^{-1} \mathbf{x}_i\right)^{-1} \left(\sum_{i=1}^{n} \mathbf{x}_i^\top \widehat{\Sigma}^{-1} \mathbf{y}_i\right) \sim N\left(\boldsymbol{\beta}, \left(\sum_{i=1}^{n} \mathbf{x}_i^\top \Sigma^{-1} \mathbf{x}_i\right)^{-1}\right). \tag{9.3}$$

An estimate of the asymptotic variance can be obtained by replacing Σ with one of its consistent estimate, for example the MLE $\widehat{\Sigma}$.

9.2.1.3   Restricted maximum likelihood (REML)

We may estimate the variance Σ without estimating β. Let

$$
Y = \begin{pmatrix} \mathbf{y}_1 \\ \mathbf{y}_2 \\ \vdots \\ \mathbf{y}_n \end{pmatrix}, \quad
X = \begin{pmatrix} \mathbf{x}_1 \\ \mathbf{x}_2 \\ \vdots \\ \mathbf{x}_n \end{pmatrix}, \quad
V = \begin{pmatrix} \Sigma & & & \\ & \Sigma & & \\ & & \ddots & \\ & & & \Sigma \end{pmatrix},
$$

and $A = X\left(X^\top V^{-1} X\right)^{-1} X^\top V^{-1}$, then it is straightforward to verify that A is idempotent $A^2 = A$ and it follows that AY and $(I - A)Y$ are independent. Thus, $Y = AY + (I - A)Y$ is an orthogonal decomposition of Y, and the density of Y is the product of the density functions of AY and $(I-A)Y$. Note that both AY and $(I-A)Y$ are multivariate normal with singular covariance matrices, and thus their density functions are defined on two subspaces that decompose $\mathbf{R}^{mn}$. The density of $(I - A)Y$ only involves Σ; hence, we can maximize it to estimate Σ without estimating β. This is called the *restricted (or residual, or reduced) maximum likelihood* (REML) estimate. Based on the REML estimate of the variance, one may estimate the parameters β in the mean function using (9.2). This estimate of β based on REML estimate of Σ may be slightly different from the MLE of β by maximizing the log-likelihood l_n with respect to both β and Σ simultaneously. In general, if the number of parameters p is large, REML is preferred over MLE for estimating variance parameters. Asymptotically, REML and MLE estimates are equivalent.

Example 9.1

Consider a study with repeated measurements under different experimental conditions. If the experimental condition is the only predictor in the study, then one may model the repeated measures as

$$
Y_{ij} = \mu_j + \varepsilon_{ij}, \ \boldsymbol{\epsilon}_i = (\varepsilon_{i1} \ \varepsilon_{i2} \ \cdots \ \varepsilon_{im}) \sim N(\mathbf{0}, \Sigma), \tag{9.4}
$$

where μ_j, $j = 1, \ldots, m$, is the effects of the jth condition (the mean response for the jth condition), and Σ is the covariance matrix for the repeated measures. If repeated outcomes are in different scales due to, say measuring different constructs or under different conditions such as weight and height and air travel vs. car travel, we generally do not impose any constraint on Σ and assume an unstructured form for Σ. MLE may be applied for inference and Wald test for linear hypothesis about the regression parameters may be performed based on (9.3). For example, we may assess whether there are differences across the different conditions by testing the null: $\mu_1 = \cdots = \mu_m = 0$. If the first four visit time points for the DOS are treated as different constructs, then the overall test of whether there are differences in the HamD scores across these four visit time points has a chi-square statistic of 5.58 with df 3, resulting in a p-value of 0.0010. Thus, we conclude that there are significant differences among the four time points.

For repeated outcomes on the same scales such as in longitudinal studies, simpler structures may be assumed for the covariance matrix Σ. For example, if any pair of outcomes involve similar constructs or conditions, then an exchangeable covariance structure may be a reasonable assumption for Σ. In such cases, the marginal variances at different conditions are the same and the correlations between any two different conditions are the same. This common correlation is called *intraclass correlation coefficient (ICC)*, a commonly used quantitative measure of variability of clustered data, describing how strongly units in the same cluster resemble each other. Based on the unstructured covariance, one may assess the exchangeable covariance assumption by testing both the homogeneity of the marginal variance and the homogeneity of the correlation. Under the exchangeable covariance assumption, the overall test of whether there are differences in the HamD scores across these four visit time points for the DOS has a chi-square statistic of 6.57 with df 3 with a p-value of 0.0002, yielding the similar conclusion as using the unstructured covariance. ⬜

In this example, there are no other factors other than the repeated conditions (time). The analysis of variance (ANOVA) has been generalized to the analysis of repeated measures in simple study designs with only a few other factors (Hand and Taylor, 1987). Inference can be similarly carried out by decomposing the sum of squares as in ANOVA. Under multivariate normality for the repeated measures, we can use a generalization called *Multivariate ANOVA* (MANOVA). In Repeated Measure ANOVA (RMANOVA), we further make the so-called *sphericity* assumption that the variances of the differences between all possible pairs of within-subject conditions are equal. The statistics in MANOVA and RMANOVA follow F-distributions, which provide the basis for inference without relying on asymptotic theory. As in linear regression, these ANOVAs are special cases of multivariate linear regression models. Since these ANOVAs only apply to very simple study designs and cannot handle missing data, they are not widely used in modern research studies and thus will not be discussed further. There are excellent textbooks on these classic models and interested readers may consult, for example, Hand and Taylor (1987), for details.

9.2.2 Linear mixed-effects models

Modeling of the covariance can be very complicated if the study design is not balanced and/or the covariance may depend on some covariates. The linear mixed-effects model (LMM) for continuous responses, a direct extension of the classical linear regression, addresses the issue of the covariance among repeated responses by modeling the within-subject covariance using latent variables, or *random effects*. Conditioning on such random effects, or the within-subject variability, the individual responses become independent, as the remaining variability only reflects differences across the different subjects. Standard statistical models such as GLM can then be applied to model the between-subject variability.

9.2.2.1 Motivation

Consider a longitudinal study with n subjects. Again, assume a set of fixed assessment times across all subjects, indexed by $t = 1, 2, \ldots, m$. If the mean $E(y_{it})$ is a linear function of time t, the classic linear model for y_{it} at each time t is

$$y_{it} = \beta_0 + \beta_1 t + \widetilde{\epsilon}_{it}, \quad \widetilde{\epsilon}_{it} \sim N\left(0, \sigma_t^2\right), \quad 1 \leq i \leq n, \quad 1 \leq t \leq m. \tag{9.5}$$

Although independent across i for each t, the $\widetilde{\epsilon}_{it}$'s are generally dependent across t for each ith subject, i.e., $Cov(\widetilde{\epsilon}_{is}, \widetilde{\epsilon}_{it}) \neq 0$ for any $1 \leq s < t \leq m$. LMM explicitly models this within-subject correlation using latent variables, or random effects.

For example, (9.5) models the mean $E(y_{it})$ as a function of time, $\mu_t = \beta_0 + \beta_1 t$. As each individual's outcomes y_{it} deviate from this mean response, the idea of random effects is to use a set of latent variables to represent such differences. In this particular case, as μ_t is determined by the intercept β_0 and slope β_1, we can use two latent variables, b_{i0} and b_{i1}, to fully capture the deviation of each individual's responses y_{it} from the mean by

$$y_{it} \mid \mathbf{b}_i = \beta_0 + \beta_1 t + b_{i0} + b_{i1} t + \epsilon_{it}, \tag{9.6}$$

$$\epsilon_{it} \sim \text{i.i.d. } N\left(0, \sigma^2\right), \quad 1 \leq i \leq n, \quad 1 \leq t \leq m,$$

where $\mathbf{b}_i = (b_{i0}, b_{i1})^\top$. Since the dependence among the $\widetilde{\epsilon}_{it}$'s in (9.5) is created by the repeated individual responses, by modeling such individual-level responses, $\beta_0 + \beta_1 t + b_{i0} +$

$b_{i1}t$, rather than the population mean, $\beta_0 + \beta_1 t$, we remove such within-subject dependence, and the error terms ϵ_{it} in (9.6) are independent.

Note that the use of different notation for the error terms between (9.5) and (9.6) is intentional, since ϵ_{it} is the error in modeling the outcome y_{it} using a subject-specific model, $E(y_{it} \mid \mathbf{b}_i) = \beta_0 + \beta_1 t + b_{i0} + b_{i1}t$, whereas $\widetilde{\epsilon}_{it}$ includes the additional between-subject variation $b_{i0} + b_{i1}t$.

By letting $\mathbf{b}_i$ vary across the subjects, we obtain a linear mixed-effects model, with the fixed-effect, $\mu_t = \beta_0 + \beta_1 t$, depicting the population mean, and the random-effect, $b_{i0} + b_{i1}t$, portraying the deviation of each individual's response from the population average. Given that the number of random effects $\mathbf{b}_i$ is the same as the sample size, we generally need to impose restrictions on $\mathbf{b}_i$. In many applications, $\mathbf{b}_i$ are assumed to follow some parametric distribution, with the multivariate normal being the most popular choice. For example, we may assume $(b_{i0} \ b_{i1}) \sim N(\mathbf{0}, D)$ in (9.6) with $D = \begin{pmatrix} d_{11} & d_{12} \\ d_{12} & d_{22} \end{pmatrix}$. Thus, in addition to $\boldsymbol{\beta}$, inference about D is also often of interest, as d_{11} (d_{22}) measures the variability of individual's intercept (slope).

9.2.2.2 Linear mixed-effects models

If the random effects take care of all the correlation among repeated outcomes, then the residuals become independent and the LMM has the following form:

$$y_{it} = \mathbf{x}_{it}^\top \boldsymbol{\beta} + \mathbf{z}_{it}^\top \mathbf{b}_i + \epsilon_{it}, \ \mathbf{b}_i \sim \text{i.i.d.} N(\mathbf{0}, D), \ \boldsymbol{\epsilon}_i \sim \text{i.i.d.} N(0, diag(\sigma_t^2)), \mathbf{b}_i \perp \boldsymbol{\epsilon}_i,$$

for $1 \leq t \leq m, 1 \leq i \leq n$, where $\mathbf{x}_{it}^\top \boldsymbol{\beta}$ is the fixed, $\mathbf{b}_i$ or $\mathbf{z}_{it}^\top \mathbf{b}_i$ the random effects, and "$\perp$" indicates the independence. Different parameters σ_t^2 can take care of the potential heteroskedasticity of the residuals across the time points. For growth-curve analysis, i.e., modeling the change of y_{it} over time as in the case of the example in (9.6), $\mathbf{z}_{it}$ is often set equal to $\mathbf{x}_{it}$. It may be unrealistic to find random effects that take care of all within-subject variability. In some studies, we may still need to model the covariance structure of the error terms even after introducing the random effects. A general LMM thus has the following form:

$$y_{it} = \mathbf{x}_{it}^\top \boldsymbol{\beta} + \mathbf{z}_{it}^\top \mathbf{b}_i + \epsilon_{it}, \mathbf{b}_i \sim \text{i.i.d.} N(\mathbf{0}, D), \boldsymbol{\epsilon}_i \sim \text{i.i.d.} N(\mathbf{0}, \Sigma), \mathbf{b}_i \perp \boldsymbol{\epsilon}_i. \tag{9.7}$$

In general, the covariance structure of the error terms will be much simplified after random effects are included. By letting $\epsilon_{it}' = \mathbf{z}_{it}^\top \mathbf{b}_i + \epsilon_{it}$, the LMM (9.7) can be written in an equivalent multivariate regression format without random effects as

$$y_{it} = \mathbf{x}_{it}^\top \boldsymbol{\beta} + \epsilon_{it}', \epsilon_i' = (\epsilon_{i1}', ..., \epsilon_{im}')^\top \sim \text{i.d.} N(\mathbf{0}, \mathbf{z}_i D \mathbf{z}_i^\top + \Sigma),$$

where $\mathbf{z}_i = (\mathbf{z}_{i1}, ..., \mathbf{z}_{im})^\top$. Note that the covariance matrix depends on covariate $\mathbf{z}_i$, thus the model can be applied in heteroskedasticity cases. In the special cases where there is only the random intercept ($z_{it} \equiv 1, b_i \sim N(0, \nu^2)$), and $\Sigma = \sigma^2 I_m$, the correlation of the response follows the exchangeable structure with common correlation $\rho = \frac{\nu^2}{\nu^2 + \sigma^2}$.

9.2.3 Inference of LMM

LMMs can be written equivalently in the form of multivariate linear regression. Thus, both the MLE and REML discussed above can be applied for inference. Based on estimated parameters, we can also estimate subject-specific random effects. Since random effects often reflect unobserved subject-specific characteristics, estimates of random effects facilitate our understanding of subject-specific features. For example, random effects may reflect unobserved genetic merits in animal breeding, and their estimation is critical in calculating of the estimated breeding values (Mrode, 2014).

9.2.3.1 Inference for Fixed Effects

For a linear hypothesis about the fixed effects

$$H_0 : K\beta = b, \tag{9.8}$$

the classical triplets, Wald, score, and likelihood ratio tests, can all be applied. For example, since $\widehat{\beta} \sim N\left(\beta, \frac{1}{n}\Sigma_\beta\right)$, by exactly the same derivation as that for the Wald statistic (4.15) in Chapter 4, we obtain the *Wald statistic*

$$Q_n^2 = n\left(K\widehat{\beta} - b\right)^\top \left(K\widehat{\Sigma}_\beta K^\top\right)^{-1} \left(K\widehat{\beta} - b\right) \tag{9.9}$$

which follows asymptotically a chi-square distribution χ_l^2 with $l = \text{rank}(K)$. This test relies on the asymptotic theory; more precisely, the covariance Σ_β is substituted by a consistent estimate of Σ_β and the variation associated with the estimate is ignored in the asymptotic inference. Hence, it may not perform well for small samples.

For small samples, F-distributions may be applied to better approximate the distributions of the test statistics. However, except in some simple situations such as those in MANOVA, the statistics do not follow F-distributions exactly. There are some methods proposed to find an approximate denominator degrees of freedom (Kenward and Roger, 1997, Schaalje et al., 2002, Schluchter and Elashoff, 1990, Satterthwaite, 1946, Kenward and Roger, 2009, Fai and Cornelius, 1996). Note that the numeric degrees of freedom is the rank of K. For example, Satterthwaite's approach obtains the degrees of freedom by trying to match the first two moments of $K\widehat{\Sigma}_\beta K^\top / K\Sigma_\beta K^\top$ with those of a chi-square random variable (Satterthwaite, 1946). However, there are no general guidelines as to which method should be preferred, although different approaches may show strengths in different situations. If there is a concern on which approach should be used, we recommend that the results from different approaches be compared.

9.2.3.2 Inference for random components

Inferences for random components usually involve two parts. The first concerns the estimate and inference for the covariance parameters of the random effects and the residuals, and the second addresses the estimate of the subject-specific random effects.

Inference for Covariance Parameters The covariance parameters can be estimated by the REML method. We are often interested in testing whether the covariance structure follows some specific pattern. For example, if the variance of a random effect is zero, then it actually means that the random effect does not exist. Thus, one can consider a model with the random effect and test whether the corresponding variance is zero. One may also be interested in testing whether some correlation parameters are equal to each other. This involves testing contrast hypotheses about the covariance parameters.

In theory, the three common approaches, Wald, score, and likelihood ratio tests, all can be applied to test covariance parameters. However, the Wald test usually converges very slowly and is in general not recommended, unless the sample size is very large. The likelihood ratio test is generally recommended; the likelihood-ratio test statistic is defined as twice the difference in magnitude of the log-likelihoods under the model as well as the model under the null hypothesis, and it follows an asymptotic chi-squared distribution with degrees of freedom equal to the dimension difference between the two parameter spaces.

Estimation of Random Effects The subject-specific random effects can be estimated, as proposed by Henderson (1950), based on the fact that

$$(\mathbf{b}_i, \mathbf{y}_i - \mathbf{x}_i\,\boldsymbol{\beta} - \mathbf{z}_i\mathbf{b}_i) \sim i.d.N\left(\mathbf{0}, \begin{pmatrix} D & \\ & \Sigma \end{pmatrix}\right).$$

Thus, the log "likelihood" for $(\mathbf{b}_i, \mathbf{y}_i - \mathbf{x}_i\,\boldsymbol{\beta} - \mathbf{z}_i\mathbf{b}_i)$ is

$$-\frac{1}{2}\log|D| - \frac{1}{2}\mathbf{b}_i^\top (D)^{-1}\mathbf{b}_i + \left(-\frac{1}{2}\log|R_i| - \frac{1}{2}(\mathbf{y}_i - \mathbf{x}_i\,\boldsymbol{\beta} - \mathbf{z}_i\mathbf{b}_i)^\top (\Sigma)^{-1}(\mathbf{y}_i - \mathbf{x}_i\,\boldsymbol{\beta} - \mathbf{z}_i\mathbf{b}_i)\right).$$

Taking the derivative with respect to $\mathbf{b}_i$, we have the "score equation"

$$-D^{-1}\mathbf{b}_i + \mathbf{z}_i^\top \Sigma^{-1}(\mathbf{y}_i - \mathbf{x}_i\,\boldsymbol{\beta} - \mathbf{z}_i\mathbf{b}_i) = \mathbf{0}.$$

Hence,

$$\mathbf{b}_i = \left(\mathbf{z}_i^\top \Sigma^{-1}\mathbf{z}_i + D^{-1}\right)^{-1}\mathbf{z}_i^\top \Sigma^{-1}(\mathbf{y}_i - \mathbf{x}_i\boldsymbol{\beta}). \tag{9.10}$$

Substituting $\boldsymbol{\beta}$ and Σ with their respective consistent estimates, we obtain

$$\widehat{\mathbf{b}}_i = \left(\mathbf{z}_i^\top \widehat{\Sigma}^{-1}\mathbf{z}_i + D^{-1}\right)^{-1}\mathbf{z}_i^\top \widehat{\Sigma}^{-1}\left(\mathbf{y}_i - \mathbf{x}_i\widehat{\boldsymbol{\beta}}\right).$$

This estimate of the random effect is known as the *Best Linear Unbiased Predictor* (BLUP) for the random effects (Henderson, 1975, Robinson, 1991).

Since we can estimate the individual random effects, we are able to calculate subject-specific fitted values by including the random effects. Thus, there are two versions of fitted values for an LMM, one with and one without the estimated random effects. The one without the random effects, $\mathbf{x}_i^\top \widehat{\boldsymbol{\beta}}$, are the *population average* or *marginal* fitted values, and the one with the random effects, $\mathbf{x}_i^\top \widehat{\boldsymbol{\beta}} + \mathbf{z}_i^\top \widehat{\mathbf{b}}_i$, are the subject-specific or conditional fitted values.

Example 9.2

Consider the following simple random intercept model without any covariate:

$$y_{it} = \mu + \alpha_i + \varepsilon_{it}, \ 1 \le i \le n, 1 \le t \le n_i,$$

where $\alpha_i \sim N(0, \nu^2)$ and $\varepsilon_{it} \sim N(0, \sigma^2)$ are all independent, and n_i is the number of repeated measures for the ith subject. Then, $\mathbf{z}_i = 1$ and $\mathbf{x}_{it}^\top \boldsymbol{\beta} = \mu$ in (9.10), and we have

$$\widehat{\alpha}_i = \left(\frac{n_i}{\sigma^2} + \frac{1}{\nu^2}\right)^{-1}\frac{n_i}{\sigma^2}(\overline{y}_{i\cdot} - \mu),$$

where $\overline{y}_{i\cdot} = \frac{1}{n_i}\sum_{t=1}^{n_i} y_{it}$ is the mean response for the ith subject. Thus, the estimate of the random effect is based on the observed deviation of the subject average $(\overline{y}_{i\cdot})$ from the population mean(μ), but shrunk by a factor $\lambda = \frac{n_i}{\sigma^2}/\left(\frac{n_i}{\sigma^2} + \frac{1}{\nu^2}\right)$. If ν^2 is held fixed, then as σ^2 goes to 0 the *shrinkage factor* λ goes to 1 and $\widehat{\alpha}_i = \lambda(\overline{y}_{i\cdot} - \mu)$ goes to $\overline{y}_{i\cdot} - \mu$; in this limiting case, the random effect accounts for all the variation of the observed mean deviation from the population for the ith subject. If σ^2 is held fixed, then as ν^2 goes to 0, the shrinkage factor λ goes to 0 and $\widehat{\alpha}_i = \lambda(\overline{y}_{i\cdot} - \mu)$ goes to 0; in this limiting case, there is no random effect and the observed subject deviation from the population mean is totally random. Between the two limiting situations, we have $0 < \lambda < 1$, and the estimated random effect, or the predicted subject-specific latent deviation from the population, is smaller than the observed deviation. In other words, the BLUP estimate shrinks the observed difference from the population toward zero. The shrinkage in the prediction results from the fact that the observed overall average deviation from the population mean consists of two independent components, the random effect and the mean random errors; the random effect is only part of the observed overall average deviation (see Problem 9.9). ⬜

9.2.4 Clustered studies

LMMs can also be applied to clustered data by using random effects for the clusters. For example, for clustered data arising from nested studies such as multi-center trials, it is often of interest to see whether there is any site effect. We can readily use the LMM above to examine this issue.

Consider modeling treatment difference at a posttreatment assessment in a multi-center, randomized trial with two treatment conditions. Let y_{ij} denote some response of interest from the jth subject within the ith site. Then, an appropriate LMM may be given by

$$y_{ij} = \beta_0 + \beta_1 x_{ij} + b_i + \epsilon_{ij}, b_i \sim \text{i.i.d. } N\left(0, \sigma_b^2\right), \epsilon_{ij} \sim \text{i.i.d.} N\left(0, \sigma^2\right), b_i \perp \epsilon_{ij},$$

where x_{ij} indicates the treatment received by subject j at site i, and b_i denotes the (random) site effect. We can formally assess whether there is any significant site effect by testing the null $H_0 : \sigma_b^2 = 0$; if the null is not rejected, we can simplify the model by removing b_i.

When the number of sites is small such as two or three, we may want to model potential site differences using fixed effects. Otherwise, it is more sensible to model site differences using random effects, since it saves more degrees of freedom for testing hypotheses concerning the fixed effects. But, more importantly, site differences in this case are typically of no particular interest, especially in multi-center trials where multiple sites are usually utilized to increase the sample size of the study.

If there are repeated measurements in the above study, then we will have two different levels of clustering. Repeated measurements from the same patient form clusters by patients, while patients within the same center form clusters at the center level. We may use multilevel mixed-effect models, with random effects at both the center level and the patient level to deal with the different levels of clustering. It is straightforward to generalize the idea of random effect to clustered studies with more than two levels.

9.2.4.1 Longitudinal and clustered data

Longitudinal data may be considered as a special kind of clustered data; all repeated measures from the same subject form a cluster. Thus, there is a natural temporal order for the repeated measures in longitudinal data, and this temporal order is critical in modeling the temporal trend and covariance structures for longitudinal data. In contrast, there is no special order among observations within the same cluster for non longitudinal general clustered data. As a result, exchangeable error terms are the natural choice for such clustered data.

Longitudinal studies are usually planned to have the same number of measures for everyone, although unbalanced clusters often occur because of missing values due to missed assessments, study dropout, etc. Thus, the cluster sizes are usually comparable in longitudinal studies. However, cluster sizes may be quite different in non longitudinal clustered data. For example, doctors may have very different numbers of patients. Sometimes, the sizes themselves may be important measures and should be taken into consideration in data analysis.

9.2.4.2 Cluster-level variables

For analysis of clustered data, it is important to consider *contextual* variables that describe properties of the clusters. Unlike longitudinal studies where characteristics at cluster levels, i.e., the subjects, are usually directly measured, some efforts may be needed to obtain cluster level covariates for general clustered data. This is because we often want to include compositional variables that are aggregated individual level variables. For example, the performance of a school may be measured by taking the average of some standard test scores of all students. We may also need contextual variables from separate resources. For example, the average income of the community to which a school (cluster) belongs may be obtained from publicly available databases outside of the school. We need to pay special attention to such contextual variables as they may not be directly available from study subjects. For inference, all the approaches discussed above for longitudinal data apply.

9.2.4.3 Contextual and individual-effects

In regression analysis of clustered data, we are usually interested in explaining variability in individual-level dependent variables using combinations of individual- and cluster-level independent variables (contextual-effects models). Thus, the predictors can be individual- or cluster-level variables. Effects of cluster-level variables, or contextual variables, are termed *contextual effects*. We may also consider interactions between individual and cluster-level variables.

For cross-sectional studies, centering predictors helps to facilitate interpretation of parameters and reduces potential multi-collinearity resulting from including higher-order or interaction terms of the predictors (e.g., quadratic or interaction terms of such predictors), but the contextual meaning and the parameter of a predictor remains the same regardless of whether a predictor is centered. However, this is generally not the case when centering predictors in clustered data analysis. For clustered data, it may be important to center a predictor at the individual level by cluster-level means and medians. For example, work hours is an important predictor of psychological well-being for employees (Bliese and Halverson, 1996). It is important to center work hours for each subject by employer-level (cluster) means or medians, especially when comparing work hours for subjects within the same employer. Since different clusters may have different averages, estimates from such centered predictors will have different interpretations from those with uncentered ones. Of course, one may also include cluster-level predictor, the employer-wise average of work hours, to study between-cluster regression relations (Bliese and Halverson, 1996).

9.2.5 Power for longitudinal and clustered studies

Longitudinal study designs are often adopted to increase "sample size" because they feature repeated assessments. When used to model stationary relationships, i.e., the same relationship over time, between a response and explanatory variables, repeated measures do increase power. However, the amount of power gained by repeated measures depends on how repeatedly assessed variables are correlated. In particular, one should not expect the same increment as achieved by directly increasing the sample size. For example, the much smaller standard errors for the estimates based on the pooled data compared to their GEE counterparts in the DOS (see the table in Example 9.8) show that the amount of power gain by repeated measures is much less than by including an equivalent number of new subjects.

To appreciate the effect of correlation on power, consider first a scenario involving high correlations among the repeated measures. In this case, there is a large amount of information shared by the repeated assessments, and thus increasing the number of assessments does not yield much new additional information. On the other hand, if such within-subject correlation is small, information will be significantly increased from additional assessments. For simple randomized longitudinal studies comparing mean differences with exchangeable correlation structure and time-independent error term variance, a longitudinal study with n subjects and m assessments is equivalent to a cross-sectional study with sample size $n\frac{m}{1+(m-1)\rho}$, where ρ is the ICC and $1+(m-1)\rho$ a *design effect*. Thus, power gains depend on the design effect, which is determined by the ICC. In the extreme case of $\rho = 1$, there is no increase in power. As ρ decreases toward 0, the equivalent sample size $n\frac{m}{1+(m-1)\rho}$

increases to nm. Thus the within-subject correlation plays an important role in power gains for longitudinal studies. It is important to weigh in logistical factors such as cost and burden of assessment before deciding to add extra assessments for the purposes of gaining additional power. Readers interested in this and related power issues may consult relevant publications for more detailed discussions of this topic (Frison and Pocock, 1992, Tu et al., 2004, 2006, 2007).

9.3 Generalized Linear Mixed-Effects Models

GLMs can be extended to a longitudinal data setting to model categorical and count responses. In the case of linear models, we add random effects $z_{it}^{\top} b_i$ to account for individuals' deviations from the (population) mean response and then model the outcome y_{it} at each time t conditional on the random effects using the standard linear regression. The mixed-effect-based mean $E\left(y_{it} \mid \mathbf{x}_i, \mathbf{z}_i, \mathbf{b}_i\right) = \mathbf{x}_{it}^{\top}\boldsymbol{\beta} + \mathbf{z}_{it}^{\top}\mathbf{b}_i$ represents the individual, rather than the population mean $\mathbf{x}_{it}^{\top}\boldsymbol{\beta}$. Hence, given $\mathbf{x}_{it}, \mathbf{z}_{it}$, and $\mathbf{b}_i$, y_{it} have a much simplified covariance structure and become independent across the assessment times if the covariance of the repeated measures is totally accounted for by the random effects.

The separation of the deterministic and random components in GLMs enables us to easily apply the same idea of adding random effects to the linear predictors in modeling categorical and count outcomes. For a categorical or count response y_{it}, consider a vector of covariates $\mathbf{x}_{it}$ ($\mathbf{z}_{it}$) for the fixed (random) effects of interest from the ith subject at time t in a longitudinal study with n subjects and m assessment times $(1 \le i \le n, 1 \le t \le m)$. For each subject, we first include the random effects $\mathbf{z}_{it}^{\top}\mathbf{b}_i$ in the linear predictor and then model y_{it} conditional on the linear predictor using a generalized linear model, i.e.,

$$y_{it} \mid \mathbf{x}_i, \mathbf{z}_i, \mathbf{b}_i \sim \text{i.i.d. } f\left(\mu_{it}\right), \quad g\left(\mu_{it}\right) = \eta_{it} = \mathbf{x}_{it}^{\top}\boldsymbol{\beta} + \mathbf{z}_{it}^{\top}\mathbf{b}_i, \quad (9.11)$$

for $1 \le i \le n$ and $1 \le t \le m$, where $g\left(\cdot\right)$ is a link function, $\mathbf{b}_i$ denotes the random effects, and $f\left(\mu\right)$ some probability distribution function with mean μ. For simplicity, we assume y_{it} given $\mathbf{x}_{it}, \mathbf{z}_{it}$, and $\mathbf{b}_i$ are independent across the different time points, although more complicated models may be applied. As in the case of LMM, $\mathbf{b}_i$ is often assumed to follow a multivariate normal distribution $\mathbf{b}_i \sim N\left(\mathbf{0}, D\right)$, although other types of more complex distributions such as mixtures of normals may also be specified. Thus, the main conceptual difference between the *generalized linear mixed-effects model* (GLMM) and its predecessor LMM is the link function $g\left(\mu\right)$.

9.3.1 Binary response

For a binary response y_{it}, it is natural to set $f\left(\mu_{it}\right) = Bernoulli\left(\mu_{it}\right)$ in (9.11). Thus, a GLMM with the logit link function for a binary response is given by

$$y_{it} \mid \mathbf{x}_i, \mathbf{z}_i, \mathbf{b}_i \sim \text{i.d. } Bernoulli\left(\mu_{it}\right), \quad \mu_{it} = \frac{\exp(\mathbf{x}_{it}^{\top}\boldsymbol{\beta} + \mathbf{z}_{it}^{\top}\mathbf{b}_i)}{1 + \exp(\mathbf{x}_{it}^{\top}\boldsymbol{\beta} + \mathbf{z}_{it}^{\top}\mathbf{b}_i)}, \quad (9.12)$$

where $\mathbf{b}_i$ denotes the random effects. GLMMs with the probit and complementary log-log link functions can be similarly defined by replacing the link functions. The use of link functions also makes it easy to apply modeling techniques for LMM to GLMM. For example, if we model the trajectory of y_{it} over time using a linear function of t with a bivariate normal random effect for the mean and slope, the GLMM becomes:

$$y_{it} \mid \mathbf{x}_i, \mathbf{z}_i, \mathbf{b}_i \sim \text{i.d. } Bernoulli\,(\mu_{it})\,, \text{logit}\,(\mu_{it}) = \eta_{it} = \beta_0 + \beta_1 t + b_{i0} + b_{i1} t,$$
$$\mathbf{b}_i = (b_{i0}, b_{i1}) \sim \text{i.i.d. } N\,(\mathbf{0}, D)\,, \quad 1 \leq i \leq n, \quad 1 \leq t \leq m. \tag{9.13}$$

As in the LMM case, $\beta_0 + \beta_1 t$ describes the (linear) change over time for the population as a whole, while $b_{i0} + b_{i1} t$ accounts for individuals' differences from this population average. Note that unlike LMM the trajectory of the mean μ_{it} of y_{it} under (9.13) is not a linear function of time, despite the fact that η_{it} is. This implies a critical difference, compared with LMM, when we interpret the fixed effect.

Note that in GLMM (9.12), we assumed that all the repeated outcomes are independent, conditioning on both fixed and random effects. If this is not the case and they are still correlated, then the correlation among the repeated binary outcomes from the same subjects may cause overdispersion as we discussed in Chapter 4. In such cases, a scaled variance may be applied to take care of the overdispersion issue.

9.3.2 Maximum likelihood inference

Since GLMMs are parametric models, conceptually we may apply the MLE method for parameter estimation and inference. Given the distribution f and the link g in (9.11), the log-likelihood is readily derived based on the independence of the repeated measurements upon conditioning on $\mathbf{x}_{it}$, $\mathbf{z}_{it}$ and $\mathbf{b}_i$. For example, for a binary response y_{it} in (9.13), this log-likelihood is given by:

$$l = \sum_{i=1}^{n} \log \left[\int_{\mathbf{b}_i} \prod_{t=1}^{m} \left[\mu_{it}^{y_{it}} \left(1 - \mu_{it}\right)^{1 - y_{it}} \right] \phi\,(\mathbf{b}_i \mid \mathbf{0}, \Sigma_b)\, d\mathbf{b}_i \right], \tag{9.14}$$

where $\phi\,(\mathbf{b}_i \mid \mathbf{0}, \Sigma_b)$ denotes the probability density function of a multivariate normal with mean $\mathbf{0}$ and variance Σ_b. Similar to the LMM case, we can perform Wald and likelihood ratio tests based on the asymptotic theory.

Although simple in appearance, it is actually quite a daunting task to compute the MLE, mainly because of the integration of the random effects in the likelihood. In the LMM case, the integration of the random effects can be carried out analytically, or in closed form, to obtain the equivalent multivariate regression without random effects. However, this is impossible for (9.14) except in special simple cases. Different numeric methods have been developed to calculate the integrals in (9.14). Common approaches include the Laplacian approximation which approximates the integrands with special forms, and Gaussian quadrature, a common numeric integration method (Pinheiro and Chao, 2006). Instead of trying to calculate the integrals in the likelihood, one may also apply pseudo models based on the linearization technique (Schabenberger and Gregoire, 1996). Different approaches often result in different estimates. Implementations of the same approach in different software packages may also produce quite different estimates (Zhang et al., 2011).

9.3.2.1 Parameter interpretation

The parameters in the fixed effect component can be interpreted by holding all other covariates and all random effects fixed. For example, if a logit link function is adopted, the parameters for the fixed effects similarly have an interpretation in terms of odds ratio. More precisely, let $\mathbf{x}_t = (x_{1t}, \tilde{\mathbf{x}}_t)$ and β_1 denote the coefficient of x_{1t} in (9.12), then the odds of response at $x_{1t} = a$ is $\exp\left(\beta_0 + \tilde{\mathbf{x}}^\top \tilde{\boldsymbol{\beta}} + \beta_1 a + \mathbf{z}_{it}^\top \mathbf{b}_i\right)$, where $\tilde{\mathbf{x}}$ $(\tilde{\boldsymbol{\beta}})$ denotes the vector $\mathbf{x}$ (the parameter vector $\boldsymbol{\beta}$) with the component of x_{1t} (β_1) removed. The odds of response after one unit increase in this covariate, i.e., $x_1 = a + 1$, with the remaining components of $\mathbf{x}$ held the same is $\exp\left(\beta_0 + \widetilde{\mathbf{x}}^\top \widetilde{\boldsymbol{\beta}} + \beta_1(a+1) + \mathbf{z}_{it}^\top \mathbf{b}_i\right)$. Thus, the odds ratio of response per unit increase in x_1 is $\frac{\exp\left(\beta_0 + \mathbf{x'}^\top \boldsymbol{\beta'} + \beta_1(a+1) + \mathbf{z}_{it}^\top \mathbf{b}_i\right)}{\exp\left(\beta_0 + \mathbf{x'}^\top \boldsymbol{\beta'} + \beta_1 a + \mathbf{z}_{it}^\top \mathbf{b}_i\right)} = \exp(\beta_1)$, if all other covariates in the fixed effect components $\widetilde{\mathbf{x}}$, all the random effects $\mathbf{b}_i$ (not observed), and the covariates $\mathbf{z}_{it}$, are held fixed.

Since the random effects $\mathbf{b}_i$ are not observed, this odds ratio interpretation must be subject specific; the parameter can only be interpreted as the log odds ratio for the same subjects had they only increased the corresponding covariate in one unit. The parameter does not have the log odds ratio interpretation at the population level. In other words, the parameters in the fixed effect component are different from the parameters for the corresponding effect at the population level.

What can we say about the odds ratio at the population level? For a given covariate value $\mathbf{x}$ and $\mathbf{z}$, the response likelihood at the population level needs to integrate out all the random effects, i.e., $\Pr(y_{it} = 1 | \mathbf{x}_i, \mathbf{z}_i) = \int_{\mathbf{b}_i} \mu_{it} \phi(\mathbf{b}_i \mid \mathbf{0}, \Sigma_b) \, d\mathbf{b}_i$. Thus, the log odds of response at a specific covariate value cannot be expressed simply as a linear function of the covariates. As a result the odd ratios when a covariate increases by one unit (and all others are held fixed) in general are different from the exponential of its coefficient; in fact the odds ratio is not even a constant, as it changes with the value of the covariate. Below, we use some numeric examples to illustrate this. See also Problem 9.18 for an approximate relation.

Consider a simple case with one univariate, time invariant covariate x and random intercept. We have $\mu_{it} = \frac{\exp(\beta_0 + \beta_1 x_i + r_i)}{1 + \exp(\beta_0 + \beta_1 x_i + r_i)}$, with $r_i \sim N(0, \nu^2)$. The population level odds of response at $x_1 = a$ is $\frac{\Pr(y_{it} = 1 | x_1 = a)}{\Pr(y_{it} = 0 | x_1 = a)} = \frac{\Pr(y_{it} = 1 | x_1 = a)}{1 - \Pr(y_{it} = 1 | x_1 = a)}$, where $\Pr(y_{it} = 1 | x_1 = a) = \int_{-\infty}^{\infty} \frac{\exp(\beta_0 + \beta_1 x_i + r_i)}{1 + \exp(\beta_0 + \beta_1 x_i + r_i)} \frac{1}{\sqrt{2\pi}} \exp\left(-\frac{r_i^2}{2\nu^2}\right) dr_i$ depends on the variance of the random effect r_i, ν^2. When x_i increases by one unit, the corresponding OR actually changes with the value of x_i. For example, when $\nu^2 = 1$ the populational level OR at level $x = 1$ vs. level $x = 0$ is 2.2974, but the OR at level $x = 2$ vs. level $x = 1$ is 2.3646. As x goes to infinity, the population-level OR at level $x + 1$ vs. level x converges to $\exp(\beta_1)$.

If the variance of the random intercept is very small, then the random effect may be ignored, in which case β_1 would approximately be the $\log(OR)$. On the other hand, if the variance is very large, then the $\log(OR)$ is close to 1, indicating that that x is no longer predictive.

9.3.2.2 Predicted values

As in LMMs, we can estimate the random effects using maximum likelihood methods. Based on the estimated fixed and random effects, we can compute the fitted values, the linear predictor or the predicted probability. If the random effect is included, $\mathbf{x}_{it}^\top \widehat{\boldsymbol{\beta}} + \mathbf{z}_{it}^\top \widehat{\mathbf{b}}_i$ is the subject specific fitted value for the linear predictor. Since the expectation of the random effect is zero, the fitted linear predictor without the random effect, $\mathbf{x}_{it}^\top \widehat{\boldsymbol{\beta}}$, is the population average of the linear predictor.

For predicted probabilities for a binary response, we need to apply the link function to the linear predictor. The fitted value with the random effect included is $\frac{\exp(\mathbf{x}_{it}^\top \widehat{\boldsymbol{\beta}} + \mathbf{z}_{it}^\top \widehat{\mathbf{b}}_i)}{1 + \exp(\mathbf{x}_{it}^\top \widehat{\boldsymbol{\beta}} + \mathbf{z}_{it}^\top \widehat{\mathbf{b}}_i)}$, the predicted probability of the outcome being 1 at time t for the subject. Note that without including the random effect, one may compute another fitted value, $\frac{\exp(\mathbf{x}_{it}^\top \widehat{\boldsymbol{\beta}})}{1 + \exp(\mathbf{x}_{it}^\top \widehat{\boldsymbol{\beta}})}$. We must be mindful that this is not the population level fitted value for the probability of $y_{it} = 1$, although they are related (see Problem 9.18).

Example 9.3

In Example 4.8, we checked how demographic information and medical burden predicted the depression outcome (dichotomized) using the baseline information of the DOS. Here, we assess their relation by applying a GLMM with the same fixed effect component and an added random intercept to the longitudinal data. More precisely, we apply the following GLMM:

$$\text{logit}(\mu_{it}) = \beta_0 + b_{i0} + \beta_1 Age_i + \beta_2 Gender_i + \beta_3 Edu_i + \beta_4 CIRS_{it} \quad (9.15)$$
$$+ \beta_5 MS1_i + \beta_6 MS2_i + \beta_7 Race_i,$$

where μ_{it} is the probability of being depressed for the ith subject at time t. In this model, the added random intercept is to account for correlations among the repeated measures. We use all subjects who completed the first four assessments. The estimates of the fixed effects are summarized in the following table:

Effect	Estimate	Standard Error	DF	t Value	Pr > |t|
Intercept	8.0289	7.1936	218	1.12	0.2656
Age	−0.1955	0.08533	674	−2.29	0.0223
Gender	1.9050	1.0947	674	1.74	0.0823
Education	−0.2903	0.2204	674	−1.32	0.1881
CIRS	0.6427	0.1126	674	5.71	< .0001
MS1	−2.2403	1.3298	674	−1.68	0.0925
MS2	−0.03240	1.4599	674	−0.02	0.9823
Race	2.5224	3.7859	674	0.67	0.5055

Based on the estimates, both gender and medical burden are significantly associated with the depression outcome at the 5% significance level. The estimated coefficient of 0.6427 for CIRS indicates that for the same person with other variates and the random effect held unchanged, the log of the OR of being depressed per unit increase in CIRS is 0.6427. However, this is not the log of the marginal odds ratio.

The estimated variance of the random intercept is 34.0344 with standard error 8.1031 and a p-value < 0.0001. If the random intercept is removed, i.e., if we assume the repeated measures are independent, then the estimated likelihood will decrease by more than 100. Thus, the likelihood ratio test also confirms the significance of the random effect. ▯

Note that there are some missing values in the data due to dropouts of some patients in the middle of the study. Thus, the estimates in the table above may be biased if the selected subgroup of subjects with completed data during the four-year study period is not representative of the initial study group at baseline. To examine this potential bias, it is necessary to consider effects of missing data on the estimates. As the latter is quite a daunting task, we will devote an entire chapter (Chapter 11) to discussing this common and complex problem.

9.3.3   Polytomous responses

We can also add random effects to the linear predictor to model the correlation among the repeated measures for polytomous responses. Note that as there are multiple linear predictors, we may include random effects in each of them. For example, for a nominal response y_{it} with levels $1, ..., J$, we may have the following GLMM

$$y_{it} \mid (\mathbf{x}_{it}, \mathbf{z}_{it}, \mathbf{b}_i) \sim \text{i.d. } MN(1, \boldsymbol{\pi}_{it}), \quad \log\left(\frac{\pi_{itj}}{\pi_{itJ}}\right) = \mathbf{x}_{it}^\top \boldsymbol{\beta}_j + \mathbf{z}_{it}^\top \mathbf{b}_{ij},$$

$$\mathbf{b}_i = (\mathbf{b}_{i1}, ..., \mathbf{b}_{iJ-1}), \quad \mathbf{b}_i \sim \text{i.i.d. } N(\mathbf{0}, D), \quad j = 1, ..., J-1, \tag{9.16}$$

where $\pi_{itj}(\mathbf{x}) = \Pr(y_{it} = j)$, $\mathbf{b}_{ij}$ denotes the random effects, and $\mathbf{x}_{it}$ and $\mathbf{z}_{it}$ are the covariates for the fixed and random effects, respectively. Note that the Jth level is used as the referent in (9.16); however, as the outcome is nominal and there is no intrinsic order among the outcome levels, the model should be invariant with respect to the choice of the reference. As discussed in Chapter 5, this is not an issue for the fixed effect component as different parameters $\boldsymbol{\beta}_j$ are specified for different levels. For the random effects, this means we should also specify different random effects $\mathbf{b}_{ij}$, i.e., it is improper to assume a common random effect $\mathbf{b}_{ij} = \mathbf{b}_i$. Further, the covariance structure of $\mathbf{b}_i = (\mathbf{b}_{i1}, ..., \mathbf{b}_{iJ-1})$ should also be invariant with respect to the choice of the reference. For example, it is in general not appropriate to assume that random effects corresponding to different levels are independent as this property is not invariant with respect to the choice of the reference level (see Problem 9.11). Unstructured covariance may be a natural choice from the point of view of modeling. However, since inference under unstructured covariance requires integration with multi-dimensional normal density functions, this choice of covariance may involve very complicated computations.

When the response variable is ordinal, we may have more flexibility as in the cross-sectional data cases. For example, it may be reasonable to assume a single random effect across different levels. If we specify an adjacent logistic model similar to (9.16), we may have the following GLMM

$$y_{it} \mid \mathbf{x}_i, \mathbf{z}_i, \mathbf{b}_i \sim \text{i.i.d. } MN(1, \boldsymbol{\pi}_{it}), \quad \log\left(\frac{\pi_{itj}}{\pi_{it(j+1)}}\right) = \mathbf{x}_{it}^\top \boldsymbol{\beta}_j + \mathbf{z}_{it}^\top \mathbf{b}_{ij}, \tag{9.17}$$

where $\mathbf{b}_{ij}$ denotes the random effects. When both $\boldsymbol{\beta}_j$ and $\mathbf{b}_{ij}$ are different across the levels j, this model is actually equivalent to model (9.16); the only difference is the arrangement of the pairs to be included in the model. However, as in the cross-sectional data case, it may be meaningful to make some or all the coefficients $\boldsymbol{\beta}_j$ and $\mathbf{b}_{ij}$ the same across the different levels j. If common $\boldsymbol{\beta}$ and $\mathbf{b}_i$ are assumed, we obtain the much simplified model:

$$y_{it} \mid \mathbf{x}_i, \mathbf{z}_i, \mathbf{b}_i \sim \text{i.i.d. } MN(1, \boldsymbol{\pi}_{it}), \quad \log\left(\frac{\pi_{itj}}{1 - \pi_{itj}}\right) = \mathbf{x}_{it}^\top \boldsymbol{\beta} + \mathbf{z}_{it}^\top \mathbf{b}_i. \tag{9.18}$$

Given the computational complexity of GLMMs, this simpler model also simplifies computations of model estimates when used to fit ordinal data.

Interpretations of the parameter $\boldsymbol{\beta}$ are similar to the cross-sectional data case, except that when a nonlinear link function is applied, $\boldsymbol{\beta}$ should be interpreted by fixing the random effects. For example, if a common coefficient $\boldsymbol{\beta}$ is assumed across different levels for a cumulative logistic model, then this common coefficient implies the proportional odds only for subjects with the same random effects across different levels. Similarly, one can assume different coefficients to test whether the proportional odds assumption is reasonable.

Example 9.4

In Example 9.3, we applied a GLMM to model the dichotomized depression diagnosis for the DOS data. Since the original depression diagnosis in the data set is an ordinal with three levels, now we apply a proportional odds model with a random intercept. Parameter estimates are summarized in the following table:

Effect	Estimate	Standard Error	DF	t Value	Pr > |t|
Intercept1	10.1137	8.7785	218	1.15	0.2505
Intercept2	14.1740	8.8499	218	1.60	0.1107
Age	−0.2466	0.1024	673	−2.41	0.0163
Gender	1.5327	1.2101	673	1.27	0.2058
Education	−0.3793	0.2740	673	−1.38	0.1667
CIRS	0.5468	0.08864	673	6.17	< .0001
MS1	−2.8241	1.6365	673	−1.73	0.0849
MS2	0.1150	1.7631	673	0.07	0.9480
Race	2.4697	4.6801	673	0.53	0.5979

Compared with the corresponding analysis on the pooled data (ignoring the correlation among repeated measures), the point estimates are quite different and the p-values are generally higher in the GLMM models when accounting for the correlation among repeated measures. For example, under the GLMM model, education is not significant anymore. □

9.3.4 Count responses

Generalized linear mixed effect models can also be developed for count responses by adding random effects to the linear predictors. For example, assuming a log link function and Poisson distributions, we may have the following mixed-effects Poisson regression model:

$$y_{it} \mid \mathbf{x}_i, \mathbf{z}_i, \mathbf{b}_i \sim \text{i.d. } Poisson\left(\mu_{it}\right), \quad \mu_{it} = \exp(\mathbf{x}_{it}^\top \boldsymbol{\beta} + \mathbf{z}_{it}^\top \mathbf{b}_i), \tag{9.19}$$

for $1 \leq i \leq n, \quad 1 \leq t \leq m$, where $\mathbf{b}_i \sim N\left(\mathbf{0}, D\right)$. The repeated outcomes are assumed to be independent conditioning on the random effects. If the random effects are not sufficient to account for all the correlation, we may have overdispersion as discussed in Chapter 6, and the Poisson distribution may be replaced with one allowing for overdispersion such as negative binomial (NB) or generalized Poisson (GP) distributions.

Interpretation of parameters in the fixed-effect component is similar to that of the Poisson regression in Chapter 6, except for controlling for the random effects. Thus, the exponentiated parameters represent the ratio of mean responses per unit increase of the independent variable for the same person with all the other fixed and random effects unchanged. Thus, the interpretation is subject specific. As in the binary case, the parameters are not exactly the marginal effect at the population level.

Example 9.5
For the Catheter Study data, we model the number of UTIs with gender and treatment as the predictors using a Poisson log-linear regression model. We assume that given the covariates and random intercept, the repeated responses are independent. The parameter estimates of the fixed effects are summarized in the following table (pooled data analysis is also included for comparison purposes):

Parameter	Random Intercept Model			Pooled Model		
	Estimate	SE	Pr > |t|	Estimate	SE	Pr > |t|
Intercept	−1.9372	0.1987	< .0001	−1.4566	0.1135	< .0001
Gender	0.1835	0.1986	0.3557	0.1054	0.1227	0.3903
Group	0.2117	0.2015	0.2938	0.1788	0.1228	0.1454

The point estimates for the fixed effects are different compared to the pooled data analysis, which is expected because of the nonlinear link function and correlated outcomes. However, directions of associations all agree between the two models, although all of them are nonsignificant. Further, the pooled data analysis generally yields smaller SE and smaller p-values. Compared with the pooled data analysis, the only difference is that the GLMM has the additional random intercept. A test of the random effect shows that the random intercept is quite significant (p-value < 0.0001). Thus the repeated measures are indeed correlated and do have an impact on the estimation. □

9.3.4.1 Mixed-effect zero-modified models

We can also add random effects to linear predictors to obtain mixed-effects models for zero-modified longitudinal data. For example, if a Poisson distribution is assumed for the data before zero-truncation, then we can add random effects to the linear predictor and have a zero-truncated Poisson (ZTP) mixed-effects model

$$y_{it} \mid \mathbf{x}_i, \mathbf{z}_i, \mathbf{b}_i \sim \text{ZTP}(\mu_{it}), \quad \log(\mu_{it}) = \mathbf{x}_{it}^\top \boldsymbol{\beta} + \mathbf{z}_{it}^\top \mathbf{b}_i. \tag{9.20}$$

As before, we assume that given $\mathbf{x}_i, \mathbf{z}_i, \mathbf{b}_i$, the repeated responses y_{it} are independent, and $\mathbf{b}_i \sim N(\mathbf{0}, D)$. The parameter of a variable in the fixed-effects is similarly interpreted by holding all other fixed- and random-effects constant, in which case its exponentiated value indicates the ratio of the mean responses per unit increase of the variable before zero-truncation.

For zero-inflated models, there are two components. We can add random effects to either or both components. For example, by including random effects in both linear predictors, we may have the following ZIP mixed-effects model

$$y_{it} \mid \mathbf{x}_i, \mathbf{z}_i, \mathbf{b}_i \sim \text{ZIP}(\mu_{it}, \rho_{it}), \quad \log(\mu_{it}) = \mathbf{v}_{it}^\top \boldsymbol{\beta}_v + (\mathbf{z}_{it}^c)^\top \mathbf{b}_i^c,$$

$$\text{logit}(\rho_{it}) = \mathbf{u}_{it}^\top \boldsymbol{\beta}_u + (\mathbf{z}_{it}^z)^\top \mathbf{b}_i^z, \quad \mathbf{b}_i = (\mathbf{b}_i^c, \mathbf{b}_i^z) \sim N(\mathbf{0}, D), \tag{9.21}$$

where $\mathbf{u}_{it}$ and $\mathbf{v}_{it}$ are two subsets of the vector $\mathbf{x}_{it}$, $\mathbf{b}_i^c$ is the random effects for the count, and $\mathbf{b}_i^z$ is the random effects for the inflated-zero component. Given $\mathbf{x}_i, \mathbf{z}_i, \mathbf{b}_i^c$, and $\mathbf{b}_i^z$, the repeated responses y_{it} are assumed independent and following ZIP. The parameter of a variable in the fixed effect is similarly interpreted by holding all other fixed- and random-effects fixed. So, for each variable in the fixed-effects of the count component, the exponentiated parameter is the ratio of the mean response per unit increase of the variable for the at-risk subgroup. Likewise, the exponentiated parameter in the fixed-effects of the zero-inflated component is the log odds ratio of being structural zeros per unit increase of the variable for the whole population.

Example 9.6

We model the number of UTIs in the Catheter Study data with zero-inflated Poisson regression, with gender and treatment as the predictors in the Poisson component, and gender and a random intercept for the zero-inflation component. We assume that given the independent variables and random intercept, the repeated outcomes are independent. Shown in the table below are the estimates of the fixed effects.

Parameter	Estimate	SE	DF	t Value	Pr > |t|	95 %	CI
Poisson Component							
Intercept	−0.5669	0.1622	192	−3.49	0.0006	−0.8868	−0.2469
Group	0.01765	0.1528	192	0.12	0.9081	−0.2837	0.3190
Gender	−0.2164	0.1855	192	−1.17	0.2449	−0.5823	0.1495
Inflated-zero Component							
Intercept	0.4812	0.4872	192	0.99	0.3245	−0.4797	1.4422
Gender	−1.3887	0.7907	192	−1.76	0.0806	−2.9483	0.1709
ν	2.8104	0.4484	192	6.27	< .0001	1.9259	3.6949

The p-value for Gender in the inflated-zero component is under 0.1, indicating that females are less likely to be "structural zeros" than males (the reference level for Gender in the model). The variance of the random intercept (ν^2) is significantly different from 0, confirming that the repeated count responses are correlated. ⬜

9.4 Marginal Models for Longitudinal Data

Instead of explicitly modeling the correlation, we may deal with the correlation using robust inference. In marginal models, we only assume the marginal distribution for the outcome at each time point. For example, we may simply assume a linear regression for a continuous response at each time point:

$$y_{it} \sim N\left(\mathbf{x}_{it}^{\top}\boldsymbol{\beta},\ \sigma^2\right), \text{ or } y_{it} = \mathbf{x}_{it}^{\top}\boldsymbol{\beta} + \epsilon_{it}, \quad \epsilon_{it} \sim N\left(0,\ \sigma^2\right).$$

Such models are called *marginal models* because the distributional assumption for y_{it} does not involve the responses at other time points. GLMs such as logistic models may also be applied for categorical responses. Unlike linear mixed-effects models, inference for marginal models relies only on the mean response, which we will discuss shortly.

For a longitudinal study with n subjects and m assessment times, let y_{it} be a response and $\mathbf{x}_{it}$ a set of explanatory variables of interest from the ith subject at time t as before. A general marginal model may be specified for the response y_{it} at each time t by the mean response:

$$y_{it} \mid \mathbf{x}_{it} \sim i.d.f\left(\mu_{it}\right), \quad g\left(\mu_{it}\right) = \mathbf{x}_{it}^{\top}\boldsymbol{\beta}, \quad 1 \le t \le m, \quad 1 \le i \le n, \tag{9.22}$$

where $\boldsymbol{\beta}$ is a $p \times 1$ vector of parameters of interest, f denotes the conditional distribution of y_{it} given $\mathbf{x}_{it}$, and g is a link function such as the logit link for a binary and logarithm function for a count response. Under (9.22), the repeated responses are linked to the explanatory variables by modeling the marginal response y_{it} at each assessment time by a generalized linear model discussed in Chapters 4–6 for cross-sectional data. However, no assumption is made about the correlation across the different time points. We discussed distribution-free inference for generalized linear models when applied to cross-sectional data using estimating equations (EEs). Although applicable to y_{it} at each time t, we cannot apply this approach to estimate $\boldsymbol{\beta}$ using all y_{it} concurrently. Without the assumption of a joint distribution, one cannot even compute the correlations among the repeated y_{it}'s. Thus, to extend the EE to the current longitudinal data setting, we must first find a way to somehow link the y_{it}'s together to provide a single estimate of $\boldsymbol{\beta}$ for the marginal model in (9.22). We discuss next how to address this fundamental issue.

9.4.1 Generalized estimation equations

We used a set of estimating equations to improve the robustness of inference when the response y_i given $\mathbf{x}_i$ does not follow the assumed parametric model from cross-sectional data. This approach yields valid inference regardless of the data distribution, if the log of the conditional mean $E\left(y_i \mid \mathbf{x}_i\right)$ is correctly modeled by a linear predictor and the sample size is sufficiently large. However, to apply this idea to the marginal models within the current context of longitudinal data, we must address the technical issues to deal with the more complex nature of correlated outcomes across the repeated assessments.

For each y_{it} at time t, we can apply the estimating equations to the cross-sectional data, yielding

$$\mathbf{w}_t = \sum_{i=1}^{n} \mathbf{x}_{it}^{\top} V_{it}^{-1} (\mathbf{x}_{it}) (y_{it} - \mu_{it}) = \sum_{i=1}^{n} D_{it} V_{it}^{-1} (\mathbf{x}_{it}) S_{it} = \mathbf{0}, \qquad (9.23)$$

where $S_{it} = y_{it} - \mu_{it}$, $D_{it} = \frac{\partial \mu_{it}}{\partial \boldsymbol{\beta}}$, and $V_{it} (\mathbf{x}_{it})$ is a function of $\mathbf{x}_{it}$. If (9.23) models stationary relationships, then it may be estimated separately for each time t. However, this will result in m different estimates of the parameter $\boldsymbol{\beta}$. Thus, it may be preferable to obtain a single estimate using all the data. If (9.23) models temporal changes which involve multiple time points, then it cannot be estimated separately at all. If we simply treat the repeated measures as independent and pool the data together, we may have the following EE for the pooled data:

$$\mathbf{w} = \sum_{t=1}^{m} \mathbf{w}_t = \sum_{t=1}^{m} \sum_{i=1}^{n} \mathbf{x}_{it}^{\top} V_{it}^{-1} (\mathbf{x}_{it}) (y_{it} - \mu_{it}) = \sum_{t=1}^{m} \sum_{i=1}^{n} D_{it} V_{it}^{-1} (\mathbf{x}_{it}) S_{it} = \mathbf{0}. \qquad (9.24)$$

However, inference based on this pooled EE treating all the repeated outcomes as independent is invalid.

As discussed in Chapter 6, (9.23) may yield the maximum likelihood estimate of $\boldsymbol{\beta}$ if y_{it} is a member of the exponential family distributions and $V_{it} (\mathbf{x}_{it}) = Var (y_{it} \mid \mathbf{x}_{it})$. Further, if $E (y_{it} \mid \mathbf{x}_{it}) = \mu_{it}$, the estimating equations still provide consistent estimates of $\boldsymbol{\beta}$ even when $V_{it} (\mathbf{x}_{it}) \neq Var (y_{it} \mid \mathbf{x}_{it})$. The latter feature lays the foundation for extending (9.23) to the longitudinal data setting.

To this end, let $\mathbf{y}_i = (y_{i1}, \dots, y_{im})^{\top}$, $\boldsymbol{\mu}_i = (\mu_{i1}, \dots, \mu_{im})^{\top}$ and $\mathbf{x}_i = (\mathbf{x}_{i1}^{\top}, \dots, \mathbf{x}_{im}^{\top})^{\top}$. In analogy to (9.23), we define a set of *generalized estimating equations* (GEE) by

$$\mathbf{w} = \sum_{i=1}^{n} D_i V_i^{-1} (\mathbf{x}_i) (\mathbf{y}_i - \boldsymbol{\mu}_i) = \sum_{i=1}^{n} D_i V_i^{-1} (\mathbf{x}_i) S_i = \mathbf{0}, \qquad (9.25)$$

where $D_i = \frac{\partial}{\partial \boldsymbol{\beta}} \boldsymbol{\mu}_i = \left(\frac{\partial \mu_{i1}}{\partial \boldsymbol{\beta}}, \dots, \frac{\partial \mu_{im}}{\partial \boldsymbol{\beta}} \right)$, $S_i = \mathbf{y}_i - \boldsymbol{\mu}_i$, and $V_i (\mathbf{x}_i)$ is a matrix (size $m \times m$) function of $\mathbf{x}_i$. The GEE above is identical in form to the one in (9.23) for cross-sectional data analysis, except for the obvious difference in the dimension of each quantity. Like its cross-sectional data counterpart, (9.25) yields consistent estimates of $\boldsymbol{\beta}$ as long as the estimating equations are unbiased. If the explanatory variable $\mathbf{x}_i$ is time-invariant, the unbiasedness of the GEE is ensured by $E (y_{it} \mid \mathbf{x}_{it}) = \mu_{it}$ in (9.22) if the mean response is correctly specified (see Section 1.4.4).

The choice of $V_i (\mathbf{x}_i)$ may affect efficiency; the optimal efficiency is achieved when $V_i (\mathbf{x}_i)$ is the true covariance matrix of the response. However, any matrix function can be used for $V_i (\mathbf{x}_i)$ without affecting the validity of the GEE estimate, as long as the resulting GEE (9.25) is solvable. Since $V_i (\mathbf{x}_i)$ does not necessarily need to be the true covariance matrix of the response, it is generally referred to as the *working covariance*. Therefore, as GEE inference mainly relies on the mean response, it can be applied to models that are specified with the mean response only without any specific marginal distributions.

9.4.1.1 Working correlation

The working covariance matrix can be specified through a working correlation and marginal variances. In most applications, we set

$$V_i = A_i^{\frac{1}{2}} R (\boldsymbol{\alpha}) A_i^{\frac{1}{2}}, \quad A_i = \mathrm{diag} (v_{it}), \quad v_{it} = Var (y_{it} \mid \mathbf{x}_{it}), \qquad (9.26)$$

where $R(\boldsymbol{\alpha})$ denotes a *working* correlation matrix parameterized by $\boldsymbol{\alpha}$, and $\text{diag}(v_{it})$ a diagonal matrix with the marginal variance v_{it} on the tth diagonal. For discrete outcomes, the variance is often a known function of the mean response. We may use this known function together with a dispersion parameter to mimic possible overdispersion. Thus, we may in general specify $Var(y_{it} \mid \mathbf{x}_{it}) = \phi v(\mu_{it})$, where $v(\mu_{it})$ is a known function for the variance, such as $v(\mu_{it}) = \mu_{it}(1 - \mu_{it})$ for Bernoulli, and ϕ is a scale parameter.

The term "working correlation" is used to emphasize the fact that $R(\boldsymbol{\alpha})$ needs not be the true correlation matrix of $\mathbf{y}_i$ (Liang and Zeger, 1986). Compared to its cross-sectional counterpart in (9.23), GEE involves an additional specification of this working correlation matrix $R(\boldsymbol{\alpha})$. The simplest choice is $R = \mathbf{I}_m$ (the $m \times m$ identity matrix). With working independence correlation, the correlated components of $\mathbf{y}_i$ are treated as if they were independent. It is readily checked that the GEE (9.25) with the independent working correlation is the same as the EE (9.24) based on the pooled data. Thus, it produces the same point estimate of the parameter $\boldsymbol{\beta}$ as that based on the pooled data. However, unlike (9.24), the GEE inference accounts for the within-subject correlation through robust inference (see Section 9.4.1.2), thereby yielding valid inference.

Usually we use more structured correlation matrices to reflect the study design. For example, if the correlation among repeated measures decays exponentially in elapsed time, one may set $R(\rho) = AR_{m,1}(\rho)$, the first-order autoregressive correlation. Thus, the working correlation $R(\boldsymbol{\alpha})$ in general involves an unknown vector of parameters $\boldsymbol{\alpha}$, and the GEE in (9.25) is a function of both $\boldsymbol{\beta}$ and $\boldsymbol{\alpha}$. We suppress its dependence on the latter parameters deliberately to emphasize that the set of equations is used to obtain the estimate of the parameter vector of interest $\boldsymbol{\beta}$. If $\boldsymbol{\alpha}$ is known, such as under the working independence model, (9.25) is free of this parameter and is readily solved for $\boldsymbol{\beta}$. Otherwise, we need to estimate the working correlation.

Commonly used working correlation matrices include the independent, exchangeable (compound symmetry), and first-order autoregressive correlation matrices. Most software packages also allow users to specify their own working correlation matrix based on the data. The content and design of a study often suggests appropriate choices of the working correlation structure. For example, if there is no intrinsic order among the correlated outcomes such as individual responses within a cluster in a therapy study, the exchangeable correlation is appropriate. In general, if the design does not suggest any particular model, we may let the data drive our selection. For example, we may perform some exploratory analyses about the correlation and choose an analytic model closely approximating its empirical counterpart. If no clear pattern emerges from the exploratory work, we may choose the *unstructured working correlation* model, which leaves the $\frac{1}{2}m(m-1)$ parameters completely unconstrained. We may also compare estimates of $\boldsymbol{\beta}$ based on different correlation structures to check sensitivity to misspecification. We discuss selection of working correlation matrices based on information criteria in Section 9.5.

Example 9.7

For longitudinal data analysis, a primary interest is to assess temporal changes. For example, we would like to compare the HIV knowledge score changes in the Sexual Health study. One would expect that the knowledge will increase as time goes on, and our initial exploratory analysis confirms such a temporal trend (see Figure 9.1). However, since only the intervention group was delivered educational information regarding the transmission and prevention of HIV, we would like to know if there are any differences between the two groups. Let z_i be the treatment indicator ($=1$ for intervention and $= 0$ for control group). We use x_{it}, $t = 1, 2$, to index the time; at the baseline ($t = 1$), $x_{i1} \equiv 0$ and at 3 months post treatment ($t = 2$), $x_{i2} \equiv 1$. Consider the following marginal model for the HIV knowledge score y_{it} at time t:

$$E\left(y_{it} \mid x_{it}, z_i\right) = \beta_0 + \beta_1 x_{it} + \beta_2 z_i + \beta_3 x_{it} z_i, \quad 1 \leq i \leq n, \quad 1 \leq t \leq 2. \tag{9.27}$$

It follows that the mean HIV knowledge score changes from baseline to 3 months post treatment are β_1 for the control group and $\beta_1 + \beta_3$ for the intervention group. Thus, β_3 represents the difference between the two groups. Like the DOS, there are some missing values. Here, we used the subset of subjects with HIV knowledge scores at both time points available. Since there are only two time points, the exchangeable and $AR_2(1)$ structures are the same. The parameter estimates (standard errors) of β_3 based on GEE with independent and exchangeable working correlation matrices are both 12.7519 (1.5193), with p-values < 0.0001 (we will discuss the inference procedures shortly). Thus, we conclude that there is a significant difference in HIV knowledge change between the two treatment groups. Note that if the repeated measurement are assumed independent, the point estimate for β_3 is the same, but the estimated standard error is 2.6269, which is actually much larger than the estimate of 1.5193 under GEE. This may not be surprising as (9.27) models the interaction, so treating the repeated measurements as independent does no artificially inflate the sample size. In fact, β_3 measures the difference in change scores $y_{i1} - y_{i0}$ between the two treatment groups. Because the two repeated measures are positively correlated, $Var\left(y_{i1} - y_{i0}\right) < Var\left(y_{i1}\right) + Var\left(y_{i0}\right)$. Thus, treating the repeated measures as independent actually artificially increased the variance in this example and may result in false non significant conclusions.

When there are only two time points as in this example, it is easy to write down the GEEs explicitly. For a general longitudinal study with two time points $(t = 1, 2)$ and covariate $\mathbf{x}$, assume that $E(y_{it} \mid \mathbf{x}_{it}) = \mu_{it} = \boldsymbol{\beta} \mathbf{x}_{it}^\top$ and $Var(y_{it} \mid \mathbf{x}_{it}) = \sigma^2$ $(t = 1, 2)$. Then, $D_i = \left(\mathbf{x}_{i1}\ \mathbf{x}_{i2}\right)$ and $A_i = \begin{pmatrix} \sigma & 0 \\ 0 & \sigma \end{pmatrix} = \sigma I_2$, where I_2 is the 2×2 identity matrix. Thus, the GEE is given by

$$\sum_{i=1}^{n} \left(\mathbf{x}_{i1}\ \mathbf{x}_{i2}\right) \left[\sigma I_2 R\left(\boldsymbol{\alpha}\right) \sigma I_2\right]^{-1} \begin{pmatrix} y_{i1} - \boldsymbol{\beta} \mathbf{x}_{i1}^\top \\ y_{i2} - \boldsymbol{\beta} \mathbf{x}_{i2}^\top \end{pmatrix} = \begin{pmatrix} 0 \\ 0 \end{pmatrix}, \tag{9.28}$$

where $R\left(\boldsymbol{\alpha}\right)$ is the working correlation. If the independent working correlation is selected, the above reduces to

$$\sum_{i=1}^{n} \left[\mathbf{x}_{i1}\left(y_{i1} - \boldsymbol{\beta} \mathbf{x}_{i1}^\top\right) + \mathbf{x}_{i2}\left(y_{i2} - \boldsymbol{\beta} \mathbf{x}_{i2}^\top\right)\right] = \mathbf{0}. \tag{9.29}$$

Under an exchangeable working correlation $R\left(\boldsymbol{\alpha}\right) = \begin{pmatrix} 1 & \rho \\ \rho & 1 \end{pmatrix}$, (9.28) yields

$$\sum_{i=1}^{n} \left[\left(\mathbf{x}_{i1} - \rho \mathbf{x}_{i2}\right)\left(y_{i1} - \boldsymbol{\beta} \mathbf{x}_{i1}^\top\right) + \left(\rho \mathbf{x}_{i1} - \mathbf{x}_{i2}\right)\left(y_{i2} - \boldsymbol{\beta} \mathbf{x}_{i2}^\top\right)\right] = \mathbf{0}. \tag{9.30}$$

If the covariate is time invariant, these two GEEs (9.29) and (9.30) are actually equivalent. However, this is not the case if the covariate is time varying. Further, because of the nonzero working correlation, (9.30) involves the product of $\mathbf{x}_{i2} y_{i1}$ and $\mathbf{x}_{i1} y_{i2}$. Thus, while (9.22) is sufficient to guarantee the unbiasedness of (9.29), it is not the case for (9.30) if $\mathbf{x}_{i1}$ and $\mathbf{x}_{i2}$ can be different (see Section 9.4.1.4 for more detailed discussion). ⬚

Different choices of $R\left(\boldsymbol{\alpha}\right)$ do generally give rise to different estimates of $\boldsymbol{\beta}$. However, regardless of the choices of the working correlation structure, estimates obtained from (9.29) and (9.30) are all consistent under the assumptions. Actually, the estimates are asymptotically normal under some minor assumptions on $R(\alpha)$. We discuss such nice asymptotic properties of GEE estimates next.

9.4.1.2 Inference

To ensure asymptotic normality, we require that $\widehat{\alpha}$ be $\sqrt{n}$-consistent. Since the working correlation need not equal the true correlation, a consistent $\widehat{\alpha}$ means that it converges in probability to some vector of constants α_0, as the sample size goes to infinity. Thus, a $\sqrt{n}$-consistent $\widehat{\alpha}$ is an estimate such that $\sqrt{n}(\widehat{\alpha} - \alpha_0)$ is bounded in probability (see Kowalski and Tu (2008)). In practice, $\widehat{\alpha}$ are usually moment estimates which are asymptotically normal and thus are $\sqrt{n}$-consistent.

For a $\sqrt{n}$-consistent $\widehat{\alpha}$, the GEE estimate $\widehat{\beta}$ is asymptotically normal, with the asymptotic variance given by

$$\Sigma_{\beta} = B^{-1} E\left(G_i S_i S_i^{\top} G_i^{\top}\right) B^{-\top}, \quad B^{-\top} = \left(B^{-1}\right)^{\top}, \tag{9.31}$$

where $G_i = D_i V_i^{-1}$ and $B = E\left(\frac{\partial(G_i S_i)}{\partial \beta}\right)$ (see Problem 9.13). A consistent estimate of Σ_{β} is given by

$$\widehat{\Sigma}_{\beta} = \frac{1}{n} \widehat{B}^{-1} \sum_{i=1}^{n} \left(\widehat{G}_i \widehat{S}_i \widehat{S}_i^{\top} \widehat{G}_i^{\top}\right) \widehat{B}^{-\top}, \tag{9.32}$$

where $\widehat{B}_i$, $\widehat{G}_i$, and $\widehat{S}_i$ denote the estimated versions of the corresponding parameters obtained by replacing β and α with their respective estimates.

For the model in Example 9.7, it is easy to check that (9.29) has the same form as the EE in (9.24) based on the pooled data. However, the asymptotic variance of the GEE estimate in (9.31) accounts for the correlations within the components of $\mathbf{y}_i$, whereas the EE in (9.24) assumes independence among the repeated observations. When the working correlation is correctly specified, i.e., it is the same as the actual correlation among the components of $\mathbf{y}_i$ given $\mathbf{x}_i$, $A_i R(\alpha) A_i = Var(\mathbf{y}_i \mid \mathbf{x}_i)$, the corresponding GEE is efficient (see Tsiatis (2006)) and (9.31) reduces to

$$\begin{aligned}
\Sigma_{\beta} &= B^{-1} D_i \left(A_i R(\alpha) A_i\right)^{-1} Var(\mathbf{y}_i \mid \mathbf{x}_i) \left(A_i R(\alpha) A_i\right)^{-1} D_i^{\top} B^{-\top} \\
&= B^{-1} D_i \left(A_i R(\alpha) A_i\right)^{-1} D_i^{\top} B^{-\top}.
\end{aligned}$$

Inference based on Wald test is straightforward following the asymptotic normality of $\widehat{\beta}$. The score test described in Section 4.2.2.1 can be similarly developed for GEE (Rotnitzky and Jewell, 1990). For small sample sizes, the Wald test is often anticonservative, i.e., with inflated type I error rates, and the score test may be used to obtain better estimates (Rotnitzky and Jewell, 1990, Zhang et al., 2011). Note that as explained in Section 4.2.2.1 in Chapter 4, score tests only rely on the estimates under the null hypothesis and thus a consistent estimate of parameters for the full model is not required for the test. Note also that the likelihood ratio test is not applicable because of the lack of a likelihood function for distribution-free marginal models.

9.4.1.3 Estimation of working correlation matrix

With an estimated parameter $\widehat{\beta}$, we can estimate α for a specified working correlation based on the residuals. For example, consider modeling a binary response y_{it} using the marginal model in (9.22) with a logistic link:

$$E(y_{it} \mid \mathbf{x}_{it}) = \mu_{it}, \quad \log\left(\frac{\mu_{it}}{1 - \mu_{it}}\right) = \mathbf{x}_{it}^{\top} \beta, \quad 1 \le i \le n, \quad 1 \le t \le m. \tag{9.33}$$

Since $Var(y_{it} \mid \mathbf{x}_{it}) = \mu_{it}(1 - \mu_{it})$, $A_i = \text{diag}(\mu_{it}(1 - \mu_{it}))$. If $\boldsymbol{\alpha}$ is unknown, we need to substitute an estimate in place of $\boldsymbol{\alpha}$ in (9.25) before solving the equations.

To illustrate, consider the unstructured working correlation matrix, $R(\boldsymbol{\alpha}) = [\rho_{st}]$, with $\rho_{st} = Corr(y_{is}, y_{it})$. We may estimate ρ_{st} by the Pearson correlation estimates:

$$\widehat{\rho}_{st} = \frac{\sum_{i=1}^{n} (r_{is} - \overline{r}_{\cdot s})(r_{it} - \overline{r}_{\cdot t})}{\sqrt{\sum_{i=1}^{n} (r_{is} - \overline{r}_{\cdot s})^2 \sum_{i=1}^{n} (r_{it} - \overline{r}_{\cdot t})^2}}, \quad r_{it} = y_{it} - \mathbf{x}_{it}^{\top}\widehat{\boldsymbol{\beta}}, \quad \overline{r}_{\cdot t} = \frac{1}{n}\sum_{i=1}^{n} r_{it}. \quad (9.34)$$

Thus, based on an initial estimate of $\boldsymbol{\beta}$, we may obtain an estimate of the working correlation. Based on the estimated working correlation, we may solve (9.25) to obtain an updated estimate of $\boldsymbol{\beta}$. The procedure continues until both estimates of $\boldsymbol{\beta}$ and $\boldsymbol{\alpha}$ converge.

Although the consistency of the GEE estimate is guaranteed regardless of the choice of the correlation matrix, the efficiency of such estimates does rely on the different choices of the correlation matrix. In general, if the sample size is large and number of repeated assessments is small, efficiency may not be a major concern and the working independence model may be sufficient. Otherwise, we may use more structured alternatives to improve efficiency. The GEE estimate achieves its optimal efficiency if the working correlation structure is the same as the true correlation structure.

9.4.1.4 Time-varying covariates

If the covariates $\mathbf{x}_i$ are time invariant, then the unbiasedness of the GEE (9.25) is ensured by $E(y_{it} \mid \mathbf{x}_{it}) = \mu_{it}$ in (9.22), i.e., if the mean response is correctly specified (see Section 1.4.4) because $E(y_{it} \mid \mathbf{x}_{it}) = \mu_{it}$ is equivalent to $E(y_{it} \mid \mathbf{x}_i) = \mu_{it}$ as $\mathbf{x}_{it} = \mathbf{x}_i$. It is straightforward to show the unbiasedness of the GEE by the theorem of iterated conditional expectations:

$$E\left(D_i V_i^{-1}(\mathbf{x}_i)(\mathbf{y}_i - \boldsymbol{\mu}_i)\right) = E\left[E\left(D_i V_i^{-1}(\mathbf{x}_i)(\mathbf{y}_i - \boldsymbol{\mu}_i) \mid \mathbf{x}_i\right)\right] \quad (9.35)$$
$$= E\left[D_i V_i^{-1}(\mathbf{x}_i) E(\mathbf{y}_i - \boldsymbol{\mu}_i \mid \mathbf{x}_i)\right] = \mathbf{0}.$$

However, this argument in general fails if $\mathbf{x}_i$ is time varying as $E(y_{it} \mid \mathbf{x}_{it}) = \mu_{it}$ is not equivalent to $E(y_{it} \mid \mathbf{x}_i)$ in general. Thus, the condition $E(y_{it} \mid \mathbf{x}_{it}) = \mu_{it}$ may not be sufficient to guarantee the unbiasedness of the GEE if some or all components of $\mathbf{x}_i$ are time varying. To clearly see this, we may carry out the matrix multiplication in the GEE and obtain

$$\sum_{i=1}^{n}\sum_{s=1}^{m}\sum_{t=1}^{m} \frac{\partial \mu_{is}}{\partial \boldsymbol{\beta}} \left[V_i^{-1}(\mathbf{x}_i)\right]_{s,t} (y_{it} - \mu_{it}) = \mathbf{0}, \quad (9.36)$$

where $\left[V_i^{-1}(\mathbf{x}_i)\right]_{s,t}$ is the (s,t) term of V_i^{-1}. If s and t in (9.36) can be different and $\mathbf{x}_{is}$ and $\mathbf{x}_{it}$ are also different, then the above iterative conditional expectation argument does not work as $\frac{\partial \mu_{is}}{\partial \boldsymbol{\beta}}$ and $y_{it} - \mu_{it}$ are functions of $\mathbf{x}_{is}$ and $\mathbf{x}_{it}$, respectively. Thus, the condition $E(y_{it} \mid \mathbf{x}_{it}) = \mu_{it}$ does not guarantee the unbiasedness of the GEE in general. However, if $V_i(\mathbf{x}_i)$ is diagonal, i.e., if we choose the independent working correlation, then $V_i^{-1}(\mathbf{x}_i)$ is also diagonal with the tth diagonal term $v^{-1}(\mu_{it})$. Since s and t in (9.36) must be the same for diagonal terms, the GEE (9.36) reduces to

$$\sum_{i=1}^{n}\sum_{t=1}^{m} \frac{\partial \mu_{it}}{\partial \boldsymbol{\beta}} v^{-1}(\mu_{it})(y_{it} - \mu_{it}) = \mathbf{0}. \quad (9.37)$$

Now, each summand only involves covariates at a single time point, hence the theorem of iterated conditional expectations can again be applied to show that the GEE is unbiased.

Alternatively, we may assume a stronger condition, $E\left(y_{it} \mid \mathbf{x}_i\right) = \mu_{it} = g^{-1}\left(\mathbf{x}_{it}^\top \boldsymbol{\beta}\right)$, for the unbiasedness of the GEE. This stronger condition is referred to as the *full covariate conditional mean* (FCCM) assumption in Diggle et al. (2002). If, conditioning on $\mathbf{x}_{it}$, all the historical (all $\mathbf{x}_{is}$ for $s < t$) and future covariates (all $\mathbf{x}_{is}$ for $s > t$) are independent of y_{it}, then $E\left(y_{it} \mid \mathbf{x}_{it}\right) = \mu_{it}$ in (9.22) implies the FCCM (see Problem 9.16), and hence the unbiasedness of the GEE. The independence of the response with the historical covariates may be approximately true if sufficient information is included in the covariate at each time point. The conditional independence of the future covariates with the response requires that none of the covariates can be adapted based on prior responses. Such variables are usually referred to as *exogenous*. Examples of time-varying exogenous covariates in clinical studies may include deterministic functions of time itself such as age and environmental variables such as weather, which are not affected by health outcomes of individuals in the studies. Covariates that are not exogenous are called *endogenous*. When there are endogenous covariates, the stronger condition of FCCM may likely fail. In such cases, it is generally recommended to use the independent working covariance for valid inferences (Pepe et al., 1994). Note that endogenous covariates may also be an issue in mixed effects models (Qian et al. 2020).

9.4.2   Binary responses

A marginal model for a longitudinal binary response y_{it} can also be easily developed under the GLM framework. The marginal model is specified for y_{it} at each time t by a GLM:

$$\Pr\left(y_{it} \mid \mathbf{x}_{it}\right) = p_{it}, \quad g\left(p_{it}\right) = \mathbf{x}_{it}^\top \boldsymbol{\beta}, \quad 1 \le t \le m, \quad 1 \le i \le n, \tag{9.38}$$

where $\boldsymbol{\beta}$ is a $p \times 1$ vector of parameters of interest and g is a link function such as the logistic function. Thus, the repeated responses are linked to the explanatory variables by modeling the marginal response y_{it} at each assessment time by a GLM. If the logit link is applied for a binary response, the coefficient of a covariate is still interpreted as the logarithm of the odds ratio between two subjects with a one-unit difference in the covariate, all other covariates being the same.

In the GEE (9.25), we may still try to specify $V_i\left(\mathbf{x}_i\right) = Var\left(\mathbf{y}_i \mid \mathbf{x}_i\right)$ through a working correlation, together with the marginal variance $Var\left(y_{it} \mid \mathbf{x}_{it}\right) = p_{it}\left(1 - p_{it}\right)$. However, for discrete responses, there may be some constraints on the elements of the correlation matrix. For example, correlations for binary responses must satisfy a set of *Frechet bounds* given by their marginal distributions (see, for example, Shults et al. (2009)). Although not necessary for ensuring consistency of GEE estimates, the use of an $R\left(\boldsymbol{\alpha}\right)$ meeting the requirement of *Frechet bounds* may yield more accurate and efficient estimates for small and moderate samples.

9.4.2.1   Alternating logistic regressions

Instead of specifying a working correlation, Carey et al. (1993) proposed to study the association among repeated binary outcomes through their odds ratios. Based on all the pair-wise odds ratios, as well as the marginal probabilities, the covariance of the repeated outcomes can be calculated; this approach has the advantage that they are not constrained by the outcome means. Specifically, Carey et al. (1993) proposed to study the covariance of the repeated binary outcomes through modeling the odds ratio with logistic models. Assume

$$\gamma_{ijk} = \log\left[\mathrm{OR}\left(y_{ij}, y_{ik}\right)\right] = \mathbf{z}_{ijk}^\top \boldsymbol{\alpha}, \quad 1 \le j < k \le m \tag{9.39}$$

where OR $(y_{ij}, y_{ik}) = \frac{\Pr(y_{ij}=1|y_{ik}=1)/\Pr(y_{ij}=0|y_{ik}=1)}{\Pr(y_{ij}=1|y_{ik}=0)/\Pr(y_{ij}=0|y_{ik}=0)}$ is the odds ratio between the outcomes at times j and k, $\boldsymbol{\alpha}$ is a $q \times 1$ vector of regression parameters, and $\mathbf{z}_{ijk}$ is a fixed set of covariates. The special case $\gamma_{ijk} = 0$ corresponds to no association (independent). For the estimation of (9.39), Carey et al. (1993) proposed another estimating equation. Let $\mu_{ij} = \mathrm{pr}\,(y_{ij} = 1)$ and $\nu_{ijk} = \mathrm{pr}\,(y_{ij} = 1, y_{ik} = 1)$. Then by Diggle (1992),

$$\mathrm{logit}\,[\Pr\,(y_{ij} = 1 \mid y_{ik} = y_{ik})] = \gamma_{ijk} y_{ik} + \log\left(\frac{\mu_{ij} - \nu_{ijk}}{1 - \mu_{ij} - \mu_{ik} + \nu_{ijk}}\right).$$

Thus, (9.39) can be fitted by regressing the binary response y_{ij} on $y_{ik} z'_{ijk}$ with offset $\log\left(\frac{\mu_{ij} - \nu_{ijk}}{1-\mu_{ij}-\mu_{ik}+\nu_{ijk}}\right)$. Since there are m choose two pairs of the binary responses and their models (9.39) may have common parameters, we may estimate them together. Let

$$\zeta_{ijk} = E\,(y_{ij} \mid y_{ik}) = \mathrm{logit}^{-1}\left\{\gamma_{ijk} y_{ik} + \log\left(\frac{\mu_{ij} - \nu_{ijk}}{1 - \mu_{ij} - \mu_{ik} + \nu_{ijk}}\right)\right\}$$

$$R_{ijk} = y_{ij} - E\,(y_{ij} \mid y_{ik} = y_{ik}) = y_{ij} - \zeta_{ijk},$$

and $\boldsymbol{\zeta}_i$ and $\mathbf{R}_i$ be the corresponding vectors with elements ζ_{ijk} and $R_{ijk}, 1 \leq j < k \leq m$, $\mathbf{T}_i = \partial \boldsymbol{\zeta}_i / \partial \boldsymbol{\alpha}$, and $\mathbf{S}_i$ the diagonal matrix with diagonal element $\zeta_{ijk}\,(1 - \zeta_{ijk})$. Then, $\boldsymbol{\alpha}$ can be estimated by an estimating equation

$$U_{\boldsymbol{\alpha}} = \sum_{i=1}^{m} \mathbf{T}_i^{\top} \mathbf{S}_i^{-1} \mathbf{R}_i = \mathbf{0}. \tag{9.40}$$

Note that this EE cannot be estimated independently as the offset term in ζ_{ijk} involves estimates of μ_{ij} and ν_{ijk}, which depends on the estimation of $\boldsymbol{\beta}$. The method of *alternating logistic regression* estimates $\boldsymbol{\alpha}$ and $\boldsymbol{\beta}$ by alternately solving (9.25) and (9.40) until convergence (Carey et al., 1993).

As concerns working correlation matrices, the log odds ratio regression structure in (9.39) does not affect the validity of inference. As long as the mean response is correctly specified, the GEE estimate is consistent regardless of the specification of (9.39). However, it does affect efficiency, and thus it would be helpful if a structure close to the reality is specified. An exchangeable odds ratio structure is commonly used (with $z_{ijk} \equiv 1$), in which case the log odds ratio is a constant for all clusters i and pairs (j, k). On the other hand, we may fully parameterize the clusters, using different parameters for each unique pair (j, k) within the cluster.

Example 9.8

Consider the following marginal model for binary depression diagnosis over time using the DOS data

$$\mathrm{logit}\,(\mu_{it}) = \beta_0 + \beta_1 Age_i + \beta_2 Gender_i + \beta_3 Edu_i + \beta_4 CIRS_{it} \tag{9.41}$$
$$+ \beta_5 MS1_i + \beta_6 MS2_i + \beta_7 Race_i.$$

In other words, we remove the random intercept from the GLMM in Example 9.3. For space consideration, we show in the table below only the estimates of Gender and CIRS, based on the pooled data analysis treating all the repeated measures as independent as well as GEE estimates using the working independent (Ind), exchangeable (Exch), and autoregressive $AR_4\,(1)$ (AR) correlation matrices, and exchangeable and fully parameterized clusters alternating logistic regression.

Method	Gender			CIRS		
	Estimate	SE	p-value	Estimate	SE	p-value
Pooled data	0.5288	0.1671	0.0016	0.2094	0.0257	< 0.0001
GEE (Ind)	0.5288	0.2951	0.0732	0.2094	0.0401	< 0.0001
GEE (Exch)	0.5275	0.2931	0.0719	0.1499	0.0297	< 0.0001
GEE (AR)	0.5162	0.2860	0.0711	0.1567	0.0278	< 0.0001
ALT(Exch)	0.5053	0.2890	0.0804	0.1630	0.0267	<.00001
ALT(Full)	0.4966	0.2851	0.0815	0.1626	0.0269	<.00001

The coefficients of Gender and CIRS have the same signs as those in Example 9.3, thereby both the marginal and mixed-effect models indicating the same direction of the effects of gender and CIRS on depression. However, the point estimates are quite different. The difference reflects the distinctive paradigms underlying the two modeling approaches. We discuss this fundamental issue in more detail later (see Section 9.4.5).

The pooled data analysis provides the same point estimates as those based on the GEE with the independent working correlation, but much underestimated variances. All the GEE estimates are comparable to each other, with the alternating logistic regressions providing the smallest standard errors, especially for CIRS. The results seem to suggest that the alternating logistic regression may indeed more authentically model the correlation, thereby providing more efficient estimates. Among the three working correlations, $AR_4(1)$ models provide smallest standard errors, very close to those by the alternating logistic regressions. However, since CIRS is time varying and likely to depend on the previous depression outcome, the FCCM assumption may be violated. Thus, the results based on the independent working correlation structure may be preferred. ⬜

9.4.3 Polytomous responses

For an ordinal response, if a score is assigned to each of the response levels, then the marginal model in (9.22) may be applied to model the mean scores. Otherwise, it is generally not sensible to compute the mean of a categorical variable response with more than two levels, and the marginal model in (9.22) does not apply. Recall that in Chapter 4 we discussed how to record the information in a categorical outcome with a series of binary indicators. Specifically, we can use a set of $K-1$ indicators to represent the outcomes of a multinomial variate with K categories. We can apply the same idea to extend the marginal model in (9.22) to a K-level categorical response for regression analysis.

Consider a K-level categorical response w_{it}. Define a set of $K-1$ longitudinal binary indicators $\mathbf{y}_{it} = \left(y_{i1t}, \ldots, y_{i(K-1)t}\right)^{\top}$, with $y_{ikt} = 1$ if $w_{it} = k$ and $y_{ikt} = 0$ otherwise $(1 \leq k \leq K-1)$. If we apply the generalized logit model in (5.6) to $\mathbf{y}_{it}$ at each time t, we obtain the following longitudinal version of this model:

$$E\left(y_{ikt} \mid \mathbf{x}_{it}\right) = \mu_{ikt} = \frac{\exp\left(\gamma_k + \boldsymbol{\beta}_k^{\top}\mathbf{x}_{it}\right)}{1 + \sum_{k=1}^{K-1}\exp\left(\gamma_k + \boldsymbol{\beta}_k^{\top}\mathbf{x}_{it}\right)}, \quad k = 1, \ldots, K-1. \qquad (9.42)$$

In the above, the γ_k's and $\boldsymbol{\beta}_k$'s have the same interpretations as in the cross-sectional data setting. The above model is parameterized by $\boldsymbol{\theta}$, representing the collection of γ_k's and $\boldsymbol{\beta}_k$'s in (9.42):

$$\boldsymbol{\theta} = \left(\boldsymbol{\gamma}^{\top}, \boldsymbol{\beta}^{\top}\right), \quad \boldsymbol{\gamma} = (\gamma_1, \ldots, \gamma_{K-1})^{\top}, \quad \boldsymbol{\beta} = \left(\boldsymbol{\beta}_1^{\top}, \ldots, \boldsymbol{\beta}_{K-1}^{\top}\right)^{\top}.$$

To extend (9.25) to provide inference about θ, let

$$\mathbf{y}_{it} = \left(y_{i1t}, \ldots, y_{i(K-1)t}\right)^\top, \quad \boldsymbol{\mu}_{it} = \left(\mu_{i1t}, \ldots, \mu_{i(K-1)t}\right)^\top, \quad \mathbf{y}_i = \left(\mathbf{y}_{i1}^\top, \ldots, \mathbf{y}_{im}^\top\right)^\top,$$

$$\boldsymbol{\mu}_i = \left(\boldsymbol{\mu}_{i1}^\top, \ldots, \boldsymbol{\mu}_{im}^\top\right)^\top, \quad \mathbf{x}_i = \left(\mathbf{x}_{i1}^\top, \ldots, \mathbf{x}_{im}^\top\right)^\top, \quad D_i = \frac{\partial}{\partial \boldsymbol{\beta}} \boldsymbol{\mu}_i, \quad S_i = \mathbf{y}_i - \boldsymbol{\mu}_i. \quad (9.43)$$

For each t, $\mathbf{y}_{it}$ has a multinomial distribution with mean $\boldsymbol{\mu}_{it}$ and variance A_{itt} given by

$$A_{itt} = \begin{pmatrix} \mu_{i1t}\left(1 - \mu_{i1t}\right) & \cdots & -\mu_{i1t}\mu_{i(K-1)t} \\ \vdots & \ddots & \vdots \\ -\mu_{i1t}\mu_{i(K-1)t} & \cdots & \mu_{i(K-1)t}\left(1 - \mu_{i(K-1)t}\right) \end{pmatrix}$$

Thus, we set $V_i = A_i^{\frac{1}{2}} R\left(\boldsymbol{\alpha}\right) A_i^{\frac{1}{2}}$ with $R\left(\boldsymbol{\alpha}\right)$ and A_i defined by

$$R\left(\boldsymbol{\alpha}\right) = \begin{pmatrix} \mathbf{I}_{K-1} & R_{12} & \cdots & R_{1m} \\ R_{12}^\top & \mathbf{I}_{K-1} & \cdots & R_{2m} \\ \vdots & \cdots & \ddots & \vdots \\ R_{1m}^\top & R_{2m}^\top & \cdots & \mathbf{I}_{K-1} \end{pmatrix}, \quad A_i = \text{diag}_t\left(A_{itt}\right) \quad (9.44)$$

where R_{jl} are some $(K-1) \times (K-1)$ matrices parameterized by $\boldsymbol{\alpha}$. For inference about θ, define the GEE in the same form as in (9.25) except for using this newly defined V_i above, with D_i and S_i as in (9.43). As in the case of a binary response, similar Frechet bounds exist for the correlation matrix, although the working correlation $R\left(\boldsymbol{\alpha}\right)$ need not satisfy such bounds. For efficiency considerations, Heagerty and Zeger (1996) generalized the alternating logistic regression technique to the ordinal cases for categorical responses. Alternatively, Touloumis et al. (2013) proposed a two-stage GEE approach to deal with the issue; they estimate the working association via local odds ratio parameterization of the marginal odds ratios of all the pairs of response levels at different time points at the first stage and then solve the GEE at the second stage. Their approach can be applied to both ordinal and nominal responses.

Example 9.9
The response variable in Example 9.8 was obtained by dichotomizing a three-level depression diagnosis. If a proportional odds model with independent working correlation is applied, we obtain the estimates below.

Effect	Estimate	Standard Error	95% Confidence Limits		Z	Pr > $\lvert Z \rvert$
Intercept1	2.1941	2.0803	−1.8832	6.2714	1.05	0.2916
Intercept2	3.1862	2.0833	−0.8971	7.2695	1.53	0.1262
Age	−0.0710	0.0251	−0.1203	−0.0218	−2.83	0.0047
Gender	0.4100	0.2864	−0.1513	0.9712	1.43	0.1522
Education	−0.0699	0.0581	−0.1838	0.0440	−1.20	0.2290
CIRS	0.2164	0.0368	0.1442	0.2886	5.87	< .0001
MS1	−0.6352	0.3488	−1.3188	0.0484	−1.82	0.0686
MS2	0.0076	0.3983	−0.7731	0.7883	0.02	0.9848
Race	0.6106	1.2989	−1.9351	3.1563	−0.47	0.6383

Compared with the corresponding GLMM analysis in Example 9.4, the general conclusions are similar. However, as in the binary case the point estimates are quite different. This is a common issue between nonlinear GLMM and GEE; see Section 9.4.5 for a discussion on this fundamental issue. □

9.4.4   Count responses

For marginal Poisson and over-dispersed Poisson regression models, we model the logarithm of the mean as a linear function of the covariates. For example, the marginal Poisson regression model is specified as

$$y_{it} \mid \mathbf{x}_{it} \sim Poisson(\mu_{it}), \quad \log(\mu_{it}) = \mathbf{x}_{it}^{\top}\boldsymbol{\beta}, \qquad (9.45)$$

where $\boldsymbol{\beta}$ is a $p \times 1$ vector of parameters of interest. The parameter $\boldsymbol{\beta}$ can be interpreted similarly as in the cross-sectional case in terms of the mean responses of the whole population.

The general theory of GEE can be applied for inference. Since $E(\mathbf{y}_i \mid \mathbf{x}_i) = \boldsymbol{\mu}_i = \exp(\mathbf{x}_{it}^{\top}\boldsymbol{\beta})$, it follows that $D_i = \frac{\partial}{\partial \boldsymbol{\beta}}\boldsymbol{\mu}_i = \mathbf{x}_{it}^{\top}\boldsymbol{\mu}_i$. For the working covariance V_i, we set $V_i = A_i^{\frac{1}{2}} R(\boldsymbol{\alpha}) A_i^{\frac{1}{2}}$ as before, where $R(\boldsymbol{\alpha})$ is the working correlation matrix parameterized by $\boldsymbol{\alpha}$ and $A_i = \mathrm{diag}(v_{it})$ with v_{it} being the variance at the tth time point. For Poisson regression, this variance equals the mean μ_{it}. Thus, if independent working correlation is utilized, the GEE has a very simple format

$$\mathbf{w}_n = \sum_{i=1}^{n} D_i V_i^{-1}(\mathbf{x}_i)(\mathbf{y}_i - \boldsymbol{\mu}_i) = \sum_{i=1}^{n} \mathbf{x}_i^{\top}(\mathbf{y}_i - \boldsymbol{\mu}_i) = \mathbf{0}. \qquad (9.46)$$

The GEE inference is still valid in cases of overdispersion as long as the mean response is correctly specified. However, one may apply marginal distributions for overdispersed count response such as the NB in place of the Poisson distribution in (9.45) to improve efficiency. For example, a marginal NB regression model may be specified as

$$y_{it} \mid \mathbf{x}_i \sim NB(\mu_{it}, \alpha'), \quad \log(\mu_{it}) = \mathbf{x}_{it}^{\top}\boldsymbol{\beta}, \qquad (9.47)$$

where α' is the dispersion parameter of the NB distribution. Since the mean response has the same format as that of the Poisson regression, the difference is only between the two working variances. For a marginal NB distribution at each time point, the marginal variance $v(\mu_{it}) = \mu_{it} + \alpha'\mu_{it}^2$. Thus, $A_i = \mathrm{diag}\left(\sqrt{\mu_{it} + \alpha'\mu_{it}^2}\right)$. Hence the GEE for NB regression is

$$\mathbf{w}_n = \sum_{i=1}^{n} D_i A_i^{-\frac{1}{2}} R^{-1}(\boldsymbol{\alpha}) A_i^{-\frac{1}{2}}(\mathbf{x}_i)(\mathbf{y}_i - \boldsymbol{\mu}_i) = \mathbf{0}. \qquad (9.48)$$

This GEE not only has parameter $\boldsymbol{\alpha}$ in the working correlation, but also contains the dispersion parameter α' in A_i. As described in Section 6.4.2.4, we may first estimate the dispersion parameter α' using MLE and then treat the estimated α' as the true value when estimating all the remaining parameters as described in the regular theory of GEE. As part of the working model assumptions, the choice of α' does not affect the validity of the GEE inference.

Example 9.10

We applied a mixed-effect model to model UTIs in the Catheter Self Management Study in Example 9.5. Here, we apply a marginal Poisson log-linear regression model. Shown in the table below are GEE estimates using the independent (Ind), exchangeable (Exch), autoregressive (AR), and unstructured (unstr) working correlation matrices.

Parameter	Estimate	SE	95%	CI	z	Pr > |z|
Independent Working Correlation						
Intercept	−1.4566	0.1702	1.7902	1.1229	−8.56	< .0001
Group	0.1788	0.1876	−0.1890	0.5465	0.95	0.3407
Gender	0.1054	0.1877	−0.2625	0.4732	0.56	0.5745
Exchangeable Working Correlation						
Intercept	−1.4156	0.1734	−1.7554	−1.0758	−8.17	< .0001
Group	0.1252	0.1875	−0.2422	0.4926	0.67	0.5042
Gender	0.1118	0.1821	−0.2451	0.4687	0.61	0.5391
First-Order Autoregressive Working Correlation						
Intercept	−1.4413	0.1700	−1.7745	−1.1080	−8.48	< .0001
Group	0.1699	0.1870	−0.1965	0.5364	0.91	0.3634
Gender	0.0941	0.1829	−0.2644	0.4526	0.51	0.6070
Unstructured Working Correlation						
Intercept	−1.4342	0.1756	−1.7782	−1.0901	−8.17	< .0001
Group	0.1621	0.1873	−0.2050	0.5293	0.87	0.3868
Gender	0.0972	0.1816	−0.2588	0.4532	0.54	0.5925

From the output, we can see that all the four GEE estimates were similar. Comparatively, the exchangeable working correlation produced the most different estimates from the others, while the AR and unstructured working correlation models had the closest estimates to each other.

$\square$

9.4.4.1 GEE for zero-modified outcomes

For zero-modified outcomes, the mean response in general is no longer the exponential of a linear predictor; thus some modifications are necessary in setting up the GEE. For zero-truncated responses, we often assume a marginal log-linear Poisson or overdispersion distribution in the absence of zero-truncation. For example, using ZTP, we get the following marginal ZTP regression models:

$$y_{it} \mid \mathbf{x}_{it} \sim ZTP(\mu_{it}), \quad \log(\mu_{it}) = \mathbf{x}_{it}^\top \boldsymbol{\beta}. \tag{9.49}$$

Since μ_{it} is the mean of a count response in the absence of zero-truncation, the marginal mean for the zero-truncated response is $\mu'_{it} = \frac{\mu_{it}}{1-\exp(-\mu_{it})}$. GEE for ZTP outcomes should be developed based on this mean response.

For zero-inflated responses, we can replace the marginal Poisson distribution in (9.22) with marginal zero-inflated distributions such as ZIP and zero-inflated negative binomial (ZINB) to obtain marginal models for zero-inflated responses. For example, using ZIP, we obtain the following marginal ZIP regression model:

$$y_{it} \mid \mathbf{x}_{it} \sim ZIP(\mu_{it}, \rho_{it}), \quad \text{logit}(\rho_i) = \mathbf{u}_{it}^\top \boldsymbol{\beta}_u, \quad \log(\mu_{it}) = \mathbf{v}_{it}^\top \boldsymbol{\beta}_v, \quad 1 < i < n,$$

where $\mathbf{u}_{it}$ and $\mathbf{v}_{it}$ are two subsets of the vector $\mathbf{x}_{it}$. The parameters can be similarly interpreted as those in the cross-sectional model.

As we commented in Section 6.5.1.3, the mean of the count response alone is not sufficient to model the mixture. So, we may add a component, the indicator of being a zero, in addition to the raw count response at each time point to facilitate the analysis (Tang et al., 2015). Instead of directly developing a system of GEE for inference, Hall and Zhang (2004) proposed the Expectation-Solution algorithm of Rosen et al. (2000) for the inference.

9.4.5 Comparison of GLMM with marginal models

For a longitudinal outcome, we have two approaches to model its trajectory over time and how the temporal changes are associated or predicted by other variables. On the one hand, we have the marginal model in (9.22) that completely ignores the within-subject correlation in the front end of model specification, but accounts for the correlated repeated outcomes at the back end of inference using the GEEs. On the other hand, the GLMM tackles this within-subject correlation directly at the time of model specification by introducing random effects to create independent individual responses (conditional on the random effects), making it possible to apply standard models such as GLM to such correlated outcomes and associated maximum likelihood for inference. The immediate consequence of the difference between the two approaches is that GLMM can provide an estimated trajectory for each individual, while the marginal approach cannot. For example, by estimating $\mathbf{b}_i$ for each subject, we can use the estimated $\widehat{\mathbf{b}}_i$ to construct a model-based individual trajectory $\widehat{\mu}_{it} = \exp\left(\mathbf{x}_{it}^\top \widehat{\boldsymbol{\beta}} + \mathbf{z}_{it} \widehat{\mathbf{b}}_i\right)$.

Another important implication is the different interpretation of the parameters $\boldsymbol{\beta}$ between the two models. From the marginal model in (9.22), we obtain

$$E\left(y_{it} \mid \mathbf{x}_{it}\right) = g^{-1}\left(\eta_{it}\right) = g^{-1}\left(\mathbf{x}_{it}^\top \boldsymbol{\beta}_m\right), \tag{9.50}$$

where g^{-1} denotes the inverse of g and $\boldsymbol{\beta}_m$ is the parameter vector under the marginal model. We can also compute $E\left(y_{it} \mid \mathbf{x}_{it}\right)$ for GLMM from (9.13) and get (see Problem 9.17)

$$E\left(y_{it} \mid \mathbf{x}_{it}\right) = E\left(g^{-1}\left(\eta_{it}\right) \mid \mathbf{x}_{it}\right), \tag{9.51}$$

where $\eta_{it} = \mathbf{x}_{it}^\top \boldsymbol{\beta}_e + \mathbf{z}_{it} \mathbf{b}_i$ and $\boldsymbol{\beta}_e$ is the parameter vector under the GLMM. Note that we used different symbols for the parameter $\boldsymbol{\beta}$ between GLMM and GEE because, except for the identity link, the two right-hand sides in (9.50) and (9.51) are generally different (see Problems 9.18 and 9.19). For example, for the identity link $g(y) = y$, it is readily shown that for the GLMM

$$E\left(g^{-1}\left(\eta_{it}\right) \mid \mathbf{x}_{it}\right) = \mathbf{x}_{it}^\top \boldsymbol{\beta}_e. \tag{9.52}$$

Also, for the marginal model, $g^{-1}\left(\mathbf{x}_{it}^\top \boldsymbol{\beta}_m\right) = \mathbf{x}_{it}^\top \boldsymbol{\beta}_m$ under the identity link. It then follows from (9.50), (9.51), and (9.52) that

$$E\left(g^{-1}\left(\eta_{it}\right) \mid \mathbf{x}_{it}\right) = \mathbf{x}_{it}^\top \boldsymbol{\beta}_e = g^{-1}\left(\mathbf{x}_{it}^\top \boldsymbol{\beta}\right) = \mathbf{x}_{it}^\top \boldsymbol{\beta}_m.$$

Thus, $\boldsymbol{\beta}_e = \boldsymbol{\beta}_m$ for the identity link and the parameters from the marginal and GLMM models have the same interpretation under either modeling approach.

Other than the identity link, $\boldsymbol{\beta}_e$ and $\boldsymbol{\beta}_m$ are generally different. We must be mindful about such a difference, since it has serious ramifications about the interpretation of $\boldsymbol{\beta}$. For example, for the logit link g,

$$E(g^{-1}\left(\eta_{it}\right) \mid \mathbf{x}_{it}) = E\left(\frac{\exp\left(\mathbf{x}_{it}^\top \boldsymbol{\beta}_e + \mathbf{z}_{it} \mathbf{b}_i\right)}{1 + \exp\left(\mathbf{x}_{it}^\top \boldsymbol{\beta}_e + \mathbf{z}_{it} \mathbf{b}_i\right)} \mid \mathbf{x}_{it}\right) \tag{9.53}$$

$$\neq \frac{\exp\left(\mathbf{x}_{it}^\top \boldsymbol{\beta}_e\right)}{1 + \exp\left(\mathbf{x}_{it}^\top \boldsymbol{\beta}_e\right)} = g^{-1}\left(\mathbf{x}_{it}^\top \boldsymbol{\beta}_e\right).$$

Thus, $\boldsymbol{\beta}$ in the fixed effect of GLMM is not identical to $\boldsymbol{\beta}$ in the marginal model. In particular, (9.53) shows that $\boldsymbol{\beta}_e$ does not have the familiar log odds ratio interpretation as its counterpart $\boldsymbol{\beta}_m$ for the marginal model. Although the two parameter vectors are related to each other in some special cases (see Problem 9.18), the relationship between the two in general can be quite complex (see Zhang et al. (2011)).

Example 9.11

We applied the marginal model in Example 9.8 and GLMM with a random intercept in Example 9.3 to the DOS data. Although the overall conclusions are similar, i.e., medical burden and gender are associated with depression, the point estimates are quite different, reflecting the different interpretations of the parameters from the two models. As mentioned, β_m from the marginal model maintains the log odds ratio interpretation for the population as in the standard logistic model for cross-sectional data, but β_e from the GLMM is more difficult to interpret. ▯

9.5 Model Diagnosis

Compared to the cross-sectional case, model evaluation for longitudinal data is much more complicated because of the correlation among the repeated measures. As in the case of cross-sectional data analysis, residual plots that depict the difference between observed and fitted values may reveal their systematic differences, indicating some type of lack of fit such as incorrect specification of the link function and/or the linear predictor. However, commonly used goodness-of-fit tests such as Pearson chi-square and deviance test statistics for cross-sectional studies can no longer be applied directly to longitudinal data, because of correlated outcomes due to repeated assessments. In this section, we discuss some common goodness-of-fit statistics for model evaluation and criteria for model selection for longitudinal data.

9.5.1 Marginal models

In general, plots of residuals are very helpful in model diagnosis. Scatter plots of residuals against fitted linear predictors or a specific covariate are commonly used. For continuous outcomes, the residuals should fluctuate randomly above and below the horizontal line at zero if the assumptions for the GEEs are satisfied. Since only the mean response is required, heteroskedasticity and lack of symmetry around zero in general are not issues for GEEs. However, residual plots for categorical responses are usually not as useful. As mentioned in Chapter 4, a plot of the residuals vs. linear predictors will have all the points fall on two deterministic curves and thus may not be informative. In such cases, we usually aggregate the residuals in some way to assess if there is any systematic deviation from what would be expected.

9.5.1.1 Chi-square statistic for binary responses

For binary responses, the plots may not be as informative because of the discrete nature of the responses. Residual-based statistics such as Pearson chi-square statistics can be used to assess model fit within the current context. Let π_{it} be the marginal mean of y_{it}, i.e., $\Pr(y_{it} = 1 \mid \mathbf{x}_{it})$, and $\widehat{\pi}_{it}$ be the fitted value. Pan (2002) generalized the Pearson chi-square statistic as follows:

$$G = \sum_{i=1}^{K} \sum_{t=1}^{m} \frac{(y_{it} - \widehat{\pi}_{it})^2}{\widehat{\pi}_{it}(1 - \widehat{\pi}_{it})}.$$

Pan (2002) showed that G has an approximately normal distribution with mean nm, as $n \to \infty$ (m is bounded). The p-value can thus be obtained by referring G to the normal distribution.

Pan (2002) also suggested an unweighted version of the chi-square statistic,

$$U = \sum_{i=1}^{K} \sum_{t=1}^{n_i} (y_{it} - \hat{\pi}_{it})^2,$$

and developed its asymptotic distribution (see Pan (2002) for details).

9.5.1.2   Cumulative sum tests

Lin et al. (2002) proposed graphical and numerical methods for model assessment based on the cumulative sums of residuals. This sum can be cumulative over certain coordinates such as a covariate, linear predictors, or some related aggregates of residuals. For example, the cumulative sum over the jth covariate is defined as

$$W_j(x) = \frac{1}{\sqrt{n}} \sum_{i=1}^{n} \sum_{t=1}^{m} I(x_{itj} \leq x) r_{it}, \qquad (9.54)$$

where r_{it} is the residual and x_{itj} is the jth covariate of the ith subject at the tth time. This cumulative sum based on a covariate can be used to assess the functional form for the covariate because $W_j(x)$ approximately follows a zero-mean Gaussian process if the GEE model is correct. Based on a large number of realizations generated from the Gaussian processes by Monte Carlo simulation, one can visually assess and formally test the plausibility of the model by the Kolmogorov-type supremum statistic $S_j \equiv \sup_x |W_j(x)|$. The p-value is the percent of times this statistic over all simulated samples is larger than the one based on the observed sample.

9.5.1.3   Covariate space partitions

Another common approach is to partition the covariate space into different regions and test whether such discretized covariates are significant (Barnhart and Williamson, 1998). Consider the marginal model in (9.22). Partition the covariate space (all covariates $\mathbf{x}_{ij}$) into M distinct regions, where $\mathbf{x}_{ij}$ can be time independent or dependent. Let $\mathbf{w}_{ij} = (w_{ij1}, \ldots, w_{ijM-1})$ denote the corresponding indicator vector for the discretized $\mathbf{x}_{ij}$ where $w_{ijk} = 1$ if $\mathbf{x}_{ij}$ belongs to the kth region, and 0 otherwise (the last region is treated as the reference). By adding $\mathbf{w}_{ij}$ as additional variables in (9.22), we have

$$E(y_{it} \mid \mathbf{x}_{it}) = \mu_{it}, \quad g(\mu_{it}) = \mathbf{x}_{it}^{\top} \boldsymbol{\beta} + \mathbf{w}_{ij}^{\top} \boldsymbol{\gamma}.$$

If the original model fits the data well, the additional variables $\mathbf{w}_{ij}$ should not have a significant contribution, and thus the null $H_0 : \boldsymbol{\gamma} = \mathbf{0}$ will not be rejected. If $\boldsymbol{\gamma}$ is significantly different from 0, then the model does not fit well. We may apply the Wald or score statistic for the test. A major difficulty with the application of the approach to real data is how to create the partitions of the covariate space. As discussed in Chapter 4, one effective and popular approach is to divide the covariate space based on the fitted values, as exemplified by the Hosmer–Lemeshow test. Thus, we may use similar methods to create partitions of the covariate space (Horton et al., 1999).

9.5.1.4   Information criteria for GEE

The techniques introduced above may be applied to assess the quality of a single model. In practice, we often need to compare two models, especially in model selection. For nested models, one may test whether the effects not included in the reduced model are significant using score or Wald tests. Information criteria, applicable to both nested and non nested models, are also commonly used alternatives. Since no distribution is assumed for GEE for marginal models, likelihood is not defined and hence likelihood-based information criteria, such as Akaike's information criterion (AIC) and Bayesian information criterion (BIC) introduced in Chapter 6 for model selection, cannot be directly applied. One way to deal with this is to construct a likelihood based on the estimating equations, or a quasi-likelihood, as a basis for computing such likelihood-based statistics.

Recall that we discussed quasi-likelihood functions for scaled variances (Section 6.3.2.2) for cross-sectional data. Let $q(y_{it}, \mu_{it})$ be the quasi-likelihood at time t, where μ_{it} and $\phi v(\mu_{it})$ are the mean and variance of y_{it} under a (cross-sectional) GLM. Under an independent working correlation matrix, the quasi-likelihood of the longitudinal data is $Q(\boldsymbol{\beta}) = \sum_{i=1}^{n} \sum_{t=1}^{m} q(y_{it}, \mu_{it})$. A quasi-likelihood under other types of working correlation models can also be constructed.

Let $\widehat{\boldsymbol{\beta}}_R$ be the GEE estimate of $\boldsymbol{\beta}$ under a given working correlation model R, and define a quasi-likelihood information criterion, QIC_u, as

$$QIC_u(R) = -2Q(\widehat{\boldsymbol{\beta}}_R) + 2p, \tag{9.55}$$

where p denotes the number of parameters in the model. We can use the QIC_u to help select competing models the same way as the AIC by choosing the one with the minimum $QIC_u(R)$. In addition to model selection, a modified version of (9.55) can also be used for selecting the working correlation by replacing p with $trace(\widehat{\Omega}_I \widehat{V}_R^{-1})$, where $\widehat{\Omega}_I$ is the sandwich variance estimate and $\widehat{V}_R$ is the model-based counterpart under working independence, evaluated at $\widehat{\boldsymbol{\beta}}_R$, i.e.,

$$QIC(R) = -2Q(\widehat{\boldsymbol{\beta}}_R) + 2trace(\widehat{\Omega}_I \widehat{V}_R^{-1}). \tag{9.56}$$

The working correlation with the smaller $QIC(R)$ is preferred (Pan, 2001).

Example 9.12

Consider the marginal model studied in Example 9.8. We divide the subjects into ten groups of comparable sizes according to the fitted linear predictors (or equivalently the fitted probabilities). Select one group as the reference level; then we may use indicator variables of the other nine groups, $\mathbf{I} = (x_1, \ldots x_9)^{\top}$, in the following model to test model fit:

$$\text{logit}\,(\mu_{it}) = \beta_0 + \beta_1 Age_i + \beta_2 Gender_i + \beta_3 Edu_i + \beta_4 CIRS_{it} + \beta_5 MS1_i \tag{9.57}$$
$$+ \beta_6 MS2_i + \beta_7 Race_i + \boldsymbol{\beta}_8^{\top} \mathbf{I}_{ij}$$

The p-value for testing $\boldsymbol{\beta}_8 = \mathbf{0}$ is 0.7243 if independent correlation is used for the working correlation. Thus, there is not sufficient evidence for the lack of fit in this case.

The QIC and QIC_u for the model above are 1122.0750 and 1077.1444, while those for the one in Example 9.8 are 1107.5489 and 1069.0662, respectively. Hence, the original model in Example 9.8 is actually better than the one with additional predictors. $\quad\Box$

9.5.2 Generalized linear mixed-effect models

Many of the techniques for marginal models can also be applied to GLMMs. There are two components in the GLMM, the fixed and the random effects. Thus, in addition to the overall goodness of fit, we may further assess each of the two components separately.

9.5.2.1 Overall goodness of fit

For GLMMs there are two versions of raw residuals, based on subject-specific conditional fitted values and population marginal fitted values. For LMM (9.7), the marginal residual vector for the ith subject is defined as $\mathbf{r}_{im} = \mathbf{y}_i - \mathbf{x}_i^\top \widehat{\boldsymbol{\beta}}$ and the conditional residual vector for the ith subject is defined as $\mathbf{r}_{ic} = \mathbf{y}_i - \mathbf{x}_i^\top \widehat{\boldsymbol{\beta}} - \mathbf{z}_{it}^\top \widehat{\mathbf{b}}_i$. The marginal residual may be applied to assess the overall goodness of fit. Since $\mathbf{y}_i - \mathbf{x}_i^\top \boldsymbol{\beta} \sim \text{i.d.} N\left(\mathbf{0}, \mathbf{z}_i D \mathbf{z}_i^\top + \sigma^2 I_m\right)$, the error terms from the subject may be correlated with different variation; thus it is not convenient to use them directly for model assessment. We may use the Cholesky decomposition of the covariance matrix to transform them into i.i.d. variables. Let C_i be the unique lower triangular matrix such that $C_i C_i^\top = Var\left(\mathbf{y}_i\right) = \mathbf{z}_i D \mathbf{z}_i^\top + R$. Then $C_i^{-1} \boldsymbol{\epsilon}_i \sim \text{i.i.d.} \ N\left(\mathbf{0}, I_m\right)$. Thus, we may compare the scaled residuals, $\mathbf{r}'_{im} = \widehat{C}_i^{-1} \mathbf{r}_{im}$, with the standard normal distribution. For example, a QQ plot may be applied to visually assess if the scaled residuals follow the standard normal distribution.

For binary responses, residual plots are not as informative. In such cases, we may use Pearson chi-square statistics. If inference is based on the Taylor linearization-based pseudo-model, we can also use a generalized chi-square statistic to assess the overall goodness of fit. This statistic is defined as

$$X^2 = \sum_{i=1}^{n} \mathbf{r}_i^\top V_i^{-1} \mathbf{r}_i,$$

where $\mathbf{r}_i$ is the residual vector for the ith subject based on the linearized pseudo-model and V_i is the estimated marginal variance based on the estimated mean responses. As in the case of the Pearson chi-square test, $\frac{X^2}{df}$ produces the usual residual dispersion estimate, where the degrees of freedom, df, is the total number of observations minus the number of parameters in the fixed-effect component.

9.5.2.2 Goodness of fit for fixed-effects

The most commonly used approach for assessing a fixed-effect is to test the significance of the relevant terms to determine whether it should be included. Thus, we can incorporate it with the step-wise model selection procedure discussed in Chapter 7 for model selection. Alternatively, we may follow the procedures described above by partitioning the covariate space into several regions and test whether the indicators for such discretized variables are significant. If the fixed-effect is adequate, those additional terms should not be significant.

Pan and Lin (2005) generalized the cumulative residual sum method to assess functional forms of the fixed effects for GLMMs. The cumulative residual sums are defined similarly as in (9.54) using the marginal residuals. The cumulative residual sums may be applied to assess the functional form of a covariate or link function, depending on whether these sums are defined over the specific covariate or all fixed effects in the linear predictor.

9.5.2.3 Assessing random component

Since the random component is typically not of primary interest for most applications, the effect of misspecification of the random component on inference has not received much attention. However, some recent studies have shown that the random component can also be critical to model fitting; considerable bias may arise if the random component is misspecified. Using simulations with a random-intercept logistic model, Litière et al. (2008) showed that the MLEs are inconsistent in the presence of misspecification. If the variability of the random-effects is small, the bias for parameters in the fixed effects is generally small. However, estimates of variation of the random-effects are always subject to considerable bias when the random-effects distribution is misspecified. Hence, it is still important to assess assumptions on the random component.

For computational convenience, multivariate normality is commonly assumed for the random-effects in practice. For LMMs, usually the errors are further assumed to be multivariate normal and independent with the random-effects. Thus the overall random component for the marginal error is again multivariate normal and the QQ plot for the scaled residuals discussed above may be applied to assess the overall normality of the random component.

It is more complicated to assess the multivariate normality of the random-effects for categorical and count responses. Although some approaches have been developed for formally testing misspecifications of random-effects, none is yet available from popular software packages. In practice, without a formal assessment of the parametric distribution assumed by the random-effects, one may fit both the parametric and the corresponding marginal models and compare their estimates for the fixed effects. If big discrepancies emerge, then the random-effects distribution may be misspecified. Alternatively, one may compare two candidate models using information criteria which we will discuss next.

9.5.2.4 Information criteria for GLMM

Since GLMM is a parametric approach, common goodness-of-fit statistics such as AIC, BIC, and likelihood ratio tests are readily applied to examine model fit and conduct model selection. The AIC is defined as $AIC = 2k - 2l\left(\widehat{\theta}\right)$, where k is the number of parameters in the model and $l\left(\widehat{\theta}\right)$ is the estimated maximum likelihood. Note that both ML and REML may be applied. If we focus on the random component and apply the restricted likelihood method, then only parameters in the covariance component are counted in k. On the other hand, if maximum likelihood is applied, then all parameters in both the mean structure and the random component will be counted in k. Vaida and Blanchard (2005) argued that both versions of AICs may not be appropriate if the focus is not at the population level. They called these AICs marginal and proposed *conditional AICs* if the focus is on the random effects (Vaida and Blanchard, 2005).

As in the cross-sectional data cases, corrections for AICs have been proposed in case the sample sizes are small. The bias-corrected AIC, AICC, is defined as $AICC = 2k\frac{mn}{mn-k-1} - 2l\left(\widehat{\theta}\right)$, where nm is the total number of observations and k is the number of parameters defined above according to the estimating methods.

For the BIC, a popular definition is

$$BIC = k\log\left(n\right) - 2l\left(\widehat{\theta}\right),$$

where n is the total number of subjects minus the number of the parameters in the fixed-effects component for REML and the number of subjects for ML approach. This definition may not be good in all situations. For example, linear models are a special case of the LMM, so if the errors are independent the sample size would be mn. To distinguish parameters in the random- and fixed-effects components, Delattre et al. (2014) proposed the following BIC,

$$BIC = -2l\left(\widehat{\theta}\right) + dim(\theta_R)\log n + dim(\theta_F)\log\left(mn\right),$$

where θ_F and θ_R denote the parameters in the fixed-effects and the random components (both random-effects and error distributions). As illustrated in Chapter 7, the BIC imposes a large penalty for estimation of each additional covariate, often yielding oversimplified models. For this reason, the AIC is more commonly used for model selection.

Note that both the AIC and BIC above are defined in terms of the likelihood of the data. Thus, we cannot use the estimates based on pseudo-models for the calculation. In fact, if different GLMMs are applied, the pseudo-responses generated for the same subjects may be very different. Thus, it is possible that between two nested models the broader model has a smaller pseudo-likelihood. Hence, information criteria cannot be defined for pseudo-models.

Example 9.13

Consider the following model with a random intercept for the DOS:

$$HAMD_{ij} = \beta_0 + \beta_1 gender_i + \beta_2 Age_i + \beta_3 education_i + \beta_4 CIRS_{ij}$$
$$+ \beta_5 visit1_{ij} + \beta_6 visit2_{ij} + \beta_7 visit3_{ij} + \alpha_i + \varepsilon_{ij}, \alpha_i \sim N(0, \nu^2),$$
$$\boldsymbol{\epsilon}_i = (\varepsilon_{i1}, \varepsilon_{i2}, \varepsilon_{i3}, \varepsilon_{i4}) \sim N(0, \sigma^2 I_4), \alpha_i \perp \boldsymbol{\epsilon}_i, \ j = 1, 2, 3, 4$$

where $visit1_{ij}, visit2_{ij}$, and $visit3_{ij}$ are the indicators for the first three visits. There are eight parameters in the mean structure and two parameters in the random component (the variance of the random intercept and the common variance of the errors). The fit statistics are summarized in the following table.

Methods	−2×Log-Likelihood	d	n	AIC	AICC	BIC
Res Likelihood	5287.1	2	229 − 8 = 221	5291.1	5291.1	5297.9
Likelihood	5275.1	10	229	5295.1	5295.3	5329.4

▯

The information criteria are commonly applied for comparison purposes. For two models that are applied to the same data set, the one with the smaller value is considered better. Since only the same type of information criterion on the same data set may be compared, the restricted likelihood-based AIC on one model cannot be compared with the likelihood-based AIC on a different model.

Example 9.14

Let us assess the GLMM (9.15) in Example 9.3, similarly dividing the subjects into ten groups of comparable sizes according to their fitted linear predictors in the fixed-effect component. Note that the fitted linear predictor with the estimated random effects included should not be used. Select one group as the reference level; then we may use indicator variables of the other nine groups, $\mathbf{I} = (x_1, \ldots x_9)^\top$, in the following model to test the model fit:

$$\text{logit}\,(\mu_{it}) = \beta_0 + b_{i0} + \beta_1 Age_i + \beta_2 Gender_i + \beta_3 Edu_i + \beta_4 CIRS_{it} \qquad (9.58)$$
$$+ \beta_5 MS1_i + \beta_6 MS2_i + \beta_7 Race_i + \boldsymbol{\beta}_8^\top \mathbf{I}_{it}.$$

The p-value for testing the null $H_0 : \boldsymbol{\beta}_3 = \mathbf{0}$ is 0.5996. Thus, the additional predictor does not significantly improve the model fit, and the original model (9.15) is adequate.

The information criteria for these two models are summarized in the following table:

Model	−2×Log-Likelihood	d	n	AIC	AICC	BIC
(9.15)	600.31	9	225	618.31	618.52	649.06
(9.58)	597.85	18	225	633.85	634.63	695.34

Note that there are several missing values in marital status, so the numbers of subjects included in the models are 225. All the different information criteria indicate that the original model (9.15) is actually better than the one with the covariate indicators added. ▯

Exercises

9.1 (a) The longitudinal data set "v4c" is in the vertical format; transform it into the horizontal format.

   (b) Transform the data set you obtained in part (a) back into the vertical format.

9.2 Plot the mean/SD of HIV knowledge of adolescent girls at baseline and three months post treatment stratified by treatment for the Sexual Health study.

9.3 Assess the trend of depression during the study for the DOS by plotting the individual profile of a random sample of the patients.

For the Problems 9.4, 9.5 and 9.10, we use data from a study on depression treatment stigma (DTS study), where subjects were randomized into two treatment groups ("treatment"= 1 for collaborative-care depression care management (DCM), and "treatment" = 0 for care as usual(CAU)) (Chen et al., 2015). Age, gender (sex = 1 for females and sex = 2 for males), education (edu = 1 for 0 years of education, 2 for 1–5 years of education, 3 for 6–8 years of education, 4 for 9–12 years of education, and 5 for 13 or more years of education), marriage status (marriage = 1 for married, 2 for divorced, 3 for unmarried, and 4 for widowed), medical burden (CIRS), HamD score for depression (HAMD_0), and stigma for depression treatment (TS_0) were measured at recruitment. Depression outcomes were measured every three months (HAMD_3, HAMD_6, HAMD_9, and HAMD_12) up to one year after the baseline. Stigma for depression treatment was measured every six months (TS_0, TS_6, and TS_12).

9.4 Perform some exploratory analysis on the DTS study described above.

   (a) Compute the mean and standard deviation of the HamD scores for the two treatment groups at each time point.

   (b) Treat repeated outcomes in HamD scores as if they were independent and fit the linear model with treatment and measurement time (as a continuous variable) as well as their interaction as the predictors. Is DCM, compared with CAU, more effective in treating depression?

   (c) Treatment stigma (TS) scores take values 0–3, with higher scores indicating more negative stigma. Compute the proportions of each level of TS scores for the two treatment groups at each time point.

   (d) Treat repeated outcomes in TS scores as if they were independent and fit a cumulative logistic model with treatment and assessment time (as a continuous variable) as well as their interaction as the predictors. Is DCM, compared with CAU, more effective in treating stigma associated with depression treatment?

9.5 For the DTS study, use subjects with all five assessments in HamD scores in the CAU group for this question. The intraclass correlation coefficient among the repeated measures in Ham-D scores can be estimated by fitting the following mixed-effects model:

$$HamD_{ij} = \mu + r_i + \varepsilon_{ij}, \ j = 0, 3, 6, 9, 12,$$

where $HamD_{ij}$ is the HamD score for the ith subject measured j months after baseline, μ is the population average of the scores, $r_i \sim N(0, \nu^2)$ is the random effect for the ith subject, $\varepsilon_{ij} \sim N(0, \sigma^2 I_4)$, and $r_i \perp (\varepsilon_{i0}, \varepsilon_{i3}, \varepsilon_{i6}, \varepsilon_{i9}, \varepsilon_{i12})$.

(a) Write the model in an equivalent multivariate regression form without random effects.

(b) Give an estimate of μ based on the mixed-effects model. Is it significantly different from 10?

(c) Estimate the intraclass correlation coefficient among the repeated HamD scores.

(d) Write down the corresponding marginal model and use the GEE to estimate μ. Is it significantly different from 10?

9.6 Recorded in the table below are testing scores of ten students in math, physics, and chemistry tests. We may treat the test scores from the three different subjects as repeated measures and use the following mixed-effects model:

$$MathScore_i = \mu_{Math} + r_i + \varepsilon_{i,math}$$
$$PhysicsScore_i = \mu_{Physics} + r_i + \varepsilon_{i,physics}$$
$$ChemistryScore_i = \mu_{Chemistry} + r_i + \varepsilon_{i,chemistry},$$

where $r_i \sim N(0, \nu^2)$ is the random effect, $(\varepsilon_{i,math}, \varepsilon_{i,physics}, \varepsilon_{i,chemistry}) \sim N(0, \sigma^2 I_3)$, and $r_i \perp (\varepsilon_{i,math}, \varepsilon_{i,physics}, \varepsilon_{i,chemistry})$.

Student	Math	Physics	Chemistry
1	93	94	97
2	89	90	88
3	96	92	97
4	83	84	81
5	80	74	81
6	71	77	80
7	96	93	92
8	87	93	94
9	77	75	74
10	79	84	76

(a) Write the model in an equivalent multivariate regression form without random-effects.

(b) Based on the model, is there any significant difference among the three subject test scores?

(c) Estimate the intraclass correlation coefficient of the testing scores and interpret its meaning within the context of test scores from the three subjects.

9.7 Fit the DOS data (data "v4c") with the following mixed-effects model with random intercept (r_i):

$$HAMTTL_{ij} = \beta_0 + \beta_1 gender_i + \beta_2 CIRS_{ij} + \beta_3 visit1_{ij} + \beta_4 visit2_{ij} + \beta_5 visit3_{ij}$$
$$+ r_i + \varepsilon_{ij}, \boldsymbol{\epsilon}_i = (\varepsilon_{i1}, \varepsilon_{i2}, \varepsilon_{i3}, \varepsilon_{i4}) \sim N(0, \sigma^2 I_4), r_i \sim N(0, \nu^2), \boldsymbol{\epsilon}_i \perp r_i.$$

(a) Write the model in the multivariate regression form without random-effects.

(b) Under the model, does the covariance matrix change with the subjects?

(c) Are fixed-effects significant at the 0.05 type I error level?

(d) Are there significant differences among the first three visits?

(e) Find estimates for σ^2 and ν^2.

(f) Estimate the predicted mean values for subjects with the same gender and CIRS scores as the last subject "ZIL8618D" in the data set.

(g) Estimate the covariance matrix for the last subject "ZIL8618D."

9.8 Add the random-effects of CIRS to the model in Problem 9.7 and fit the DOS data (data "v4c") with the following mixed-effects model:

$$HAMTTL_{ij} = \beta_0 + \beta_1 gender_i + \beta_2 CIRS_{ij} + \beta_3 visit1_{ij} + \beta_4 visit2_{ij} + \beta_5 visit3_{ij}$$

$$+ r_i + s_i CIRS_{ij} + \varepsilon_{ij}, \boldsymbol{\epsilon}_i \sim N(0, \sigma^2 I_4), (r_i, s_i) \sim N(\boldsymbol{0}, \begin{pmatrix} \nu^2 & \tau\mu\nu \\ \tau\mu\nu & \mu^2 \end{pmatrix}), \boldsymbol{\epsilon}_i \perp (r_i, s_i).$$

(a) Under the model, does the covariance matrix change with the subjects?

(b) Are fixed-effects significant at the 0.05 type I error level?

(c) Are there significant differences among the first three visits?

(d) Find estimates for τ, σ^2, ν^2, and μ^2.

(e) Estimate the predicted mean values for subjects with the same gender and CIRS scores at the last subject "ZIL8618D" in the data set.

(f) Estimate the covariance matrix for the last subject "ZIL8618D."

9.9 Suppose $x \sim N\left(0, \nu^2\right), y \sim N\left(0, \sigma^2\right)$, and x is independent of y. Given $x + y = 1$, find the MLE for x.

9.10 In this question we develop a regression model to assess the treatment effect for stigma in the DTS study, controlling for demographics and baseline measurements. We will use the cumulative logit link functions for the four-level stigma outcome and apply some model selection criteria to trim the independent variables.

(a) Perform the backwards model selection for marginal models using treatment, visit time, age, gender, marriage status, education, baseline HamD, baseline TS, and CIRS as covariates, with the elimination threshold for p-values set at 20%. Treat treatment, gender, marriage status, and baseline TS as categorical and the others as continuous variables. Since the purpose is to assess treatment effect, which occurs over time, treatment and its interaction with time will be kept in the model. Summarize the selection process (which factors are eliminated at each step and why).

(b) Is there a significant treatment effect based on the model obtained in part (a)?

(c) For the marginal model in part (a), identify subjects with the most influence on the coefficient of the interaction of treatment effects and time.

(d) Based on the marginal model obtained in part (a), develop a generalized mixed-effects model using the same fixed-effects and adding a random intercept. Is there significant treatment effect based on the generalized mixed-effects model?

(e) Identify subjects with the most influence on the estimate of treatment effects based on the model in part (d). Compare the results with those from part (c).

9.11 Consider the following simple random intercept generalized logit model for a categorical response with three levels 1–3:

$$y_{it} \mid \mathbf{b}_i \sim \text{i.d. } MN\left(1, \boldsymbol{\pi}_{it}\right), \quad \log\left(\frac{\pi_{itj}}{1 - \pi_{itj}}\right) = b_{ij}, \; j = 1, 2. \tag{9.59}$$

Show that if the first level is selected as the reference, then the model can be equivalently expressed as

$$y_{it} \mid \mathbf{b}_i \sim \text{i.d. } MN\left(1, \boldsymbol{\pi}_{it}\right), \quad \log\left(\frac{\pi_{itj}}{1 - \pi_{itj}}\right) = b'_{ij}, \ j = 2, 3. \qquad (9.60)$$

If b_{i1} and b_{i2} are independent, then b'_{i2} and b'_{i3} in general become dependent.

9.12 Generalize the model considered in Example 4.11 to a marginal model for the longitudinal DOS data and compare the findings with that in Example 4.11.

9.13 Prove (9.31).

9.14 Redo Problem 9.12 using the three-category depression outcome as the response together with a cumulative logit link function.

9.15 Use the technique discussed in Section 9.4.4.1 to develop a GEE approach for zero-inflated Poisson model for count responses in longitudinal studies.

9.16 Show that $\mathbf{x}_i \perp y_{it} \mid \mathbf{x}_{it}$ and $E\left(y_{it} \mid \mathbf{x}_{it}\right) = \mu_{it}$ imply the FCCM.

9.17 Show the identities in (9.51) and (9.52) (see Problem 1.2).

9.18 Consider the GLMM in (9.11) with a logit link. Show that

(a) $E\left(y_{it} \mid \mathbf{x}_{it}, \mathbf{z}_{it}\right) \approx \Phi\left(\frac{\mathbf{x}_{it}^\top \boldsymbol{\beta}}{\sqrt{c^2 + \mathbf{z}_{it}^\top \Sigma_b \mathbf{z}_{it}}}\right)$, where $c = \frac{15\pi}{16\sqrt{3}}$ and $\Phi\left(\cdot\right)$ denotes the cumulative distribution function of the standard normal with mean 0 and variance 1 (see Johnson et al. (1994)).

(b) $\Phi\left(\frac{\mathbf{x}_{it}^\top \boldsymbol{\beta}}{\sqrt{c^2 + \mathbf{z}_{it}^\top \Sigma_b \mathbf{z}_{it}}}\right) \approx \text{logit}^{-1}\left(\left(1 + c^{-2}\mathbf{z}_{it}^\top \Sigma_b \mathbf{z}_{it}\right)^{-\frac{1}{2}} \mathbf{x}_{it}^\top \boldsymbol{\beta}\right)$.

(c) if $\mathbf{z}_{it} = \mathbf{z}_t$, i.e., $\mathbf{z}_{it}$ is independent of individual characteristics, then $\boldsymbol{\beta}_M \approx \left(1 + c^{-2}\mathbf{z}_t^\top \Sigma_b \mathbf{z}_t\right)^{-\frac{1}{2}} \boldsymbol{\beta}$, where $\boldsymbol{\beta}_M$ is the parameter of the corresponding marginal model.

9.19 Consider the GLMM in (9.11), with a log link. Show that

(a) $E\left(y_{it} \mid \mathbf{x}_{it}, \mathbf{z}_{it}\right) = E\left[\exp\left(\frac{1}{2}\mathbf{z}_{it}^\top \Sigma_b \mathbf{z}_{it}\right) \mid \mathbf{x}_{it}\right] \exp\left(\mathbf{x}_{it}^\top \boldsymbol{\beta}\right)$.

(b) if $\mathbf{z}_{it}$ is independent with $\mathbf{x}_{it}$, then $E\left(y_{it} \mid \mathbf{x}_{it}\right) = \exp\left(\gamma_0 + \mathbf{x}_{it}^\top \boldsymbol{\beta}\right)$, where $\gamma_0 = \log\left[E\left(\exp\left(\frac{1}{2}\mathbf{z}_{it}^\top \Sigma_b \mathbf{z}_{it}\right)\right)\right]$.

(c) if $\mathbf{z}_{it}$ is a subvector of $\mathbf{x}_{it}$, say $\mathbf{x}_{it} = \left(\mathbf{z}_{it}^\top, \mathbf{w}_{it}^\top\right)^\top$, then

$$E\left(y_{it} \mid \mathbf{x}_{it}, \mathbf{z}_{it}\right) = \exp\left[\mathbf{w}_{it}^\top \boldsymbol{\beta}_w + \mathbf{z}_{it}^\top \left(\boldsymbol{\beta}_z + \frac{1}{2}\Sigma_b \mathbf{z}_{it}\right)\right].$$

9.20 Construct a generalized linear mixed-effects model for the longitudinal DOS data with the fixed-effects component similar to that in Problem 9.12 and a random intercept and assess the model fit.

9.21 Construct a generalized linear mixed-effects model for the Sexual Health study data with the fixed-effects component similar to that in Example 9.7 and random slopes of the time effect and assess the model fit.

9.22 Assess the models in Problems 9.12 and 9.14.

9.23 Use the longitudinal Catheter Self Management Study data for this question ("intake.csv" contains demographic and baseline information, and "catheter.csv" contains follow-up measurements). We model the binary response for having any UTIs during the last two months with gender, age, education, and treatment as predictors. Fit the following models and test whether there is a significant group difference.

(a) Model the binary UTI severity outcome with the following logistic model with random intercept

$$\text{logit}\,(\pi_{ij}) = \alpha_i + \beta_0 + \beta_1 Gender_i + \beta_2 Age_i + \beta_3 Education_i + \beta_4 Group_i,$$

where π_{ij} is the probability of having UTIs during the last two months for the ith subject at the jth assessment, *Gender* is the indicator variable for female, and $\alpha_i \sim N(0, \sigma^2)$ is the random intercept.

(b) Change the link function in (a) to probit and clog-log.

(c) Change the GLMM in (a) to GEE for the marginal model

$$\text{logit}\,(\pi_{ij}) = \beta_0 + \beta_1 Gender_i + \beta_2 Age_i + \beta_3 Education_i + \beta_4 Group_i,$$

with the independent working correlation.

(d) Change the link function in (c) to probit and clog-log.

9.24 In this question, we change the response in Problem 9.23 to a three-level response by grouping the counts of UTIs into three levels, 0, 1, and ≥ 2. Fit the following models and test whether there is a significant group difference.

(a) Model the ordinal three-level UTI count outcome with the following cumulative model with random intercept

$$\text{logit}\,(\gamma_{ij}) = \alpha_i + \beta_0 + \beta_1 Gender_i + \beta_2 Age_i + \beta_3 Education_i + \beta_4 Group_i,$$

$j = 1$ and 2, where γ_{i1} and γ_{i2} are the probabilities of having no UTI and having at most one UTI, respectively, and $\alpha_i \sim N(0, \sigma^2)$ is the random intercept.

(b) Change the link function in (a) to probit and cloglog.

(c) Change the GLMM in (a) to GEE for the marginal model

$$\text{logit}\,(\gamma_{ij}) = \beta_0 + \beta_1 Gender_i + \beta_2 Age_i + \beta_3 Education_i + \beta_4 Group_i,$$

$j = 1$ and 2, with the independent working correlation.

(d) Change the link function in (c) to probit and cloglog.

.

Chapter 10

Evaluation of Instruments

In this chapter, we focus on assessing diagnostic and screening instruments. Such tools are commonly used in clinical research studies, ranging from physical health to psychological well-being to social functioning. Since early detection of diseases often leads to early intervention and speedier recovery while false diagnoses unnecessarily expose individuals to potentially harmful treatments (Bach et al., 2007), it is important to assess accuracy of such instruments so that informative decisions can be made. When the true status is available, we can assess accuracy by comparing test results directly with the known true status. The receiver operating characteristic (ROC) curve is commonly used when the true status is binary such as presence or absence of a disease. For a continuous or ordinal outcome (with a reasonably wide range so it can be treated as a continuous), we may use concordance correlation coefficients.

In mental health and psychosocial research, many instruments are designed to measure conceptual constructs characterized by certain behavioral patterns, and as such it is generally more difficult to objectively evaluate their ability to predict a behavioral criterion, or *criterion validity*. Further, as such latent constructs are typically multi-faceted, they generally require multiple sets of facet-specific and concept-driven items (questions) to measure them. In this case, we must ensure the coherence, or *internal consistency*, of the items, so that they work together to capture the different aspects of a facet and even a set of related facets, or *domain*, of the construct. Another important consideration for instrument evaluation is its *test-retest reliability*. As measurement errors are random, test results or instrument scores will vary from repeated administrations to the same individual. Large variations will create errors for disease diagnosis and cause problems for replicating research findings. As validity does not imply reliability, and vice versa, this issue must be addressed separately.

Many instruments in psychosocial studies yield discrete scores, which are typically analyzed by methods rooted in statistical models for continuous outcomes such as Pearson correlation and linear regression. Such scales may be approximately modeled by these methods, provided that they have a reasonably wide range in their outcomes. However, it is important to apply distribution-free inference to obtain accurate inference, because of the discrete nature of the outcome and departures from mathematical distributional models for continuous outcomes such as normality for most such instruments.

In Section 10.1, we focus on the ability of a diagnostic test to detect a disease of interest, or *diagnostic-ability* of the test. In Section 10.2, we focus on assessing the added value in prediction or diagnostic-ability when a new risk factor is added to a base model. In Section 10.3, we devote our attention to the study of criterion validity under a gold standard (continuous or ordinal outcome with a reasonably large range). In Section 10.4, we take up the issue of internal reliability. We conclude this chapter with a discussion on test-retest reliability.

10.1 Diagnostic-Ability

In this section, we consider the assessment of a diagnostic test T, when the true disease status D, a binary outcome with 1 (0) for diseased (nondiseased), is known. If T is binary, then we may use sensitivity and specificity defined in Chapter 2 to assess the diagnostic-ability of the test. For the case of an ordinal or continuous T, we may assess the accuracy of the test by studying the distribution of the variable for the diseased and nondiseased groups, using regression methods discussed in Chapter 4 with the disease status D as the response variable. However, a more popular approach to modeling the relationship between T and D is the ROC curve.

10.1.1 Receiver operating characteristic curves

For continuous- and ordinal-valued diagnostic tests, it is important to dichotomize the outcome for diagnostic purposes, especially for clinical research. For example, the body mass index (BMI) is a continuous outcome indicating the amount of body fat based on comparing the weight with the height of an individual. To use BMI as a screening tool for obesity or pre-obese conditions, it is convenient to discretize the continuous outcome into easy-to-interpret categories. For example, 25 is commonly used as a cut-point for overweight for adults, with a BMI above 25 indicating overweight. However, as BMI is a heuristic proxy for human body fat, which varies from person to person depending on the physique of an individual, it is not 100% correlated with the actual amount of body fat in the person. Thus, it is quite possible that a person with a BMI above 25 is still fit and normal. To reduce the chance to erroneously label a person as being overweight, one may increase the cut-point to make the criterion more stringent, or specific, for overweight. Doing so, however, would increase the error in the other direction by mislabeling overweight subjects as being nonoverweight. The ROC curve aims to balance the sensitivity and specificity by considering all possible cut-points.

10.1.1.1 Continuous ROC curves

First consider the case of a continuous T. We may use a cut-point c to generate a binary test; the test is positive if $T > c$, and negative otherwise. Here, we assume that a larger test score implies a higher likelihood of being diseased. The same considerations apply if the disease is indicated by a smaller test score.

Let $Se(c) = \Pr(T > c \mid D = 1)$ and $Sp(c) = \Pr(T \leq c \mid D = 0)$ be the sensitivity and specificity at the cut-point c. The ROC curve is the plot of all points, $(1 - Sp(c), Se(c))$, in the xy-plane when c ranges over all real numbers. The ROC curve shows many important properties of a diagnostic test by providing a visual display of the relationship between the test sensitivity and specificity. Let $F_k(t) = \Pr(T \leq t \mid D = k)$, $k = 1, 0$, be the cumulative distribution function (CDF) of T for the diseased ($D = 1$) and nondiseased ($D = 0$) subjects. Then $Se(c) = 1 - F_1(c)$ and $Sp(c) = F_0(c)$. It follows immediately from the properties of CDFs that ROC curves increase from $(0,0)$ to $(1,1)$. The sensitivity and specificity change in opposite directions as the cut-point varies, indicating that the performance of a test is determined jointly by both indices. As noted earlier, we can reduce the rate of false diagnosis of nonoverweight people by increasing the cut-point of BMI. However, the improved specificity does not translate into more accurate diagnoses, as doing so will increase the chance of misdetecting overweight people, undermining the sensitivity of this screening tool.

If T is based on blind guessing, then $F_0 = F_1$, in which case the ROC curve is just the diagonal line from $(0,0)$ to $(1,1)$. For an informative test where a larger test score implies a higher likelihood of being diseased, the ROC curve should lie above this *diagonal reference line*, and the ratio of the density functions $f_1(t)/f_0(t)$ should be an increasing function of t, where $f_k(t) = F_k'(t)$ $(k = 0, 1)$. Since $-f_1(t)/f_0(t)$ is the slope of the tangent line of the ROC curve at the cut-point t (see Problem 10.4), it follows that the ROC curve should have decreasing tangent slopes, forming a concave shape. Concavity is a characteristic of *proper* ROC curves, and ROC curves without the property are called *improper*.

We may model ROC curves through distributions of T. For example, if assuming $T \sim N(0,1)$ $(N(\alpha, \sigma^2))$ for the nondiseased (diseased) subjects, then for all cut-points c,

$$Se(c) = 1 - \Phi\left(\frac{c - \alpha}{\sigma}\right) \text{ and } Sp(c) = \Phi(c), \quad c \in R, \tag{10.1}$$

where Φ denotes the CDF of $N(0,1)$. Hence, the corresponding ROC curve takes the parametric form $\left(1 - \Phi(c), 1 - \Phi(\frac{c-\alpha}{\sigma})\right)$ $(-\infty < c < \infty)$, or more succinctly:

$$y = 1 - \Phi\left(\frac{\Phi^{-1}(1-x) - \alpha}{\sigma}\right) = \Phi\left(\frac{\Phi^{-1}(x) + \alpha}{\sigma}\right), \quad 0 \le x \le 1. \tag{10.2}$$

Shown in Figure 10.1 is a plot of such *binormal* ROC curves for several different pairs of (α, σ). Due to the artifact of the binormal model, a binormal ROC curve is always improper, if the two normal distributions for the diseased and nondiseased group have different variances (see Problem 10.5). For example, the ROC curve corresponding to $\sigma = 2$ (see Figure 10.1) is not concave near the upper right corner, and part of the curve falls under the diagonal reference line. However, this typically occurs in a very small region near the corners of the ROC and thus may not be a serious issue in practice.

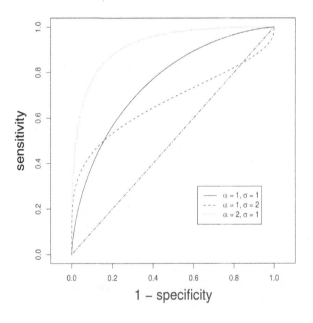

FIGURE 10.1: Binormal ROC curves.

Note that the ROC curve is invariant under monotone transformations (see Problem 10.2); thus, the assumption of standard normal for the nondiseased group does not impose any restriction on the application, if a monotone transformation of the test outcome is allowed. This invariance implies that the binormal ROC curve is not restricted to normally distributed test outcomes and is applicable to a much wider class of distributions that can be transformed to normal distributions under some monotone transformation. This property has important implications for later discussion of inference about ROC curves.

Note that positive (negative) predictive value, or PPV (NPV), is often used in place of or as a supplement to sensitivity and specificity, especially when screening for rare diseases. Since sensitivity and specificity are characteristics of a test among diseased and non-diseased groups, they do not depend on disease prevalence. In many studies, especially those of rare diseases, we may want to minimize false positive or false negative tests, which not only depend on sensitivity and specificity, but also on disease prevalence. Thus PPV and NPV are useful for optimizing the performance of the test for a given situation. For example, PPV is quite sensitive to small changes in specificity if the disease prevalence is extremely low, such as breast cancer for young women and HIV for low-risk populations. To see this, consider the rate of false positive, or false positive predictive value (FPPV):

$$FPPV(c) = 1 - PPV(c) = \frac{(1-p)(1-Sp(c))}{p \cdot Se(c) + (1-p)(1-Sp(c))},$$

where p is the disease prevalence. For any fixed cut-point c, FPPV(c) approaches 1 as p decreases to 0. Thus, even for a test with very high specificity Sp(c), we will have a high FPPV when p is really small (Tu et al., 1992, 1994). This fact explains the primary reason why screening is not recommended for a rare disease or a population at extremely low risk for such a disease, such as mammographic screening for breast cancer for women younger than 40. In such cases, it is important to calibrate the test to achieve the highest specificity possible.

10.1.1.2   Discrete ROC curves

For many instruments, especially those based on questionnaires, the outcome is often not continuous, but rather discrete ordinal. The cut-point in this case can no longer vary over all the real numbers.

Consider an instrument with m possible levels, $v_1 < v_2 < \cdots < v_m$, as its outcomes. There are a total of $m+1$ pairs of (Se_j, Sp_j):

$$Se_j = \sum_{l \geq j} p_1(v_l), \quad Sp_j = \sum_{l < j} p_2(v_l), \quad j = 1, \ldots, m,$$

$$Se_{m+1} = 0, \quad Sp_{m+1} = 1,$$

where $p_k(v) = \Pr(T = v \mid D = k)$ $(k = 0, 1)$ denotes the probability distribution function of the ordinal test outcome T from the diseased $(k = 1)$ and nondiseased $(k = 0)$ group. These $m+1$ points collectively characterize the diagnostic-ability of the instrument. The ROC curve for the ordinal-valued test is obtained by sequentially connecting the points $(1 - Sp_j, Se_j)$ by line segments, starting from $(0,0)$ and ending at $(1,1)$. If the test is binary, then the ROC curve actually consists of the line segments from $(0,0)$ to $(1 - Sp, Se)$ to $(1,1)$, in which case the ROC curve reduces to the specificity and sensitivity of the binary test.

10.1.1.3   Optimal cut-point

Finding an optimal cut-point for a diagnostic test is a great way to balance the two test characteristics, especially for clinical purposes. A simple criterion for the optimal threshold is to maximize the sum of sensitivity and specificity, or the *Youden index*, defined as $Sp+Se-1$. Depending on diagnostic objectives, available resources, and logistic considerations, optimal cut-points are also often sought to minimize the cost of false positive and/or negative diagnoses, if such information is available.

10.1.2 Inference

Without any assumption on the distribution of the test outcome T, one may estimate the ROC curve empirically using observed proportions. If test scores can be reasonably modeled by parametric distributions such as the normal, inference can be based on the maximum likelihood estimate (MLE) to achieve greater efficiency.

10.1.2.1 Empirical estimate

Let t_{1i} (t_{0j}) denote the (continuous) test outcome from the ith (jth) subject of a sample consisting of n_1 diseased (n_0 nondiseased) subjects ($1 \leq i \leq n_1$, $1 \leq j \leq n_0$). For a given cut-point c, we can readily estimate the sensitivity and specificity by

$$\widehat{Se}(c) = \frac{1}{n_1} \sum_{i=1}^{n_1} I_{\{t_{1i} \geq c\}}, \quad \widehat{Sp}(c) = \frac{1}{n_0} \sum_{j=1}^{n_0} I_{\{t_{0j} < c\}},$$

where $I_{\{t \geq c\}}$ is an indicator with the value 1 for $t \geq c$ and 0 otherwise.

Now rearrange the $n_1 + n_0$ pooled outcomes t_{1i} and t_{0j} in an ascending order and denote the distinct points by $x_1 < \cdots < x_m$, where m ($\leq n_1 + n_0$) is the total number of such values. The ordered set $\{x_i; 1 \leq i \leq m\}$ divides the real line into $m + 1$ intervals, namely, $(-\infty, x_1)$, $[x_1, x_2)$, $\ldots$, $[x_m, \infty)$. As the estimates $\widehat{Se}(c)$ and $\widehat{Sp}(c)$ do not change when c varies within each interval, there are $m+1$ distinct pairs $\left(1 - \widehat{Sp}(c), \widehat{Se}(c)\right)$. The *empirical ROC curve* is the piece-wise linear curve connecting the points defined by such pairs. As the sample sizes n_1 and n_0 increase, more distinct values will be observed in the sample, yielding a smoother and more accurate ROC estimate. As no parametric model is assumed for the test outcome or the ROC curve, the empirical ROC curve is free of any artifact.

The same ideas and procedures apply to ordinal test outcomes as well. However, as the number of distinct values in the observed sample m cannot exceed that of the different categories of the test results, say K, the empirical ROC curve contains at most K line segments, regardless of how large the sample is.

10.1.2.2 Binormal estimate

If a continuous test score follows approximately a normal distribution for both the diseased and nondiseased subjects, the resulting binormal ROC curve is totally determined by the means and variances of these normal distributions, as discussed in Section 10.1.1. In many applications, the original test outcome may not be normally distributed, or not even approximately, but it may still be reasonably modeled by such a parametric approach when rescaled under some monotone transformation. If the transforming function is known, say $g(\cdot)$, we can apply the binormal model to the transformed outcome $z = g(y)$. However, given a very limited number of analytic functions, it is likely difficult to find such a transformation in most cases. By ranking the data, Zou and Hall (2000) developed an ML rank-based estimate for binormal ROC curves; however, the method is very computationally intensive.

Binormal models can also be applied to ordinal tests, if we assume they are based on some normally distributed latent outcomes. More precisely, for an ordinal test with m

levels, $v_1 < \cdots < v_m$, suppose the m levels are the result of grouping the values of a latent continuous outcome based on a set of cut-points $c_1 < \cdots < c_{m-1}$. If the latent outcome follows the normal distribution for the diseased and nondiseased populations, then c_j's will satisfy

$$p_1(v_j) = \Phi(c_j) - \Phi(c_{j-1}),$$

$$p_0(v_j) = \Phi\left(\frac{c_j - \alpha}{\sigma}\right) - \Phi\left(\frac{c_{j-1} - \alpha}{\sigma}\right),$$

$$j = 1, \ldots, m, \quad c_0 = -\infty, \quad c_m = \infty. \tag{10.3}$$

For random samples of diseased and nondiseased subjects, suppose that there are n_{kj} with disease status k ($k = 1$ for disease and 0 for nondisease) and test outcome v_j, the likelihood under the binormal model is $L = \prod_{k=0}^{1} \prod_{j=1}^{m} p_k^{n_{kj}}(v_j)$. Note that although α and σ are of primary interest, the c_j's are also unknown. Thus L is maximized with respect to all these parameters (Dorfman and Alf, 1969). Also, although (10.3) allows us to estimate ROC curves without finding the transformation to transform the data to the normal distribution, the estimates obtained are still subject to the constraints imposed by the normal distribution. If $\sigma = 1$, i.e., the binormal model is proper, then the cumulative probit model described in Chapter 5, Section 5.3.1.2, may be applied (see Problem 10.6).

10.1.3   Areas under ROC curves

The area under the ROC curve (AUC) is a commonly used summary index for the accuracy of the continuous- or ordinal-valued test. Note that different terminology is commonly utilized in statistical learning. The AUC under an ROC curve is called a *c-statistic* or *concordance statistic*. The positive predictive value is called *precision*, and sensitivity is called *recall*.

In general, larger values of AUC indicate better performance of the test and vice versa. It is easy to check that for a binary test, AUC $= \frac{Sp + Se}{2}$, which is essentially the Youden index discussed earlier (see Problem 10.1). In the special case of a gold standard test, the curve goes from (0,0) to (0,1) to (1,1), with AUC $= 1$. At the other end of the spectrum, the ROC curve becomes the reference diagonal line, with AUC $= 0.5$, for a test based on random guessing. Thus, for an informative test, $0.5 < \text{AUC} \le 1$. In general, a test with an AUC of 0.8 or higher is considered a good test.

The AUC for a continuous or an ordinal test is given by

$$AUC = \Pr(t_1 > t_0) + \frac{1}{2}\Pr(t_1 = t_0), \tag{10.4}$$

where t_k denotes the test outcome for the diseased ($k = 1$) and nondiseased ($k = 0$) subject (see Problem 10.7), and the second term accounts for the contribution of tied observations. For the binormal ROC curve defined in (10.2), AUC $= \Phi\left(\frac{\alpha}{\sqrt{1+\sigma^2}}\right)$ (see Problem 10.8).

Given an estimated ROC curve, we can immediately obtain an estimate of the AUC by finding the area under the estimated curve. Alternatively, based on (10.4), we may also apply the Mann–Whitney–Wilcoxon statistic discussed in Chapter 2 to estimate the AUC, without relying on such an ROC curve estimate. By applying the Mann–Whitney–Wilcoxon statistic, we have

$$\widehat{AUC} = \frac{1}{n_1 n_0} \sum_{i=1}^{n_1} \sum_{j=1}^{n_0} \left[I_{\{t_{0j} < t_{1i}\}} + \frac{1}{2} I_{\{t_{0j} = t_{1i}\}} \right]. \tag{10.5}$$

The above is actually identical to the estimate of AUC obtained by computing the area under the empirical ROC curve (see Problem 10.9). Note that even for the continuous test, we may have tied test outcomes due to limited measurement precision, grouping and/or rounding, and thus we may still use the tie-corrected U-statistic in (10.5).

In many applications, multiple diagnostic tests are applied to each subject from the diseased and nondiseased groups, yielding correlated test outcomes. The AUCs of the different tests are correlated, and the theory of multivariate U-statistics can be applied to facilitate inference (DeLong et al., 1988; Kowalski and Tu, 2008).

Example 10.1

In the PPD, the subjects are administered several depression screening tools, including the SCID, the Beck Depression Inventory-II (BDI-II), and the Edinburgh postnatal depression scale (EPDS), the latter being developed specifically for postpartum depression. By treating the SCID diagnosis as the gold standard, we obtain empirical ROC curve estimates for the BDI-II and EPDS relative to the depression diagnosis (either major or minor depression), as shown in Figure 10.2.

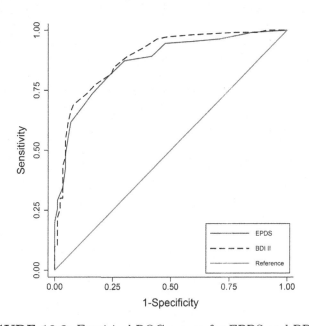

FIGURE 10.2: Empirical ROC curves for EPDS and BDI II.

The two estimated curves seem to be quite close to each other. Further both follow closely the left hand and the top border, indicating that they are quite informative for detecting depression in this particular study population.

The estimated AUCs of the two tests are 0.868 (EPDS) and 0.887 (BDI-II), with the corresponding standard errors 0.026 and 0.024. So both are excellent screening tools for depression in this population and are significantly better than blind guessing (p-values < 0.0001 for testing the null AUC $= 0.5$ for both tests).

By modeling the two AUCs using multivariate U-statistics, we obtain a difference of 0.0197 between the two AUC estimates with a standard error 0.0214. As p-value = 0.3587, we find no significant difference in diagnostic-ability for depression between the two screening tests. □

Note that in this example, we treated SCID as a gold standard test. If an imperfect test is used as "gold standard," estimates of sensitivities and specificities, as well as ROC curves, may be biased; see Reitsma et al. (2009) for a discussion on diagnostic accuracy studies with an imperfect standard.

10.2 Added Predictability

For some diseases, there are well-established risk factors and prediction models. For example, cholesterol levels, blood pressure, smoking, and diabetes are well-known risk factors, and the Framingham Risk Score for Hard Coronary Heart Disease is a commonly applied risk calculator for cardiovascular diseases (Wilson et al., 1998). When new risk factors are found, it is in general of interest to assess their added value. The new factors should be included if they significantly improve the predictability of the disease. On the other hand, including additional predictors may make the prediction model unnecessarily complicated if the added value is negligible. Thus, we often need to compare two regression models for the binary outcome, one with some well-established predictors (base model) and the other with an additional factor included (expanded model), to assess the added utility of this additional factor.

10.2.1 Difference in AUCs

For a binary regression, one can use the ROC curve for the fitted linear predictors or, equivalently, the fitted probabilities, to indicate the performance of the model. When a new predictor is added to the model, we would expect that the goodness of fit would increase, which in turn would lead to increased AUC. In fact, the increase in AUC is commonly used to evaluate the contribution of a new variable to the predictability of the model (D'Agostino et al., 1997). Note that since the AUCs are computed from the fitted values, it may not be appropriate to apply the test of DeLong et al. (1988) for inference about AUCs (Demler et al., 2012). As the linear predictors in regression models, or the coefficients, are estimated, the AUCs are not only correlated, but also subject to sampling variability in the estimated linear predictors. Thus except in the univariate cases, the AUC changes with the estimated coefficients and valid inference must account for such sampling variability. While correct inference may be performed for these U-statistics with estimated parameters (Demler et al., 2017), many software packages may simply ignore the issue; thus we need to interpret the results with caution.

A common issue with the approach of using improved AUC to assess the added value of a new predictor is that it is relatively insensitive to changes in absolute risk estimates. When the AUC of the base model is already high, any improvement may be very small even for new variables with very good diagnostic performance (Cook, 2007; Pencina et al., 2008). In such cases, Pencina et al. (2008) developed two new measures, net reclassification improvement (NRI) and integrated discrimination improvement (IDI), as alternatives to the AUC.

10.2.2 Net reclassification improvement

Net reclassification improvement (NRI) is a popular index to quantify improvement contributed by a new variable in classifying subjects (Pencina et al., 2008). Based on predefined risk categories, we can classify subjects into different risk categories according to the fitted probabilities by the base and expanded models. An "upward" risk reclassification where the risk category based on the expanded model is riskier than the base model for a diseased subject is considered an improvement. Likewise, a "downward" risk reclassification for a nondiseased subject is also considered an improvement. Changes in the other directions are considered deteriorations. For diseased and nondiseased subjects, the net improvement is the difference in the proportions of improvement and deterioration, and the NRI is the total of the NRI for both groups. Specifically, suppose that we have I predefined risk categories with cut-points $c_0 = 0 < c_1 < c_2 < ... < c_{I-1} < 1 = c_I$, then a subject with fitted probability $\widehat{p}$ is in the risk category $R_j = [c_{j-1}, c_j)$ if $c_{j-1} < \widehat{p} < c_j$. Let $n_{ij}^D \left(n_{ij}^N\right)$ be the number of subjects who are classified as in R_i based on the base model and in R_j based on the expanded model; we have the following classification-reclassification table:

	Expanded Model							
	Diseased				**Nondiseased**			
Base Model	R_1	R_2	$\cdots$	R_I	R_1	R_2	$\cdots$	R_I
R_1	n_{11}^D	n_{12}^D	$\cdots$	n_{1I}^D	n_{11}^N	n_{12}^N	$\cdots$	n_{1I}^N
R_2	n_{21}^D	n_{22}^D	$\cdots$	n_{2I}^D	n_{21}^N	n_{22}^N	$\cdots$	n_{2I}^N
$\vdots$	$\vdots$	$\vdots$	$\ddots$	$\vdots$	$\vdots$	$\vdots$	$\ddots$	$\vdots$
R_I	n_{I1}^D	n_{I2}^D	$\cdots$	n_{II}^D	n_{I1}^N	n_{I2}^N	$\cdots$	n_{II}^N

For subjects in the diagonal of each of the diseased and nondiseased tables, the classifications under both models are the same; thus they do not contribute to the NRI. However, the classifications are different for subjects off the diagonal of each table. Within each table, the subjects above and under the diagonal are "upward" and "downward" reclassified. Thus, the number of net improvements is $\sum_{i<j} n_{ij}^D - \sum_{i>j} n_{ij}^D$ for diseased and $\sum_{i>j} n_{ij}^N - \sum_{i<j} n_{ij}^N$ for nondiseased subjects. The NRI is defined as the sum of the proportions of net improvements in the diseased and nondiseased groups:

$$NRI = \frac{\sum_{\{i:Y=1\}} \left(\widehat{p}_{iE} - \widehat{p}_{iB}\right)}{n_D} + \frac{\sum_{\{i:Y=0\}} \left(\widehat{p}_{iB} - \widehat{p}_{iE}\right)}{n_N},$$

where n_D and n_N are the total number of diseased and nondiseased subjects, respectively.

If there are well-established risk categories, it is natural to use them in the calculation of the NRI. Otherwise, defining the risk categories may be rather arbitrary. The NRI may vary substantially depending on the number of the categories and the choice in risk cutoffs (Mühlenbruch et al., 2013). To overcome this issue, continuous NRI is also proposed (Pencina et al., 2011). The continuous NRI is similarly defined as the sum of proportions of net improvements in the diseased and nondiseased groups. However, as there are no risk categories for continuous NRI, we check whether the fitted probability changes and the direction of change. For example, a diseased subject with increased fitted probability will be considered an improvement. We may specify a threshold and require that the changes in the fitted probabilities be larger than the threshold to be considered as improvements/deteriorations.

Although commonly used in practice, the NRI has some potential issues. For example, Pepe et al. (2015) demonstrated through simulations that for a large test data set, using risk models derived from a training set, the NRI is likely to be positive even when the added covariate has no predictive information.

10.2.3   Integrated discrimination improvement

The NRI categorizes subjects into improvement, deterioration, or no change; thus the change magnitude is mostly ignored. For categorical NRI, regardless of how large the changes may be in the predicted probabilities for a subject, they will be ignored if they do not cross the cut-points and hence the subject will remain in the same risk category. On the other hand, a very small change in the predicted probabilities may be counted as an improvement or deterioration if it happens to be near the cut-points. The continuous NRI does not consider the size of changes, but only whether they cross a threshold. To take the change magnitude into consideration, we can add up all the changes in the predicted probabilities. Again, an increase (decrease) in the predicted probabilities for diseased is an improvement (deterioration). *Integrated discrimination improvement* (IDI) is the total net improvement amount from both diseased and nondiseased subjects:

$$IDI = \frac{\sum_{\{i:Y=1\}} \widehat{p}_{iE} - \widehat{p}_{iB}}{n_D} + \frac{\sum_{\{i:Y=0\}} \widehat{p}_{iB} - \widehat{p}_{iE}}{n_N},$$

where $\widehat{p}_{iE}$ and $\widehat{p}_{iB}$ are the predicted probabilities under the expanded and base models, respectively.

Example 10.2

In Example 10.1, we compared two depression screening tools, BDI-II and EPDS. Now let us assess the added predictability of BDI-II over EPDS. Thus, the base model has EPDS as the only predictor, and the expanded model includes both as predictors. The estimated AUC for the base model is 0.868, and the estimated AUC when BDI-II is added is 0.8998. The AUC difference is 0.0322 with SE = 0.0130 and p-value 0.0130. Thus, the improvement in AUC is significant. Note that the SE is based on the AUC that ignores the sampling variation associated with the estimated coefficients in the logistic models.

If we use two risk categories with a cut-point of 50%, we have the following classification-reclassification table:

| | EPDS + DBI-II | | | |
| | Depressed | | Nondepressed | |
EPDS	**Lower Risk**	**Higher Risk**	**Lower Risk**	**Higher Risk**
Lower Risk	12	9	62	4
Higher Risk	4	85	6	14

There are 110 diseased subjects in the data set. Among the 89 diseased subjects who are classified at the higher risk ($p \geq 0.5$) by the base model (EPDS alone), four are reclassified at lower risk by the expanded model; thus, there are four deteriorations. Among the 21 diseased subjects who are classified as at the lower risk ($p < 0.5$) by the base model (EPDS alone), nine are reclassified at higher risk by the expanded model; thus, there are nine improvements. So, the net improvement for the diseased group is $\frac{9-4}{110} = 0.04546$. Similarly, the net improvement for the nondiseased group is $\frac{6-4}{86} = 0.02326$. Hence, the estimated NRI is $0.04546 + 0.02326 = 0.06872$. The standard error may be estimated based on correlated proportions, ignoring the sampling variation associated with the parameter estimates of the two models. Alternatively, resampling methods such as bootstrap may be applied. Based on 10,000 bootstrap samples, we obtain an estimated 0.04831 for the standard error and a 95% CI for NRI $(-0.01385, 0.1739)$. If the sample proportion for depression, $\frac{110}{196}$, is used as the cut-point, then the estimate NRI is 0.08182, with an estimated SE 0.04684 and a 95% CI $(-0.01376, 0.1704)$ based on 10,000 bootstrap samples. In both cases, the NRI is not significant.

The estimated continuous NRI is 0.8419 with SE 0.1719 and a 95% CI $(0.4897, 1.1640)$. For the IDI, the point estimate is 0.07483 with SE 0.01792 and a p-value .00003. Thus, there is a significant improvement based on the continuous NRI and IDI.

⬜

10.3 Criterion Validity

In this section, we study the validity of instruments by comparing them with a gold or reference (relatively more accurate) standard. For categorical or ordinal outcomes with very limited range, Kappa is the most popular measure in that regard (see Chapter 2). If both the instrument and gold standard have a continuous or ordinal outcome with a large range, we may assess such validity by the Pearson product-moment correlation. However, we must be mindful about its limitation when it is used to examine the accuracy of the instrument.

To illustrate, consider a hypothetical study of six subjects with data from both an instrument and a gold standard. Suppose that the pairs of outcomes (y_{i1}, y_{i2}) from the gold standard (y_{i1}) and instrument (y_{i2}) from the study subjects are as follows:

$$(3,5), \ (4,6), \ (5,7), \ (6,8), \ (7,9), \ (8,10).$$

Although the outcomes hardly agree at all, the sample Pearson correlation $\widehat{\rho}_{Pearson} = 1$. The paradox results from the upward bias in the instrument, a constant difference of 2 versus the gold standard across all the subjects, which cannot be detected by the Pearson correlation. Thus, although perfectly correlated, i.e., with perfect *precision*, the instrument has poor criterion validity, or poor *accuracy*. To address this flaw of the Pearson correlation, one may use the concordance correlation coefficient (CCC) introduced next.

10.3.1 Concordance correlation coefficient

Let y_{1i} (y_{2i}) denote the outcome from the gold standard (instrument). The CCC between y_{1i} and y_{2i} is defined as

$$\rho_{CCC} = \frac{2\sigma_{12}}{\sigma_1^2 + \sigma_2^2 + (\mu_1 - \mu_2)^2}, \quad \sigma_{12} = Cov(y_{i1}, y_{i2}),$$
$$\mu_k = E(y_{ik}), \quad \sigma_k^2 = Var(y_{ik}), \quad k = 1, 2. \tag{10.6}$$

This index ρ_{CCC} ranges between -1 and 1; $\rho_{CCC} = 1$ if the two raters completely agree ($y_{1i} \equiv y_{2i}$), $\rho_{CCC} = -1$ if $\mu_1 = \mu_2$ and the two raters are completely opposite to each other's ratings with respect to the common center $\mu = \mu_1 = \mu_2$, and $\rho_{CCC} = 0$ if $\sigma_{12} = 0$ (see Problem 10.17). Further, ρ_{CCC} can be expressed as

$$\rho_{CCC} = \rho_{Pearson} C_b, \quad C_b = 2\left[\left(\frac{\sigma_1}{\sigma_2}\right) + \left(\frac{\sigma_2}{\sigma_1}\right) + \left(\frac{\mu_1 - \mu_2}{\sqrt{\sigma_1\sigma_2}}\right)^2\right]^{-1}. \tag{10.7}$$

The quantity C_b is a measure of accuracy. In general, $C_b \leq 1$, and $C_b = 1$ occurs only when $\mu_1 = \mu_2$ and $\sigma_1 = \sigma_2$ (see Problem 10.13), in which case ρ_{CCC} reduces to $\rho_{Pearson}$. As C_b is inversely related to the *bias* $\delta = \mu_1 - \mu_2$, C_b increases as the bias becomes smaller. Thus, unlike $\rho_{Pearson}$ and other popular association measures such as Spearman's ρ and Kendall's τ, ρ_{CCC} captures both the accuracy, defined by δ, and precision, defined by σ_1 and σ_2, of the instrument.

Maximum likelihood may be used for inference about ρ_{CCC} under some parametric assumptions for the joint distribution of $\mathbf{y}_i = (y_{i1}, y_{i2})^\top$. For example, under a bivariate normal $\mathbf{y}_i$, the MLE is obtained by substituting the sample means, variances, and covariance in place of their respective parameter counterparts in Lin (1989),

$$\widehat{\rho}_{CCC} = \frac{2s_{12}}{s_1^2 + s_2^2 + (\overline{y}_1. - \overline{y}_2.)^2}, \tag{10.8}$$

where $\overline{y}_1.$ and $\overline{y}_2.$ are the sample means and s_1^2, s_2^2, and s_{12} are the sample variances and covariance of the outcome $\mathbf{y}_i$. The asymptotic normal distribution of $\widehat{\rho}_{CCC}$ may be used to provide inference about ρ_{CCC}, in which case the asymptotic variance is readily evaluated (Lin, 1989).

For discrete outcomes, $\widehat{\rho}_{CCC}$ no longer has the normality-based interpretation. As a result, the asymptotic variance of the MLE based on multivariate normality generally does not yield valid inference for discrete outcomes. However, since $\widehat{\rho}_{CCC}$ in (10.8) is a function of sample moments and such estimates are robust to deviations from assumed distributions such as a bivariate normal for $\mathbf{y}_i$, $\widehat{\rho}_{CCC}$ still yields a consistent estimate. Further, $\widehat{\rho}_{CCC}$ is asymptotically normal (Kowalski and Tu, 2008), which can be used to provide inference about ρ_{CCC}, regardless of the data distribution.

Example 10.3
Risky sexual behaviors such as unprotected vaginal sex are often retrospectively assessed in sexual health and related research studies. Although convenient to administer, this approach lacks the ability to provide reliable information due in large part to recall bias, especially over an extended period of time. In the Sexual Health pilot study, adolescent girls were asked to recall their sexual activities such as unprotected vaginal sex at the end of the study (three months). To assess recall bias, these girls were also instructed to use a daily diary to record the same information. Since the information from the diary is much more accurate, it can serve as a reference standard for assessing the validity of the retrospective assessment. We may apply the Pearson correlation and CCC to compare the outcomes between the retrospective recall at three months and the daily diary.

TABLE 10.1: Pearson correlation and CCC between recall (y_{1i}) and diary (y_{2i})

μ_1 (σ_1)	μ_2 (σ_2)	$\frac{\|\mu_1 - \mu_2\|}{\sqrt{\sigma_1 \sigma_2}}$	σ_1/σ_2	$\rho_{Pearson}$ (SE)	ρ_{CCC} (SE)
13.8 (23.5)	9.1 (12.3)	0.28	1.91	0.47 (0.13)	0.38 (0.19)

Based on estimates summarized in Table 10.1, the retrospective recall y_{i1} is not only upwardly biased, but much more variable as well, as indicated by the ratio of the standard deviation (σ_1/σ_2). The large difference in spread is primarily responsible for the difference between the two coefficients. Also, the two outcomes are only moderately correlated, indicating that a large interval such as three months does affect the accuracy of reporting. □

10.4 Internal Reliability

In mental health and psychosocial research, many instruments are designed to measure conceptual constructs characterized by certain behavioral patterns, and as such it is generally more difficult to objectively evaluate their ability to predict a behavioral criterion, or *criterion validity*. Further, as such latent constructs are typically multi-faceted, they generally require multiple sets of facet-specific and concept-driven items (questions) to measure them. In this case, we must ensure the coherence, or *internal consistency*, of the items, so they can complement each other to capture the different aspects of a facet and even a set of related facets, or *domain*, of the construct.

For example, Medical Outcomes Study 36-Items Short Form Health Survey (SF-36) is an instrument developed for assessing quality of life (QOL), which encompasses a number of dimensions pertaining to an individual's emotional, social, and physical well-being, including the ability to function in the ordinary tasks of living. The 36 items, or questions, of this instrument are grouped into eight domains to capture eight related but clearly distinct constructs ranging from physical function to mental health to social function. Given the multi-faceted conceptual constructs, it is not possible to validate this instrument using a single gold standard based on some behavioral patterns or medical conditions. Thus, evaluating the validity of such an instrument is more complex, not only requiring the selection of appropriate criteria for criterion validity, but also the assessment of the items' *construct validity*, which refers to the items' ability to measure the concept of the construct such as QOL.

The development of an instrument measuring latent constructs generally starts with a formative study consisting of a small "nominal" group of subjects with the disease of interest and a focus group of experts specializing in this topic. The focus group discusses and confirms the structure of the instrument, typically consisting of facets and/or domains (broader concepts than facets) such as physical and mental health and social functioning, as in the case of SF-36 discussed in the beginning of this chapter. The nominal group then proposes a list of potential items under each facet or domain, after a careful review of the instrument structure. The items identified are further refined by the focus group through in-depth interviews of the subject in the nominal group. The selected items by the focus group then undergo some pilot testing with a small group of subjects. The data collected are analyzed to confirm and/or revise the within-facet or within-domain items for construct validity using statistical modeling tools such as exploratory and confirmatory factor analysis, and correlation and regression analysis. Readers interested in the process of instrument construction may consult relevant books for details (Streiner et al., 2015). In this section, we discuss how to assess the internal validity of the instrument to form meaningful dimensional scales to quantify such latent constructs.

The internal reliability of an instrument is concerned with the cohesion of the items within a facet or domain of the instrument to capture the various attributes of the latent construct of interest. This is different from the criterion validity of the instrument, which focuses on establishing the validity of the latent construct itself by relating the scale or subscales of the instrument with some relevant gold or reference standards. In addition to CCC and correlation measures discussed in the preceding section, regression methods may also be used to assess the relationship between instrument scales (or subscales) and gold (reference) standards.

10.4.1 Spearman–Brown Spearman-Brown Rho

Internal validity is also known as *internal consistency*. As noted earlier, it is concerned with the cohesion of a set of items (or questions) when used together to measure a latent construct of interest such as depression, social functioning, or personality. As the total item score is used as a dimensional scale for the latent construct, it is important that such item scores are coherent, or additive. For example, the 36 items of the SF-36 are divided into eight domains: Physical Function (PF), Role-Physical, Bodily Pain, General Health,Vitality, Social Function, Role-Emotional, and Mental-Health. The PF domain has ten items, probing different physical activities such as walking, running, bending, and kneeling (RAND Health). Each item is scored 1, 2, or 3, indicating the respective levels of limitation, "Limited a lot," "Limited a little," and "Not limited at all," when performing each of these activities. For the PF scale to measure the cumulative burden of performing such daily activities, the item scores must be at least positively correlated.

One way to ensure such internal consistency is to examine all between-item correlations within the facet or domain. This approach, however, is impractical for facets or domains with a large number of items. In addition, it is difficult to assess the strength of item coherence with a large number of pairwise correlations.

A formal framework for assessing internal validity is the *domain-sampling* model. Under this classic model, a measure of the latent construct of interest is obtained from a random sample of K items from a population of items that this construct underlies. The value of the latent construct is the limit of the averaged item scores y_k when the number of items approaches infinity, $y_\infty = \lim_{K \to \infty} \frac{1}{K} \sum_{k=1}^{K} y_k$, akin to the population mean of a random variable. Under the domain-sampling approach, all items share an equal amount of the common core of the construct, i.e., the product-moment (PM) correlation between the kth item and the latent construct y_∞, $p_1 = Corr\,(y_k, y_\infty)$, is a constant independent of any particular item k. This common correlation with y_∞ is called the *reliability index* p_1 for a *single* item.

Let

$$\overline{\rho}_K = \binom{K}{2}^{-1} \sum_{(k,l) \in C_2^K} \rho_{kl}, \quad \overline{\rho}_\infty = \lim_{K \to \infty} \overline{\rho}_K, \tag{10.9}$$

where $\rho_{kl} = Corr\,(y_k, y_l)$ and C_2^K denotes the set of $\binom{K}{2}$ combinations of two distinct elements (k, l) from the integer set $\{1, \ldots, K\}$. Then, it can be shown that $p_1 = \sqrt{\overline{\rho}_\infty}$ (see Problem 10.15). The limit of averaged pair-wise correlations $\overline{\rho}_\infty$ is known as the *reliability coefficient*.

The identity $Corr\,(y_k, y_\infty) = \sqrt{\overline{\rho}_\infty}$ is fundamental to classical measurement theory, as it expresses the incomputable correlation involving the latent true score y_∞ as a function of an estimable quantity $\overline{\rho}_\infty$. To minimize measurement error, the averaged item score, $\overline{y}. = \frac{1}{K} \sum_{k=1}^{K} y_k$, or equivalently the total score, $y. = \sum_{k=1}^{K} y_k$, is used in most applications. The Pearson correlation between $\overline{y}.$ and y_∞, p_K, is called the *reliability index* of K *multiple* items.

When K is large, we have approximately $p_K = \sqrt{\frac{K\overline{\rho}_\infty}{1+(K-1)\overline{\rho}_\infty}}$, with the approximation error approaching 0 as K increases (see Problem 10.15). This second important identity, the *Spearman–Brown Prophecy* formula, generalizes the first fundamental result for a single item to a measure of internal consistency of K items. Accordingly, $\rho_K = \frac{K\overline{\rho}_\infty}{1+(K-1)\overline{\rho}_\infty}$ is called the *reliability coefficient* for *a facet or domain* consisting of K items. In theory $-1 \leq \rho_K \leq 1$, but in real study applications $\rho_K \geq 0$, since facet or domain items from an internally consistent instrument are positively correlated. As a special case with $K = 1$, $\rho_1 = \overline{\rho}_\infty$ and $\rho_1 = p_1^2$, reducing to the identities for the single-item case, respectively.

Ideally, we want to select an optimal subset of size k from the K same domain items in a scale or subscale with the smallest k and largest ρ_k. In practice, such a subset may not exist, in which case we prefer a subset with a smaller size, even if it has a ρ_k that is a bit smaller than a subset with more than k items (Nunnally and Bernstein, 1994).

Similar to the validity measures introduced previously in the chapter, we may substitute sample moments to obtain consistent estimates of the reliability coefficients and indices above. For example, if y_{ik} denote the responses to a set of K items from n subjects, we can estimate the reliability coefficient $\bar{\rho}_\infty$ by an estimate of $\bar{\rho}_K$, i.e.,

$$\widehat{\bar{\rho}}_K = \binom{K}{2}^{-1} \sum_{(k,l) \in C_2^K} \widehat{\rho}_{kl},$$

where $\widehat{\rho}_{kl}$ is the Pearson correlation between y_{ik} and y_{il}. As $\widehat{\rho}_{kl}$ is consistent and $\widehat{\bar{\rho}}_K$ is a continuous function of the $\widehat{\rho}_{kl}$'s, $\widehat{\bar{\rho}}_K$ above is also a consistent estimate of $\bar{\rho}_K$ (Kowalski and Tu, 2008).

Classical methods for inference about $\bar{\rho}_K$ assume joint multivariate normality for $\mathbf{y}_i = (y_{i1}, \ldots, y_{iK})^\top$, which may not be appropriate for ordinal outcomes within the current context. Thus, for valid inference, alternatives such as U-statistic based nonparametric methods may be applied (Tu et al., 2007).

10.4.2 Cronbach coefficient alpha

The most popular index for internal consistency for a set of K items is the *Cronbach coefficient alpha*:

$$\alpha_K = \frac{K}{K-1} \frac{\sum_{k \neq l} Cov(y_k, y_l)}{\sum_{k=1}^K \sum_{l=1}^K Cov(y_k, y_l)}. \tag{10.10}$$

The Cronbach α_K is generally different from the Spearman–Brown ρ_K, unless the item variance $\sigma^2 = Var(y_k)$ is a constant independent of k (see Problem 10.16). The difference between the two may be small for instruments consisting of items with similar item variances.

The Cronbach coefficient alpha is motivated by split-half reliability, which is itself a common index for internal validity. Under the split-half approach, the items within a facet (domain) are separated into two parallel subfacets (domains), and reliability measures such as the Pearson correlation are then applied to assess the correlation between the two subfacet items scores. A major shortcoming of the split-half method is that the division of the items can be arbitrary, with different ways of splitting the items, resulting in different estimates of the reliability. For example, suppose a and b are the two subtest scores, then one of the commonly used split-half reliability indices is defined as $1 - \frac{Var(a-b)}{Var(a+b)} = \frac{4Cov(a,b)}{\sum_{k=1}^K \sum_{l=1}^K Cov(y_k, y_l)}$ (Rulon's formula). Note that $\frac{Var(a-b)}{Var(a+b)}$ is the ratio of the variance of the differences between the two half tests $(a - b)$ with the variance of the total scores $(a + b)$. The coefficient will be high if highly correlated items are separated into the two subtests. However, if we average over all the possible split options, we would essentially have the Cronbach α.

A Cronbach coefficient alpha estimates the proportion of variance in the item scores attributable to the true score variance. Like all the other validity and reliability measures, it ranges from 0 to 1 in real study applications. If the items are all independent, $Cov(y_k, y_l) = 0$ for all $(k, l) \in C_2^K$, implying $\alpha_K = 0$. At the other end of the spectrum, if the item scores are all perfectly correlated, $\alpha_K = 1$. Thus, the normalizing factor $\frac{K}{K-1}$ in (10.10) is used to scale the index to ensure that $\alpha_K = 1$ in the latter case. In general, larger values of α

indicate stronger item coherence, with an acceptable range of α_K being 0.7 or higher in most applications.

Like Spearman–Brown rho, we also select an optimal subset with the smallest number of items possible, even if it has a slightly smaller than a subset with more items (Nunnally and Bernstein, 1994). A potential caveat in applying Cronbach coefficient α in practice is the influence of the number of items K on the value of the coefficient; α_K increases toward 1 as K becomes large, all other things being equal. Thus, we must be mindful about this dependence when interpreting and comparing several scales with different numbers of items. For example, we may require a higher α_K for a scale with a larger number of items.

In addition to the overall internal reliability, Cronbach's coefficient alpha may also be used for refining items within a facet or domain. For example, if two items are highly correlated, we may consider removing one of them because of the potential redundancy in the information captured by the items, as well as the artifact on the value of α induced by a larger K. One common practice is to remove one item, compute α based on the remaining items, and repeat the process for every item in the facet or domain. The resulting α's are then compared with the one based on the original scale to inform decisions regarding whether some items should be considered for removal. In general, if α increases after an item is removed, we may consider deleting this particular item, since the lower α suggests that the item is either not well or too highly correlated with the other remaining ones.

The contribution of the items to the Cronbach coefficient α depends on the variances of the items. Sometimes, items are standardized, i.e., transformed to have (sample) mean 0 and variance 1 before being used to compute the coefficient α. For the standardized item scores, the covariance is the same as the correlation, yielding the standardized coefficient $\alpha_K = \frac{K\bar{r}}{1+(K-1)\bar{r}}$, where $\bar{r}$ is the average of all the pairwise correlations among the items. For example, some latent constructs such as intelligence quotient (IQ) do not have a natural unit that one can relate easily with familiar dimensional scales such as weight and height, and the standardization of item scores is a way to succinctly describe the distribution of the construct in a population. In the case of IQ, it is arguably easier to understand someone's level of intelligence by knowing the person's IQ percentile than the IQ score itself. However, if the scores of a scale are well interpreted, it is more informative to use the raw scores in applications. For example, the first item in the SF-36 is "In general, would you say your health is:" which is scored as 1, 2, 3, 4, and 5, representing Excellent, Very Good, Good, Fair and Poor, respectively. It is more convenient to keep the original scores, which is exactly how it is used in most applications.

When low values of α indicate that the items are not well correlated, it is likely that the items measure more than one construct and may need to be reanalyzed to examine the structure and dimensionality of the construct by factor analysis and/or related methods. It may even be necessary to regroup the items to create additional facets (domains) to characterize the construct.

10.4.3   Intraclass correlation coefficient

Another popular approach for assessing internal reliability is the *intraclass correlation coefficient* (ICC) McGraw and Wong, 1996; Shrout and Fleiss, 1979. Unlike Cronbach's coefficient alpha and the Spearman–Brown rho, the ICC explicitly models the variability of the latent construct and between-item variation in a population of interest, hence the name of intraclass correlation, based on the linear mixed-effects (LMM) model introduced in the last chapter. Thus, in addition to providing a measure of reliability, the LMM approach also yields an estimate of the latent construct for each subject sampled.

Consider again a set of K items for measuring some latent construct of interest, and let y_{ik} denote the score of the kth item from the ith subject ($1 \leq i \leq n$, $1 \leq k \leq K$). The LMM for item scores y_{ik} has the form

$$y_{ik} = \mu + \lambda_i + \epsilon_{ik}, \quad \lambda_i \sim N\left(0, \sigma_I^2\right), \quad \epsilon_{ik} \sim N\left(0, \sigma^2\right), \quad 1 \leq k \leq K. \tag{10.11}$$

As discussed in the preceding chapter, μ is the fixed effect denoting the population mean score of the construct, λ_i is the random effect representing the deviation of the ith subject's score from the population mean, and ϵ_{ik} is the difference between the kth observed item score y_{ik} and the core of the latent construct $\mu + \lambda_i$ captured by this item. It follows that the mixed effect, $y_{i\infty} = \mu + \lambda_i$, is the (latent) true score of the construct for the ith subject. By employing the mixed-effect model, we are able to distinguish the variability of the latent construct σ_I^2 from the between-item variation σ^2.

The ratio $\rho_{ICC} = \frac{\sigma_I^2}{\sigma_I^2 + \sigma^2}$, known as the ICC, describes the fraction of the variation of the latent construct relative to the total variability of the item score, with ρ_{ICC} closer to 1 (0) indicating a good (poor) internal consistency. Since $\rho_{ICC} = Corr\left(y_{ik}, y_{il}\right)$ is a constant independent of k and l ($k \neq l$) (see Problem 10.19), ρ_{ICC} is indeed a correlation.

There are close relationships between other reliability indices and the ICC ρ_{ICC} derived from the LMM in (10.11). Since $Corr\left(y_{ik}, y_{il}\right)$ is a constant independent of k and l ($k \neq l$), the average of all the pairwise correlations among the items, $\bar{r} = \rho_{ICC}$. Thus, $\alpha_K = \frac{K\rho_{ICC}}{1 + (K-1)\rho_{ICC}}$. Since $\rho_1 = \rho_{ICC}$ (see Problem 10.19), it follows that the single-item reliability index $p_1 = \sqrt{\rho_{ICC}}$, the first fundamental result for the relationship between p_1 and ρ_1 derived under the classic domain sampling model in Section 10.4.1. However, unlike the domain sampling model, ρ_1 under (10.11) is equal to ICC ρ_{ICC}, rather than the limit of averaged pair-wise Pearson correlations. This simplicity, however, is achieved at the expense of assuming a common variance σ^2 for the measurement error ϵ_{ik} in (10.11).

As σ_I^2 and σ^2 are typically estimated by moment estimates even under the normal assumption in (10.11) as in most software packages (McGraw and Wong, 1996; Lu et al., 2014, the estimate of ρ_{ICC} is still consistent for discrete ordinal outcomes. However, as inference is still based on the normal assumptions, it is generally not appropriate for ordinal outcomes within our context. Distribution-free models by replacing the normally distributed λ_i and ϵ_{ik} with variates centered at 0 should be used to provide valid inference. The theory of U-statistics can be utilized to develop inference procedures for such distribution-free models (Kowalski and Tu, 2008; Lu et al., 2014).

Example 10.4

The SF-36 has been translated into many foreign languages and used in more than 40 countries as part of the international quality-of-life assessment project (Lubetkin et al., 2003, Wang et al., 2006). A study was conducted to evaluate the performance of a Chinese version of SF-36, or CSF-36, when used to assess health-related quality of life (HRQOL) for patients with hypertension, coronary heart diseases, chronic gastritis, or peptic ulcer in mainland China (Yang et al., 2012). As noted earlier, the SF-36 instrument has eight domains, which can further be aggregated into the Physical and Mental Health subscales. In this study, there were 534 patients from the aforementioned four disease groups: 157 with hypertension, 133 with coronary heart disease, 124 with chronic gastritis, and 120 with peptic ulcer. The patients ranged in age from 16 to 86, with a mean age of 54.7.

Shown in Table 10.2 are the estimates of Cronbach coefficient α and ICC based on the original ten items, as well as the nine remaining items after one of them is removed, for the PF domain of the CSF-36. The two versions of the estimates of the coefficient α are nearly indistinguishable. Thus, removing an item from this domain has a negligible effect

TABLE 10.2: Cronbach coefficient alpha and ICC for the PF domain of the CSF-36

					Item removed					
None	**1st**	**2nd**	**3rd**	**4th**	**5th**	**6th**	**7th**	**8th**	**9th**	**10th**
			Cronbach's coefficient α $\left(\frac{\text{Original}}{\text{Standardized}}\right)$							
$\frac{0.92}{0.93}$	$\frac{0.93}{0.93}$	$\frac{0.93}{0.93}$	$\frac{0.92}{0.93}$	$\frac{0.92}{0.92}$	$\frac{0.93}{0.93}$	$\frac{0.93}{0.93}$	$\frac{0.92}{0.92}$	$\frac{0.92}{0.93}$	$\frac{0.92}{0.92}$	$\frac{0.93}{0.93}$
				Intraclass correlation						
0.58	0.60	0.58	0.59	0.59	0.58	0.60	0.58	0.58	0.59	0.59

on the value of either coefficient, supporting the fact that all items contribute about an equal amount of variability to the domain scale. Based on the results above, the PF scale has a good overall α.

It is interesting to see a large difference between the values of the α and ICC coefficients, with α values almost double those of the ICC. This is expected from the relationship $\alpha_K = \frac{K\rho_{ICC}}{1+(K-1)\rho_{ICC}}$, given the relatively low ICC and large number of items. Accordingly, one must keep this relationship in mind when applying different reliability indices in practice. ☐

10.5 Test-Retest Reliability

In addition to internal and criterion validity, it is also important to assess the test-retest reliability of the instrument. This reliability is concerned with the ability of the instrument to yield identical or similar results when administered repeatedly to the same subject at different, but closely spaced times points. The exact spacing may vary, depending on the subject matter of the construct, but the idea is to select a time window to minimize any systematic difference. For example, in the study of CSF-36 discussed in Example 10.4, each of the 534 study patients was administered the CSF-36 at the time of admission to hospital. A subsample of 197 patients was also randomly selected to take the questionnaire a second time 1–2 days after being admitted. The time window 1–2 days was too short for a sizable amount of change to take place due to treatment, but long enough to minimize systematic bias such as memory recalls.

All of the reliability and validity measures discussed above as well as other related correlation techniques such as Spearman's rho can be applied to assess test-retest reliability. As in the case of validity assessment, association measures such as the Pearson coefficient and Spearman's rho correlations are insensitive to systematic changes over the repeated administrations of the instrument and as such may not be used in situations where systematic differences are expected. For example, a second random subsample of 409 patients in the CSF-36 study was administered the questionnaire at discharge (after about two weeks of treatment) to study the sensitivity of the CSF-36 in response to changes in QOL due to treatment effects. As the QOL outcomes are likely to have been improved for the patients thanks to the treatments received, the mean of each of the domain scores at discharge would have been different from its counterpart at admission. Association measures do not assess such temporal changes, potentially yielding wrong conclusions about the test-retest reliability of the CSF-36.

Example 10.5

Consider the random subsample of 197 patients in the CSF-36 study who were asked to take the questionnaire again 1–2 days after admission. As the time is too short for patients to show a significant improvement in their quality of life, we may use the data at the two time points to assess the test-retest reliability of the translated instrument.

Shown in Table 10.3 are the estimates, standard errors, and 95% confidence intervals of Pearson correlation, CCC, and ICC between the two assessment times for the PF domain based on the subset of 197 subjects. The three indices yield quite similar values, and the high values for all the three coefficients indicate good test-retest reliability for this particular domain of the CSF-36.

TABLE 10.3: Pearson correlation, CCC, and ICC between admission and 1–2 days post-admission to hospital for the PF domain

Index	$\rho_{Pearson}$	ρ_{CCC}	ρ_{ICC}
Estimate (SE)	0.76 (0.022)	0.75 (0.031)	0.76 (0.051)
95% CI	(0.72, 0.80)	(0.69, 0.81)	(0.66, 0.86)

The near identical estimates across the three indices provide a strong indication that there is very little "drift" in the domain outcome from admission to post-admission to the hospital, as expected for such a short period, since otherwise the estimated ρ_{CCC} would have been substantially smaller than its counterpart $\rho_{Pearson}$ as noted in Example 10.3. □

Exercises

10.1 Show that for a binary diagnostic test, AUC = $\frac{1}{2}$ (sensitivity + specificity).

10.2 Show that the ROC curve is invariant under monotone transformation of the test variable.

10.3 Verify (10.1) and (10.2).

10.4 Let S be a curve in the two-dimensional x-y plane defined parametrically by $x = F(t)$ and $y = G(t)$, where F and G are smooth functions. Show that the slope of the tangent line at an interior point t_0 of S is $G'(t_0)/F'(t_0)$. Use this fact to derive the slopes for ROC curves.

10.5 Show that a binormal ROC curve is improper if the two normal distributions for diseased and nondiseased have different variances.

10.6 Show that for the binormal model in (10.3), if properness is further assumed, then this model reduces to the cumulative probit model with the disease status as the predictor and the ordinal test result as the response.

10.7 Let t_k be the test outcome for the diseased $(k = 1)$ and nondiseased $(k = 0)$ subject. Show

    (a) $\text{AUC} = \Pr(t_1 \geq t_0)$ if t_k is continuous;

    (b) $\text{AUC} = \Pr(t_1 > t_0) + \frac{1}{2}\Pr(t_1 = t_0)$ if t_k is discrete.

10.8 Express the AUC of a binormal ROC curve in terms of the means and variances of the two underlying normal distributions.

10.9 Show that the estimate in (10.5) equals the area under the empirical ROC curve.

10.10 In assessing the accuracy of HAM-D for the DOS, treat the SCID diagnosis of depression as a gold standard to

    (a) Estimate the ROC curve;

    (b) Estimate the AUC;

    (c) Which cut-points would you suggest based on the data?

10.11 For the DOS, we would like to manually calculate two two-category NRIs when medical burden is added to HAM-D in the prediction of depression, using a) 0.5 and b) the sample depression rate as the cut-point. For each cut-point, answer the following questions:

    (a) Generate the classification-reclassification table separately for depressed and nondepressed subjects.

    (b) What are the numbers of improvements and deteriorations for depressed and nondepressed subjects?

    (c) Estimate the two-category NRIs.

10.12 For the DOS, we would like to assess the added value of the medical burden after HAM-D in the prediction of depression. In Problem 10.11, we computed two-category NRIs. Here, we would like to compute some further indices.

    (a) Assess the difference in the AUCs under the ROC curves;

    (b) Estimate the continuous NRI, without a threshold;

    (c) Estimate the continuous NRI, with a threshold 0.05;

    (d) Estimate the IDI.

10.13 Verify (10.7) and show

    (a) $C_b \leq 1$;

    (b) $\rho_{CCC} = \rho_{Pearson}$, if and only if $\mu_1 = \mu_2$ and $\sigma_1 = \sigma_2$.

10.14 For the Sexual Health pilot study, compute CCC and ICC between the diary and retrospective recall outcomes for the number of instances of unprotected vaginal sex.

10.15 For the domain sampling model described in Section 10.4.1, show

    (a) $p_1 = \sqrt{\bar{\rho}_\infty}$;

    (b) $p_K = \sqrt{\frac{K\bar{\rho}_\infty}{1+(K-1)\bar{\rho}_\infty}} + o(1)$, where $o(1)$ is a higher order term with $o(1) \to 0$ as $K \to \infty$.

10.16 For the domain sampling model described in Section 10.4.1, show

    (a) If $Var(y_k) = \sigma^2$ is a constant, the Spearman–Brown ρ_K and Cronbach coefficient alpha α_K are identical;

    (b) If $Cov(y_k, y_l) \geq c > 0$ for all $1 \leq k, l \leq K$, then $\lim_{K \to \infty} \alpha_K = 1$;

    (c) Choose a setting where $\sigma_k^2 = Var(y_k)$ is a function of k and compare the estimates of ρ_K and α_K using Monte Carlo simulation with a sample size 5,000.

10.17 Show that CCC ranges between -1 and 1 and identify the scenarios in which CCC takes the value 1, -1, and or 0.

10.18 Show that the moment-based estimate $\widehat{\rho}_{CCC}$ in (10.8) is consistent.

10.19 Let y_{ik} be a continuous outcome for the kth instrument from the ith subject ($1 \leq i \leq n$, $1 \leq k \leq K$). Assume that y_{ik} follows the LMM in (10.11). Let $y_{i\infty} = \mu + \lambda_i$. Show

    (a) $\rho_{ICC} = \rho_1 = Corr(y_{ik}, y_{il})$ is a constant independent of k and l ($k \neq l$).

    (b) $\rho_1 = Corr(y_k, y_{i\infty}) = \sqrt{\rho_1}$.

10.20 Estimate the reliability index and the Cronbach coefficient alpha for each of the eight domains of CSF-36 based on the study described in Example 10.4. Assess whether each item is coherently associated with the other remaining items.

10.21 Assess the test-retest reliability for each of the domains of CSF-36 based on the study described in Example 10.5.

Chapter 11

Analysis of Incomplete Data

An important issue we have intentionally avoided so far is the problem of incomplete data. It is common that we may not be able to collect all the data we intend to collect for a variety of reasons. For example, patients may refuse to provide sensitive information such as sexual abuse or may not even know the answers to some questions such as family health history, yielding missing values of the pertinent variables. Missing values are also a more common phenomenon in modern longitudinal clinical trials. In such studies, patients are followed up for a period of time, and some may miss a certain number of visits and/or even drop out of the study completely, creating missing data for the outcomes to be assessed at the missed visits. Simply ignoring such missing values may produce seriously biased inference.

In Section 11.1, we first describe some common reasons for missing values and how bias arises if missing values are ignored. We then delve into missing-value mechanisms and associated statistical models in Section 11.2. We discuss statistical models to address the impact of the different missing-value mechanisms on inference for affected outcomes in Section 11.3 and illustrate their applications under different missing data circumstances in Section 11.4.

11.1 Incomplete Data and Associated Impact

The occurrence of missing values is a common phenomenon in modern clinical trials as well as observational studies. In this section, we elaborate on some common types of missing data to help develop an appreciation of the mechanisms underlying the different types of missing values.

11.1.1 Observational missing

The most common situation involving missing values is our inability to obtain the values of the variables of interest as planned. In clinical trials, patients may refuse to provide sensitive information such as income and sexual orientation. They may also be truly clueless about the information being asked for. For example, family health history is a common question in many health studies, especially those focusing on genetic risk and/or protective factors for the diseases of interest. It is not uncommon that some subjects do not have such information, especially for the ones not living with them. This problem is further compounded by technological advances. For example, with increased use of web-based assessment tools, data collection methods, and cost-efficient storage devices, investigators are tempted to collect as much information that may potentially be relevant to study objectives as possible, further increasing the chance of missing data.

A common and important missing data phenomenon in modern longitudinal studies is that subjects may drop out of the study prematurely. As we described in Chapter 9, patients are followed up for a period of time in longitudinal studies, making it almost impossible to avoid this issue, even in well-designed and well-executed clinical trials. Subjects may quit study or not show up at follow-up visits due to problems with transportation, weather conditions, health status, relocation, etc. For example, in the PPD, one of the key factors determining whether postpartum mothers will participate and stay in the study is whether childcare is provided. In clinical trials, missing data may also be the result of patients' deteriorated or improved health conditions due to treatment-related complications, treatment responses, etc. Some of these reasons are clearly treatment related while others are not. Although in some studies attempts are made to continue to collect information on such patients, information about the patients after they drop out is often missing, threatening the validity of the data analysis if missing values are simply ignored.

11.1.2 Missing by design

The concept of missing values also arises from a variety of other contexts, which may not appear to be a missing-value problem. For example, when studying the accuracy of a new diagnostic test, it may happen that not all subjects who are administered the new test have their true disease status confirmed, since gold standard tests are sometime too expensive and time-consuming to perform for everyone tested, and some may even involve mentally and/or physically intrusive procedures such as surgery. In such a situation, subjects with negative test outcomes may be less likely to receive a gold standard evaluation than those with positive test outcomes. In some applications, some subjects may be too ill to undergo a more intrusive gold standard evaluation. In all such cases, the decision on whether or not to verify the subject's true disease status depends on the person's diagnostic outcome, health condition, and related prognosis, giving rise to missing data for the confirmatory test outcome.

In survey studies, complex sampling designs are often employed to obtain a sample with a well-balanced representation of the composition of the study population with regard to demographic and other study-defined parameters such as the purpose of the survey, cost, and feasibility. For example, if a particular racial/ethnicity or mental/physical group accounts for a very small fraction of a targeted population, we may need to oversample this group so that it is sufficiently represented in the sample for reliable inference. In modern clinical trials, simple randomization that assigns patients to treatment conditions with equal probability is too often insufficient to address confounding bias, and more structured randomization methods such as stratified block randomization, urn randomization, and adaptive sampling are typically used to balance the distribution of covariates across different treatment conditions. In such cases, subjects are not randomly sampled from the target population or randomly assigned to treatment groups with an equally likely chance, but rather with different sampling probabilities to achieve desired characteristics called for by the study design.

A common feature of these samples is that subjects are no longer sampled with an equally likely chance. As a result, standard statistical methods introduced in the previous chapters do not apply. By interpreting the varying sampling probabilities as the result from filtering subjects sampled with an equally likely chance by some missing data mechanism, methods based on the missing data concept can be applied to facilitate inference (see Section 11.4 for details). Note that the missing value has a different interpretation within the current context, since it refers to missing subjects, rather than missing outcomes from the subject as in the prior settings.

11.1.3   Counterfactual missing

The concept of missing values may also be applied to the counterfactual framework, a popular paradigm for studying causality, to facilitate inference, especially with observational studies (Rubin, 1976). For example, consider a study to examine the effect of certain type of exposure (either a risk factor such as smoking or a protective factor like an intervention) on some health outcome of interest. Each patient in the study could potentially have two outcomes: one that occurs if he/she has the exposure, and the other that results if he/she does not have the exposure. Since a patient can only have one exposure status, the two outcomes only exist conceptually, with only one of the two actually being observed in practice. Note that unlike the missingness caused by logistics and health-related reasons as in the clinical trial studies, one of the potential outcomes is always missing. Nonetheless, methods for missing values can still be applied to facilitate analysis (see Section 11.4.2).

11.1.4   Impact of missing values

A naive approach, which is still commonly practiced in some areas of research, is simply ignoring missing values and proceeding with the subsample with complete data. Indeed, we have used such a *listwise deletion* procedure in all the analyses of the previous chapters. However, as the resulting complete-data subsample is typically not representative of the study population for which inference is desired, simply ignoring this issue generally yields biased estimates. We use a hypothetical example involving diagnostic tests to illustrate this point. This example is also used in Section 11.3 to motivate the development of methods for missing values. Thus, analysis results from some of the examples in the prior chapters with a relatively large amount of missing data may not be correct, and a portion of these will be reanalyzed in this chapter.

Example 11.1
Suppose that we are interested in estimating the prevalence of a disease based on a sample of 1000 subjects randomly selected from the population of interest. Summarized in Table 11.1(a) is the information about the number of diseased subjects and test results from a screening test. Now suppose that 50% of those tested negative have their test results confirmed, with the results shown in Table 11.1(b).

TABLE 11.1:   A hypothetical study of a diagnostic test

	Test			Test	
Diseased	Positive	Negative	Diseased	Positive	Negative
Yes	450	50	Yes	450	25
No	100	400	No	100	200
(a) Complete Data			(b) Verified Subsample		

If the remaining subjects without their negative tests validated are deleted, the naive estimate of prevalence based on the subset in Table 11.1(b) is $\frac{475}{775} = 61.3\%$, much higher than 50% obtained based on the original complete data in Table 11.1(a).   ☐

The cause of bias is the selection of a subset of subjects for verification of their disease status. While the initial random sample is representative of the target population, the subsample of subjects selected for the confirmation of disease status is not. Such a phenomenon due to selection bias has been noted before. For example, in Chapters 3 and 4 we discussed Simpson's paradox, which is the result of basing analysis on unrepresentative samples of the study population. To remove such bias or reduce its effect, we need to understand how the missing value arises and its relationship to selection bias.

11.2   Missing Data Mechanism

Since the reasons for missing values vary and the validity of inference in the presence of missing values depends on how the missing values arise, it is important to make plausible assumptions and model the missing data mechanisms accordingly. Such assumptions allow statisticians to ignore the multitude of reasons for missing data and focus instead on these missing data models when addressing their impact on inference. In this section, we introduce three statistical models, or mechanisms, with increased generality for missing data, which together characterize the impact of missing data on inference under all scenarios of missing data.

11.2.1   Missing completely at random

Consider a study with n subjects, and let $\mathbf{y}_i = (y_{i1}, \ldots, y_{im})^\top$ be an $m \times 1$ vector of responses for the ith subject. Let $\mathbf{r}_i = (r_{i1}, \ldots, r_{im})^\top$ be a vector of indicators; $r_{it} = 1$ if the corresponding component of $\mathbf{y}_i$, y_{it}, is observed, and $r_{it} = 0$ otherwise. The simplest scenario is that the event of missing, or missingness, of any component of $\mathbf{y}_i$ does not depend on any of the variables of interest, observed or otherwise, or simply,

$$\mathbf{y}_i \perp \mathbf{r}_i.$$

This is called the *missing completely at random* (MCAR) mechanism (Rubin, 1976). For example, in clinical trials, missing data at follow-up visits due to reasons unrelated to studies such as patient's relocation and conflict of schedule generally fall into this category. This model corresponds to a lay person's notion of random missing, i.e., the missing data mechanism has no influence whatsoever on any of the patient's outcomes. Thus, under MCAR the subjects who are completely observed have exactly the same distribution as those with missing values for every outcome of the study. This invariance property with respect to the distribution of the outcome may be used to empirically check this assumption.

Consider first a simple situation where missing values are confined to a single component of $\mathbf{y}_i$, say y_m. The sample can then be divided into two subgroups, with one consisting of those with this variable observed, and hence the complete data, and the other formed by the remaining subjects with missing values for this variable. Under MCAR, the two groups should have the same distribution with respect to each of the other variables $y_1, \ldots, y_{m-1}$. We may apply methods for comparing two independent groups such as contingency table methods for categorical variables and t tests or Mann–Whitney–Wilcoxon tests for continuous variables to examine this defining property of MCAR.

If missing data occur for more than one variable, we may generalize the approach above by comparing variables that are not subject to missing, or always observed, across different missing data patterns. For example, if missing values only occur for two variables, there are at most four missing patterns: missing for both, missing for only one of the variables, and missing for none of the variables. We can then compare the groups defined by the patterns to see whether each outcome has the same distribution across the groups. Such an approach is limited because it relies on those variables that are not subject to missing, which may be few in practice. It also suffers from the multiple comparison issue, especially when comparing a large number of missing data patterns (see Chapter 3, Section 3.2, for a discussion about the issue of multiple comparison).

A more efficient alternative is to use all available data (Little, 1988; Chen and Little, 1999). Suppose there are K missing data patterns. Under MCAR, the K different groups, derived based on the different patterns, follow the same distribution. In particular, they all should have the same mean with respect to each of the variables. The idea is to compare the sample means based on the observed values for each pattern with the corresponding means estimated based on the entire sample.

Let $\widehat{\boldsymbol{\mu}}$ ($\widehat{\Sigma}$) be an estimate of the population mean $\boldsymbol{\mu}$ (variance matrix Σ) of $\mathbf{y}_i$ under MCAR based on all observed data. For subjects with missing values, only part of the components of $\mathbf{y}_i$ are observed. Let $\overline{\mathbf{y}}_{obs,j}$ be the sample average of the observed variables in the jth pattern. If MCAR holds, $\overline{\mathbf{y}}_{obs,j}$ should be close to the subvector $\widehat{\boldsymbol{\mu}}_j$ of $\widehat{\boldsymbol{\mu}}$ defined by the observed components of $\mathbf{y}_i$ under the jth pattern. The variation of $\overline{\mathbf{y}}_{obs,j}$ can be measured by $\frac{1}{m_j}\widehat{\Sigma}_j$, where $\widehat{\Sigma}_j$ is the submatrix of $\widehat{\Sigma}$ corresponding to $\widehat{\boldsymbol{\mu}}_j$ and m_j is the number of subjects in the jth missing pattern. Thus, we may use the following statistic,

$$M = \sum_{j=1}^{K} m_j \left(\overline{\mathbf{y}}_{obs,j} - \widehat{\boldsymbol{\mu}}_j\right) \widehat{\Sigma}_j^{-1} \left(\overline{\mathbf{y}}_{obs,j} - \widehat{\boldsymbol{\mu}}_j\right)^{\top}, \tag{11.1}$$

to test MCAR, where the sum runs over the K missing data patterns.

In the above, estimates of $\boldsymbol{\mu}$ and Σ may be obtained from maximum likelihood by assuming multivariate normality or estimating equations (Little, 1988; Chen and Little, 1999). The latter may also be applied directly to test MCAR, even in the regression setting, which is discussed next. In both cases, the statistic M in (11.1) has an asymptotic chi-square distribution with $\sum_{j=1}^{K} p_j - p$ degrees of freedom, where p denotes the length of $\mathbf{y}_i$ and p_j the dimension of $\mathbf{y}_i$ observed for the jth pattern.

11.2.2 Missing at random

MCAR is a strong assumption and may not be satisfied in many real study applications. In clinical trials, patients may be lost to follow up because of deteriorated or improved health conditions, and thus the probability of the missed visit does depend on the missing outcome. Simply ignoring the missing data generally yields biased estimates. On the other hand, the dependence on the missing outcome makes it impossible to model the missing data probability directly as a function of the missing outcome. Thus, plausible assumptions are needed to enable modeling of missingness based on observed outcomes.

The *missing at random* (MAR) assumption posits a mechanism that is completely determined by the observed components $\mathbf{y}_{i,obs}$ of $\mathbf{y}_i$. Under MAR, the missingness is related with, but becomes independent of, the missing value, when conditioned upon $\mathbf{y}_{i,obs}$, i.e.,

$$\mathbf{y}_i \perp \mathbf{r}_i \mid \mathbf{y}_{i,obs}. \tag{11.2}$$

For example, suppose that only the last component of $\mathbf{y}_i$, y_{im}, is subject to missing; i.e., $\widetilde{\mathbf{y}}_{im} = \left(y_{i1}, \ldots, y_{i(m-1)}\right)^{\top}$ is always observed. Then, it is MAR if the missingness of y_{im}, which may depend on y_{im}, becomes independent of y_{im} after controlling for $\widetilde{\mathbf{y}}_{im}$. In this case, $y_{im} \perp r_{im} \mid \mathbf{y}_{i,obs}$, where $\mathbf{y}_{i,obs} = \widetilde{\mathbf{y}}_{im}$ and r_{im} is the missing data indicator for y_{im}.

In general, missing values may occur in any component of $\mathbf{y}_i$ and thus can create rather complex relationships between the missingness and observed values, especially for large m, making modeling such relationships quite a daunting task. For example, for $m = 3$, if missing values can occur in both y_{i2} and y_{i3}, the missingness of y_{i2} (y_{i3}) may depend on y_{i1} (y_{i2}) only, or y_{i3} (y_{i2}) only, or both y_{i1} and y_{i3} (y_{i1} and y_{i2}).

The *monotone missing data pattern (MMDP)* is often used to facilitate applications of MAR in practice. Under this assumption, a subject with missing values in a component, y_{it}, implies that the values for all subsequent components, y_{is} ($t < s \leq m$), are also missing. With the additional constraint of MMDP, only the very first missing component of $\mathbf{y}_i$, say y_{ik}, needs to be considered, in which case the defining MAR condition in (11.2) becomes $\mathbf{y}_{i,obs} = \widetilde{\mathbf{y}}_{ik}$. In other words, MAR under MMDP assumes that the missingness of y_{it} depends only on the preceding components, y_{il} ($1 \leq l < t$). Thus, MMDP allows us to apply the same modeling considerations for the special case of missing the last component y_{im} to the general case involving any missing component of $\mathbf{y}_i$.

For regression analysis with cross-sectional study data, we model the response of interest y_i by conditioning on a set of regressors, or explanatory variables, $\mathbf{x}_i$, which is assumed to be observed. By setting $\mathbf{y}_{i,obs} = \mathbf{x}_i$, the MAR condition in (11.2) becomes $y_i \perp r_i \mid \mathbf{x}_i$, where r_i is the indicator for missingness of y_i. Because of this and the fact that $\mathbf{x}_i$ is also conditioned upon when modeling y_i, MAR in this special case is often referred to as MCAR.

For longitudinal data analysis, we identify t as the visit number and y_{it} as the response at time t. Let $\widetilde{\mathbf{y}}_{it} = \left(y_{i1}, \ldots, y_{i(t-1)}\right)^{\top}$. Then, specific for longitudinal data, the MAR condition in (11.2) takes the form $y_{it} \perp r_{it} \mid \widetilde{\mathbf{y}}_{it}$. For regression analysis, let $\mathbf{x}_{it}$ denote a vector of regressors at time t (always observed), and the corresponding MAR condition becomes $y_{it} \perp r_{it} \mid \mathbf{y}_{it,obs}, \mathbf{x}_{it}$. Thus, the assumption of MAR in this case also involves the regressors.

MMDP is particularly a natural choice for modeling missing data in longitudinal studies, because study dropouts follow such a missing-data pattern. Further, it is straightforward to model MAR under MMDP, which we discuss next.

11.2.2.1 Modeling of missing mechanism

If there is only one component subject to missing such as y_{im} as in the example above, then there are only two missing-value patterns defined by the values of the missing value indicator r_i. Methods described in Chapter 4 can be applied to model the missingness in this special case. For example, we may model the probability of missing y_{im} using the following logistic regression:

$$\text{logit} \left(\Pr(r_i = 1 \mid \mathbf{y}_i)\right) = \text{logit} \left(\Pr(r_i = 1 \mid \widetilde{\mathbf{y}}_{im})\right) = \boldsymbol{\alpha}^{\top} \widetilde{\mathbf{y}}_{im}, \tag{11.3}$$

where $\widetilde{\mathbf{y}}_{im} = \left(y_{i1}, \ldots, y_{i(m-1)}\right)^{\top}$ is always observed.

If missing values occur in more than one component, there will be more than two missing-data patterns, which can be expressed by a multinomial response defined by the corresponding missing data indictors. Models for multinomial responses such as the generalized logit model may be used to model the different missing-value patterns. However, this approach is quite restrictive, since only those variables that are always observed can be included as predictors. Under MMDP, this problem is much simplified, since only the two missing-value patterns associated with the very first missing visit need to be considered for each of the components of $\mathbf{y}_i$ involving missing values. We illustrate the modeling process using a longitudinal study, with $\mathbf{y}_i$ denoting the repeated responses over time and $\mathbf{x}_i$ the vector of static, or time-invariant covariates (always observed).

Let $\widetilde{\mathbf{y}}_{it} = \left(y_{i1}, \ldots, y_{i(t-1)}\right)^{\top}$ be a vector containing all responses up to the $(t-1)$th visit ($1 \leq t \leq m$). Under MMDP and MAR, the missingness of any component y_{it} of $\mathbf{y}_i$ is determined by the outcomes from prior visits, i.e.,

$$\Pr(r_{it} = 1 \mid \mathbf{x}_i, \mathbf{y}_i) = \Pr\left(r_{it} = 1 \mid \mathbf{x}_i, \tilde{\mathbf{y}}_{it}\right), \quad 1 \le t \le m.$$

Since no missing value occurs at baseline $t = 1$, $\Pr(r_{i1} = 1) = 1$. It is readily checked that for $t \ge 2$

$$\Pr\left(r_{it} = 1 \mid \mathbf{x}_i, \tilde{\mathbf{y}}_{it}\right) = \prod_{j=2}^{t} \Pr(r_{ij} = 1 \mid \mathbf{x}_i, \tilde{\mathbf{y}}_{ij}, r_{i(j-1)} = 1).$$

We can model each transition probability, $\Pr(r_{ij} = 1 \mid \mathbf{x}_i, \tilde{\mathbf{y}}_{ij}, r_{i(j-1)} = 1)$, using a logistic regression akin to (11.3) or other models for binary responses discussed in Chapter 4, i.e.,

$$\Pr(r_{it} = 1 \mid \mathbf{x}_i, \mathbf{y}_i, r_{i(t-1)} = 1) = f_t\left(\mathbf{x}, \tilde{\mathbf{y}}_{it}; \boldsymbol{\alpha}_t\right), \tag{11.4}$$

where $f_t\left(\mathbf{x}_i, \tilde{\mathbf{y}}_{it}; \boldsymbol{\alpha}_t\right)$ denotes the model for the probability of $r_{it} = 1$ given $(\mathbf{x}_i, \tilde{\mathbf{y}}_{it})$ parameterized by $\boldsymbol{\alpha}_t$.

Example 11.2
In Example 11.1, if the subjects to be administrated with a gold standard test were chosen in a completely random fashion, then the missingness of the true disease status would be MCAR. We can empirically check this by comparing the test results between the two groups defined by the decision to verify their test outcomes. A chi-square test applied to the 2×2 table for the groups rejects the MCAR assumption (p-value < 0.00001). For this hypothetical example, we know that the decision to verify the disease status is based on the test results, and thus conceptually the missingness of a gold standard diagnosis follows MAR. We discuss how to make valid inference based on such a missing-value mechanism in Section 11.3.   ▯

11.2.3 Missing not at random

If $\mathbf{r}_i$ and $\mathbf{y}_i$ are still related despite conditioning on $\mathbf{y}_{i,obs}$, then it is called *missing not at random* (MNAR). In such situations, regression models such as (11.3) do not sufficiently model the missing mechanism, since missing values themselves must be included as predictors. For example, consider a longitudinal data vector $\mathbf{y}_i$ in a nonregression setting and suppose that only y_{im} is subject to missing. Under MNAR, we may use the following to model the missing mechanism:

$$\Pr(r_{im} = 1 \mid \mathbf{y}_i^m, y_{im}) = f\left(\boldsymbol{\alpha}^\top \mathbf{y}_i^m + \gamma y_{im}\right), \tag{11.5}$$

where f is a known function parameterized by $\boldsymbol{\alpha}$ and γ. However, the above in general is not estimable without further assumptions, since it involves a missing predictor y_{im}.

MNAR occurs if some variables upon which the missing data mechanism depends are not collected or included in the study. For example, if the missing value mechanism below depends on $\mathbf{z}_i$, but not all components of $\mathbf{z}_i$ are observed or included, then it is MNAR, and biased inference may result if MAR is modeled by the available components:

$$\Pr(r_{im} = 1 \mid \mathbf{y}_i^m, \mathbf{z}_i) = g\left(\boldsymbol{\alpha}^\top \mathbf{y}_i^m + \boldsymbol{\gamma}^\top \mathbf{z}_i\right), \tag{11.6}$$

where g is a known function parameterized by $\boldsymbol{\alpha}$ and $\boldsymbol{\gamma}$.

In practice, if a sufficient number of covariates are included and appropriate models are applied, MAR should be a reasonable model for most applications. Thus, a feasible and popular approach is to assume MAR and then assess the impact of the assumptions on inference. We discuss such sensitivity analysis in Section 11.3.5.

Example 11.3
In Example 11.1, the missing-value mechanism would be MNAR if the test results were not collected or included in the study. By including the test result as a covariate, it becomes MAR. In this particular example, we know the true missing value mechanism by design. In general, it is not possible to ascertain MAR based on the observed data. ⬜

11.3 Methods for Incomplete Data

A naive approach to dealing with missing values is simply ignoring them. This is valid only under MCAR (see Problem 11.4). Thus, if MCAR is plausible, we can discard subjects with missing values and base inference only on the remaining subgroup of subjects with completely observed data. On the other hand, if only a very small fraction of subjects have missing values, we may also dismiss such subjects, since inference based on those without missing values is likely to be close to the one based on the complete data. Thus, biased estimates may arise only when MCAR is untenable and the amount of missing values is substantial. In this section, we discuss common approaches for addressing missing values, including maximum likelihood, inverse probability weighting, and imputation methods.

11.3.1 Maximum likelihood method

Let $\mathbf{y}_i$ be a $m \times 1$ vector of responses of interest and $\mathbf{r}_i$ be the corresponding missing data indicator vector, i.e., $r_{ij} = 1$ if y_{ij} is observed and 0 otherwise. If a parametric model is posited for the joint distribution $h(\mathbf{y}, \mathbf{r})$ of $\mathbf{y}_i$ and $\mathbf{r}_i$, then inference can be based on maximum likelihood (ML). This approach typically proceeds in one of two ways, depending on how the joint distribution is factored and modeled.

If $h(\mathbf{y}, \mathbf{r})$ is factored into the following product of a marginal and a conditional distribution,

$$h(\mathbf{y}, \mathbf{r}) = f(\mathbf{y}_i; \boldsymbol{\beta})g(\mathbf{r}_i \mid \mathbf{y}_i; \boldsymbol{\alpha}), \tag{11.7}$$

we need to model the marginal $f(\mathbf{y}_i; \boldsymbol{\beta})$ and conditional $g(\mathbf{r}_i \mid \mathbf{y}_i; \boldsymbol{\alpha})$ distribution. Alternatively, $h(\mathbf{y}, \mathbf{r})$ can be expressed as the product of a different marginal $g'(\mathbf{r}_i \mid \boldsymbol{\alpha}')$ and conditional $f'(\mathbf{y}_i \mid \mathbf{r}_i; \boldsymbol{\beta}')$, and modeling $h(\mathbf{y}, \mathbf{r})$ can proceed by specifying these alternative marginal and conditional distributions. Under the latter *pattern mixture* model approach, the response of interest, $f'(\mathbf{y}_i \mid \mathbf{r}_i; \boldsymbol{\beta}')$, is modeled through a mixture based on the different missing-value patterns defined by $g'(\mathbf{r}_i \mid \boldsymbol{\alpha}')$. Below, we focus on the factorization in (11.7), since this *selection* model, so named because the probability $g(\mathbf{r}_i \mid \mathbf{y}_i; \boldsymbol{\alpha})$ reflects a selection process, provides a simpler alternative to address MAR.

Under MAR, $g(\mathbf{r}_i \mid \mathbf{y}_i; \boldsymbol{\alpha}) = g(\mathbf{r}_i \mid \mathbf{y}_i^o; \boldsymbol{\alpha})$, where $\mathbf{y}_i^m$ ($\mathbf{y}_i^o$) denotes the missing (observed) component of $\mathbf{y}_i$. It follows that

$$f(\mathbf{y}_i^o, \mathbf{r}_i \mid \mathbf{x}_i) = \int f(\mathbf{y}_i^m, \mathbf{y}_i^o; \boldsymbol{\beta})g(\mathbf{r}_i \mid \mathbf{y}_i^m, \mathbf{y}_i^o; \boldsymbol{\alpha})d\mathbf{y}_i^m$$

$$= g(\mathbf{r}_i \mid \mathbf{y}_i^o; \boldsymbol{\alpha}) \int f(\mathbf{y}_i^m, \mathbf{y}_i^o; \boldsymbol{\beta})d\mathbf{y}_i^m$$

$$= g(\mathbf{r}_i \mid \mathbf{y}_i^o; \boldsymbol{\alpha})f(\mathbf{y}_i^o; \boldsymbol{\beta}). \tag{11.8}$$

The log-likelihood based on the joint observations $(\mathbf{y}_i^o, \mathbf{r}_i)$ has been separated into two parts with one involving $\boldsymbol{\alpha}$ and the other containing $\boldsymbol{\beta}$. Thus, if $\boldsymbol{\alpha}$ and $\boldsymbol{\beta}$ are disjoint, inference about the regression model of interest $f(\mathbf{y}_i; \boldsymbol{\beta})$ can simply be based on the observed-data likelihood $f(\mathbf{y}_i^o; \boldsymbol{\beta})$. In other words, missing data can be "ignored," if interest centers on modeling $\mathbf{y}_i$. For this reason, MAR is often called *ignorable missing*.

It should be emphasized, however, that if $\boldsymbol{\alpha}$ and $\boldsymbol{\beta}$ are not disjoint, inference based on $f(\mathbf{y}_i^o; \boldsymbol{\beta})$ may be incorrect. In practice, it is generally difficult to validate this disjoint assumption, creating a potential weakness for applications of the selection model.

Example 11.4

In Example 11.1, since both the disease status and test result are binary, their joint distribution follows a multinomial with four categories. Let $p_{kj} = \Pr(d_i = k$ and $t_i = j)$ for $k, j = 0, 1$, where $d_i = 1$ (0) for the diseased (nondiseased) status and $t_i = 1$ (0) for the positive (negative) test result. Since the missingness of d_i depends on the observed value of the test t_i, it is MAR and the observed-data likelihood is

$$(p_{11} + p_{01})^{m_1} (p_{10} + p_{00})^{m_0} p_{11}^{n_{11}} p_{10}^{n_{10}} p_{01}^{n_{01}} p_{00}^{n_{00}}, \tag{11.9}$$

where n_{kj} is the number of subjects with observed $d_i = k$ and $t_i = j$ $(k, j = 0, 1)$, and m_j is the number of subjects with d_i missing and observed $t_i = j$ $(j = 0, 1)$.

The MLE for the prevalence is 0.5, with an estimated standard error of 0.017. Note that in this simple example, the MLE and associated asymptotic variance estimate are readily evaluated. But, as discussed in Chapter 9, computation of estimates based on likelihood-based methods is generally quite complex, even for complete data. □

Because missing values can be ignored under MAR, this mechanism is commonly assumed in real data applications. However, as argued in Section 11.2.2.1, it is important to assess the plausibility of this missing-value model within the particular context of the study, since appearances can be deceiving, and mechanisms that appear MAR may actually follow MNAR.

11.3.2   Imputation methods

One straightforward approach for dealing with missing values is to simply fill them in. In the simplest case, a number is imputed for each missing value, and the resulting "complete" data are then analyzed as if it were really completely observed. This *single imputation* procedure can normally produce valid point estimates under correctly specified imputation assumptions. For example, under MCAR, the mean of the missing value is the same as the population mean. Thus, we may fill in the missing value of a variable with its sample mean based on the observed data. However, since each missing value is imputed with only one number, the variability of the original data is underestimated, yielding potentially false significant findings. For example, by imputing the missing values of the disease status in Example 11.1 with the observed means from the corresponding test results, we obtain a complete data set as shown in Table 11.1(a), which can be used to obtain the correct estimate of the probability, but not the variance of the estimate, as the sample size is artificially inflated.

11.3.2.1   Mean score methods

One way to address the problem of underestimated variance resulting from single imputation is to account for sampling variability through estimating equations (EEs). We illustrate this procedure within the context of regression analysis.

Consider a regression model:

$$E\left(y_i \mid \mathbf{x}_i\right) = f\left(\mathbf{x}_i; \boldsymbol{\beta}\right), \quad 1 \le i \le n, \tag{11.10}$$

where $\boldsymbol{\beta}$ is a vector of parameters of interest. Without any missing value, the following EEs can be used to provide inference about $\boldsymbol{\beta}$:

$$\sum_{i=1}^{n} G(\mathbf{x}_i)\left[y_i - f\left(\mathbf{x}_i; \boldsymbol{\beta}\right)\right] = \mathbf{0}, \tag{11.11}$$

where $G(\mathbf{x}_i)$ is some known (matrix) function of $\boldsymbol{\beta}$. Now suppose that the response y_i is subject to missing, but the missingness is conditionally independent of y_i given $\mathbf{x}_i$ and $\mathbf{z}_i$, i.e.,

$$r_i \perp y_i \mid \mathbf{x}_i, \mathbf{z}_i, \quad 1 \le i \le n, \tag{11.12}$$

where $\mathbf{z}_i$ is some other vector of covariates and, along with $\mathbf{x}_i$, is always observed. For example, in longitudinal studies, if y_i (r_i) represents a response (associated missing-data indicator) at the current visit and $\mathbf{z}_i$ consists of all the observed responses prior to y_i, (11.12) is the MAR condition discussed in Section 11.2.2.

If the condition in (11.12) only involves $\mathbf{x}_i$, then the missing data mechanism is MCAR, as noted in Section 11.2.2. In this special case, the naive EE that simply ignores those subjects with missing values,

$$\sum_{i=1}^{n} r_i G(\mathbf{x}_i)\left[y_i - f\left(\mathbf{x}_i; \boldsymbol{\beta}\right)\right] = \mathbf{0}, \tag{11.13}$$

is unbiased, thereby yielding consistent estimates of $\boldsymbol{\beta}$. In general, if additional information $\mathbf{z}_i$ is also needed to ensure (11.12), then (11.13) is no longer unbiased (see Problem 11.4).

To correct the bias, we impute missing values based on the information from both $\mathbf{x}_i$ and $\mathbf{z}_i$, since as posited in (11.12), the missingness of y_i only depends on these variables. Suppose the dependence is described by a second model

$$E\left[y_i \mid (\mathbf{x}_i, \mathbf{z}_i)\right] = g\left(\mathbf{x}_i, \mathbf{z}_i; \boldsymbol{\gamma}\right), \tag{11.14}$$

where $g\left(\mathbf{x}_i, \mathbf{z}_i; \boldsymbol{\gamma}\right)$ is some known function parameterized by $\boldsymbol{\gamma}$. This second model (11.14) may be estimated using either ML or EEs based on the observed data. Regardless of the approach used, we solve a set of equations of the following form:

$$\sum_{i=1}^{n} r_i H(\mathbf{x}_i, \mathbf{z}_i)\left[y_i - g\left(\mathbf{x}_i, \mathbf{z}_i; \boldsymbol{\gamma}\right)\right] = \mathbf{0}, \tag{11.15}$$

where $H(\mathbf{x}_i, \mathbf{z}_i)$ is a known (matrix) function of $\boldsymbol{\gamma}$. Given an estimate of $\widehat{\boldsymbol{\gamma}}$ of $\boldsymbol{\gamma}$, we can use the estimated mean $g\left(\mathbf{x}_i, \mathbf{z}_i; \widehat{\boldsymbol{\gamma}}\right)$ to impute the missing y_i and then estimate $\boldsymbol{\beta}$ using (11.11) based on the completed data. The consistency of the estimate $\widehat{\boldsymbol{\beta}}$ is guaranteed by the unbiasedness of the EE, since (11.11) can be reexpressed as

$$\sum_{i=1}^{n} G(\mathbf{x}_i)\left[r_i y_i + (1 - r_i)\, g\left(\mathbf{x}_i, \mathbf{z}_i; \boldsymbol{\gamma}\right) - f\left(\mathbf{x}_i; \boldsymbol{\beta}\right)\right] = \mathbf{0}, \tag{11.16}$$

which is unbiased (see Problem 11.5). We may also obtain the asymptotic variance of $\widehat{\boldsymbol{\beta}}$ from the above EE. However, since the equations in (11.16) require an estimate of $\boldsymbol{\gamma}$, this variance again underestimates the true sampling variability. To address this underestimation, we may combine the two estimating equations in (11.15) and (11.16) for simultaneous inference about $\boldsymbol{\beta}$ and $\boldsymbol{\gamma}$.

The above is known as the *mean score* method, since the imputation is carried out to complete the missing y_i in the score, or EEs (11.11). This approach may also be applied to missing covariate $\mathbf{x}_i$ if y_i is not missing and $f(\cdot)$ is linear. Readers interested in this may consult Pepe et al. (1994) and Reilly and Pepe (1995) for details.

For the mean score method to produce valid inference, it is critical that the prediction model (11.14) is correctly specified. In addition, inference may become quite complex computationally when applying the mean score procedure to real study data, as separate variance estimates may be needed depending on the situations at hand.

Example 11.5

Without missing values, the prevalence $p = \Pr(d_i = 1)$ in Example 11.1 can be estimated by the EE:

$$\sum_{i=1}^{n} (d_i - p) = 0.$$

However, when there are missing values in d_i, the naive EE $\sum_{i=1}^{n} r_i (d_i - p) = 0$ may become biased. We may use observed test results t_i to predict those missing values in d_i to correct the bias.

Assume that d_i depends on t_i and other covariates $\mathbf{x}_i$ through the following relationship:

$$\Pr(d_i = 1 \mid \mathbf{x}_i, t_i) = g\left(\mathbf{x}_i, t_i; \boldsymbol{\beta}\right), \quad 1 \leq i \leq n. \tag{11.17}$$

By using the above to impute the missing values in d_i, we can apply the mean score method for valid inference. In Example 11.1, since both d_i and t_i are binary with no other covariate and d_i is observed if $t_i = 1$, (11.17) is determined by the proportion $\alpha = \Pr(d_i = 1 \mid t_i = 0)$, which can be estimated by the observed sample proportion under the MAR assumption, $r_i \perp d_i \mid t_i$. Since the observed sample proportion can be written as the solution to the EE, $\sum_{i=1}^{n} r_i(1 - t_i)(d_i - \alpha) = 0$, valid inference for the prevalence can be obtained by the following combined EE:

$$\sum_{i=1}^{n} [r_i d_i + (1 - r_i)\alpha - p] = 0,$$

$$\sum_{i=1}^{n} r_i(1 - t_i)(d_i - \alpha) = 0.$$

The mean score estimate for the prevalence is 0.5, with an estimated standard error of 0.0172. This point estimate is the same as that obtained by imputing α for the missing d_i. However, the standard error estimate corrects the downward bias in our earlier estimate 0.0158 obtained from single imputation. ⬜

11.3.2.2 Multiple imputation

Multiple imputation is another popular approach to overcome the issue of underestimation of variation under single imputation (Rubin, 1978). Instead of a single value, this approach assigns multiple plausible numbers to each missing value, yielding multiple completed data sets to provide information for correcting the downward bias in the variance estimate. Multiple imputation consists of three major steps:

(1) Data imputation. Based on some prediction models, each missing value is filled in with several, say m (imputation size), plausible numbers, generating m complete data sets;

(2) Analysis of multiply imputed data sets. Apply complete-data models such as those described in the previous chapters to each completed data set. For the parameter of interest, say $\boldsymbol{\beta}$, we obtain m estimates $\widehat{\boldsymbol{\beta}}_j$ and associated variance estimate $\widehat{\mathbf{u}}_j$, one from each of the completed-data analysis ($1 \leq j \leq m$);

(3) Synthesis of results in (2). The multiple imputation estimate of $\boldsymbol{\beta}$ is $\widehat{\boldsymbol{\beta}} = \frac{1}{m} \sum_{j=1}^{m} \widehat{\boldsymbol{\beta}}_j$, but the variance estimate of $\widehat{\boldsymbol{\beta}}$, $\widehat{V}_{\boldsymbol{\beta}} = \overline{W} + \left(1 + \frac{1}{m}\right) B$, has two components: the within-imputation sample variance estimated by $\overline{W} = \frac{1}{m} \sum_{j=1}^{m} \widehat{\mathbf{u}}_j$ and the between-imputation sample variance estimated by $B = \frac{1}{m-1} \sum_{j=1}^{m} \left(\widehat{\boldsymbol{\beta}}_j - \widehat{\boldsymbol{\beta}}\right)^{\top} \left(\widehat{\boldsymbol{\beta}}_j - \widehat{\boldsymbol{\beta}}\right)$. The adjustment using the between-imputation variance is actually quite intuitive, if one recalls the formula for conditional variances (see Problem 1.2).

For a scalar β, inference can be based on the approximation

$$\frac{1}{\sqrt{v}} \left(\widehat{\beta} - \beta\right) \sim t_\nu, \quad \nu = (m-1) \left(1 + \frac{\overline{W}}{(1 + \frac{1}{m})B}\right)^2,$$

where t_ν denotes the t-distribution with ν degrees of freedom (Rubin, 1987). The larger the imputation size m, the more accurate the t approximation. However, a choice of $m = 20$ seems to suffice for most practical purposes. A similar F-statistic is available when $\boldsymbol{\beta}$ is a vector (see Schafer (1997)).

For very large m, $\frac{1}{m}$ is close to 0, and $\widehat{\Sigma}_{\boldsymbol{\beta}}$ is approximately $\overline{W} + B$. Thus, the variance estimate $\overline{W} = \widehat{\mathbf{u}}$ from single imputation underestimates the true variability by an amount B.

The most important step of MI is data imputation; the quality of the imputed data translates directly to the validity of inference. In general, variables known to be predictive of the missingness of the response should be included in the imputation model. Likewise, when imputing missing covariates, it is important to include the response as a predictor.

For imputing missing values, we start with a model appropriate for the type of response. For example, for a binary y_i, we may model the probability of $y_i = 1$, or mean of y_i, given x_i using any of the models for binary responses discussed in Chapter 4 such as logistic regression as follows:

$$\Pr\left(y_i = 1 \mid \mathbf{x}_i\right) = f(\mathbf{x}_i; \boldsymbol{\alpha}), \tag{11.18}$$

where f is a known function parameterized by $\boldsymbol{\alpha}$ with the exact form depending on the specific model used. Since the missing y_i is (conditionally) independent of any other variables given $\mathbf{x}_i$ according to (11.18), it follows from the discussion in Section 11.2.2 that the missing y_i within the context of the imputation model above follows MCAR and can be readily fitted based on the observed y_i.

Let $\widehat{\boldsymbol{\alpha}}$ and $\widehat{V}_{\boldsymbol{\alpha}}$ be an estimate of $\boldsymbol{\alpha}$ and associated variance estimate. We assume that $\widehat{\boldsymbol{\alpha}}$ has approximately a normal distribution, $N\left(\boldsymbol{\alpha}, \widehat{V}_{\boldsymbol{\alpha}}\right)$, which is not a strong assumption since most estimates such as ML have an asymptotic normal distribution. In the imputation step, we first sample m copies of the parameter vector $\boldsymbol{\alpha}_j$ from $N\left(\widehat{\boldsymbol{\alpha}}, \widehat{V}_{\boldsymbol{\alpha}}\right)$ ($1 \leq j \leq m$). For each sampled $\boldsymbol{\alpha}_j$, we compute the fitted probability, $f(\mathbf{x}_i; \boldsymbol{\alpha}_j)$, for each subject with a missing y_i, followed by simulating a Bernoulli response based on $f(\mathbf{x}_i; \boldsymbol{\alpha}_j)$ to replace the missing value, thereby completing the data.

Example 11.6

To illustrate multiple imputation, consider again the hypothetical data in Example 11.1. We first impute the missing d's using the logistic model

$$\text{logit}\,(\Pr(d = 1 \mid t)) = \alpha_0 + \alpha_1 t,$$

where $\boldsymbol{\alpha} = (\alpha_0, \alpha_1)^\top$ is the vector of parameters. By fitting the logistic model based on the observed data, we obtain the MLE $\widehat{\boldsymbol{\alpha}} = (-2.0794, 0.5835)^\top$ and associated variance estimate $\widehat{V}(\widehat{\boldsymbol{\alpha}}) = \begin{pmatrix} 0.0450 & -0.0450 \\ -0.0450 & 0.0572 \end{pmatrix}$. With the imputation model in place, we can start the multiple imputation process by following the three steps above. In this simple example, we set $m = 10$.

Step 1. By sampling from $N(\widehat{\boldsymbol{\alpha}}, \widehat{V}(\widehat{\boldsymbol{\alpha}}))$ ten times, we obtain ten values $\boldsymbol{\alpha}_j = (\alpha_{j0}, \alpha_{j1})^\top$ of $\boldsymbol{\alpha}$ as follows:

Sample	1	2	3	4	5	6	7	8	9	10
$-\alpha_{j0}$	2.37	2.09	2.16	1.54	2.15	2.22	1.99	2.06	2.07	2.20
α_{j1}	3.97	3.64	3.55	2.93	3.78	3.67	3.52	3.69	3.69	3.74

For each $\boldsymbol{\alpha}_j$, we generate a Bernoulli d_i for each subject with missing d_i, with the probability of success given by

$$\Pr(d_i = 1) = \begin{cases} \dfrac{\exp(\alpha_{j0} + \alpha_{j1})}{1 + \exp(\alpha_{j0} + \alpha_{j1})} & \text{if } t_i = 1, \\[2mm] \dfrac{\exp(\alpha_{j0})}{1 + \exp(\alpha_{j0})} & \text{if } t_i = 0, \end{cases} \qquad 1 \le i \le n.$$

Note that in this example, missingness of d_i only happens when $t_i = 0$. This results in ten complete data sets.

Step 2. For each of the ten complete data, we can estimate the prevalence $\widehat{p}_j$ and its variation $\widehat{v}_j$. The ten complete samples as well as the estimates are summarized in the following table:

Sample	1	2	3	4	5	6	7	8	9	10
cell (0,0)	433	422	423	410	427	428	424	420	431	427
cell (1,0)	42	53	52	65	48	47	51	55	44	48
$\widehat{p}_j$	0.49	0.50	0.50	0.52	0.50	0.50	0.50	0.50	0.49	0.50
$\widehat{v}_j \times 10^4$	2.50	2.50	2.50	2.50	2.50	2.50	2.50	2.50	2.50	2.50

Step 3. Combine the results obtained in Step 2. The point estimate of the prevalence would be $\frac{1}{10}\sum_{j=1}^{10}\widehat{p}_j = 0.501$. The variance of the estimate is $B = \frac{1}{10}\sum_{j=1}^{10}\widehat{v}_j + (1 + \frac{1}{10})var(\widehat{p}_j) = 0.0002962$, which gives standard error 0.0172. ▯

Note that, in general, different estimates will result if the procedure is repeated, due to variation in sampling in the imputation step using Monte Carlo simulations. Such variability is generally not an issue. However, if the significance of a predictor is straddling the borderline such as when the p-value is close to 0.05, different conclusions may be reached between the repeated runs.

11.3.3   Inverse probability weighting

The inverse probability weighting (IPW) approach has a long history in the analysis of sample survey data (Horvitz and Thompson, 1952). The basic idea is to treat each sampled subject as a representative of a group of subjects to account for the unsampled ones by the representability of each subject sampled when estimating population characteristics. For example, an observed subject with a 20% likelihood of being sampled represents a group of $1/0.2 = 5$ similar subjects, including the one sampled. The representativeness of each observed subject is then used as a weight to construct estimates of population-level parameters.

When applied to address missing data within our context, consider the model in (11.10) and suppose the missing value mechanism is modeled by

$$\Pr(r_i = 1 \mid \mathbf{x}_i, \mathbf{z}_i) = \pi_i(\mathbf{x}_i, \mathbf{z}_i; \boldsymbol{\alpha}), \tag{11.19}$$

where $\pi_i(\cdot)$ is some function parameterized by a vector of parameters $\boldsymbol{\alpha}$. Then the following EE:

$$\frac{1}{n} \sum_{i=1}^{n} \frac{r_i}{\pi_i} G(\mathbf{x}_i) [y_i - f(\mathbf{x}_i; \boldsymbol{\beta})] = \mathbf{0}, \tag{11.20}$$

is unbiased (see Problem 11.6), yielding consistent estimates of $\boldsymbol{\beta}$.

In some studies such as the two-phase design in Example 11.1, the probability π_i may be known by design. In this case, we may immediately make inference based on the EE in (11.20). However, in most applications, we need to specify a form for π_i and estimating $\boldsymbol{\alpha}$ in (11.19). Following the discussion in Section 11.3.2 for inference using the mean score method, we may combine the score or estimating equations for the model in (11.20) with the EE in (11.19) for simultaneous inference about $\boldsymbol{\beta}$ and $\boldsymbol{\alpha}$. Note that even when π_i is known as in studies with two-phase designs, the estimated version is often preferred over the true π_i because it may fit the observed data better (see Problem 11.1).

For IPW to provide inference at the population-level, each subject must have a positive probability of being observed, i.e., $\pi_i > 0$. In other words, the subgroups that comprise the study population must have their representatives observed. For extremely small π_i's, the inverses of such π_i can become quite large, causing IPW estimates obtained from (11.20) to be highly volatile. To ensure good behaviors of the IPW estimates, we require that the π_i's be bounded away from 0:

$$\pi_i > c > 0, \quad 1 \le i \le n,$$

where c is some positive constant.

Under the IPW approach, we do not need to model the relationship between the variables subject to missing and those always observed, as in the case of the mean score method. However, like the latter, biased estimates will result if $\pi_i(\mathbf{x}_i, \mathbf{z}_i; \boldsymbol{\beta})$ is not correctly specified.

11.3.4   Doubly robust estimate

One may combine the mean score and IPW approaches to obtain a *doubly robust* estimate (Robins et al., 1994; Robins and Rotnitzky, 1995), in the sense that the estimate of $\boldsymbol{\beta}$ is consistent if at least one of the prediction and missing mechanism models is correctly specified (see Problem 11.9).

Example 11.7
Following the IPW approach, estimates of disease prevalence in Example 11.1 can be obtained by the EE $\frac{1}{n} \sum_{i=1}^{n} \frac{r_i}{\pi_i} (d_i - p) = 0$, where $\pi_i = 1$ (0.5) if $t_i = 1$ (0) are known by design. By solving the EE above, we obtain an estimate of prevalence $= \frac{450 + 2 \times 25}{550 + 2 \times 225} = 0.5$ and associated standard error estimate 0.0172. In this simple example, we obtain the same estimate of prevalence from the mean score and IPW methods. In general, differences are

expected between the two estimates, albeit both will yield valid inference if the model for the missing mechanism and the prediction model are correctly specified. ⬜

11.3.5 Sensitivity analysis

Although plausible for most studies and popular in practice, MAR itself is not testable by the observed data, making it impossible to rule out the possibility of MNAR in a given study. When the validity of MAR becomes a serious concern, it is time to think about assessing the impact of MAR on inference by performing *sensitivity analysis*. The general principle of such analysis is to specify an MNAR model such as the one in (11.5) that includes MAR as a special case, and then assess how inference will change as the missing data mechanism deviates from MAR under the posited MNAR model.

Specifically, we vary γ in (11.5) over a range of selected values and make inference about the parameters of interest for each γ under MNAR using methods such as ML. If a small deviation from MAR (small γ) generates a significant change in inference, then inference based on MAR may not be reliable. Otherwise, MAR is a reasonably good approximation to the true missing data mechanism for the range of γ considered.

For sensitivity analysis to be meaningful and informative, we must select a range of γ to reflect the potential degree of violation of MAR for the study at hand. In that respect, it is critical to solicit input from experts familiar with the content and subject matter of the study. We illustrate the ideas with a simple example. Readers interested in this topic may consult the relevant literature such as Molenberghs and Kenward (2007) and Daniels and Hogan (2008).

Example 11.8

We have applied MLE, mean score, and IPW to the data in Example 11.1 under MAR. How will the MAR assumption affect the inference? To carry out a sensitivity analysis in this example, we assume the following MNAR mechanism for observing subjects with a positive test:

$$\text{logit}(\pi_i) = c_0 + \gamma d_i,$$

where the parameter γ represents the (log) odds ratio of observing a diseased vs. a nondiseased subject. Thus, if $\gamma > 0$ ($\gamma < 0$), the diseased subject with a positive test is more (less) likely to be observed. When $\gamma = 0$, the MNAR model reduces to MAR. In this simple example, it is easy to compute the ML estimate.

TABLE 11.2: Estimates of prevalences under different γ's

γ	-1	$-2/3$	$-1/3$	0	$1/3$	$2/3$	1
$\hat{p}$	0.5321	0.5191	0.5084	0.5	0.4935	0.4886	0.4849
SE	0.0193	0.0185	0.0177	0.0172	0.0167	0.0164	0.0162

Summarized in Table 11.2 are the estimates of prevalences for a range of γ, which show a monotone decreasing function of γ. This decreasing trend reflects the increased likelihood of observing diseased subjects in the sample. Thus, each increment of γ increases the representativeness of the diseased population by the observed disease subjects, which in turn reduces the weight placed on the diseased subjects observed, thereby resulting in smaller estimates of prevalence. ⬜

11.4 Applications

In this section, we illustrate the methods for missing values introduced in the last section with some additional applications. From the range of examples considered below, it should be clear that the methods discussed above are sufficiently general to address missing data in most situations where MAR is plausible. Through adaptions and modifications, these examples can be tailored to a wide spectrum of applications. The reader is in a better position to decide as to which approach is best suited for a particular application at hand.

11.4.1 Verification bias of diagnostic studies

In assessing the accuracy of diagnostic tests, gold standard tests may not be administrated to every subject due to concerns of cost, health risk, and other pertinent issues, yielding missing data for the true disease status for a subset of the subjects. Since the decision for administration of the gold standard is often made based on the test results and other related patient characteristics and prognoses, the missing mechanism is generally not MCAR. Thus, *verification bias* may result if missing values are simply ignored, as illustrated by Example 11.1.

Let $(\mathbf{x}_i, t_i, d_i)$ be an i.i.d. sample from a population, with $\mathbf{x}_i$ denoting a covariate vector, t_i a binary test result, and d_i the true disease status ($1 \le i \le n$). Assume that the covariate $\mathbf{x}_i$ and test result t_i are always observed, but d_i is subject to missing, with the missing status indicated by v_i ($= 1$ if d_i is observed, and 0 otherwise). If the decision to administer the gold standard test is based on the observed covariates $\mathbf{x}_i$ and the test result t_i, a quite plausible and commonly adopted assumption in the literature on this topic (Alonzo et al., 2003), then $v_i \perp d_i \mid (\mathbf{x}_i, t_i)$, i.e., the missingness of d_i is of MAR type.

Suppose we have a prediction model

$$\Pr(d_i = 1 \mid \mathbf{x}_i, t_i) = f(\mathbf{x}_i, t_i; \boldsymbol{\alpha}), \tag{11.21}$$

where f is a function defined by the parameter $\boldsymbol{\alpha}$. Once the specific form of f is established, such as based on input from experts familiar with the disease under consideration, we can apply the mean score method to estimate the sensitivity of the test, $Se = \Pr(t = 1 \mid d = 1)$, yielding

$$\widehat{Se}_{MS} = \frac{\sum_{i=1}^n [d_i v_i t_i + f(\mathbf{x}_i, t_i; \widehat{\boldsymbol{\alpha}})(1 - v_i)t_i]}{\sum_{i=1}^n [d_i v_i + f(\mathbf{x}_i, t_i; \widehat{\boldsymbol{\alpha}})(1 - v_i)]},$$

where $\widehat{\boldsymbol{\alpha}}$ is an estimate for $\boldsymbol{\alpha}$ based on model (11.21).

Alternatively, if the probability of d_i being observed, $\pi_i = \Pr(v_i = 1 \mid \mathbf{x}_i, t_i)$, is known, we may estimate Se by IPW method:

$$\widehat{Se}_{IPWK} = \frac{\sum_{i=1}^n v_i t_i d_i \pi_i^{-1}}{\sum_{i=1}^n v_i d_i \pi_i^{-1}}. \tag{11.22}$$

Since π_i is generally unknown (except in rare cases such as studies with two-phase designs), we need to model and estimate π_i in order to use the IPW estimate in (11.22). Suppose a plausible model is given by

$$\Pr(v_i = 1 \mid \mathbf{x}_i, t_i) = \pi_i = g(\mathbf{x}_i, t_i; \boldsymbol{\gamma}), \tag{11.23}$$

where g is a known function defined by the parameter $\boldsymbol{\gamma}$. The model in (11.23) does not involve any missing value and can be readily estimated by standard methods. Denoting by

$\widehat{\gamma}$ some estimate of γ, we immediately obtain an estimate $\widehat{\pi}_i = g(\mathbf{x}_i, t_i; \widehat{\gamma})$ of π_i and hence an estimate $\widehat{Se}_{IPW}$ of Se:

$$\widehat{Se}_{IPW} = \frac{\sum_{i=1}^{n} v_i t_i d_i \widehat{\pi}_i^{-1}}{\sum_{i=1}^{n} v_i d_i \widehat{\pi}_i^{-1}}. \tag{11.24}$$

Similar approaches can be applied to estimate specificity. By connecting point estimates of sensitivity and specificity, we can obtain empirical ROC curves and use them to estimate AUC. Closed variance formula may be difficult to obtain, but the computation may be facilitated by resampling methods such as bootstrap.

Alternatively, we may directly estimate AUC using the theory of U-statistics. Within the context of verification bias, it is not possible to separate the data into two groups according to disease status because of missing values. However, since

$$AUC = \Pr(T_i < T_j \mid D_i < D_j) = \frac{\Pr(T_i < T_j \text{ and } D_i < D_j)}{\Pr(D_i < D_j)}, \tag{11.25}$$

we may estimate the AUC under missing data by computing a consistent estimate of the right-side of (11.25) using missing-data methods. For example, under the assumption of (11.23), we can apply the IPW approach to obtain the following estimate:

$$\widehat{AUC} = \frac{\sum_{i \neq j} \frac{v_i}{\widehat{\pi}_i} \frac{v_j}{\widehat{\pi}_j} \left[I(t_i < t_j) + \frac{1}{2} I(t_i = t_j) \right] I(d_i < d_j)}{\sum_{i \neq j} \frac{v_i}{\widehat{\pi}_i} \frac{v_j}{\widehat{\pi}_j} I(d_i < d_j)},$$

where $I(\cdot)$ is a set indicator function. The AUC estimate above equals the area under the empirical estimated ROC curve constructed based on connecting the IPW estimates of sensitivity and specificity (He et al., 2009).

Example 11.9

In DOS, there is no missing value in either HAM-D or depression diagnosis (SCID) at baseline, so we may directly assess the accuracy of HAM-D in diagnosis of depression. For illustrative purposes, we create a subset of subjects with missing SCID to resemble missing data arising from a two-phase design as in the hypothetical study in Example 11.1. In this way, we can use estimates from the complete data for comparing the performance between different methods. For everyone in this subset, HAM-D is available, but SCID is observed by the following selection probability based on the subject's age, and HAM-D and CIRS scores:

$$\pi = 0.1 + 0.3 I \,(\text{HAM-D} > 7) + 0.3 I \,(\text{CIRS} > 7) + 0.3 I \,(\text{Age} < 75). \tag{11.26}$$

Under this model, 283 of the 742 patients (38.1%) are selected for SCID verification of depression diagnosis.

We apply both mean score and IPW for inference about the sensitivity of HAM-D in detecting depression. For mean score, we use the following predict model:

$$\Pr(\text{Depressed}) = \alpha_0' + \alpha_1' \text{HAMD} + \alpha_2' \text{CIRS} + \alpha_3' \text{Age}, \tag{11.27}$$

while for IPW, we model the selection probability as

$$\pi = \alpha_0 + \alpha_1 I \,(\text{HAM-D} > 7) + \alpha_2 I \,(\text{CIRS} > 7) + \alpha_3 I \,(\text{Age} < 75). \tag{11.28}$$

This is the same model as the one in (11.28), but with the known coefficients replaced by unknown parameters to be estimated based on the observed data. To illustrate the effect of model misspecification on inference, we also consider a second model for π given by

$$\pi = \alpha_0 + \alpha_1 \text{HAMD} + \alpha_2 \text{CIRS} + \alpha_3 \text{Age}. \tag{11.29}$$

Although involving the same set of predictors, the above is a wrong model for the selection probability π, but useful for examining the performance of IPW when the weight is incorrectly modeled. By applying mean score and IPW in the respective settings, we can make inference about the sensitivity and specificity of HAM-D for detecting depression.

Shown in the following table are estimates of sensitivity and associated standard errors if subjects with HAM-D > 7 are considered positive.

Methods	Full	Naive	MS	IPWK	IPWE1	IPWE2
Sensitivity	0.824	0.894	0.807	0.827	0.827	0.857
SE	0.025	0.029	0.046	0.042	0.038	0.043

The column "IPWK" indicates the IPW estimate based on the true value of the selection probability π in (11.26), while those labeled "MS," "IPWE1," and "IPWE2" show the mean score estimate based on (11.27), and the two IPW estimates based on the correct (11.28) and incorrect (11.29) model for the missing mechanism. To help assess the performance of the different methods, the table also contains the estimated sensitivity based on the original complete data (under "Full") and the observed incomplete data (under "Naive"). Compared to the upwardly biased Naive estimate, the mean score and IPW (IPWK and IPWE1) estimates are considerably closer to the full estimate. In this particular example, IPWK and IPWE1 seem to have outperformed their mean score counterpart, which may not be surprising since either exact selection probabilities or exactly the same model for missingness of SCID as the one that generates the missing values is used in these two IPW approaches, rather than an approximation as in the case of mean score. The upward bias in the IPWE2 estimate of sensitivity shows that IPW is sensitive to the quality of models for missing-value mechanisms. But despite this weakness, IPWE2 is still better than the naive estimate, which completely ignores the missing values.

We may want to assess model fit when modeling missing-value mechanisms. The model selection techniques discussed in Chapter 7 can be applied to aid in the selection of appropriate models.   □

11.4.2   Causal inference of treatment effects

We have discussed many methods in this book to study correlation, or association, between two or more variables. However, none applies if we want to go beyond association to infer a causal relationship among correlated outcomes. A primary reason for such a difficulty is confounding factors, observed or otherwise. Unless such factors are all identified and controlled for, association observed cannot be attributed to causation. For example, if patients in one treatment have a higher rate of recovery from a disease of interest than those in another treatment, we cannot generally conclude that the first treatment is more effective, since the difference could simply be due to differential disease severity between the two treatment groups. Alternatively, if those in the first treatment group are in better healthcare facilities and/or have easier access to some efficacious adjunctive therapy, we could also see a difference in recovery between the two groups.

The most popular approach for controlling for confounding is randomization. Indeed, randomization is generally regarded as the gold standard for inferring causal relations, since, unlike all other alternatives, it takes the guesswork out of identifying elements of confounding by self-regulating such factors across different treatment groups. For example, if the subjects are randomized into the two treatment conditions in the above example, any difference observed between the two groups must be ascribed to the differential effect of the treatments.

In some studies, however, it is not feasible or even ethical to apply randomization. For example, it is simply not possible to study the effect of smoking on health using a randomized trial. Instead, such studies attempt to explicitly identify and control for factors of confounding to infer causal relationships. The case-control design discussed in Chapter 4 is a prime example of this type of nonrandomization-based strategies for causal inference. Despite their popularity in practice, however, such "retrospective" study designs generally do not completely get rid of selection bias. This fundamental limitation of nonrandomization-based methods, although rather obvious on intuitive grounds, is actually quite difficult to demonstrate analytically. In fact, such an inquiry raises a more fundamental question as to how treatment effects are defined in the first place. After all, randomization is the means by which to control for confounding, rather than to define the notion of causal effect.

The concept of *counterfactual outcome*, the underpinnings of the modern causal inference paradigm, addresses the gap in classic statistical inference to answer this fundamental question (Rubin, 1976). The idea is that for every patient, there is a potential outcome for each treatment condition received, and the treatment effect is defined by the difference between the outcomes in response to the respective treatments from the same individual. Thus, treatment effect is defined for each subject based on his/her differential responses to different treatments, rather than averaged responses over a group of subjects as often conceived, thereby free of any confounder. The counterfactual outcome not only addresses the lack of a conceptual basis for causal analysis of data from nonrandomized studies, but also provides a building block for developing complex study designs and associated inference methods.

For notational brevity, consider two treatment conditions, and let $y_{i,j}$ denote the potential outcome for the ith subject under the jth treatment ($1 \leq i \leq n$, $1 \leq j \leq 2$). We observe only one of the two outcomes, $y_{i,1}$ or $y_{i,2}$, depending on the treatment received by the patient. The difference between $y_{i,1}$ and $y_{i,2}$ can be attributed to the differential effect of the treatments, since there is absolutely no other confounder in this case. However, as one of $y_{i,1}$ and $y_{i,2}$ is always unobserved, standard statistical methods cannot be applied, but methods for missing data can be used to facilitate inference.

Under this new paradigm, the mean response $E(y_{i,1} - y_{i,2})$, albeit unobserved, represents the effect of treatment. Let z_i be an indicator for the first treatment; then $y_{i,1}$ ($y_{i,2}$) is observable only if $z_i = 1$ (0). Under simple randomization, the assignment of treatment is random and free of any selection bias, yielding

$$E(y_{i,j}) = E(y_{i,j} \mid z_i = j), \quad 1 \leq i \leq n.$$

The above shows that missing values in the counterfactual outcomes $y_{i,j}$ are of the MCAR type and can thus be completely ignored. It follows from the above that $E(y_{i,j})$ can be estimated based on the observed component of each subject's counterfactual outcomes corresponding to the assigned treatment.

Selection bias may arise when applying the above to studies employing more complex randomization schemes or no randomization at all (e.g., observational studies). Let $\mathbf{x}_i$ be a vector of covariates upon which treatment assignments are based. Then the missing mechanism for the unobserved outcome no longer follows MCAR, but rather an MAR as defined by

$$(y_{i,1}, y_{i,2}) \perp z_i \mid \mathbf{x}_i. \tag{11.30}$$

Although nonrandom unconditionally, the assignment is random given the covariates $\mathbf{x}_i$; thus

$$E(y_{i,1} \mid \mathbf{x}_i) - E(y_{i,2} \mid \mathbf{x}_i) = E(y_{i,1} \mid z_i = 1, \mathbf{x}_i) - E(y_{i,2} \mid z_i = 0, \mathbf{x}_i).$$

Hence, we can apply any of the missing data approaches discussed in Section 11.3 to estimate the effect of treatment. For example, the IPW estimate of $E(y_{i,1} - y_{i,2})$ is a weighted average of the treatment effects over the different values of $\mathbf{x}_i$, with the weights defined by the distribution of $\mathbf{x}_i$.

Within the context of causal inference, the MAR condition in (11.30) is known as the *strongly ignorable treatment assignment* assumption (Rosenbaum and Rubin, 1983). Although treatment assignments for the whole study subjects do not follow simple randomization, the ones within each of the strata defined by the distinct values of the distribution of $\mathbf{x}_i$ do. Thus, $E(y_{i,1} \mid \mathbf{x}_i)$ and $E(y_{i,2} \mid \mathbf{x}_i)$ can be estimated within each group using the corresponding sample means. The overall treatment effect can then be estimated by averaging these subgroup means weighted by the group sizes. The approach may not result in reliable estimates or simply does not work, if some groups have a small or even 0 number of subjects for one or both treatment conditions. This can occur if the overall sample size is relative small, and/or the number of distinct values of $\mathbf{x}_i$ is large such as when $\mathbf{x}_i$ contains continuous components. The propensity score partition discussed below helps facilitate the creation of strata in such situations.

Let $\pi(\mathbf{x}_i) = \Pr(z_i = 1 \mid \mathbf{x}_i)$. It follows from the condition in (11.30) that

$$E(y_{i,1}) = E\left[\frac{z_i y_{i,1}}{\pi(\mathbf{x}_i)}\right] \text{ and } E(y_{i,2}) = E\left[\frac{(1 - z_i)y_{i,2}}{1 - \pi(\mathbf{x}_i)}\right]. \tag{11.31}$$

Hence, IPW may be applied to correct selection bias. The probability of treatment assignment conditional on the observed covariate $\mathbf{x}_i$, $\pi(\mathbf{x}_i)$, is called the *propensity score* (Rosenbaum and Rubin, 1983). Conditioning on any given score, the counterfactual outcomes are independent of the treatment assignment, i.e., for any $e \in (0, 1)$,

$$E(y_{i,k} \mid z_i = 0, \pi_i = e) = E(y_{i,k} \mid z_i = 1, \pi_i = e) = E(y_{i,k} \mid \pi_i = e), \quad k = 1, 2. \tag{11.32}$$

This follows directly from (11.30), using the iterated conditional expectation argument (see Rosenbaum and Rubin (1983, 1984) and Rosenbaum (2002)). The propensity score is usually unknown and typically estimated using a logistic regression model or discriminant analysis.

The propensity score π_i is a scalar-valued function regardless of the dimension and type (e.g., continuous or categorical) of $\mathbf{x}_i$. We may partition the range of the estimated propensity score based on the observed sample to create five to ten groups of comparable size, akin to the Hosmer–Lemeshow goodness-of-fit test. For each group, the treatment effect is estimated by the corresponding sample means, and again the overall treatment effect can then be estimated by averaging these group means weighted by the group sizes. Simulation studies show that such a partition seems to be sufficient to remove 90% of the bias (Rosenbaum and Rubin, 1984).

As a method based on post-hoc adjustment, propensity score inherits the same limitations as other retrospective designs such as case-control studies and thus can only balance observed covariates, not unobserved confounders. Despite this fundamental limitation, we may be able to minimize selection bias in practice by attempting to collect and include in the model for propensity scores all potential confounders. As noted earlier, there is simply no other alternative when studying certain causal relationships such as smoking on health.

Example 11.10

By treating the two groups with known and missing disease status as two treatment groups, we may also apply the propensity score stratification method to Example 11.9. By dividing the subjects into ten groups of about equal size according to the propensity scores π_i, we obtain an estimate of sensitivity 0.813, with a standard error 0.048, if the model in (11.28) is used to estimate the propensity score. The sensitivity estimate (standard error) changes to 0.838 (0.033) if propensity scores are estimated based on (11.29). Since models for missing mechanisms are only used to divide subjects into subgroups, propensity-score-based methods seem to provide more robust estimates, as compared to other approaches that rely on estimates of missing-value models such as IPW (see He and McDermott (2012)). ⬚

11.4.3 Longitudinal data with missing values

Missing values are a common issue for longitudinal studies. Under ignorable missing mechanisms, i.e., missing data that follow MCAR and MAR, generalized mixed-effect models discussed in Chapter 9 may be applied. However, as noted in Section 11.3.1, such parametric models yield consistent estimates under MAR only when the distribution assumptions are met. Although the generalized estimating equation (GEE) provides robust estimates without imposing any constraint on the distribution of the response, this semiparametric alternative is less robust when it comes to dealing with missing data, as it only guarantees consistent estimates under MCAR. By applying the principle of IPW, we can develop weighted GEE (WGEE) to provide valid inference under MAR.

Consider a longitudinal study with n subjects and m assessments. Let y_{it} be a response and $\mathbf{x}_{it}$ a vector of independent variables of interest as in Chapter 9. Consider the distribution-free regression model in (9.22), which for convenience is copied below:

$$E\left(y_{it} \mid \mathbf{x}_{it}\right) = \mu_{it}, \quad g\left(\mu_{it}\right) = \mathbf{x}_{it}^{\top}\boldsymbol{\beta}, \quad 1 \le t \le m, \quad 1 \le i \le n, \tag{11.33}$$

where $g\left(\cdot\right)$ is some known link function and $\boldsymbol{\beta}$ is a $p \times 1$ vector. The consistency of the GEE estimate is ensured by the unbiasedness of a set of estimating equations as discussed in Chapter 9. In the presence of missing y_{it}, the EEs no longer remain unbiased, except when the missing value follows MCAR (see Problem 11.4).

11.4.3.1 Weighted generalized estimating equations

Let r_{it} denote a binary indicator with the value 1 if y_{it} is observed and 0 otherwise. We assume no missing value at baseline $t = 1$ so that $r_{i1} \equiv 1$ $(1 \le i \le n)$. Let

$$\pi_{it} = \Pr\left(r_{it} = 1 \mid \mathbf{x}_i, \mathbf{y}_i\right), \quad \Delta_{it} = \frac{r_{it}}{\pi_{it}}, \quad \Delta_i = \text{diag}\left(\Delta_{i1}, ..., \Delta_{im}\right), \tag{11.34}$$

where $\text{diag}(\Delta_{i1}, ..., \Delta_{im})$ denotes an $m \times m$ diagonal matrix with Δ_{it} on the tth diagonal. Assuming π_{it} are known and $\pi_{it} \ge c > 0$ $(1 \le i \le n, 2 \le t \le m)$, the IPW-based WGEE for the class of models in (11.33) have the form

$$\mathbf{w}_n\left(\boldsymbol{\beta}\right) = \sum_{i=1}^{n} G_i\left(\mathbf{x}_i\right) \Delta_i S_i = \sum_{i=1}^{n} G_i\left(\mathbf{x}_i\right) \Delta_i \left(\mathbf{y}_i - \mathbf{h}_i\right) = \mathbf{0}, \tag{11.35}$$

where $G_i\left(\mathbf{x}_i\right)$, S_i, y_i, and h_i are all defined the same way as in (9.25) of Chapter 9. It is readily checked that $E\left(\mathbf{w}_n\left(\boldsymbol{\beta}\right)\right) = \mathbf{0}$, and hence the EEs in (11.35) are unbiased, yielding consistent estimates.

As noted in Section 11.3.3, we assume $\pi_{it} \ge c > 0$ so that all subgroups with missing values are represented in the sample observed and there are no extremely large weight to make the inference unstable. Also, G_i may depend on some parameter vector $\boldsymbol{\alpha}$, and for most applications, $G_i\left(\mathbf{x}_i\right) = \left(\frac{\partial}{\partial\boldsymbol{\beta}}\Delta_i\mathbf{h}_i\right) V_i^{-1} = D_i\Delta_i V_i^{-1}$, where D_i and V_i are defined the same way as for the unweighted GEE in (9.26).

As in the case of unweighted GEE discussed in Chapter 9, the consistency of the estimate from (11.35) is independent of the type of estimates of $\boldsymbol{\alpha}$ used, but the asymptotic normality of the estimate $\widehat{\boldsymbol{\beta}}$ is ensured when $\boldsymbol{\alpha}$ is substituted by some $\sqrt{n}$-consistent estimate. Specifically, if $\widehat{\boldsymbol{\alpha}}$ is $\sqrt{n}$-consistent, then the estimate $\widehat{\boldsymbol{\beta}}$ obtained by solving the WGEE (11.35) is asymptotically normal with the asymptotic variance Σ_β given by (see Kowalski and Tu (2008)):

$$\Sigma_\beta = B^{-1} E\left(G_i \Delta_i S_i S_i^\top \Delta_i G_i^\top\right) B^{-\top}, \quad B = E\left(G_i \Delta_i D_i^\top\right). \qquad (11.36)$$

A consistent estimate of Σ_β can be obtained using sample moment substituted by $\sqrt{n}$-consistent estimates of β and α.

The probability π_{it} is generally unknown. To model and estimate this weight function, let $\widetilde{\mathbf{x}}_{it} = (\mathbf{x}_{i1}^\top, \ldots, \mathbf{x}_{i(t-1)}^\top)^\top$ and $\widetilde{\mathbf{y}}_{it} = (y_{i1}, \ldots, y_{i(t-1)})^\top$ $(2 \leq t \leq m)$, denoting the vectors containing the explanatory and response variables prior to time t, respectively. Then $H_{it} = \{\widetilde{\mathbf{x}}_{it}, \widetilde{\mathbf{y}}_{it}; 2 \leq t \leq m\}$ represents the observed data prior to time t $(2 \leq t \leq m, 1 \leq i \leq n)$. Under MAR, we have

$$\pi_{it} = \Pr\left(r_{it} = 1 \mid \mathbf{x}_i, \mathbf{y}_i\right) = \Pr\left(r_{it} = 1 \mid H_{it}\right). \qquad (11.37)$$

Under the MAR assumption for MMDP discussed in Section 11.2.2, y_{it} is observed only if all y_{is} prior to time t are observed, allowing us to model the one-step transition probability of the occurrence of missing data and then compute π_{it} as a function of such transition probabilities.

Let $p_{it} = E\left(r_{it} = 1 \mid r_{i(t-1)} = 1, H_{it}\right)$ denote the one-step transition probability from observing the response at $t-1$ to t $(2 \leq t \leq m)$. We can readily model p_{it} using models for binary responses such as a logistic regression below:

$$\begin{aligned} \text{logit } (p_{it}) &= \text{logit} \left[E\left(r_{it} = 1 \mid r_{i(t-1)} = 1, H_{it}\right)\right] \qquad (11.38) \\ &= g_t\left(\gamma_t, \widetilde{\mathbf{x}}_{it}, \widetilde{\mathbf{y}}_{it}\right), \quad 2 \leq t \leq m, \quad 1 \leq i \leq n, \end{aligned}$$

where $g_t\left(\gamma_t, \widetilde{\mathbf{x}}_{it}, \widetilde{\mathbf{y}}_{it}\right)$ is some function of $(\gamma_t, \widetilde{\mathbf{x}}_{it}, \widetilde{\mathbf{y}}_{it})$. To transition from p_{it} to π_{it}, we use the following identity:

$$\pi_{it}(\gamma) = p_{it} \Pr\left(r_{i(t-1)} = 1 \mid H_{i(t-1)}\right) = \prod_{s=2}^{t} p_{is}(\gamma_s), \qquad (11.39)$$

where $\gamma = \left(\gamma_1^\top, \ldots, \gamma_m^\top\right)^\top$. Thus, we can estimate π_{it} from the modeled p_{it} in (11.38) using the above relationship.

By substituting estimates of π_{it} into (11.35), we obtain consistent estimates of the parameters of interest β. As discussed in Section 11.3.2, the asymptotic variance obtained from (11.35) generally underestimates the sampling variability of the estimate of β as the variability in the estimated version of π_{it} is ignored. Alternatively, we may combine (11.35) with the estimating (or score) equations from estimating γ for simultaneous inference about both β and γ.

Example 11.11

We now refit the model in Example 9.8 by adding the subjects with missing visits to the subset of those with complete data on the first four visits in the DOS. We use age, gender, education, medical burden (CIRS), HAMD, and depression diagnosis (Depd) at the previous visit to predict the dropout at the current visit, i.e., for a subject observed at visit $j-1$, the probability of this same subject at the next visit j is modeled by

$$\begin{aligned} \text{logit}(p_{ij}) &= \beta_0^j + \beta_1^j Age_i + \beta_2^j Gender_i + \beta_3^j Education + \beta_4^j CIRS_{ij-1} \qquad (11.40) \\ &+ \beta_5^j HAMD_{ij-1} + \beta_6^j Depd_{ij-1}, 2 \leq j \leq 4, \quad 1 \leq i \leq n. \end{aligned}$$

By (11.39), the probability of the ith subject being observed at visit k is given by $\pi_{ik} = \Pi_{j=2}^{k} p_{ij}$ $(2 \leq k \leq 4)$. Note that a small fraction of subjects also have intermittent missing values (visits) prior to dropout. For these subjects, we dropped all data following the first missing visit to obtain a data set in full compliance with the MMDP pattern.

Shown in the following table are the estimates of the same regression coefficients for the model in Example 9.8, but obtained with a WGEE based on weighting each subject in the data according to the model in (11.40).

Parameter	Estimate	SE	95% CI		Z	Pr > | Z |
Intercept	1.4674	2.1101	−2.6682	5.6031	0.70	0.4868
Age	−0.0511	0.0250	−0.1001	−0.0021	−2.04	0.0410
Gender	0.4424	0.2909	−0.1277	1.0124	1.52	0.1283
Education	−0.0912	0.0600	−0.2087	0.0263	−1.52	0.1282
CIRS	0.2001	0.0370	0.1277	0.2726	5.41	< .0001
MS1	−0.4333	0.3661	−1.1509	0.2842	−1.18	0.2366
MS2	0.0390	0.4325	−0.8086	0.8866	0.09	0.9282
Race	1.1286	1.4483	−1.7100	3.9672	0.78	0.4358

Comparing with the earlier GEE analysis based on the subset of completely observed subjects, the conclusions are the same. However, the p-values are smaller here, indicating improved power by including all available data. ⬜

11.4.3.2 Multiple imputation

If we can use covariates and observed responses to predict missing outcomes, we may also apply multiple imputation (MI) to address missing values. We illustrate an application of MI within the context of GEE to correct the bias of GEE estimates under MAR.

Example 11.12
In Example 9.7, we assessed the effect of a HIV prevention intervention on HIV knowledge using only the subset of subjects with complete data at both baseline and three months posttreatment. We can also apply GEE to all subjects, including those who have only baseline data. Shown in Table 11.3 are the GEE estimates of β_3 for the same regression model in (9.27) using only the subset with complete data (GEE(C)) and using all available data ignoring missing values (GEE).

TABLE 11.3: GEE, WGEE, and MI-GEE estimates of β_3

Method	Estimate	Standard Error	95% Confidence Limits		Pr > |Z|
GEE(C)	12.7519	1.5193	9.7740	15.7297	<.0001
GEE	12.9438	1.5365	9.9323	15.9553	<.0001
WGEE	12.6380	1.4984	9.7012	15.5748	<.0001
MI-GEE	12.6416	1.4293	9.8402	15.4429	<.0001

In general, both approaches are valid only when missing values at posttreatment follow MCAR. Since there are considerable difference between these two GEE estimates, this may not be the case for the example. We now apply WGEE and MI to correct the bias in the GEE estimates.

Let y_{it} denote the knowledge outcome from the ith subject at time t, with $t = 1$ (2) indicating baseline (posttreatment) and r_i is the indicator of whether y_{i2} is observed. Our WGEE is based on the following MAR assumption:

$$r_i \perp y_{i2} \mid Age_i, Treatment\,(z_i)\,, y_{i1}, CESD_{i1}. \tag{11.41}$$

More precisely, we assume the following model for the missing mechanism at posttreatment:

$$\text{logit}(\Pr(y_{i2} \text{ observed})) = \alpha_0 + \alpha_1 Age_i + \alpha_2 z_i + \alpha_3 y_{i1} + \alpha_4 CESD_{i1}. \tag{11.42}$$

Note that for model (11.42), there is no missing value issue as all the covariates and the responses are observed. The model for the missing mechanism (11.43) may be refined depending on input from experts experienced with HIV prevention intervention research, especially with the particular study population of the study. Model selection procedures discussed in Chapter 7 may be applied to compare competing models. The WGEE estimates based on the above weight model are also shown in Table 11.3.

Alternatively, we may apply MI to correct the bias in the GEE estimate. Suppose we have the following model for imputing missing y_{i2}:

$$y_{i2} \sim \gamma_0 + \gamma_1 Age_i + \gamma_2 z_i + \gamma_3 y_{i1} + \gamma_4 CESD_{i1}. \tag{11.43}$$

Note that there are missing values in the response variable in (11.43). However, the covariates in (11.43) are always observed. Under the MAR assumption (11.41), the missingness of y_{i2} actually follows MCAR for (11.43) in the sense of regression analysis as discussed in Section 11.2.2. Thus, the model (11.43) can be readily estimated based on the subset of subjects with observed y_{i2}. However, because of the missing values, it will be more difficult to assess the prediction model (11.43). Thus, input from experts with experience in the subject field is especially important.

We can use (11.43) to impute the missing y_{i2}, apply GEE to the imputed completed data set, repeat the process multiple times, and combine the GEE estimates to obtain valid inference. Also shown in Table 11.3 are the results from MI-GEE based on ten multiply imputed data sets. Note that different MI-GEE estimates will result if the procedure is repeated because of the randomness of the imputed values.

Without external information about the missing mechanism and the prediction model, it would be difficult to compare the estimates based on the data set itself. However, the difference between the two GEE estimates ignoring missing values (GEE(C) and GEE) indicate that the missing knowledge outcome at posttreatment does not follow MCAR. This is confirmed by the estimate of the missing mechanism; age is significantly associated with the missingness of posttreatment knowledge (p-value 0.0317). Thus, the MCAR assumption is unlikely to be true, and these two GEE approaches ignoring missing values may be biased.   □

Beunckens et al. (2008) compared MI-based GEE (MI-GEE) with WGEE through simulated data and found that MI-GEE is less biased and more precise in small and moderate samples, thereby suggesting that MI-GEE would be preferable over WGEE in practice. Of course, the performance of MI-GEE depends on the quality of imputing (prediction) models. Likewise, the performance of WGEE is defined by the quality of models for the missing mechanism. Ultimately, it is our degree of confidence in the model for the missing mechanism that determines which approach to use for a particular application.

Note that MI-GEE may not be appropriate when missing values are the result of subject dropout, as it often results in missing values in all time-varying covariates in addition to the response. On the other hand, MI-GEE can be applied to deal with intermittent missing data, although it often requires very strong assumptions such as multivariate normality. In addition, MI-GEE is easier to implement and available in common statistical packages.

11.4.4   Survey studies

Analysis of survey study data requires some special attention because of their complex sampling designs employed to obtain a sample representative of the population of interest.

To ensure that certain groups of interest are well represented in the sample, it is necessary to sample subjects in multiple stages. As a result, the subjects sampled are generally clustered with varying selection probabilities, violating the usual i.i.d. assumption upon which most standard statistical methods are developed. IPW can again be used to address the special sampling features underlying such data.

11.4.4.1 Survey sampling design

Stratified sampling is probably the most popular multi-stage procedure to collect data from a targeted population. In this "top-down" approach, nested sampling is performed, starting with sampling clusters, or *Primary Sampling Units* (PSUs), such as counties and houses, and ending at the bottom level by sampling the individual subjects. By carefully selecting stratification criteria, we can obtain a sample representing well all subpopulations of interest across the strata and homogeneous within each stratum.

For example, to sample people in the United States, we may start by sampling the states, constituting the PSUs. Within each state sampled, we may sample counties, followed by households from within each sampled county and individual persons from within each family sampled. At each stage, we may oversample some specific subgroups such as minorities so that all different groups of interest are well represented to permit reliable inference.

The Behavioral Risk Factor Surveillance (BRFSS) is an annual survey about behavioral risks to health among adults in the United States (Centers for Disease Control and Prevention (CDC), 2010). The survey is stratified by states, a natural choice given that the state health departments are the ones to collect data. Within each state, telephone numbers are sampled through random-digit dialing. The PSUs here are the blocks of numbers with the same first eight digits. Most states sample phone numbers according to the presumed density of known telephone household numbers, with numbers in dense strata sampled at a higher rate. Next, one adult is randomly sampled from each selected household.

11.4.4.2 Sampling weights

For a subject to be selected in a multi-stage sampling study, the unit to which the subject belongs at each level must be sampled. Thus, the overall sampling probability of the subject is the product of the sampling probabilities of all such units. Most survey studies report *sampling weight*, the inverse of the sampling probability for each subject sampled. As discussed in Section 11.3.3, the sampling weight of a subject represents the size of the subgroup of subjects expressed by the subject sampled.

Nonresponse may occur at any of the levels of multi-stage sampling. For example, a common type of nonresponse is "out of scope," which occurs if the house sampled is unoccupied, such as a vacation home, or the subject selected is unreachable, such as those in the military. Even if the house selected is occupied, the person contacted may simply turn down the request for participation or provide incomplete information. We must redistribute the weights from such nonresponses over their sampled counterparts to preserve the total sampling weights.

Discrepancy between sampling weights and population size may also occur due to inclusion/exclusion and other study-related criteria. For example, adults without a household number in the BRFSS survey were excluded by design. Thus, the total sampling weight is generally not equal to the size of the targeted population, and *post-stratification adjustment* is necessary to correct such deviations. For example, in BRFSS, adjustments were made to force the sum of the weighted frequencies to equal the estimated state population, with the final sampling weights FINALWT given by

$$\text{FINALWT} = \text{STRWT} \times \text{1OVERNPH} \times \text{NAD} \times \text{POSTSTRAT},$$

where STRWT is the inverse of the sampling fraction of each phone number, 1OVERNPH is the inverse of the number of residential telephone numbers in the respondent's household, NAD is the number of adults in the respondent's household, and POSTSTRAT is the post-stratification adjustment based on age, sex, and race/ethnicity.

11.4.4.3 Analysis of survey data

By applying IPW, we can infer population-level parameters based on surveyed subjects and associated sampling weights. Consider a target population with N subjects, and let y_i be the value of the outcome of interest from the ith subject sampled with a sampling probability π_i $(i = 1, \ldots, n)$. The IPW, or Horvitz–Thompson, estimates of the population total T and mean μ of y_i are given by

$$T = \sum_{i=1}^{n} \frac{y_i}{\pi_i} \quad \text{and} \quad \widehat{\mu} = \frac{1}{N} \sum_{i=1}^{n} \frac{y_i}{\pi_i}. \tag{11.44}$$

Both estimates are unbiased with respect to their respective parameters. The population size N is assumed known in the above, but can be estimated by $\sum_{i=1}^{n} \frac{1}{\pi_i}$ if unknown. The latter is a special case of T above when $y_i \equiv 1$ $(1 \leq i \leq n)$. The estimate $\widehat{\mu}$ obtained by substituting N with $\sum_{i=1}^{n} \frac{1}{\pi_i}$ may not be unbiased, but remain consistent.

The IPW approach can also be applied to regression analysis. However, if all the variables used to construct the sampling weights are included as covariates, it is not necessary to apply the sampling weights in the regression model, because their effects on estimates of parameters are already accounted for by the covariates. Since weighting typically reduces efficiency, the unweighted version may be preferred in such situations. However, in many survey studies, details about the calculations of sampling weights may not be provided, in which case IPW must be used to address the selection bias from multi-stage sampling.

Inference for survey data is compounded by correlated responses arising from nested sampling in multi-stage designs, destroying the independence among the sampled subjects as required by most statistical methods. For example, if a household is not sampled in BRFSS, then no adult in that household will be selected. The fundamental assumption is violated even under the simplest sampling procedure due to the finite size of the population (see Problem 11.13). Thus, we must consider covariance between y_i and y_j when computing the variance of the estimate in (11.44). For example, the variance of $\widehat{\mu}$ is given by (see also Problem 11.14):

$$\sigma_\mu^2 = Var\left(\frac{1}{N} \sum_{i=1}^{n} \frac{y_i}{\pi_i}\right) = \frac{1}{N^2}\left[E\left(\frac{y_i y_j}{\pi_i \pi_j}\right) - \left(\sum_{i=1}^{n} y_i\right)^2\right]$$

$$= \frac{1}{N^2} \sum_{i,j=1}^{N}\left(\frac{\pi_{ij} y_i y_j}{\pi_i \pi_j} - y_i y_j\right), \tag{11.45}$$

where π_{ij} is the probability that the ith and jth subjects are sampled simultaneously. In the case of simple sampling, $\pi_i = \frac{n}{N}$ and $\pi_{ij} = \frac{n(n-1)}{N(N-1)}$ $(i \neq j)$, and the above variance formula simplifies to $\frac{N-n}{N} \frac{Var(y)}{n}$. The *finite population correction* $\frac{N-n}{N}$ reflects the finite nature of the population, which is not ignorable if a significant proportion of the population is sampled. For example, for census studies, the entire population is sampled, and so all the π's are equal to 1, in which case this correction factor reduces to 0, indicating no variability in the estimate.

Note that we may speak of "variance" of estimate even with census data when performing regression analysis. This is because our primary interest is whether there is essential association between two variables and more variables. In other words, we want to know if such associations occur beyond chance. From this standpoint, the entire population in a survey study is viewed as a (random) sample from a superpopulation of infinite sample size, for which inference is intended. Thus, asymptotic theories may still be applied to facilitate the computation of variance estimates and p-values. Resampling methods may also be used, especially for moderate sample and population sizes. Interested readers may consult books devoted to survey studies for details such as Cochran (2007), Korn and Graubard (1999), and Thompson (2002).

Example 11.13

Consider a question on behavioral risks to health surveyed in the 2010 BRFSS: "Was there a time in the past 12 months when you needed to see a doctor but could not because of cost?" The variable MEDCOST in the data of the survey, downloadable from the CDC website, shows that 14.6% of the Americans adults said "yes" to this question. As the answer should be related with income, a chi-square test was applied to the two-way table formed by this and the income variable income2, yielding a p-value < 0.0001. The result seems to support the hypothesized relationship between the likelihood of seeing a doctor and income. ▯

Exercises

11.1 Suppose for an i.i.d. sample of size n, the disease status, d_i, is MCAR with the probability of each d_i being observed given by $\pi = 0.75$.

(a) Show that $\frac{1}{n} \sum_{d_i \text{ observed}} \frac{d_i}{\pi}$ is a consistent estimate of population prevalence, $\Pr(d_i = 1)$ (IPW estimate with known probabilities).

(b) Show that $\frac{1}{n} \sum_{d_i \text{ observed}} \frac{d_i}{\widehat{\pi}} = \frac{\sum_{d_i \text{ observed}} d_i}{\sum_{d_i \text{ observed}} 1}$ is a consistent estimate of the population prevalence, where $\widehat{\pi} = \frac{\sum_{d_i \text{ observed}} 1}{n}$ is an estimate of π (IPW estimate with estimated probabilities).

(c) Compare the variances of the two estimates in (a) and (b) to confirm that the use of estimated probability π improves efficiency, provided that the model for π is correct.

11.2 For the Sexual Health study, check whether the missingness of three-month posttreatment HIV knowledge is MCAR.

11.3 For Example 11.4, we are interested in the sensitivity and specificity of the test.

(a) Compute the MLEs of sensitivity and specificity and their asymptotic variances based on the likelihood (11.9).

(b) Another way to parametrize the distribution is to use $\Pr(t = 1)$, $\Pr(d = 1 \mid t = 1)$ (PPV), and $\Pr(d = 0 \mid t = 0)$ (NPV). Write down the likelihood and compute the MLEs of sensitivity and specificity and their asymptotic variances.

(c) Compare the estimates in (a) and (b). They should be the same, since MLE does not depend on how the model is parameterized.

11.4 Prove that the estimating equations in (11.13) are unbiased under MCAR, but are generally biased without the stringent MCAR assumption.

11.5 Show that the estimating equations (11.16) are unbiased.

11.6 Prove that the estimating equations (11.20) are unbiased.

11.7 Use mean score, IPW, and MI methods to estimate the sensitivity and specificity of the test in Example 11.1.

11.8 Use the simulated DOS baseline data (with missing values in depression diagnosis) to assess the accuracy of HAM-D in diagnosis of depression.

   (a) Estimate the ROC curve using mean score, IPW, and MI methods.

   (b) Estimate the AUC based on the empirical ROC curve obtained in (a) for each of the mean score, IPW, and MI methods.

   (c) Compare the results obtained using the three different methods in parts (a) and (b).

11.9 Consider the prevalence in Example 11.1. Suppose we have a model for the missing mechanism $\pi_i = \Pr(r_i = 1)$ and one for the disease process $\widetilde{d}_i = \Pr(d_i = 1)$

   (a) Verify that the estimating equation

$$\sum_{i=1}^{n} \left[\frac{r_i}{\pi_i} (d_i - p) + \left(1 - \frac{r_i}{\pi_i} \right) \left(\widetilde{d}_i - p \right) \right] = 0 \qquad (11.46)$$

   is unbiased if the model for π_i is correct, i.e., $\pi_i = \Pr(r_i = 1)$, and hence the estimate of prevalence based on (11.46) is consistent.

   (b) Rewrite the estimating equation in (11.9) as

$$\sum_{i=1}^{n} \left[\left(\widetilde{d}_i - p \right) + \frac{r_i}{\pi_i} \left(d_i - \widetilde{d}_i \right) \right] = 0. \qquad (11.47)$$

   Check that the estimating equation is also unbiased, provided that the disease model is correct, i.e., $\widetilde{d}_i = \Pr(d_i = 1)$. Hence, the estimate of prevalence is consistent if either π_i or d_i are correctly modeled.

11.10 Show that the estimating equations in (11.35) are unbiased.

11.11 Prove (11.32).

11.12 Apply WGEE by changing the dichotomized depression diagnosis in the model for the missing mechanism in Example 11.11 to the original three-level scale and compare the results from the two versions of the depression diagnosis variable.

11.13 For a simple random sample from a population of size N, the subjects are not sampled independently because of the finite size of the population.

   (a) Show that the probability of being sampled for each subject is $\frac{n}{N}$, where n is the number of subjects in the sample.

   (b) If sampled sequentially, then conditioning on the first one being sampled, the probability of sampling each of the remaining subjects is $\frac{n-1}{N-1}$.

   (c) Use (a) and (b) to show that the subjects in the simple random sample are not independent.

11.14 Compute the variance of $\widehat{\mu}$ in (11.45) via the following steps.

(a) Suppose the N subjects of the population are labeled from 1 to N. The ith subject with outcome y_i is sampled with the probability π_i. Show that $\widehat{\mu} = \frac{1}{N} \sum_{i=1}^{N} \frac{r_i y_i}{\pi_i}$, where $r_i = 1$ if the ith subject is sampled and 0 otherwise.

(b) Show that $Cov\left(\frac{r_i y_i}{\pi_i}, \frac{r_j y_j}{\pi_j}\right) = \begin{cases} \frac{y_i^2}{\pi_i^2}\pi_i(1 - \pi_i) & \text{if } i = j, \\ \frac{\pi_{ij} y_i y_j}{\pi_i \pi_j} & \text{if } i \neq j. \end{cases}$

(c) Show that $Var(\widehat{\mu}) = \frac{1}{N^2} \sum_{i,j}^{N} \left(\frac{\pi_{ij} y_i y_j}{\pi_i \pi_j} - y_i y_j\right)$.

11.15 Use the BRFSS 2010 survey to

(a) Estimate the proportion of persons who are "employed for wages";

(b) Check whether employment and income are related.

References

Agresti, A. (2002). *Categorical data analysis* (2nd ed.). New York: Wiley-Interscience.

Agresti, A. (2007). *An introduction to categorical data analysis*. Hoboken, NJ: John Wiley & Sons.

Agresti, A. and B. A. Coull (1998). Approximate is better than exact for interval estimation of binomial proportions. *The American Statistician 52*, 119–126.

Albert, A. and J. Anderson (1984). On the existence of maximum likelihood estimates in logistic regression models. *Biometrika 71*(1), 1–10.

Alonzo, T. A., M. S. Pepe, and T. Lumley (2003). Estimating disease prevalence in two-phase studies. *Biostatistics 4*, 313–326.

Aranda-Ordaz, F. J. (1981). On two families of transformations to additivity for binary response data. *Biometrika 68*(2), 357–363.

Armitage, P. (1955). Tests for linear trends in proportions and frequencies. *Biometrics 11*(3), 375–386.

Austin, P. C. and E. W. Steyerberg (2013). Predictive accuracy of risk factors and markers: A simulation study of the effect of novel markers on different performance measures for logistic regression models. *Statistics in Medicine 32*(4), 661–672.

Bach, P., J. Jett, U. Pastorino, M. Tockman, S. Swensen, and C. Begg (2007). Computed tomography screening and lung cancer outcomes. *JAMA: The Journal of the American Medical Association 297*(9), 953–961.

Barnhart, H. and J. Williamson (1998). Goodness-of-fit tests for GEE modeling with binary responses. *Biometrics 54*(2), 720–729.

Belsley, D. A., E. Kuh, and R. Welsch (1980). *Regression Diagnostics*: Identifying influential data and sources of collinearity. Hoboken, NJ: John Wiley & Sons.

Beunckens, C., C. Sotto, and G. Molenberghs (2008). A simulation study comparing weighted estimating equations with multiple imputation based estimating equations for longitudinal binary data. *Computational Statistics and Data Analysis 52*(3), 1533–1548.

Birch, M. (1963). Maximum likelihood in three-way contingency tables. *Journal of the Royal Statistical Society. Series B (Methodological) 25*(1), 220–233.

Birch, M. (1964). The detection of partial association, I: The 2×2 case. *Journal of the Royal Statistical Society, Series B 26*(3), 13–324.

Bliese, P. D. and R. R. Halverson (1996). Individual and nomothetic models of job stress: An examination of work hours, cohesion, and well-being 1. *Journal of Applied Social Psychology 26*(13), 1171–1189.

Bliss, C. (1935). The calculation of the dosage-mortality curve. *Annals of Applied Biology 22*(1), 134–167.

Bollen, K. A. and R. W. Jackman. Regression diagnostics: An expository treatment of outliers and influential cases. In J. Fox and J. S. Long (Eds.), *Modern methods of data analysis*. Newbury Park, CA: SAGE.

Bowker, A. H. (1948). A test for symmetry in contingency tables. *Journal of the American Statistical Association 43*(244), 572–574.

Box, G., G. Jenkins, and G. Reinsel (2008). *Time series analysis: Forecasting and control* (4th ed.). Hoboken, NJ: John Wiley & Sons.

Bradley, R. A. (1954). Rank analysis of incomplete block designs. II. Additional tables for the method of paired comparisons. *Biometrika 41*, 502–537.

Bradley, R. A. (1955). Rank analysis of incomplete block designs. III. Some large-sample results on estimation and power for a method of paired comparisons. *Biometrika 42*, 450–470.

Bradley, R. A. and M. E. Terry (1952). Rank analysis of incomplete block designs. I. The method of paired comparisons. *Biometrika 39*, 324–345.

Brant, R. (1990). Assessing proportionality in the proportional odds model for ordinal logistic regression. *Biometrics*, Vol. 46, 1171–1178.

Breslow, N., N. Day, and J. Schlesselman (1982). Statistical methods in cancer research. Volume 1—The analysis of case-control studies. *Journal of Occupational and Environmental Medicine 24*(4), 255.

Brown, A. D., T. T. Cai, and A. Dasgupta (2001). Interval estimation for a binomial proportion. *Statistical Science 16(2)*, 101–133.

Brown, M. and J. Benedetti (1977). Sampling behavior of test for correlation in two-way contingency tables. *Journal of the American Statistical Association 72*(358), 309–315.

Burnham, K. and D. Anderson (2002). *Model selection and multimodel inference: A practical information-theoretic approach*. New York: Springer Verlag.

Carey, V., S. L. Zeger, and P. Diggle (1993). Modelling multivariate binary data with alternating logistic regressions. *Biometrika 80*(3), 517–526.

Carmeron, A. and P. Trivedi (1986). Econometric models based on count data: Comparisons and application of some estimators. *Journal of Applied Econometrics 46*, 29–53.

Casella, G., R. Berger, and R. Berger (2002). *Statistical inference*. Pacific Grove, CA: Duxbury Press.

Caserta, M. T., T. G. O'Connor, P. A. Wyman, H. Wang, J. Moynihan, W. Cross, X. Tu, and X. Jin (2008). The associations between psychosocial stress and the frequency of illness, and innate and adaptive immune function in children. *Brain, Behavior, and Immunity 22*(6), 933–940.

Centers for Disease Control and Prevention (CDC) (2010). *Behavioral Risk Factor Surveillance System Survey Data*. U.S. Department of Health and Human Services, Centers for Disease Control and Prevention.

Chaudron, L., P. Szilagyi, W. Tang, E. Anson, N. Talbot, H. Wadkins, X. Tu, and K. Wisner (2010). Accuracy of depression screening tools for identifying postpartum depression among urban mothers. *Pediatrics 125*(3), e609 –e617.

Chen, H. and R. Little (1999). A test of missing completely at random for generalised estimating equations with missing data. *Biometrika 86*(1), 1–13.

Chen, S., Y. Conwell, J. He, N. Lu, and J. Wu (2015). Depression care management for adults older than 60 years in primary care clinics in urban china: A cluster-randomised trial. *The Lancet Psychiatry 2*(4), 332–339.

Cheng, S. and J. S. Long (2007). Testing for IIA in the multinomial logit model. *Sociological Methods & Research 35*(4), 583–600.

Chernoff, H. and E. L. Lehmann (1954). The use of maximum likelihood estimates in χ^2 tests for goodness of fit. *The Annals of Mathematical Statistics 25*, 579–586.

Clopper, C. and E. S. Pearson (1934). The use of confidence or fiducial limits illustrated in the case of the binomial. *Biometrika 26*, 404–413.

Cochran, W. (2007). *Sampling techniques*. New Delhi: Wiley-India.

Cochran, W. G. (1954). Some methods for strengthening the common χ^2 tests. *Biometrics 10*, 417–451.

Cohen, J. (1960). A coefficient of agreement for nominal scales. *Educational and Psychological Measurement 20*(1), 37–46.

COMBINE Study Research Group (2006). Combined pharmacotherapies and behavioral interventions for alcohol dependence – The COMBINE study: A randomized controlled trial. *JAMA: The Journal of the American Medical Association 295*(17), 2003–2017.

Consul, P. and M. M. Shoukri (1985). The generalized poisson distribution when the sample mean is larger than the sample variance. *Communications in Statistics-Simulation and Computation 14*(3), 667–681.

Conway, R. W. and W. L. Maxwell (1962). A queuing model with state dependent service rates. *Journal of Industrial Engineering 12*(2), 132–136.

Cook, N. R. (2007). Use and misuse of the receiver operating characteristic curve in risk prediction. *Circulation 115*(7), 928–935.

Copas, J. (1989). Unweighted sum of squares test for proportions. *Journal of the Royal Statistical Society: Series C (Applied Statistics) 38*(1), 71–80.

Cox, D. (1972). Regression models and life-tables. *Journal of the Royal Statistical Society. Series B (Methodological)*, Vol. 34, 187–220.

Cox, D. and D. Oakes (1984). *Analysis of survival data*. London: Chapman and Hall/CRC.

Cox, D. R. (1958). Two further applications of a model for binary regression. *Biometrika 45*(3/4), 562–565.

Cramér, H. (1946). *Mathematical methods of statistics*. Princeton Mathematical Series, vol. 9. Princeton, NJ: Princeton University Press.

Cui, X. J., J. M. Lyness, W. Tang, X. Tu, and Y. Conwell (2008). Outcomes and predictors of late-life depression trajectories in older primary care patients. *The American Journal of Geriatric Psychiatry 16*(5), 406–415.

D'Agostino, R., J. Griffith, C. Schmid, and N. Terrin (1997). Measures for evaluating model performance. In *Proceedings-American Statistical Association Biometrics Section*, pp. 253–258.

Daniels, M. and J. Hogan (2008). *Missing data in longitudinal studies: Strategies for Bayesian modeling and sensitivity analysis.* Boca Raton, FL: Chapman and Hall.

Dean, C. and J. Lawless (1989). Tests for detecting overdispersion in Poisson regression models. *Journal of the American Statistical Association 84*(406), 467–472.

Delattre, M., M. Lavielle, and M.-A. Poursat (2014). A note on BIC in mixed-effects models. *Electronic Journal of Statistics 8*(1), 456–475.

DeLong, E., D. DeLong, and D. Clarke-Pearson (1988). Comparing the areas under two or more correlated receiver operating characteristic curves: A nonparametric approach. *Biometrics 44*(3), 837–845.

Demler, O. V., M. J. Pencina, N. R. Cook, and R. B. D'Agostino Sr (2017). Asymptotic distribution of δ auc, nris, and idi based on theory of u-statistics. *Statistics in Medicine 36*(21), 3334–3360.

Demler, O. V., M. J. Pencina, and R. B. D'Agostino Sr (2012). Misuse of delong test to compare aucs for nested models. *Statistics in Medicine 31*(23), 2577–2587.

Deng, D. and S. R. Paul (2000). Score tests for zero inflation in generalized linear models. *Canadian Journal of Statistics 28*(3), 563–570.

Diggle, P. J., P. J. Heagerty, K.-Y. Liang, and S. L. Zeger (2002). *Analysis of longitudinal data* (2nd ed.), Volume 25 of *Oxford Statistical Science Series.* Oxford: Oxford University Press.

Dorfman, D. and E. Alf (1969). Maximum-likelihood estimation of parameters of signal-detection theory and determination of confidence intervals–Rating-method data. *Journal of Mathematical Psychology 6*(3), 487–496.

Duberstein, P., Y. Ma, B. Chapman, Y. Conwell, J. McGriff, J. Coyne, N. Franus, M. Heisel, K. Kaukeinen, S. Sörensen, X. Tu, and J. Lyness (2011). Detection of depression in older adults by family and friends: Distinguishing mood disorder signals from the noise of personality and everyday life. *International Psychogeriatrics 23*(04), 634–643.

Edwards, D. (2000). *Introduction to graphical modelling.* New York: Springer Verlag.

Edwards, D. and S. Kreiner (1983). The analysis of contingency tables by graphical models. *Biometrika 70*(3), 553–565.

Efron, B. and D. V. Hinkley (1978). Assessing the accuracy of the maximum likelihood estimator: Observed versus expected fisher information. *Biometrika 65*(3), 457–483.

Fagerland, M. W. and D. W. Hosmer (2013). A goodness-of-fit test for the proportional odds regression model. *Statistics in Medicine 32*(13), 2235–2249.

Fagerland, M. W. and D. W. Hosmer (2016). Tests for goodness of fit in ordinal logistic regression models. *Journal of Statistical Computation and Simulation 86*(17), 3398–3418.

Fagerland, M. W., D. W. Hosmer, and A. M. Bofin (2008). Multinomial goodness-of-fit tests for logistic regression models. *Statistics in Medicine 27*(21), 4238–4253.

Fai, A. H.-T. and P. L. Cornelius (1996). Approximate F-tests of multiple degree of freedom hypotheses in generalized least squares analyses of unbalanced split-plot experiments. *Journal of Statistical Computation and Simulation 54*(4), 363–378.

Famoye, F. and K. P. Singh (2006). Zero-inflated generalized poisson regression model with an application to domestic violence data. *Journal of Data Science 4*(1), 117–130.

Feinstein, A. R. and D. V. Cicchetti (1990). High agreement but low kappa: I. the problems of two paradoxes. *Journal of Clinical Epidemiology 43*(6), 543–549.

Firth, D. (1993). Bias reduction of maximum likelihood estimates. *Biometrika 80*, 27–38.

Fisher, R. (1922). On the mathematical foundations of theoretical statistics. *Philosophical Transactions of the Royal Society of London. Series A, Containing Papers of a Mathematical or Physical Character 222*, 309–368.

Freeman, G. H. and J. H. Halton (1951). Note on an exact treatment of contingency, goodness of fit and other problems of significance. *Biometrika 38*, 141–149.

Frison, L. and S. Pocock (1992). Repeated measures in clinical trials: Analysis using mean summary statistics and its implications for design. *Statistics in Medicine 11*(13), 1685–1704.

Gart, J. (1970). Point and interval estimation of the common odds ratio in the combination of 2 *times* 2 tables with fixed marginals. *Biometrika 57*(3), 471–475.

Gart, J. (1971). The comparison of proportions: A review of significance tests, confidence intervals and adjustments for stratification. *Revue de l'Institut International de Statistique/Review of the International Statistical Institute 39*(2), 148–169.

Gehan, E. (1965). A generalized wilcoxon test for comparing arbitrarily singly-censored samples. *Biometrika 52*(1–2), 203–223.

Gibbons, J. D. and S. Chakraborti (2003). *Nonparametric statistical inference* (4th ed.), Volume 168 of *Statistics: Textbooks and Monographs*. New York: Marcel Dekker.

Goodman, L. and W. Kruskal (1954). Measures of association for cross classifications. *Journal of the American Statistical Association 49*, 732–764.

Goodman, L. A. (1979). Simple models for the analysis of association in cross-classifications having ordered categories. *Journal of the American Statistical Association 74*(367), 537–552.

Goodman, L. A. (1985). The analysis of cross-classified data having ordered and/or unordered categories: Association models, correlation models, and asymmetry models for contingency tables with or without missing entries. *The Annals of Statistics 13*, 10–69.

Green, D. and J. Swets (1966). *Signal detection theory and psychophysics*. New York: Wiley.

Guikema, S. D. and J. P. Goffelt (2008). A flexible count data regression model for risk analysis. *Risk Analysis: An International Journal 28*(1), 213–223.

Hall, D. and Z. Zhang (2004). Marginal models for zero inflated clustered data. *Statistical Modelling 4*(3), 161–180.

Hand, D. J. and C. C. Taylor (1987). *Multivariate analysis of variance and repeated measures: A practical approach for behavioural scientists*. London, UK: Chapman & Hall.

Haneuse, S. (2016). Distinguishing selection bias and confounding bias in comparative effectiveness research. *Medical Care 54*(4), e23.

Harris, M. N. and X. Zhao (2007). A zero-inflated ordered probit model, with an application to modelling tobacco consumption. *Journal of Econometrics 141*(2), 1073–1099.

He, H., J. M. Lyness, and M. P. McDermott (2009). Direct estimation of the area under the receiver operating characteristic curve in the presence of verification bias. *Statistics in Medicine 28*(3), 361–376.

He, H. and M. McDermott (2012). A robust method for correcting verification bias for binary tests. *Biostatistics 13*(1), 32–47.

He, H., H. Zhang, P. Ye, and W. Tang (2019). A test of inflated zeros for poisson regression models. *Statistical Methods in Medical Research 28*(4), 1157–1169.

Heagerty, P. J. and S. L. Zeger (1996). Marginal regression models for clustered ordinal measurements. *Journal of the American Statistical Association 91*(435), 1024–1036.

Henderson, C. R. (1950). Estimation of genetic parameters. *Biometrics 6*(2), 186–187.

Henderson, C. R. (1975). Best linear unbiased estimation and prediction under a selection model. *Biometrics 31*, 423–447.

Hilbe, J. M. (2011). *Negative binomial regression*. Cambridge, UK: Cambridge University Press.

Hirji, K. (1992). Computing exact distributions for polytomous response data. *Journal of the American Statistical Association 87*(418), 487–492.

Hirji, K., C. Mehta, and N. Patel (1987). Computing distributions for exact logistic regression. *Journal of the American Statistical Association 82*(400), 1110–1117.

Hirji, K., C. Mehta, and N. Patel (1988). Exact inference for matched case-control studies. *Biometrics 44*(3), 803–814.

Hirji, K., A. Tsiatis, and C. Mehta (1989). Median unbiased estimation for binary data. *The American Statistician 43*(1), 7–11.

Holford, T., C. White, and J. Kelsey (1978). Multivariate analysis for matched case-control studies. *American Journal of Epidemiology 107*(3), 245–256.

Horton, N., J. Bebchuk, C. Jones, S. Lipsitz, P. Catalano, G. Zahner, and G. Fitzmaurice (1999). Goodness-of-fit for GEE: An example with mental health service utilization. *Statistics in Medicine 18*(2), 213–222.

Horvitz, D. G. and D. J. Thompson (1952). A generalization of sampling without replacement from a finite universe. *Journal of the American Statistical Association 47*, 663–685.

Hosmer, D. and S. Lemesbow (1980). Goodness of fit tests for the multiple logistic regression model. *Communications in Statistics–Theory and Methods 9*(10), 1043–1069.

Huber, P. J. (1964). Robust estimation of a location parameter. *The Annals of Mathematical Statistics 35*, 73–101.

Hutson, A. D. (2006). Modifying the exact test for a binomial proportion and comparisons with other approaches. *Journal of Applied Statistics 33*(7), 679–690.

Jansakul, N. and J. Hinde (2002). Score tests for zero-inflated poisson models. *Computational Statistics & Data Analysis 40*(1), 75–96.

Joe, H. and R. Zhu (2005). Generalized poisson distribution: The property of mixture of poisson and comparison with negative binomial distribution. *Biometrical Journal: Journal of Mathematical Methods in Biosciences 47*(2), 219–229.

Johnson, N., S. Kotz, and N. Balakrishnan (1994). *Continuous univariate distributions* (2nd ed.), Volume 1. New York: Wiley.

Jonckheere, A. (1954). A distribution-free k-sample test against ordered alternatives. *Biometrika 41*, 133–145.

Kalbfleisch, J. and R. Prentice (2002). *The statistical analysis of failure time data* (2nd ed.). New York: Wiley.

Kenward, M. G. and J. H. Roger (1997). Small sample inference for fixed effects from restricted maximum likelihood. *Biometrics 53*, 983–997.

Kenward, M. G. and J. H. Roger (2009). An improved approximation to the precision of fixed effects from restricted maximum likelihood. *Computational Statistics & Data Analysis 53*(7), 2583–2595.

Konishi, S. and G. Kitagawa (2008). *Information criteria and statistical modeling*. New York: Springer.

Korn, E. and B. Graubard (1999). *Analysis of health surveys*. New York: John Wiley & Sons.

Kowalski, J. and X. M. Tu (2008). *Modern applied U-statistics*. Wiley Series in Probability and Statistics. Hoboken, NJ: Wiley-Interscience.

Kruskal, W. (1952). A nonparametric test for the several sample problem. *The Annals of Mathematical Statistics 23*(4), 525–540.

Kruskal, W. and W. Wallis (1952). Use of ranks in one-criterion variance analysis. *Journal of the American Statistical Association 47*(260), 583–621.

Kuritz, S., J. Landis, and G. Koch (1988). A general overview of Mantel–Haenszel methods: Applications and recent developments. *Annual Review of Public Health 9*(1), 123–160.

Lagakos, S. W., L. M. Barraj, and V. d. Gruttola (1988). Nonparametric analysis of truncated survival data, with application to aids. *Biometrika 75*(3), 515–523.

Lamberti, J., D. Olson, J. Crilly, T. Olivares, G. Williams, X. Tu, W. Tang, K. Wiener, S. Dvorin, and M. Dietz (2006). Prevalence of the metabolic syndrome among patients receiving clozapine. *American Journal of Psychiatry 163*(7), 1273–1276.

Lancaster, H. O. (1969). *The chi-squared distribution*. New York: John Wiley & Sons.

Lauritzen, S. (1996). *Graphical Models (Oxford Statistical Science Series)*. Oxford, England: Claredon Press.

Lawless, J. (2002). *Statistical models and methods for lifetime data* (2nd ed.). New York: Wiley.

Liang, K. and S. Zeger (1986). Longitudinal data analysis using generalized linear models. *Biometrika 73*(1), 13–22.

Lin, D., L. Wei, and Z. Ying (2002). Model-checking techniques based on cumulative residuals. *Biometrics 58*(1), 1–12.

Lin, L. (1989). A concordance correlation coefficient to evaluate reproducibility. *Biometrics 45*(1), 255–268.

Litière, S., A. Alonso, and G. Molenberghs (2008). The impact of a misspecified random-effects distribution on the estimation and the performance of inferential procedures in generalized linear mixed models. *Statistics in Medicine 27*(16), 3125–3144.

Little, R. (1988). A test of missing completely at random for multivariate data with missing values. *Journal of the American Statistical Association 83*, 1198–1202.

Long, J. (1997). *Regression models for categorical and limited dependent variables.* Thousand Oaks: Sage Publications.

Long, S. and J. Freese (2006). *Regression models for categorical dependent variables using stata* (2nd ed.). College Station, TX: Stata Press.

Lu, N., T. Chen, P. Wu, D. Gunzler, H. Zhang, H. He, and X. Tu (2014). Functional response models for intraclass correlation coefficients. *Journal of Applied Statistics 41*(11), 2539–2556.

Lubetkin, E., H. Jia, and M. Gold (2003). Use of the SF-36 in low-income Chinese American primary care patients. *Medical Care 41*(4), 447–457.

Lyness, J. M., J. Kim, W. Tang, X. Tu, Y. Conwell, D. A. King, and E. D. Caine (2007). The clinical significance of subsyndromal depression in older primary care patients. *The American Journal of Geriatric Psychiatry 15*(3), 214–223.

Lyness, J. M., Q. Yu, W. Tang, X. Tu, and Y. Conwell (2009). Risks for depression onset in primary care elderly patients: Potential targets for preventive interventions. *American Journal of Psychiatry 166*(12), 1375–1383.

Mann, H. and D. Whitney (1947). On a test of whether one of two random variables is stochastically larger than the other. *The Annals of Mathematical Statistics 18*(1), 50–60.

Mantel, N. and W. Haenszel (1959). Statistical aspects of the analysis of data from retrospective studies of disease. *Journal of the National Cancer Institute 22*(4), 719–748.

Mantel, N. and W. Hankey (1975). The odds ratio of a 2×2 contingency table. *The American Statistician 29*, 143–145.

Maxwell, A. (1970). Comparing the classification of subjects by two independent judges. *The British Journal of Psychiatry 116*(535), 651–655.

McCullagh, P. (1986). The conditional distribution of goodness-of-fit statistics for discrete data. *Journal of the American Statistical Association 81*(393), 104–107.

McCullagh, P. and J. Nelder (1989). *Generalized linear models.* London: Chapman & Hall/CRC.

McGraw, K. and S. Wong (1996). Forming inferences about some intraclass correlation coefficients. *Psychological Methods 1*(1), 30–46.

McNemar, Q. (1947). Note on the sampling error of the difference between correlated proportions or percentages. *Psychometrika 12*(2), 153–157.

McNutt, L.-A., C. Wu, X. Xue, and J. P. Hafner (2003). Estimating the relative risk in cohort studies and clinical trials of common outcomes. *American Journal of Epidemiology 157*(10), 940–943.

Mehta, C., N. Patel, and R. Gray (1985). Computing an exact confidence interval for the common odds ratio in several 2×2 contingency tables. *Journal of the American Statistical Association 80*(392), 969–973.

Moghimbeigi, A., M. R. Eshraghian, K. Mohammad, and B. Mcardle (2008). Multilevel zero-inflated negative binomial regression modeling for over-dispersed count data with extra zeros. *Journal of Applied Statistics 35*(10), 1193–1202.

Molenberghs, G. and M. Kenward (2007). *Missing data in clinical studies.* Hoboken, NJ: John Wiley & Sons.

Molenberghs, G. and G. Verbeke (2005). *Models for discrete longitudinal data.* New York: Springer Verlag.

Morrison-Beedy, D., M. Carey, H. Crean, and S. Jones (2011). Risk behaviors among adolescent girls in an hiv prevention trial. *Western Journal of Nursing Research 33*(5), 690–711.

Morrison-Beedy, D., M. P. Carey, C. Y. Feng, and X. M. Tu (2008). Predicting sexual risk behaviors among adolescent and young women using a prospective diary method. *Research in Nursing and Health 31*(4), 329–340.

Mrode, R. A. (2014). *Linear models for the prediction of animal breeding values* (3rd ed.). Cabi.

Mühlenbruch, K., A. Heraclides, E. W. Steyerberg, H.-G. Joost, H. Boeing, and M. B. Schulze (2013). Assessing improvement in disease prediction using net reclassification improvement: Impact of risk cut-offs and number of risk categories. *European Journal of Epidemiology 28*(1), 25–33.

Nelder, J. and R. Wedderburn (1972). Generalized linear models. *Journal of the Royal Statistical Society. Series A (General) 135*(3), 370–384.

Neter, J., W. Wasserman, and M. H. Kutner (1990). *Applied linear statistical models: Regression, analysis of variance, and experimental designs.* Homewood, IL: Richard D. Irwin Inc.

Nunnally, J. C. and I. H. Bernstein (1994). *Psychometric theory* (3rd ed.). New York : McGraw-Hill.

Osius, G. and D. Rojek (1992). Normal goodness-of-fit tests for multinomial models with large degrees of freedom. *Journal of the American Statistical Association 87*(420), 1145–1152.

Pan, W. (2001). Akaike's information criterion in generalized estimating equations. *Biometrics 57*(1), 120–125.

Pan, W. (2002). Goodness-of-fit Tests for GEE with Correlated Binary Data. *Scandinavian Journal of Statistics 29*(1), 101–110.

Pan, Z. and D. Lin (2005). Goodness-of-fit methods for generalized linear mixed models. *Biometrics 61*(4), 1000–1009.

Pencina, M. J., R. B. D'Agostino Sr, R. B. D'Agostino Jr, and R. S. Vasan (2008). Evaluating the added predictive ability of a new marker: From area under the roc curve to reclassification and beyond. *Statistics in Medicine 27*(2), 157–172.

Pencina, M. J., R. B. D'Agostino Sr, and E. W. Steyerberg (2011). Extensions of net reclassification improvement calculations to measure usefulness of new biomarkers. *Statistics in Medicine 30*(1), 11–21.

Pepe, M. S., J. Fan, Z. Feng, T. Gerds, and J. Hilden (2015). The net reclassification index (nri): A misleading measure of prediction improvement even with independent test data sets. *Statistics in Biosciences 7*(2), 282–295.

Pepe, M. S., M. Reilly, and T. R. Fleming (1994). Auxiliary outcome data and the mean score method. *Journal of Statistical Planning and Inference 42*, 137–160.

Peterson, B. and F. E. Harrell Jr (1990). Partial proportional odds models for ordinal response variables. *Journal of the Royal Statistical Society: Series C (Applied Statistics) 39*(2), 205–217.

Peyhardi, J., C. Trottier, and Y. Guédon (2015). A new specification of generalized linear models for categorical responses. *Biometrika 102*(4), 889–906.

Pinheiro, J. C. and E. C. Chao (2006). Efficient Laplacian and adaptive Gaussian quadrature algorithms for multilevel generalized linear mixed models. *Journal of Computational and Graphical Statistics 15*(1), 58–81.

Pregibon, D. (1980). Goodness of link tests for generalized linear models. *Journal of the Royal Statistical Society: Series C (Applied Statistics) 29*(1), 15–24.

Pregibon, D. (1981). Logistic regression diagnostics. *The Annals of Statistics 9*(4), 705–724.

Qian, T., P. Klasnja, and S. A. Murphy (2020). Linear mixed models with endogenous covariates: Modeling sequential treatment effects with application to a mobile health study. *Statistical Science 35*(3), 375–390.

Quenouille, M. (1949). Approximate tests of correlation in time-series. *Journal of the Royal Statistical Society. Series B (Methodological) 11*(1), 68–84.

Quenouille, M. (1956). Notes on bias in estimation. *Biometrika 43*(3), 353–360.

Reilly, M. and M. S. Pepe (1995). A mean score method for missing and auxiliary covariate data in regression models. *Biometrika 82*, 299–314.

Reitsma, J. B., A. W. Rutjes, K. S. Khan, A. Coomarasamy, and P. M. Bossuyt (2009). A review of solutions for diagnostic accuracy studies with an imperfect or missing reference standard. *Journal of Clinical Epidemiology 62*(8), 797–806.

Ridout, M., J. Hinde, and C. G. Demétrio (2001). A score test for testing a zero-inflated poisson regression model against zero-inflated negative binomial alternatives. *Biometrics 57*(1), 219–223.

Robins, J. M. and A. Rotnitzky (1995). Semiparametric efficiency in multivariate regression models with missing data. *Journal of the American Statistical Association 90*, 122–129.

Robins, J. M., A. Rotnitzky, and L. P. Zhao (1994). Estimation of regression coefficients when some regressors are not always observed. *Journal of the American Statistical Association 89*, 846–866.

Robinson, G. K. (1991). That BLUP is a good thing: The estimation of random effects. *Statistical Science 6*, 15–32.

Rosen, O., W. Jiang, and M. Tanner (2000). Mixtures of marginal models. *Biometrika 87*(2), 391–404.

Rosenbaum, P. R. (2002). *Observational Studies*. New York: Springer.

Rosenbaum, P. R. and D. B. Rubin (1983). The central role of the propensity score in observational studies for causal effects. *Biometrika 70*, 41–55.

Rosenbaum, P. R. and D. B. Rubin (1984). Reducing bias in observational studies using sub-classification on the propensity score. *Journal of the American Statistical Association 79*, 516–524.

Rothman, K. J. (1998). *Modern epidemiology*. Philadelphia, PA: Lippincott Williams & Wilkins.

Rotnitzky, A. and N. Jewell (1990). Hypothesis testing of regression parameters in semipara-metric generalized linear models for cluster correlated data. *Biometrika 77*(3), 485–497.

Royston, P. and D. G. Altman (1994). Regression using fractional polynomials of continuous covariates: Parsimonious parametric modelling. *Journal of the Royal Statistical Society: Series C (Applied Statistics) 43*(3), 429–453.

Rozanov, Y. (1977). *Probability theory: A concise course*. Mineola, NY: Dover Publications.

Rubin, D. B. (1976). Inference and missing data. *Biometrika 63*(3), 581–592.

Rubin, D. B. (1978). Multiple imputations in sample surveys. *American Statistical Association, Proceedings of the Survey Research Methods Section*, 22–34.

Rubin, D. B. (1987). *Multiple imputation for nonresponse in surveys*. London: Chapman & Hall.

Rubin, D. B. and N. Schenker (1987). Logit-based interval estimation for binomial data using the Jeffreys prior. *Sociological Methodology 17*, 131–144.

Samuels, M. (1993). Simpson's paradox and related phenomena. *Journal of the American Statistical Association 88*(421), 81–88.

Santner, T. and D. Duffy (1986). A note on A. Albert and J. A. Anderson's conditions for the existence of maximum likelihood estimates in logistic regression models. *Biometrika 73*(3), 755–758.

Satterthwaite, F. E. (1946). An approximate distribution of estimates of variance components. *Biometrics Bulletin 2*(6), 110–114.

Schaalje, G. B., J. B. McBride, and G. W. Fellingham (2002). Adequacy of approximations to distributions of test statistics in complex mixed linear models. *Journal of Agricultural, Biological, and Environmental Statistics 7*(4), 512–524.

Schabenberger, O. and T. Gregoire (1996). Population-averaged and subjectspecific approaches for clustered categorical data. *Journal of Statistical Computation and Simulation 54*(1-3), 231–253.

Schafer, J. L. (1997). *Analysis of incomplete multivariate data*. London: Chapman & Hall.

Schluchter, M. D. and J. T. Elashoff (1990). Small-sample adjustments to tests with unbalanced repeated measures assuming several covariance structures. *Journal of Statistical Computation and Simulation 37*(1-2), 69–87.

Self, S. and K. Liang (1987). Asymptotic properties of maximum likelihood estimators and likelihood ratio tests under nonstandard conditions. *Journal of the American Statistical Association 82*(398), 605–610.

Shannon, C. E. (1948). A mathematical theory of communication. *The Bell System Technical Journal 27*, 379–423, 623–656.

Shmueli, G., T. P. Minka, J. B. Kadane, S. Borle, and P. Boatwright (2005). A useful distribution for fitting discrete data: Revival of the conway–maxwell–poisson distribution. *Journal of the Royal Statistical Society: Series C (Applied Statistics) 54*(1), 127–142.

Shrout, P. and J. Fleiss (1979). Intraclass correlations: Uses in assessing rater reliability. *Psychological Bulletin 86*(2), 420–428.

Shults, J., W. G. Sun, X. Tu, H. Kim, J. Amsterdam, J. A. Hilbe, and T. Ten-Have (2009). A comparison of several approaches for choosing between working correlation structures in generalized estimating equation analysis of longitudinal binary data. *Statistics in Medicine 28*(18), 2338–2355.

Sim, S. Z., R. C. Gupta, and S. H. Ong (2018). Zero-inflated conway-maxwell poisson distribution to analyze discrete data. *The International Journal of Biostatistics 14*(1).

Spearman, C. (1904). The proof and measurement of association between two things. *The American Journal of Psychology 15*(1), 71–101.

Spiegelhalter, D. J. (1986). Probabilistic prediction in patient management and clinical trials. *Statistics in Medicine 5*(5), 421–433.

Stevens, R. J. and K. K. Poppe (2020). Validation of clinical prediction models: What does the "calibration slope" really measure? *Journal of Clinical Epidemiology 118*, 93–99.

Steyerberg, E. W., A. J. Vickers, N. R. Cook, T. Gerds, M. Gonen, N. Obuchowski, M. J. Pencina, and M. W. Kattan (2010). Assessing the performance of prediction models: A framework for some traditional and novel measures. *Epidemiology (Cambridge, Mass.) 21*(1), 128.

Stokes, M., C. Davis, and G. Koch (2009). *Categorical data analysis using the SAS system* (2nd ed.). Cary, NC: SAS Publishing.

Storer, B. E. and J. Crowley (1985). A diagnostic for cox regression and general conditional likelihoods. *Journal of the American Statistical Association 80*(389), 139–147.

Streiner, D. L., G. R. Norman, and J. Cairney (2015). *Health measurement scales: A practical guide to their development and use.* Oxford University Press, USA.

Stuart, A. (1955). A test for homogeneity of the marginal distributions in a two-way classification. *Biometrika 42*(3), 412–416.

Stukel, T. A. (1988). Generalized logistic models. *Journal of the American Statistical Association 83*(402), 426–431.

Tang, W., N. Lu, T. Chen, W. Wang, D. D. Gunzler, Y. Han, and X. M. Tu (2015). On performance of parametric and distribution-free models for zero-inflated and over-dispersed count responses. *Statistics in Medicine 34*(24), 3235–3245.

Tang, Y. and W. Tang (2019). Testing modified zeros for poisson regression models. *Statistical Methods in Medical Research 28*(10–11), 3123–3141.

Tarone, R. E. (1985). On heterogeneity tests based on efficient scores. *Biometrika 72*(1), 91–95.

Terpstra, T. (1952). The asymptotic normality and consistency of Kendall's test against trend, when ties are present in one ranking. *Indagationes Mathematicae 14*(1952), 327–333.

Thompson, S. K. (2002). *Sampling* (2nd ed.). Wiley Series in Probability and Statistics. New York: Wiley-Interscience.

Tian, G.-L., H. Ma, Y. Zhou, and D. Deng (2015). Generalized endpoint-inflated binomial model. *Computational Statistics & Data Analysis 89*, 97–114.

Tjur, T. (2009). Coefficients of determination in logistic regression modelsa new proposal: The coefficient of discrimination. *The American Statistician 63*(4), 366–372.

Touloumis, A., A. Agresti, and M. Kateri (2013). Gee for multinomial responses using a local odds ratios parameterization. *Biometrics 69*(3), 633–640.

Train, K. E. (2009). *Discrete choice methods with simulation*. London: Cambridge University Press.

Tritchler, D. (1984). An algorithm for exact logistic regression. *Journal of the American Statistical Association 79*(387), 709–711.

Tsiatis, A. A. (2006). *Semiparametric theory and missing data*. Springer Series in Statistics. New York: Springer.

Tu, X., J. Kowalski, P. Crits-Christoph, and R. Gallop (2006). Power analyses for correlations from clustered study designs. *Statistics in Medicine 25*(15), 2587–2606.

Tu, X., J. Kowalski, J. Zhang, K. Lynch, and P. Crits-Christoph (2004). Power analyses for longitudinal trials and other clustered designs. *Statistics in Medicine 23*(18), 2799–2815.

Tu, X., E. Litvak, and M. Pagano (1992). Issues in human immunodeficiency virus (HIV) screening programs. *American Journal of Epidemiology 136*(2), 244–255.

Tu, X., E. Litvak, and M. Pagano (1994). Screening tests: Can we get more by doing less? *Statistics in Medicine 13*(19–20), 1905–1919.

Tu, X. M., C. Feng, J. Kowalski, W. Tang, H. Wang, C. Wan, and Y. Ma (2007). Correlation analysis for longitudinal data: Applications to HIV and psychosocial research. *Statistics in Medicine 26*(22), 4116–4138.

Vaida, F. and S. Blanchard (2005). Conditional Akaike information for mixed-effects models. *Biometrika 92*(2), 351–370.

Van den Broek, J. (1995). A score test for zero inflation in a poisson distribution. *Biometrics 51*(2), 738–743.

Vuong, Q. (1989). Likelihood ratio tests for model selection and non-nested hypotheses. *Econometrica 57*(2), 307–333.

Wang, W., V. Lopez, C. Ying, and D. Thompson (2006). The psychometric properties of the chinese version of the sf-36 health survey in patients with myocardial infarction in mainland china. *Quality of Life Research 15*(9), 1525–1531.

White, H. (1982). Maximum likelihood estimation of misspecified models. *Econometrica 50*(1), 1–25.

Wickens, T. (1989). *Multiway contingency tables analysis for the social sciences*. Hillsdale, NJ: Lawrence Erlbaum Associates.

Wilcoxon, F. (1945). Individual comparisons by ranking methods. *Biometrics Bulletin 1*(6), 80–83.

Wilde, M. H., J. M. McMahon, M. V. McDonald, W. Tang, W. Wang, J. Brasch, E. Fairbanks, S. Shah, F. Zhang, and D.-G. D. Chen (2015). Self-management intervention for long-term indwelling urinary catheter users: Randomized clinical trial. *Nursing research 64*(1), 24–34.

Williams, D. (1982). Extra-binomial variation in logistic linear models. *Journal of the Royal Statistical Society. Series C (Applied Statistics) 31*(2), 144–148.

Wilson, E. B. (1927). Probable inference, the law of succession, and statistical inference. *Journal of the American Statistical Association 22*, 209–212.

Wilson, P. W., R. B. D'Agostino, D. Levy, A. M. Belanger, H. Silbershatz, and W. B. Kannel (1998). Prediction of coronary heart disease using risk factor categories. *Circulation 97*(18), 1837–1847.

Woolf, B. (1955). On estimating the relation between blood group and disease. *Annals of Human Genetics 19*(4), 251–253.

Yang, Y. (2005). Can the strengths of AIC and BIC be shared? A conflict between model indentification and regression estimation. *Biometrika 92*(4), 937–950.

Yang, Z., W. Li, X. Tu, W. Tang, S. Messing, L. Duan, J. Pan, X. Li, and C. Wan (2012). Validation and psychometric properties of chinese version of sf-36 in patients with hypertension, coronary heart diseases, chronic gastritis and peptic ulcer. *International Journal of Clinical Practice 66*(10), 991–998.

Yates, J. F. (1982). External correspondence: Decompositions of the mean probability score. *Organizational Behavior and Human Performance 30*(1), 132–156.

Yau, K. K., K. Wang, and A. H. Lee (2003). Zero-inflated negative binomial mixed regression modeling of over-dispersed count data with extra zeros. *Biometrical Journal: Journal of Mathematical Methods in Biosciences 45*(4), 437–452.

Ye, P., X. Qiao, W. Tang, C. Wang, and H. He (2022). Testing latent class of subjects with structural zeros in negative binomial models with applications to gut microbiome data. *Statistical Methods in Medical Research*. Vol. 31, 2237–2254.

Ye, P., Y. Tang, L. Sun, W. Tang, and H. He (2021). Testing inflated zeros in binomial regression models. *Biometrical Journal 63*(1), 59–80.

Zelen, M. (1971). The analysis of several 2×2 contingency tables. *Biometrika 58*(1), 129–137.

Zhang, C., G.-L. Tian, and K.-W. Ng (2016). Properties of the zero-and-one inflated poisson distribution and likelihood-based inference methods. *Statistics and Its Interface 9*(1), 11–32.

Zhang, H., N. Lu, C. Feng, S. Thurston, Y. Xia, L. Zhu, and X. Tu (2011). On fitting generalized linear mixed-effects models for binary responses using different statistical packages. *Statistics in Medicine 30*, 2562–2572.

Zhang, H., Y. Xia, R. Chen, D. Gunzler, W. Tang, and X. Tu (2011). Modeling longitudinal binomial responses: Implications from two dueling paradigms. *Journal of Applied Statistics 38*, 2373–2390.

Zou, K. and W. Hall (2000). Two transformation models for estimating an ROC curve derived from continuous data. *Journal of Applied Statistics 27*(5), 621–631.

Index

Printed in the United States
by Baker & Taylor Publisher Services